Earthquake Seismology: Tools, Techniques and Instrumentation

Earthquake Seismology: Tools, Techniques and Instrumentation

Edited by **Daniel Galea**

SYRAWOOD
PUBLISHING HOUSE
New York

Published by Syrawood Publishing House,
750 Third Avenue, 9th Floor,
New York, NY 10017, USA
www.syrawoodpublishinghouse.com

Earthquake Seismology: Tools, Techniques and Instrumentation
Edited by Daniel Galea

International Standard Book Number: 978-1-68286-216-2 (Hardback)

Printed in China.

Contents

	Preface	**VII**
Chapter 1	**High-precision relocation of seismic sequences above a dipping Moho: the case of the January–February 2014 seismic sequence on Cephalonia island (Greece)** V. K. Karastathis, E. Mouzakiotis, A. Ganas and G. A. Papadopoulos	**1**
Chapter 2	**Finite-difference modelling to evaluate seismic P-wave and shear-wave field data** T. Burschil, T. Beilecke and C. M. Krawczyk	**13**
Chapter 3	**Receiver function images of the Hellenic subduction zone and comparison to microseismicity** F. Sodoudi, A. Brüstle, T. Meier, R. Kind, W. Friederich and Egelados working group	**28**
Chapter 4	**Variations of the crustal thickness in Nepal Himalayas based on tomographic inversion of regional earthquake data** I. Koulakov, G. Maksotova, S. Mukhopadhyay, J. Raoof, J. R. Kayal, A. Jakovlev and A. Vasilevsky	**45**
Chapter 5	**Volcanological aspects of the northwest region of Paraná continental flood basalts (Brazil)** F. Braz Machado, E. Reis Viana Rocha-Júnior, L. Soares Marques and A. J. Ranalli Nardy	**55**
Chapter 6	**Short-lived tectonic switch mechanism for long-term pulses of volcanic activity after mega-thrust earthquakes** M. Lupi and S. A. Miller	**70**
Chapter 7	**Petrophysical constraints on the seismic properties of the Kaapvaal craton mantle root** V. Baptiste and A. Tommasi	**82**
Chapter 8	**Review of some significant claimed irregularities in Scandinavian postglacial uplift on timescales of tens to thousands of years – earthquakes in Denmark?** S. Gregersen and P. H. Voss	**101**
Chapter 9	**New insights on the occurrence of peperites and sedimentary deposits within the silicic volcanic sequences of the Paraná Magmatic Province, Brazil** A. C. F. Luchetti, A. J. R. Nardy, F. B. Machado, J. E. O. Madeira and J. M. Arnosio	**111**

Chapter 10 **Seismic visibility of a deep subduction channel – insights from numerical simulation of high-frequency seismic waves emitted from intermediate depth earthquakes** **121**
W. Friederich, L. Lambrecht, B. Stöckhert, S. Wassmann and C. Moos

Chapter 11 **Seismogenic frictional melting in the magmatic column** **140**
J. E. Kendrick, Y. Lavallée, K.-U. Hess, S. De Angelis, A. Ferk, H. E. Gaunt, P. G. Meredith, D. B. Dingwell and R. Leonhardt

Chapter 12 **Focal mechanisms in the southern Aegean from temporary seismic networks – implications for the regional stress field and ongoing deformation processes** **150**
W. Friederich, A. Brüstle, L. Küperkoch, T. Meier, S. Lamara and Egelados Working Group

Chapter 13 **Seismic structure of the lithosphere and upper mantle beneath the ocean islands near mid-oceanic ridges** **173**
C. Haldar, P. Kumar and M. Ravi Kumar

Chapter 14 **On the complexity of surface ruptures during normal faulting earthquakes: excerpts from the 6 April 2009 L'Aquila (central Italy) earthquake (M_w 6.3)** **184**
L. Bonini, D. Di Bucci, G. Toscani, S. Seno and G. Valensise

Chapter 15 **BrO / SO_2 molar ratios from scanning DOAS measurements in the NOVAC network** **204**
P. Lübcke, N. Bobrowski, S. Arellano, B. Galle, G. Garzón, L. Vogel and U. Platt

Chapter 16 **AxiSEM: broadband 3-D seismic wavefields in axisymmetric media** **220**
T. Nissen-Meyer, M. van Driel, S. C. Stähler, K. Hosseini, S. Hempel, L. Auer, A. Colombi and A. Fournier

Permissions

List of Contributors

Preface

Every book is initially just a concept; it takes months of research and hard work to give it the final shape in which the readers receive it. In its early stages, this book also went through rigorous reviewing. The notable contributions made by experts from across the globe were first molded into patterned chapters and then arranged in a sensibly sequential manner to bring out the best results.

Seismology is a prominent scientific field that aims to understand and analyse the seismic waves and tectonic movements. It plays a crucial part in study of several interdisciplinary subjects like geophysics, earthquake engineering, etc. This text attempts to elucidate the latest instruments and techniques used in seismology. Some of the topics discussed in this extensive book are structural geology, tectonophysics, geodynamics and geomorphology which will provide a comprehensive overview of the discipline. It strives to present an exhaustive insight into the current progress and advancements in the field of seismology. Students, researchers and professionals associated with the field of seismology and allied fields will benefit alike from this book.

It has been my immense pleasure to be a part of this project and to contribute my years of learning in such a meaningful form. I would like to take this opportunity to thank all the people who have been associated with the completion of this book at any step.

Editor

1

High-precision relocation of seismic sequences above a dipping Moho: the case of the January–February 2014 seismic sequence on Cephalonia island (Greece)

V. K. Karastathis, E. Mouzakiotis, A. Ganas, and G. A. Papadopoulos

National Observatory of Athens, Institute of Geodynamics, Lofos Nymfon, P.O. Box 20048, 11810 Athens, Greece

Correspondence to: V. K. Karastathis (karastathis@noa.gr)

Abstract. Detailed velocity structure and Moho mapping is of crucial importance for a high precision relocation of seismicity occurring out of, or marginal to, the geometry of seismological networks. Usually the seismographic networks do not cover the boundaries of converging plates such as the Hellenic arc. The crustal thinning from the plate boundary towards the back-arc area creates significant errors in accurately locating the earthquake, especially when distant seismic phases are included in the analysis. The case of the Cephalonia (Ionian Sea, Greece) sequence of January–February 2014 provided an excellent example where the hypocentral precision was greatly affected by the crustal thinning from the plate boundary at the Ionian sea towards the Aegean sea. This effect was examined in detail by testing various velocity models of the region in order to determine an optimal model. Our tests resulted in the adoption of a velocity model that resembles the crustal thinning of the region. Then, a relocation procedure was performed in the Cephalonia sequence for the time period of 26 January to 15 May 2014 by applying probabilistic non-linear location algorithms. The high-precision relocation resulted in an improved spatial distribution of the seismicity with respect to the preliminary locations and provided a reliable basis to examine seismotectonic implications of the Cephalonia sequence.

1 Introduction

On 26 January (13:55:42 and 18:45:08 GMT) and 3 February 2014 (03:08:44 GMT), western Cephalonia, an island on the Ionian Sea (Greece) (Fig. 1), was ruptured by three strong earthquakes of magnitudes $M_w = 6.0$, $M_w = 5.3$, and $M_w = 5.9$ (Table 1, Fig. 2). The two strongest earthquakes caused considerable damage in buildings and infrastructure as well as several types of ground failures (rockfalls, landslides, soil liquefaction) on the Paliki peninsula, mainly in Lixouri and its surrounding villages (Papadopoulos et al., 2014; Valkaniotis et al., 2014) (Fig. 1). The peak ground acceleration (PGA) recorded in several localities at accelerometric stations, operated by the National Observatory of Athens, Institute of Geodynamics (NOAGI) (NOA web report, 2014a, b) and the Institute of Engineering Seismology and Earthquake Engineering (EPPO-ITSAK, 2014a, b) reached up to 0.56 and 0.77 *g* in Lixouri during the first and third earthquake, respectively. Since only three permanent seismic stations existed in Cephalonia, on 28 and 29 January 2014 four portable seismic stations (Fig. 2) were installed by NOAGI in the aftershock zone of western Cephalonia with the aim of improving the monitoring capabilities.

Cephalonia has been hit by many strong earthquakes in the past (Ambraseys, 2009; Papazachos and Papazachou, 2003). In the instrumental era of seismology, the most important activity was a series of very strong, lethal earthquakes that ruptured eastern and central Cephalonia, with the largest ($M_s = 7.2$) occurring on 12 August 1953. Large earthquakes occurred off the coast of western Cephalonia in 1972 and 1983 (Scordilis et al., 1985). The very high seismicity of Cephalonia is controlled by two major seismotectonic structures. The first is the right-lateral strike-slip Cephalonia Transform Fault zone (CTFZ) comprising of the NNE–SSW trending Lefkada segment in the north and the NE–SW trending Cephalonia segment in the south (Louvari

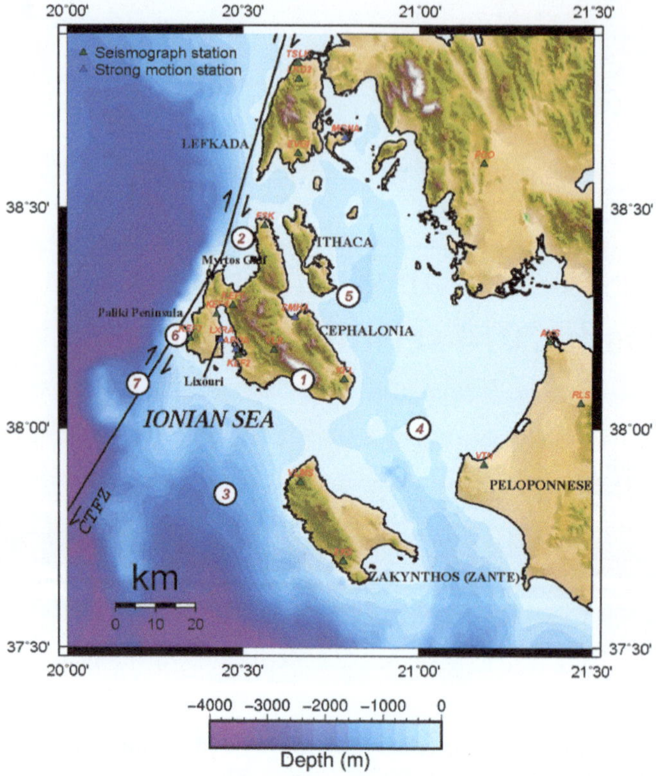

Figure 1. Location map of the study region. The Cephalonia Transform Fault zone (CTFZ) is also indicated on the map. The seismographic stations of the Hellenic Unified Seismological Network (HUSN) are depicted with green-colored triangles and the strong motion stations with blue ones. Significant earthquakes after 1900 (Papazachos et al., 2010) are shown as white circles: (1) 24 January 1912, $M = 6.8$; (2) 9 August 1953, $M = 6.4$; (3) 11 August 1953, $M = 6.8$; (4) 12 August 1953, $M = 6.3$; (5) 12 August 1953, $M = 7.2$; (6) 17 September 1972, $M = 6.3$; (7) 17 January 1983, $M = 7.0$.

et al., 1999) (Fig. 1). The strong ($M_w = 6.2$) earthquake on 14 August 2003 ruptured off the coast of western Lefkada along the Lefkada segment of CTFZ (Fig. 1; Papadopoulos et al., 2003). Recently, Papadopoulos et al. (2014) proposed, based on the spatial pattern of the Cephalonia 2014 earthquake sequence, that the Lefkada segment does not terminate at the NW side of Cephalonia, as previously thought, but extends into western Cephalonia (Fig. 1). A second major structure that controls the seismicity of the area is the northeastern subduction of the Ionian segment of the Hellenic Arc beneath Cephalonia (Sachpazi et al., 2000), thus making up a highly complex seismotectonic setting.

The January–February 2014 seismic sequence is the first one that ruptured western Cephalonia and was recorded by modern seismograph instruments. Therefore, the study of this sequence is of particular importance to better understand the seismicity of Cephalonia. To this aim we performed a high-precision relocation of the seismic sequence of more than 3300 events of magnitude range $M = 1.0$–6.0 from 26 January to 15 May 2014. Then, relocation results were utilized to interpret the seismotectonics of the 2014 activity as well as the geometry and kinematics of the major structure of the CTFZ.

2 The problem of location

The routine determination of earthquake hypocentral parameters usually suffers from significant errors.

More specifically, the main sources of errors for an accurate determination of the hypocentral parameters are (a) picking errors, (b) false identification of the seismic phases, (c) insufficient number of phases, (d) deficient azimuthal coverage of the seismographic network, and finally (e) use of non-effective seismic velocity models that are usually oversimplified (often 1-D) without adequate information for the velocity structure and the lateral velocity heterogeneities. It is a common practice that unified, regional 1-D velocity models are in use, which is also the case of NOAGI for the daily seismicity monitoring of Greece (see Fig. 3). Such velocity models deviate considerably from the local velocity structure of an area, especially outside of the area covered by the network.

Although it is feasible to derive reliable seismic velocity models for areas on a local scale through the implementation of nonlinear inversion techniques that simultaneously invert travel-time data for the hypocentral parameters and seismic velocity determination (Kissling et al., 1994; Kissling, 1995; Thurber, 1993; Koulakov, 2009), it is difficult to perform this practice on a regional scale. Additionally, information coming from crustal seismic surveys is usually useful to build only local velocity models. The crustal mapping based on gravity models suffers from intrinsic ambiguity and the resolution of the seismic velocity global models is too low (Koulakov and Sobolev, 2006).

Inaccurate hypocenter determinations become more acute in the presence of strong structural anomalies and variations that make a structure different from the horizontally layered Earth. For example, crustal thickness is strongly affected in areas situated in the vicinity of convergent plate boundaries. This is the case for the thick continental Aegean crust in the vicinity of the Hellenic subduction zone. In fact, the compressional regime along the Hellenic Arc led to folding and thin-skinned tectonics as well as to the creation of the Mediterranean ridge, which evolved to an accretionary prism, and to subsequent thickening of the crust (Underhill, 1989; Yem et al., 2011). Crustal surveys have shown that in western Greece, where the oceanic crust of the African plate is sliding beneath the Aegean area, the continental crustal thickness exceeds 40 km and becomes progressively thinner to the east (Makris, 1978; Tsokas and Hansen, 1997; Papazachos and Nolet, 1997; Tiberi et al., 2001; Karagianni et al., 2005; Pearce et al., 2012; Makris et al., 2013). At the southern Aegean Sea the crustal thickness reaches values as

Table 1. Focal parameters of the three strong earthquakes of the January–February 2014 Cephalonia seismic sequence as preliminary calculated by NOA. The revised solutions are also shown.

Date	Time	Lat (deg)	Long (deg)	M_W	M_L	h (km)	
26.01.14	13.55.42.7	38.2190 38.2102	20.5322 20.4614	6.0	5.8	16.4 16.5	Preliminary Relocated
26.01.14	18.45.08.3	38.2282 38.2514	20.4138 20.4555	5.3	5.1	16.5 12.5	Preliminary Relocated
03.02.14	03.08.44.7	38.2462 38.2734	20.3958 20.4310	5.9	5.7	11.3 4.6	Preliminary Relocated

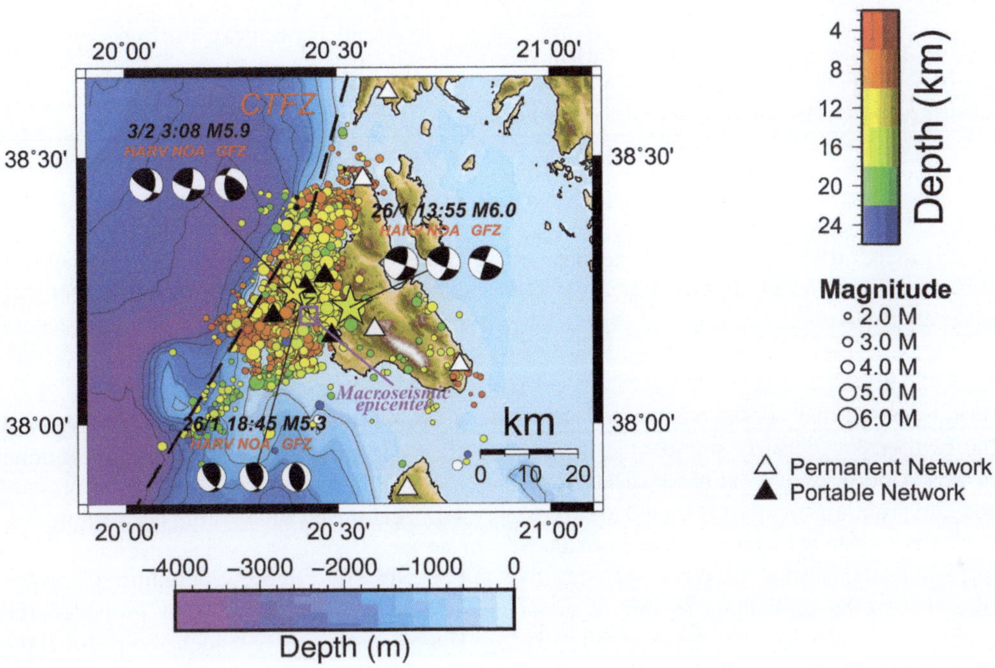

Figure 2. Map of the aftershock sequence from 26 January to 15 May 2014 as determined by the National Observatory of Athens. The moment tensor solutions for the largest events as calculated by the Global Centroid Moment Tensor Project: Harvard University, USA (HARV), National Observatory of Athens (NOA), and the German Research Centre for Geoscience (GFZ).

low as 20 km or less (Makris, 1975, 1976, 1977; Bohnhoff et al., 2001; Tirel et al., 2004). A similar pattern of crustal thinning to the east but with shallower Moho depths (28 km under Cephalonia) was proposed by Sodoudi et al. (2006). Thus, when an earthquake occurs in the thick part of the crust and the paths of the first arriving waves pass through the Moho as it progressively becomes shallower, the travel-time errors may increase considerably with the increase of the epicentral distance. In contrast, shallower events are not so strongly affected, particularly in short epicentral distances since only Pg phases are actually picked. This structure causes an asymmetrical shape to the head-wave wavefront. Due to this structure, the adoption of a 1-D velocity model (see NOA-IG model in Fig. 3) can cause systematic travel-time residuals at the event's location.

3 Analysis of seismological data

The Cephalonia 2014 seismic sequence was examined exactly in this context. The first strong earthquake on 26 January 2014 was recorded by the permanent stations of the HUSN (Hellenic Unified Seismological Network, 2015) before the installation of the portable network in Cephalonia. The focus of that event was preliminary determined by NOAGI at a location situated about 5 km NNE from the city of Argostoli and at a focal depth of $h = 21$ km (Fig. 2) (Table 1). This location clearly falls outside of the aftershock cloud as well as far from Lixouri, where the macroseismic epicenter was determined according to the field observations performed by Papadopoulos et al. (2014) (Fig. 2). The epicentral area is situated at the geometrical edge of the

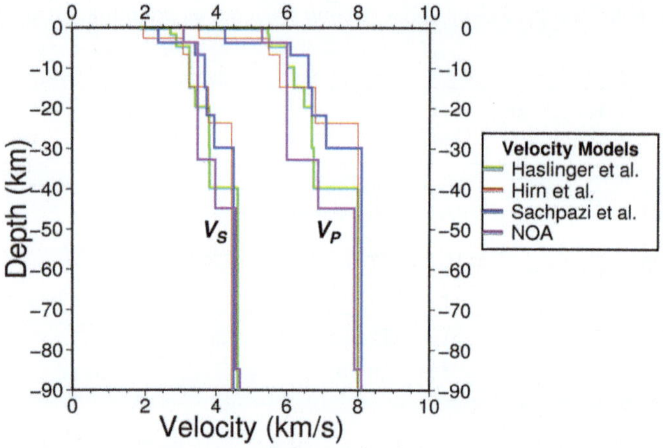

Figure 3. The seismic P- and S-wave velocity models tested for the relocation of the aftershock sequence. Green, blue, and red correspond to velocity models proposed by Haslinger et al. (1999), Sachpazi et al. (2000), and Hirn et al. (1996), respectively. The model routinely used by NOAGI is marked in purple. The S-wave model of NOAGI is based on a constant $V_p/V_s = 1.73$. For Hirn et al. (1996) and Sachpazi et al. (2000), a constant V_p/V_s ratio of 1.80 was used as proposed by previous studies (Hatzfeld et al., 1995; Le Meur et al., 1997; see also Haslinger et al., 1999). Haslinger et al. (1999) provide a separate velocity model.

permanent network, meaning that no stations were situated to the west of the earthquake focus. Consequently, the azimuthal coverage was poor, leading to unstable location solutions. In addition, the preliminary solution included several phases from significantly large epicentral distances extending up to 360 km, where the crustal thickness was significantly smaller than that in the subduction regime of western Cephalonia. The 1-D model used by NOAGI (Fig. 3), with the Moho boundary placed at 40 km, could not match Pn phase data from areas of thin crust, particularly at large epicentral distances. Therefore, significant errors were involved and consequently the epicenter of the first strong earthquake was shifted substantially to the east. Its aftershocks, however, were not shifted significantly because their small magnitudes prevented the use of many distant stations where the velocity model was inadequate. In a next paragraph it is explained more precisely how errors are introduced in the epicentral solution.

The preliminary epicenter of the strong ($M_w = 5.3$) aftershock on 26 January 2014 (18:45:08 GMT) was located in a noticeably better agreement to the macroseismic epicenter (Fig. 2), although it was calculated before the installation of the portable network in Cephalonia. This is explained by the fact that its actual focal depth was shallower than that of the first strong earthquake. It is also worth mentioning that 1 s after the first strong shock on 26 January 2014, another strong aftershock occurred. Unfortunately, the waveform of this event partially overlapped with the waveform of the first strong shock, making an accurate estimation of its local and moment magnitudes quite difficult. The estimation of its duration magnitude is $M_d = 5.0$. The existence of this event, although not reported in earthquake catalogs, justifies the accounts of local people of another shock strongly felt in Cephalonia a few seconds after the first major event.

After the installation of the portable network in western Cephalonia, the maximum azimuthal gap was decreased significantly, becoming less than 180° for most of the seismic events. The preliminary focus of the strong earthquake ($M_w = 5.9$) on 3 February 2014 was located at a shallower depth ($h = 11$ km, Table 1) but within the aftershock cloud (Fig. 2) and at a position close to its macroseismic epicenter, which again was determined to be (Papadopoulos et al., 2014) in Lixouri.

To effectively relocate the hypocenter data of the sequence one may use a velocity model resembling as much as possible the real velocity structure. An effective model can compensate for most of the systematic time residuals created at distant stations. An alternative method is to use a 1-D model to find the epicentral distance range where the effect of the inclined Moho does not considerably affect the location accuracy. This approach limits the seismic phases taken into account. It is reasonably expected that the two different approaches should lead to quite similar results.

4 Selection of a velocity model – relocation of the Cephalonia 2014 earthquake sequence

4.1 Proposed models for the region

For western Greece, including Cephalonia, few seismic velocity models have been proposed (Hirn et al., 1996; Haslinger et al., 1999; Sachpazi et al., 2000). The model proposed by Hirn et al. (1996) was directly based on the results of the crustal seismic surveys carried out in 1992 within the project STREAMERS. The profile ION-7, with an azimuth of N62, was conducted offshore between Cephalonia and Zakynthos (Zante) islands, starting from the deep Ionian basin and reaching the western Gulf of Patras for a total length of 180 km (see maps in Hirn et al., 1996). For the data acquisition, the motor vessel (M/V) *Bin Hai 511* from Geco-Prakla was used with a 36-airgun array (for processing details see also in Kokinou et al., 2005). The seismic reflection profile acquired provided useful information for the shallower structure. However, no precise information was gathered for the Moho interface. To get a rough estimate of the Moho depth, Hirn et al. (1996) performed ray-tracing modeling of the wide-angle travel-time data recorded at distant onshore stations positioned on the Greek mainland. Those stations were located only at the eastern side of the profile. Further assumptions were made for the velocity values beneath the 7 km depth. Thus, these authors discussed a model with the lower crustal interface (V_p between 5.8 and 6.8 km s^{-1}) at 15 km depth and the Moho boundary at 25 km. Since the structure in this model was almost horizontal, the 1-D model in Fig. 3 can be easily derived.

The velocity model of Haslinger et al. (1999) (Fig. 3) was built for the eastern region of Lefkada (western Greece), which as regards to the Cephalonia 2014 sequence, concentrates a high percentage of the ray paths between the earthquakes and the stations. This model was based on a passive experiment and was built as a "1-D minimum velocity model" for this region by the VELEST algorithm (Kissling et al., 1994; Kissling, 1995) and used in a following stage as the initial model in the local earthquake tomography method and SIMULPS code (Thurber, 1993; Eberhart-Phillips, 1990, 1993), implemented to calculate the 3-D crustal velocity structure. The SIMULPS code uses a linearized damped least-square inversion to solve the non-linear problem of the hypocentral location and velocity model. Because of this non-linear nature of the problem, the initial velocity model and the initial hypocenter locations in the inversion procedure should be as close as possible to their true values. The "1-D minimum velocity model" calculated by the VELEST algorithm can provide a good approximation and be used as an initial velocity model. The minimum 1-D velocity models are usually used for seismicity relocation (e.g., Lippitsch et al., 2005; Ganas et al., 2014).

The model proposed by Sachpazi et al. (2000) (Fig. 3) was also created by the VELEST algorithm in order to be used as the initial model in a 3-D local earthquake tomography to determine the velocity structure of the studied area. The procedure for the construction of a 1-D minimum velocity model is highly dependent on the selection of an initial model (Karastathis et al., 2011) and, therefore, it is usually based on the results of seismic profiles. Sachpazi et al. (2000) based their initial model on the seismic profiles presented by Hirn et al. (1996).

For the adoption of an appropriate seismic velocity model, we compared the three 1-D models mentioned above (Fig. 3). As we will see later in detail, the model that performed better was that proposed by Haslinger et al. (1999). With a vertical velocity gradient based on Haslinger et al. (1999) we constructed, tested, and adopted a 2-D velocity model with a non-horizontal Moho boundary based on Papazachos and Nolet (1997). Figure 4 shows the vertical cross section of the 2-D model.

4.2 Synthetic data

Before comparing the performance of these models, we examine the influence on the location procedure of the non-horizontal Moho boundary in the Aegean region. More specifically, in order to assess the impact of the errors imposed in the earthquake location procedure by the adoption of a simplified 1-D model in the presence of a non-horizontal Moho structure, we constructed synthetic arrival times for the adopted model using the 3-D version of the eikonal finite-difference scheme of Podvin and Lecomte (1991) and esti-

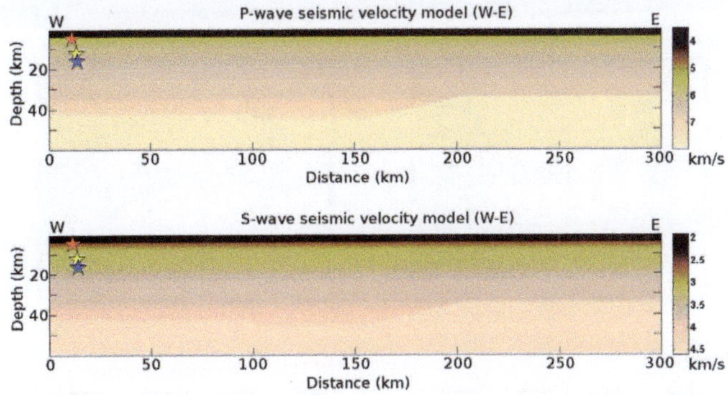

Figure 4. The 2-D velocity model tested to assess the influence of Moho structure. The Moho boundary has been based on the results of Papazachos and Nolet (1997). The position of Cephalonia is between 0 and 50 km. The hypocenters of the major events are shown with stars: blue for the event of 26 January 2014 ($M_W = 6.0$), yellow for the aftershock on 26 January 2014 ($M_W = 5.3$), and red for the event of 3 February 2014 ($M_W = 5.9$).

mated the time differences for both a horizontal and a non-horizontal Moho structure. The velocity gradient was based on Haslinger et al. (1999). The 2-D model with the non-horizontal Moho boundary is shown in Fig. 4 and the results of the comparison with the respective 1-D model are depicted in Fig. 5. As one may expect, the time difference is zero only when the first arrivals are also zero due to the Pg seismic phases. Obviously, the shallower events, with focal depth between 5 and 10 km are not, or only slightly, affected, particularly when they do not bare enough energy to travel at long epicentral distances. As a result, the majority of the shallow aftershocks remain unaffected. In contrast, the influence is higher for the deeper and stronger events, such as the first strong earthquake on 26 January 2014. For this strong earthquake three different epicenters were calculated with the use of the 1-D (with the inclusion or the exclusion of distant phases) and 2-D (with all phases) models (Fig. 6). It can be seen how the simplified 1-D velocity model affects the epicenter location when distant phases are taken into account. The error decreases when distant phases are omitted.

We concluded that, for lack of reliable knowledge regarding the structure and velocity of the Moho boundary and in the presence of poor azimuthal seismographic coverage, it is preferable to limit the range of the epicentral distances of stations used and to base the location mostly on the Pg phases.

The data processing has been performed by the NonLinLoc algorithm (Lomax et al., 2000) that follows a non-linear earthquake location method, giving a complete probabilistic solution expressed in terms of a posterior density function (PDF) (Tarantola and Valette, 1982). The function is calculated using the equal differential time (EDT) likelihood function and depicted by confidence ellipsoids. Therefore the higher the confidence of the velocity model, the smaller the ellipsoids of the event locations. This probabilistic approach

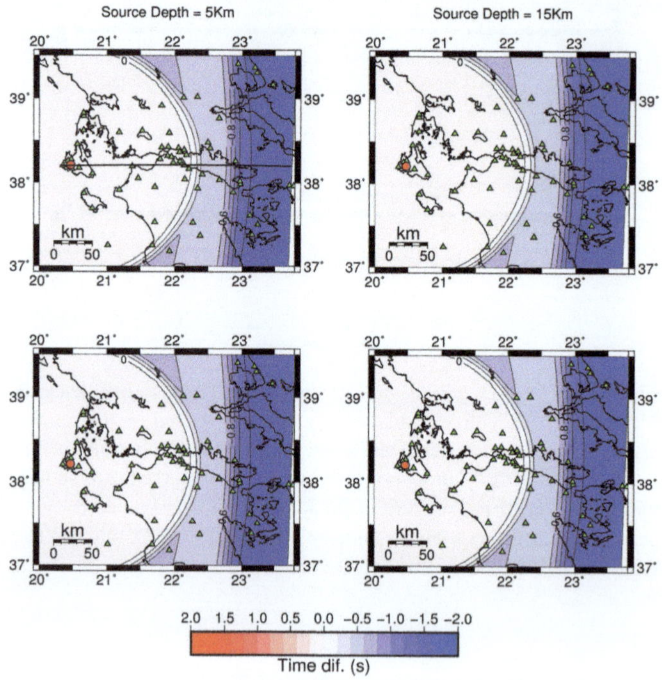

Figure 5. Time differences of P-wave (the two panels in the upper row) and S-wave (the two panels in the lower panel) arrivals between synthetic data calculated on the basis of the 1-D model and data collected using the 2-D model (based on the same 1-D model but with a non-horizontal Moho boundary). Earthquake focal depths of 5 km (left column) and 15 km (right column) are represented. The hypothetical epicenter is shown as a red circle. The errors imported in the case which does not take into account the Moho structure can be significant at long distances.

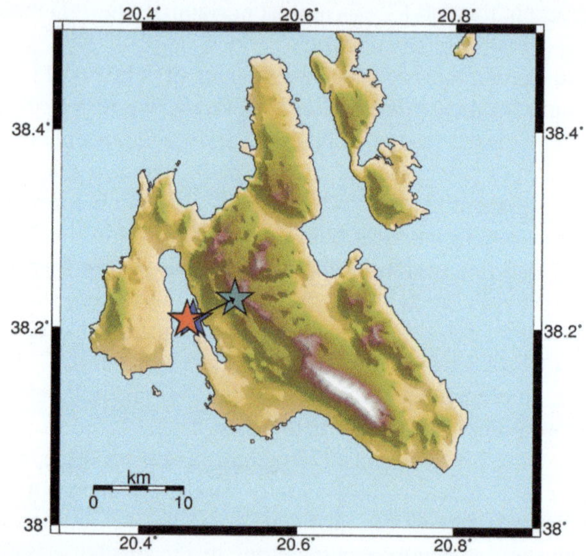

Figure 6. The epicenter of the strong earthquake ($M_W = 6.0$) on 26 January 2014 calculated with the 1-D model of Haslinger et al. (1999) (see Fig. 3) including distant phases (light blue star at the east), the same 1-D model excluding distant phases (blue star at the west), and the 2-D model (see Fig. 4) including distant phases (red star).

is characterized by strong advantages compared to linearized methods. More precisely, the EDT function provides a more reliable uncertainty estimate, especially in the presence of outliers, than the conventional least-squares L1 and L2 norms for the misfit calculation between the observed and calculated travel times. Another advantage of the method is that it is independent of origin time, so the 4-D problem of hypocenter location reduces to a 3-D search over spatial location of the hypocenter (latitude, longitude, depth). The NonLinLoc algorithm can also use 2-D and 3-D velocity models.

4.3 Real data

For the relocation of the Cephalonia 2014 aftershock sequence, we used NOAGI phase data set consisting of more than 44 000 P-wave and 24 000 S-wave arrivals for the time interval from 26 January to 15 May 2014, corresponding to more than 3300 events. Phase data from distant stations ($\Delta > 120$ km) were excluded. It should be noted, however, that small magnitude events remained unaffected since they could hardly be identified at greater epicentral distances anyway.

We compared both the 2-D and 1-D versions of the adopted model with the three models proposed by previous authors (Fig. 3). For the comparison we selected events with at least six P- and one S-wave arrivals and an azimuthal gap lower than 180°. The station delays were calculated as an average value of observed travel-time residual of the well-located events (GAP < 180°, RMS < 1 s, horizontal uncertainty ERH and vertical uncertainty ERZ < 1 km) and applied to the location procedure. Station corrections compensate for the effect of the station's local geology, which could not be taken into account by the use of a 1-D velocity model. The adopted model succeeds in producing a more compact horizontal projection (Fig. 7) and verifies that the aftershock sequence, trending NNE–SSW, covers only the western part of Cephalonia at a length of about 35 km and maximum lateral width of about 10 km (see also Table 1 for the relocated hypocenters of the three major events). It is noteworthy that the relocated aftershock area nearly coincides with the main part of the macroseismic field that is covered by the isoseismal of level V, which is also the area of ground failures produced by the strong earthquakes on 26 January and 3 February 2014 (Papadopoulos et al., 2014).

The model of Haslinger et al. (1999) performs significantly better than those of Hirn et al. (1996), Sachpazi et al. (2000), and that of NOAGI because it is derived from histograms of horizontal and vertical location uncertainties (Table 2). The vast majority (about 80 %) of the events relocated with the adopted model have a horizontal uncertainty less than 900 m. Moreover, the 52 % of the relocated events

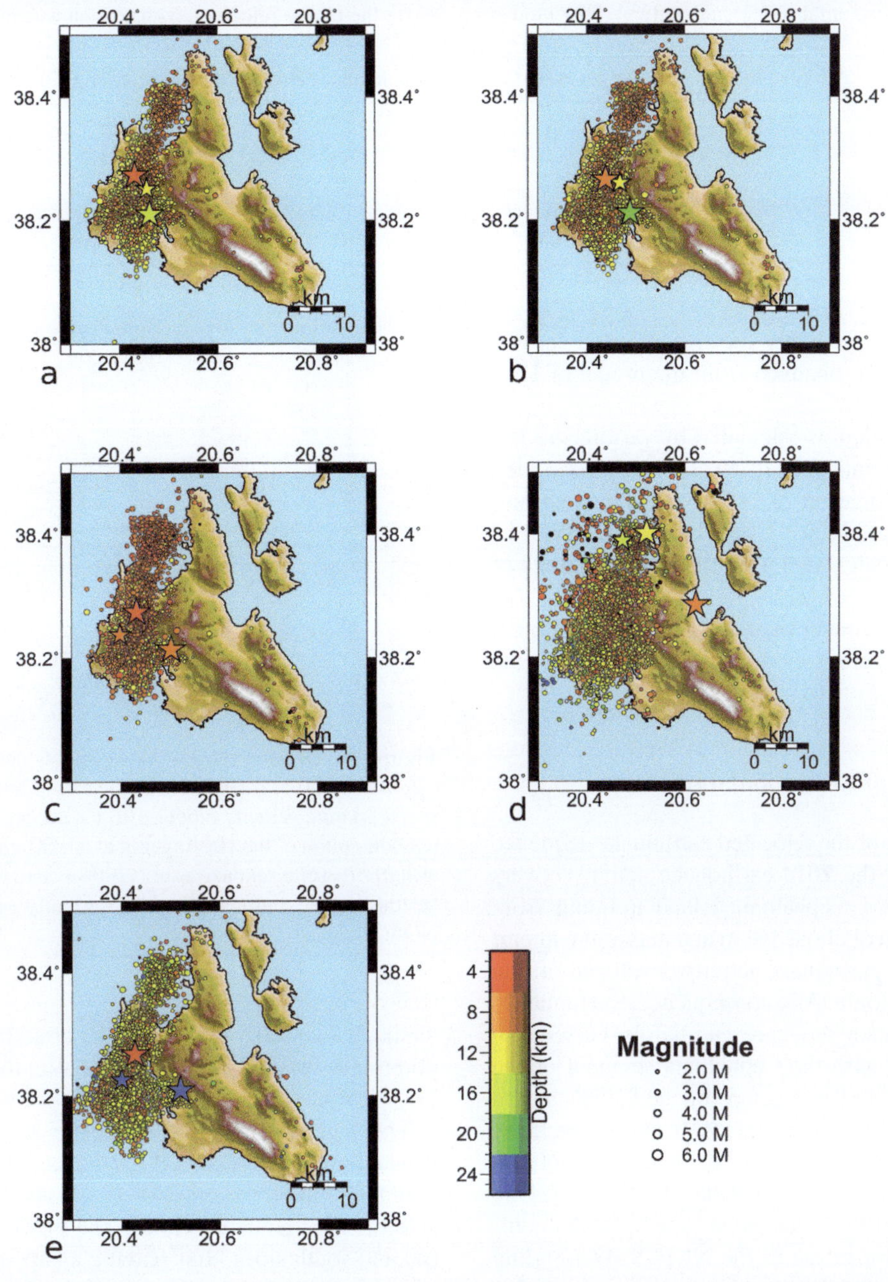

Figure 7. Relocation results by using the velocity models of **(a)** the 2-D velocity model that resembles the Moho structure, **(b)** the 1-D velocity model proposed by Haslinger et al. (1999), **(c)** the 1-D velocity model proposed by Sachpazi et al. (2000), **(d)** the 1-D velocity model proposed by Hirn et al. (1996), and **(e)** the 1-D velocity model used by NOAGI for the daily seismic monitoring.

have a horizontal uncertainty less than 600 m (Table 2). As we can also see in the same table, the original 1-D model of Haslinger et al. (1999) is not notably inferior to the other three models, which produced significantly larger uncertainties. Similar results can be seen also in Fig. 9, depicting the vertical uncertainty distribution (see also Table 2).

For the first 15 km of depth, the model of Haslinger et al. (1999) has similar velocity values with the model proposed by Hirn et al. (1996), which was produced from reliable data of seismic reflection profiles. However, there is an obvious discrepancy at depths greater than 15 km. This might be explained by the fact that for these depths, Hirn et al. (1996) used results only from ray-tracing modeling based on common receiver data only from one side (eastern part) of the seismic traverse. In contrast, Haslinger et al. (1999) did not consider an abrupt increase in the velocity structure at 15 km depth and proposed a Moho boundary at 40 km, whereas Hirn et al. (1996) used a Moho depth at 25 km. The

Table 2. Distribution of horizontal (ERH) and vertical (ERZ) uncertainties for the events relocated with several seismic velocity models.

Performance	Adopted model	Haslinger et al. (1999)	Sachpazi et al. (2000)	Hirn et al. (1996)	NOAGI
ERH < 900 m	86 %	83 %	71 %	39 %	76 %
ERH < 600 m	57 %	57 %	43 %	21 %	46 %
ERZ < 900 m	89 %	87 %	81 %	64 %	68 %
ERZ < 600 m	69 %	67 %	59 %	44 %	43 %
RMS SD	0.063	0.065	0.119	0.488	0.086

shallow Moho boundary is likely the main reason for the poor relocation results we obtained from the model of Hirn et al. (1996).

The aftershock focal depths calculated by various models also show significant variations (Fig. 8). The adopted model, as well as that of Haslinger et al. (1999), has the vast majority of focal depths between 6 and 14 km. The model of Sachpazi et al. (2000) calculated a significant percentage of the hypocenters at depths between 4 and 6 km and a very low percentage with depths greater than 12 km. No hypocenters with unrealistic depth values (< 4) were calculated by the models based on Haslinger et al. (1999).

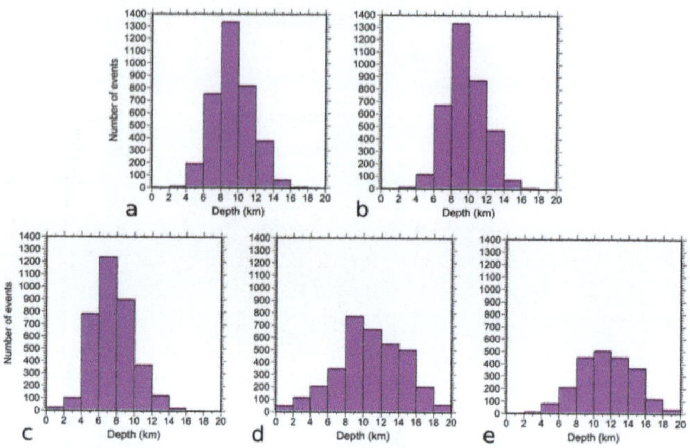

Figure 8. Distribution of hypocentral depth of the relocated Cephalonia 2014 earthquake sequence up to 15 May 2014 for different seismic velocity models: **(a)** the adopted 2-D model, which is a modification of the Haslinger et al. (1999) model; **(b)** Haslinger et al. (1999); **(c)** Sachpazi et al. (2000); **(d)** Hirn et al. (1996); and **(e)** the model routinely used in the daily seismic monitoring by NOAGI.

5 Seismotectonic implications and discussion

The spatial distribution of the relocated earthquake sequence (Fig. 7a) confirms that the 2014 earthquake activity covers only the western part of Cephalonia Island trending from NNE to SSW at a length of about 35 km and maximum lateral width of ca. 10 km. No earthquake activity developed off the coast of western Cephalonia. As a consequence, the January–February 2014 earthquake sequence can hardly be seismotectonically associated with the Cephalonia segment of the major right-lateral strike-slip structure of the CTFZ, as the latter was proposed by Louvari et al. (1999). On the contrary, the aftershock pattern implies that the 2014 activity ruptured western Cephalonia due to onshore strike-slip faulting. One possible scenario is that the activated strike-slip faults comprise the southern prolongation of the NNE–SSW trending Lefkada segment of the CTFZ. Papadopoulos et al. (2014) suggested that the Lefkada CTFZ segment does not terminate off the coast of NW Cephalonia, as proposed by previous authors (Louvari et al., 1999), but extends further into western Cephalonia. Another scenario is that the activated strike-slip faults comprise segments of a 30 km, nearly N–S trending fault zone that splits the island into a western and an eastern part. The consequence is that western Cephalonia appears as a seismotectonic block independent of eastern Cephalonia, which hosted the sources of the large 1953 earthquakes.

The space–time evolution of the 2014 sequence (Fig. 9), based on the high-precision relocated earthquake catalogue that we produced, indicates that soon after the occurrence of the first strong earthquake on 26 January 2014, the aftershock area was already well shaped. No further expansion of the aftershock area was observed either after 29 January, when the portable network was installed, or after the strong earthquake on 3 February 2014.

The Cephalonia 2014 sequence is geographically distributed into two clusters (Figs. 7a and 10). The first is small, with a length on the order of 10 km and occupying the north side of the aftershock cloud. The other extends in the central and south sides, thus leaving an apparent spatial gap between the two clusters. Papadopoulos et al. (2014) suggested that the area of the 2014 gap had already ruptured from the strong ($M_w = 5.5$) strike-slip earthquake on 25 March 2007. However, no temporal relation was found between these two clusters and the occurrence of the strong events on 26 January and 3 February 2014. The north cluster abuts but does not overlap the southern side of the aftershock area of the strong 2003 Lefkada ($M_w = 6.2$) main shock (Papadopoulos et al., 2003). Besides, the foreshock activity that preceded the first strong earthquake on 26 January 2014 by about 4 days was recorded exactly in the area of the north (small) aftershock cluster (Papadopoulos et al., 2014). This may indicate that the 2014 activity was initiated at the northern part of the aftershock area where the 2003 Lefkada activity diminished. Therefore, we observe that a shallow tectonic structure exists

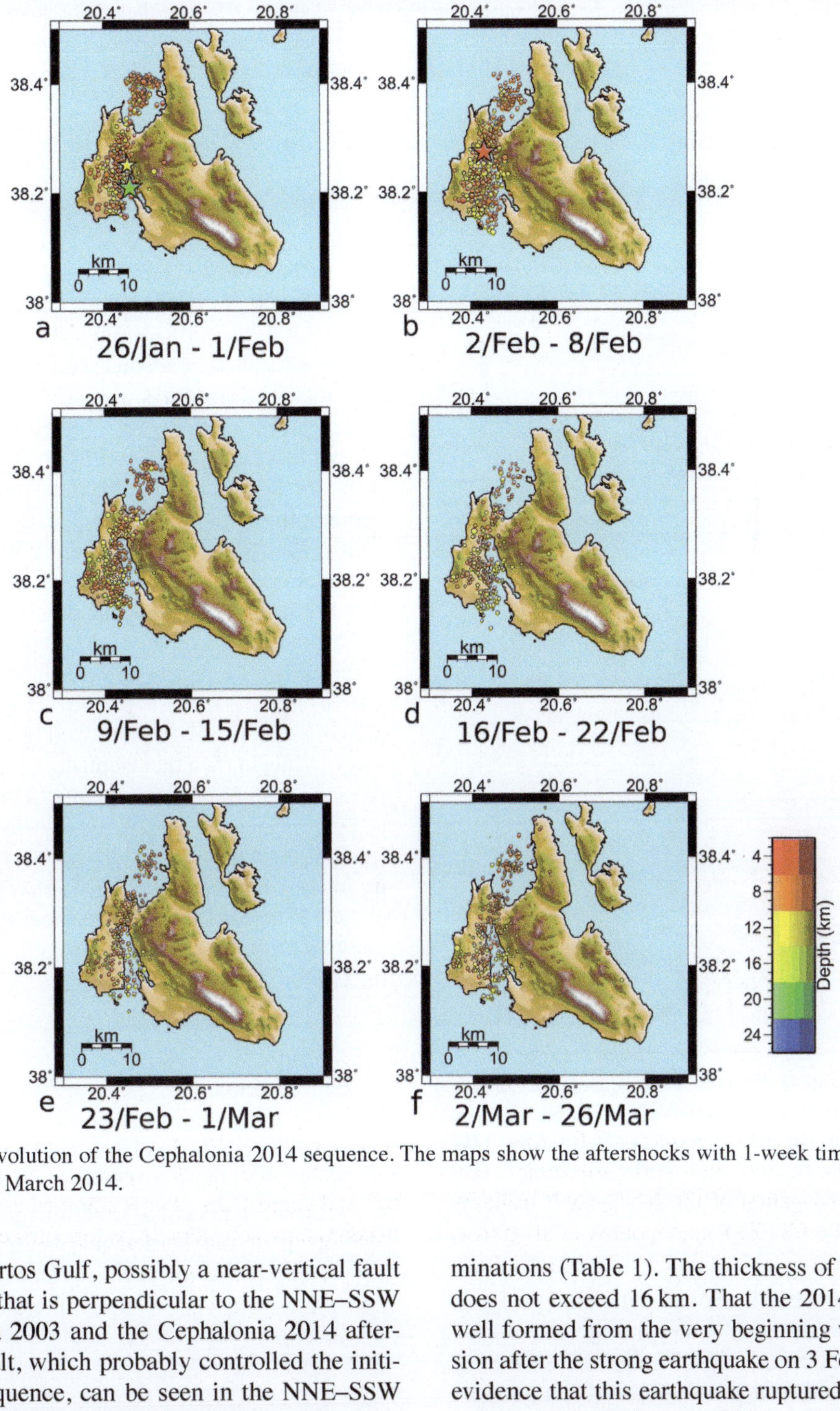

Figure 9. Space–time evolution of the Cephalonia 2014 sequence. The maps show the aftershocks with 1-week time interval **(a–f)** between 26 January 2014 and 26 March 2014.

in the area of the Myrtos Gulf, possibly a near-vertical fault striking WNW–ESE that is perpendicular to the NNE–SSW strike of the Lefkada 2003 and the Cephalonia 2014 aftershock areas. This fault, which probably controlled the initiation of the 2014 sequence, can be seen in the NNE–SSW cross section in Fig. 10c that depicts clearly the vertical geometry of the EW cross fault at Myrtos Gulf at depths of 5–12 km. Evidence also comes from the space–time evolution of the relocated 2014 sequence (Fig. 9), but further examination is needed.

With the relocation applied, the foci of the three strongest earthquakes of the sequence shifted at shallower depths, while the first strong earthquake on 26 January 2014 also shifted towards WNW with respect to the preliminary determinations (Table 1). The thickness of the seismogenic layer does not exceed 16 km. That the 2014 aftershock area was well formed from the very beginning without spatial expansion after the strong earthquake on 3 February 2014 provides evidence that this earthquake ruptured within the aftershock volume of the 26 January 2014 earthquake, the largest event of the sequence. From this point of view we may consider that the 3 February event was the strongest aftershock of the sequence that ruptured at a shallower depth and at a different fault from that of the main shock on 26 January 2014.

To further control the fault patterns associated with the 26 January and 3 February earthquakes, we constructed two respective vertical cross sections as shown in Fig. 10. One may observe that in the section corresponding to the main

Table 3. Mean horizontal and absolute source location distances between the adopted model and the other seismic velocity models.

	Haslinger et al. (1999)	Sachpazi et al. (2000)	Hirn et al. (1996)	NOAGI
Mean horizontal distance (km)	0.7	1.8	5.3	2.4
Mean absolute distance (km)	2.5	3.1	7.1	3.7

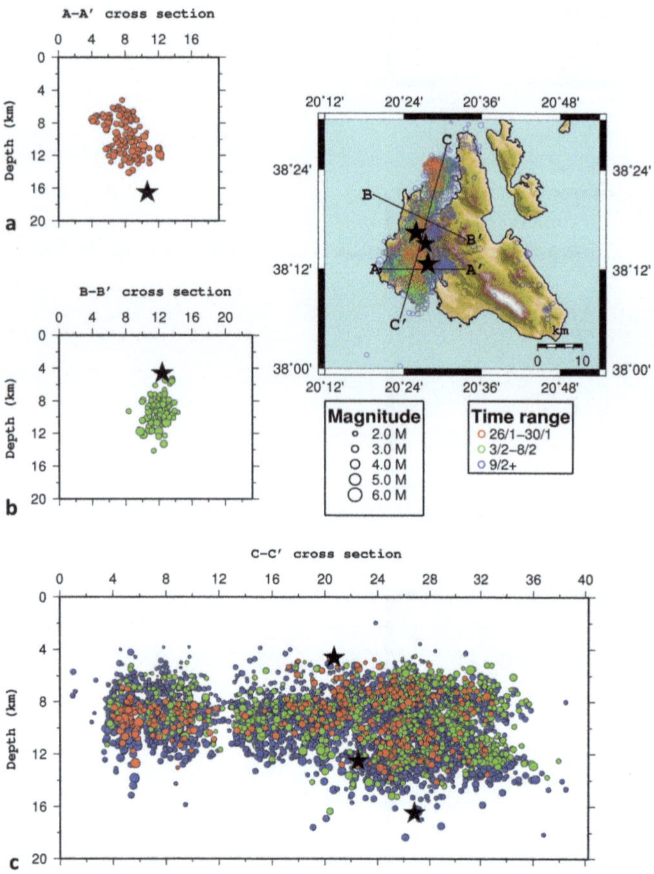

Figure 10. Vertical sections of the aftershock sequence and their location on the map: **(a)** the hypocenters (with GAP < 180°) between 26 and 30 January 2014; **(b)** the hypocenters (with GAP < 180°) between 3 and 8 February 2014; and **(c)** the hypocenters (with GAP < 180°) for the whole sequence (26 January–15 May 2014).

shock on 26 January, the aftershocks until 30 January included in a ±4 km wide zone seem arranged in a plane of nearly (+/−10°) N–S and dip of about 65° to the east (Fig. 10a). The preferred fault-plane adopted by Papadopoulos et al. (2014) and Valkaniotis et al. (2014) is of strike 23° and dip 68° to the east, which is consistent with the geometry represented by the vertical section. However, the vertical section through the hypocenter of the 3 February 2014 event (Fig. 10b) shows that the fault plane strikes nearly (+/−10°) N–S but its dip is about 65° to the west. The geometry of this fault plane is compatible with the fault plane that dips 56° to the west according to the focal mechanism computed by the GFZ (183°/56°/138°; reported on the European-Mediterranean Seismological Centre website (2015). The western dip in combination with the oblique-slip rake may result in uplift of the hanging wall (western) of the N–S fault during co-seismic motion.

Our relocation procedure suggests a different fault model than that of Karakostas et al. (2014; their Fig. 8), who suggested a right step of CTFZ, on the Paliki peninsula. Contrarily, we detected the activation of two blind strike-slip faults along the N–S axis at the Myrtos Gulf–Lixouri line, possibly overlapping with a left step. The 26 January 2014 activated fault (Fig. 10a) is in agreement with the blind fault model of Valkaniotis et al. (2014). Such a fault configuration may explain the co-seismic uplift seen on the Paliki peninsula in InSAR data (Boncori et al., 2015) because our relocation data in the epicentral region of the 3 February 2014 event point to a west-dipping fault (Fig. 10b). If this is the case, then the hanging wall of this fault moved upwards during co-seismic motion; it is known that Cephalonia region is in state of compression with maximum horizontal stress oriented at N78° E (+/−9°; Ganas et al., 2013), as determined from GPS data. In addition, Lagios et al. (2012) obtained a N–S discontinuity in their horizontal velocity field (see Fig. 4c of Lagios et al., 2012) across the Gulf of Argostoli, which may be indicative of a crustal block boundary or a large fault zone beneath the Gulf.

6 Conclusions

1. The strong earthquakes on Cephalonia on 26 January and 3 February 2014 clearly show the problem of incorrect hypocentral determination for earthquakes that occur at the borders of the seismograph network of Greece, near the region of convergence of the tectonic plates when an oversimplified 1-D velocity model is used.

2. We propose a 2-D velocity model with a non-horizontal Moho boundary that resembles the crustal thinning of the Hellenic Arc region.

3. We applied a relocation procedure in the 2014 Cephalonia sequence for the time period from 26 January to 15 May 2014 by applying probabilistic non-linear location algorithms. The thickness of the seismogenic layer of the upper (Aegean) plate does not exceed 16 km.

4. The distribution of the relocated epicenters covers only the western part of Cephalonia, trending NNE–SSW for about 35 km with a lateral width of ca. 10 km.

5. The 2014 earthquake sequence is associated with dextral, strike-slip faults on the island of Cephalonia. Moreover, a vertical section through the hypocenter of the 3 February 2014 event shows that the fault plane strikes nearly N–S but its dip is about 65° to the west.

6. Our relocated hypocentral distribution delineates a shallow, near-vertical fault existing in the area of Myrtos Gulf, striking WNW–ESE, perpendicular to both the strike of the Lefkada 2003 and the Cephalonia 2014 aftershock areas.

Acknowledgements. Thanks are extended to the acting director and the staff of the Institute of Geodynamics for the daily processing of many seismic phases as well as for the installation of the portable network in Cephalonia. Thanks are also due to the topical editor of the journal, Takaaki Taira, for his suggestions on improving the original manuscript. We would also like thank Ivan Koulakov and the anonymous reviewer for the thorough review of the manuscript and their constructive comments.

Edited by: T. Taira

References

Ambraseys, N.: Earthquakes in the Mediterranean and Middle East, Cambridge University Press, New York, 8, 4333–4355, 2009.

Bohnhoff, M., Makris, J., Papanikolaou, D., and Stavrakakis, G.: Crustal investigation of the Hellenic subduction zone using wide aperture seismic data, Tectonophysics, 343, 239–262, 2001.

Boncori, J. P. M., Papoutsis, I., Pezzo, G., Tolomei, C., Atzori, S., Ganas, A., Karastathis, V., Salvi, S., Kontoes, C., and Antonioli, A.: The February 2014 Cephalonia Earthquake (Greece): 3D Deformation Field and Source Modeling from Multiple SAR Techniques, Seismol. Res. Lett., in press, 2015.

Eberhart-Phillips, D.: Three-dimensional P and S velocity structure in the Coalinga Region, California, J. Geophys. Res., 95, 15343–15363, 1990.

Eberhart-Phillips, D.: Local earthquake tomography: earthquake source regions, in: Seismic Tomography: Theory and Practice, edited by: Iyer, H. M. and Hiahara, K., Chapman and Hall, London, 613–643, 1993.

EPPO-ITSAK: The January 26th, 2014 Earthquake in Cephalonia, Preliminary Report, by EPPO-ITSAK, available at: http://www.slideshare.net/itsak-eppo/2014-0127-kefaloniaeqpreliminaryreporta, last access: 28 January 2015, 2014a (in Greek).

EPPO-ITSAK: Strong Ground Motion of the February 3, 2014 Cephalonia Earthquake: Effect on Soil and Built Environment in Combination with the January 26, 2014 Event, available at: http://www.slideshare.net/itsak-eppo/20140203-kefaloniaeq-report-en, last access: 28 January 2015, 2014b.

European-Mediterranean Seismological Centre website: Full tensors solutions (event 20140203 03:08), http://www.emsc-csem.org/Earthquake/mtfull.php?id=357329, last access: 28 January 2015.

Ganas, A., Marinou, A., Anastasiou, D., Paradissis, D., Papazissi, K., Tzavaras, P., and Drakatos, G.: GPS-derived estimates of crustal deformation in the central and north Ionian Sea, Greece: 3-yr results from NOANET continuous network data, J. Geodynamics, 67, 62–71, 2013.

Ganas, A., Karastathis, V., Moshou, A., Valkaniotis, S., Mouzakiotis, E., and Papathanassiou, G.: Aftershock properties and seismotectonic setting of the 7 August 2013 moderate earthquake in Kallidromon Mountain, central Greece, Tectonophysics, 617, 101–113, 2014.

Haslinger, F., Kissling, E., Ansorge, J., Hatzfeld, D., Papadimitriou, E., Karakostas, V., Makropoulos, K., Kahle, H. G., and Peter, Y.: 3D crustal structure from local earthquake tomography around the Gulf of Arta (Ionian region, NW Greece), Tectonophysics, 304, 201–218, 1999.

Hatzfeld, D., Kassaras, I., Panagiotopoulos, D., Amorese, D., Makropoulos, K., Karakaisis G., and Coutant, O.: Microseismicity and strain pattern in northwestern Greece, Tectonics, 14, 773–785, 1995.

Hellenic Unified Seismological Network: Webpage of Hellenic Unified Seismic Network (H.U.S.N.) hosted in the website of Institute of Geodynamics, of National Observatory of Athens: http://www.gein.noa.gr/en/networks/husn/, last access: 28 January 2015.

Hirn, A., Sachpazi, M., Siliqi, R., Bride, J. M., Marnelis, F., Cernobori, L., and STREAMERS-PROFILES group: A traverse of the Ionian islands front with coincident normal incidence and wide-angle seismics, Tectonophysics, 264, 35–49, 1996.

Karagianni, E. E., Papazachos, C. B., Panagiotopoulos, D. G., Suhadolc, P., Vuan, A., and Panza, G. F.: Shear velocity structure in the Aegean area obtained by inversion of Rayleigh waves, Geophys. J. Int., 160, 127–143, 2005.

Karakostas, V., Papadimitriou, E., Mesimeri, M., Gkarlaouni, C., and Paradisopoulou, P.: The 2014 Kefalonia doublet (Mw 6.1 and Mw 6.0), Central Ionian Islands, Greece: Seismotectonic implications along the Kefalonia transform fault zone, Acta Geophys., 1–16, 2014.

Karastathis, V. K., Papoulia, J., Di Fiore, B., Makris, J., Tsambas, A., Stampolidis, A., and Papadopoulos, G. A.: Deep structure investigations of the geothermal field of the North Euboean Gulf, Greece, using 3-D local earthquake tomography and Curie point depth analysis, J. Volcanol. Geoth. Res., 206, 106–120, 2011.

Kissling, E.: Velest User's Guide, Institute of Geophysics, ETH Zurich, Switzerland, 26 pp., http://www.seg.ethz.ch/software/velest_guide.pdf, 1995.

Kissling, E., Ellsworth, W. L., Eberhart-Phillips, D., and Kradolfer, U.: Initial reference models in local earthquake tomography, J. Geophys. Res., 99, 19635–19646, 1994.

Kokinou, E., Kamberis, E., Vafidis, A., Monopolis, D., Ananiadis, G., and Zelelidis, A.: Deep seismic reflection data Greece: a new crustal model for the Ionian sea, J. Petrol. Geol., 28, 81–98, 2005.

Koulakov, I.: LOTOS code for local earthquake tomographic inversion: benchmarks for testing tomographic algorithms, B. Seismol. Soc. Am., 99, 194–214, 2009.

Koulakov, I. and Sobolev, S. V.: Moho depth and three-dimensional P and S structure of the crust and uppermost mantle in the Eastern Mediterranean and Middle East derived from tomographic inversion of local ISC data, Geophys. J. Int., 164, 218–235, 2006.

Lagios, E., Papadimitriou, P., Novali, F., Sakkas, V., Fumagalli, A., Vlachou, K., and Del Conte, S.: Combined Seismicity Pattern Analysis, DGPS and PSInSAR studies in the broader area of Cephalonia (Greece), Tectonophysics, 524, 43–58, 2012.

Le Meur, H., Virieux, J., and Podvin, P.: Seismic tomography of the Gulf of Corinth: a comparison of methods, Ann. Geophys.-Italy, 40, 1–24, 1997.

Lippitsch, R., White, R. S., and Soosalu, H.: Precise hypocentre relocation of microearthquakes in a high temperature geothermal field: the Torfajökull central volcano, Iceland. Geophys. J. Int, 160, 371–388, 2005.

Lomax, A., Virieux, J., Volant, P., and Berge-Thierry, C.: Probabilistic earthquake location in 3-D and layered models, in: Advances in Seismic Event Location, edited by: Thurber, C. H. and Rabinowitz, N., Kluwer Academic Publishers, Dordrecht/Boston/London, 101–134, 2000.

Louvari, E., Kiratzi, A. A., and Papazachos, B. C.: The CTF and its extension to western Lefkada Island, Tectonophysics, 308, 223–236, 1999.

Makris, J.: Crustal Structure of the Aegean Sea and the Hellenides Obtained from Geophysical Survey, J. Geophys., 41, 441–443, 1975.

Makris, J.: A dynamic model of the Hellenic arc deduced from geophysical data, Tectonophysics, 36, 339–346, 1976.

Makris, J.: Geophysical Investigations of the Hellenides, Hamburger Geophysikalische Einzelschriften, G. M. L. Wittenborn Söhne, Hamburg, Heft 34, 124 pp., 1977.

Makris, J.: The crust and upper mantle of the Aegean region from deep seismic soundings, Tectonophysics, 46, 269–284, 1978.

Makris, J., Papoulia, J., and Yegorova, T.: A 3-D density model of Greece constrained by gravity and seismic data, Geophys. J. Int., 194, 1–17, 2013.

NOA web report, internal report, available at: http://www.gein.noa.gr/Documents/pdf/Cefalonia_20140126_preliminary_web.pdf (last access: 4 June 2014), 2014a.

NOA web report, internal report, available at: http://www.gein.noa.gr/Documents/pdf/Cefalonia_20140203_preliminary_web.pdf (last access: 4 June 2014), 2014b.

Papadopoulos, G. A., Karastathis, V. K., Ganas, A., Pavlides, S., Fokaefs, A., and Orfanogiannaki, K.: The Lefkada, Ionian Sea (Greece), shock (Mw 6.2) of 14 August 2003: Evidence for the characteristic earthquake from seismicity and ground failures, Earth Planet. Space, 55, 713–718, 2003.

Papadopoulos, G. A., Karastathis, V. K., Koukouvelas, I., Sachpazi, M., Baskoutas, I., Chouliaras, G., Agalos, A., Daskalaki, E., Minadakis, G., Moschou, A., Mouzakiotis, E., Orfanogiannaki, A., Papageorgiou, A., Spanos D., and Triantafyllou, I.: The Cephalonia, Ionian Sea (Greece), sequence of strong earthquakes of January-February 2014: a first report, Res. Geophys., 4, 19–30, 2014.

Papazachos, B. C. and Papazachou C. B.: The Earthquakes of Greece, Ziti Editions, Thessaloniki, 2003 (in Greek).

Papazachos, B. C., Comninakis, P. E., Scordilis, E. M., Karakaisis, G. F., and Papazachos C. B.: A catalogue of earthquakes in the Mediterranean and surrounding area for the period 1901–2010, Publ. Geophys. Laboratory, University of Thessaloniki, Greece, 2010.

Papazachos, C. B. and Nolet, G.: P and S deep velocity structure of the Hellenic area obtained by robust nonlinear inversion of travel times, J. Geophys. Res., 102, 8349–8367, 1997.

Pearce, F. D., Rondenay, S., Sachpazi, M., Charalampakis, M., and Royden, L. H.: Seismic investigation of the transition from continental to oceanic subduction along the western Hellenic Subduction Zone, J. Geophys. Res., 117, 1978–2012, 2012.

Podvin, P. and Lecomte, I.: Finite difference computation of traveltimes in very contrasted velocity models: a massively parallel approach and its associated tools, Geophys. J. Int., 105, 271–284, 1991.

Sachpazi, M., Hirn, A., Clement, C., Laigle, M., Haslinger, F., Kissling, E., Charvis, P., Hello, Y., Lepine, J., Sapine, M., and Ansorge, J.: Western Hellenic subduction and Cephalonia transform: local earthquakes and plate transport and strain, Tectonophysics, 319, 301–319, 2000.

Scordilis, E. M., Karakaisis, G. F., Karakostas, B. G., Panagiotopoulos, D. G., Comninakis, P. E., and Papazachos, B. C.: Evidence for Transform Faulting in the Ionian Sea: the Cephalonia Island Earthquake Sequence of 1983, Pure Appl. Geophys., 123, 388–397, 1985.

Sodoudi, F., Kind, R., Hatzfeld, D., Priestley, K., Hanka, W., Wylegalla, K., Stavrakakis, G., Vafidis, A., Harjes H.-P., and Bohnhoff, M.: Lithospheric structure of the Aegean obtained from P and S receiver functions, J. Geophys. Res., 111, B12307, doi:10.1029/2005JB003932, 2006.

Tarantola, A. and Valette, B.: Inverse problems = quest for information, J. Geophys., 50, 159–170, 1982.

Thurber, C. H.: Local earthquake tomography: velocities and $V_p = V_s$ – theory, in: Seismic Tomography: Theory and Practice, edited by: Iyer, H. M. and Hiahara, K., Chapman and Hall, London, 563–583, 1993.

Tiberi, C., Diament, M., Lyon-Caen, H., and King, T.: Moho topography beneath the Corinth Rift area (Greece) from inversion of gravity data, Geophys. J. Int., 145, 797–808, 2001.

Tirel, C., Gueydan, F., Tiberi, C., and Brun, J. P.:. Aegean crustal thickness inferred from gravity inversion. Geodynamical implications, Earth Planet. Sc. Lett., 228, 267–280, 2004.

Tsokas, G. N. and Hansen, R. O.: Study of the crustal thickness and the subducting lithosphere in Greece from gravity data, J. Geophys. Res., 102, 20585–20597, 1997.

Underhill, J. R.: Late Cenozoic deformation of the Hellenide foreland, western Greece, Geol. Soc. Am. Bull., 101, 613–634, 1989.

Valkaniotis, S., Ganas, A., Papathanassiou, G., and Papanikolaou, M.: Field observations of geological effects triggered by the January-February 2014 Cephalonia (Ionian Sea, Greece) earthquakes, Tectonophysics, 630, 150–157, doi:10.1016/j.tecto.2014.05.012, 2014.

Yem, L. M., Camera, L., Mascle, J., and Ribodetti, A.: Seismic stratigraphy and deformational styles of the offshore Cyrenaica (Libya) and bordering Mediterranean Ridge, Geophys. J. Int., 185, 65–77, 2011.

2

Finite-difference modelling to evaluate seismic P-wave and shear-wave field data

T. Burschil, T. Beilecke, and C. M. Krawczyk

Leibniz Institute for Applied Geophysics, Stilleweg 2, 30655 Hannover, Germany

Correspondence to: T. Burschil (thomas.burschil@liag-hannover.de)

Abstract. High-resolution reflection seismic methods are an established non-destructive tool for engineering tasks. In the near surface, shear-wave reflection seismic measurements usually offer a higher spatial resolution in the same effective signal frequency spectrum than P-wave data, but data quality varies more strongly.

To discuss the causes of these differences, we investigated a P-wave and a SH-wave seismic reflection profile measured at the same location on the island of Föhr, Germany and applied seismic reflection processing to the field data as well as finite-difference modelling of the seismic wave field. The simulations calculated were adapted to the acquisition field geometry, comprising 2 m receiver distance (1 m for SH wave) and 4 m shot distance along the 1.5 km long P-wave and 800 m long SH-wave profiles. A Ricker wavelet and the use of absorbing frames were first-order model parameters. The petrophysical parameters to populate the structural models down to 400 m depth were taken from borehole data, VSP (vertical seismic profile) measurements and cross-plot relations.

The simulation of the P-wave wave-field was based on interpretation of the P-wave depth section that included a priori information from boreholes and airborne electromagnetics. Velocities for 14 layers in the model were derived from the analysis of five nearby VSPs ($v_P = 1600$–$2300\,\mathrm{m\,s^{-1}}$). Synthetic shot data were compared with the field data and seismic sections were created. Major features like direct wave and reflections are imaged. We reproduce the mayor reflectors in the depth section of the field data, e.g. a prominent till layer and several deep reflectors. The SH-wave model was adapted accordingly but only led to minor correlation with the field data and produced a higher signal-to-noise ratio. Therefore, we suggest to consider for future simulations additional features like intrinsic damping, thin layering, or a near-surface weathering layer. These may lead to a better understanding of key parameters determining the data quality of near-surface shear-wave seismic measurements.

1 Introduction

Near-surface geophysical methods constitute a non-destructive means to investigate the shallow subsurface. Especially engineering tasks, for instance geohazard assessment or groundwater prospecting, profit from structural and parametrical methods (e.g. Miller, 2013; Kirsch, 2008). In some cases, results of geophysical prospecting are compiled into 3-D models and can act as input for, e.g. groundwater flow modelling.

Among other geophysical methods, high-resolution reflection seismics constitute a valuable tool to extract structural information. The reflection seismic method using compressional waves (P waves) is an established method and often produces accurate results for near-surface tasks (e.g. Steeples and Miller, 1990; Rumpel et al., 2009; Jørgensen et al., 2012). In many cases, however, a higher spatial resolution is desired than P waves offer. This is one important reason, why shear waves have become popular. Since shear waves have a lower velocity than P waves, they offer shorter wavelengths, i.e. higher spatial resolution, within the same frequency band as P waves. Also, shear wave velocity information yields an additional parameter than the P-wave velocity information alone and offers more detailed studies of elastic properties, like for instance Poisson's ratio. The combination of P- and S-wave velocities can help characterize the lithology or pore fluid (e.g. Sheriff and Geldart, 1995). For shallow application, recent developments have been successful (e.g. Inazaki, 2004; Pugin et al., 2009a, b; Polom et al., 2010, 2013; Krawczyk et al., 2013). However, we experience that shear wave reflection seismics more strongly varies in quality compared with compressional wave seismics.

Seismic modelling is an important tool to evaluate seismic field data, yet not often applied for near-surface tasks. Most studies applying wave-field modelling concentrate on deep reservoirs and crustal structures (Robertsson et al., 2007). A comprehensive approach is reported by Bellefleur et al. (2012), which we use accordingly in this study. One current method to create synthetic seismograms of complex structures is finite-difference (FD) modelling (Alterman and Karal, 1968). The advantage of the FD method is the ability to choose arbitrary heterogeneous input models without fundamental restrictions. Drawbacks like high-computational requirements have become less restricting during the past years. The FD code by Bohlen (2002) has the ability to include intrinsic seismic wave absorption, an advantage with respect to the study of unconsolidated material.

Full waveform inversion with either synthetic or real field data is an evolving field in exploration seismology. It aims at the automatic subsurface model generation from seismic field data. Although there has been a lot of progress in its evolution, and it can be applied routinely for simple subsurface structures, especially in marine environments, its application for near-surface studies is still experimental (Romdhane et al., 2011; Groos et al., 2012; Plessix, 2012). This has to do with the importance of a good starting model for the inversion process but also with the large parameter space in the near surface. Also, many seismic surveys try to avoid low frequencies in order to not generate large surface waves. Yet, the low frequencies are an important prerequisite to find the global optimum in the optimization process of the inversion (e.g. Virieux and Operto, 2009; Fichtner et al., 2011). Therefore, we focus here on direct modelling.

In this paper we describe the acquisition and processing of P-wave and SH-wave reflection seismic field data along an example profile. We generated synthetic P-wave data of the same profile with a FD algorithm on the basis of the P-wave depth section interpreted with a priori geological and geophysical information. The synthetic data were compared with the field data. Main reflectors can be reproduced. To assess the lower signal-to-noise ratio of the recorded shear-wave field data, we simulated SH-wave seismograms on the basis of this model to comprehend quality variations in the SH-wave reflection field data.

2 Finite-difference modelling

Seismic wave-field modelling reveals the propagation of seismic waves within a subsurface model. So far, no exact analytical solution exists for the calculation of such wave fields in arbitrary media. Therefore, a number of approximation methods have been introduced over the years to solve the specific wave-equation (Carcione et al., 2002; Fichtner et al., 2011). One method is FD modelling that works for arbitrary media. The medium is divided into a grid small enough to represent the natural phenomena of elastic waves. Changes of elastic parameters for each grid point are approximated for defined time steps and simulate seismic waves travelling through the gridded model. Forces at any chosen location within the model excite the respective grid points and start the wave propagation. These forces are independent in time and place and constitute seismic sources. Several elastic properties can be extracted at any grid point and any time. These properties represent, e.g. the seismic response to a receiver at the surface or in a borehole. FD modelling simulates the entire wave-field and thus contains surface waves naturally. FD modelling has only recently become feasible for routine application, because of enhanced computing capabilities.

The FD software we used in this work is described in Bohlen (2002). This code, Seismic mOdelling with FInite differences called SOFI, is based on earlier work by Virieux (1986) and Levander (1988) for the elastic wave simulation but also on work by Robertsson et al. (1994) for the viscoelastic case. It allows the user to also consider intrinsic wave absorption (viscoelasticity, Q), and it provides an alternative rotated grid representation of the subsurface, based on work by Gold et al. (1997) and Saenger et al. (2000) for a more exact simulation of surface waves. Absorption is a common phenomenon in unconsolidated near-surface rock units, and surface waves pose a typical problem in near-surface seismic data processing. The consideration of these phenomena in the simulation offers a better separation of the different subsurface parameters responsible for seismic field data signatures. SOFI provides full wave-field simulations for 3-D media and 2-D modelling. The 2-D codes simulate either P and SV waves within the propagation plane or SH waves oscillating orthogonally to the propagation plane. For geometrical reasons SH waves will never convert into P or SV waves in this case. This kind of simulation is often unrealistic because the subsurface has a 3-D structure in which arbitrary wave conversion takes place. Nonetheless, it is a valuable aid for the interpretation of the SH-wave field surveys. SOFI requires v_P, v_S, and density (ρ) as input information and optionally accepts absorption models for P and S waves. The SH version does not require compressional wave parameters.

Another important feature of the code is its ability to run simulations in parallel threads on multiprocessor computers to save time. For such parallel simulation, the model can be split up into subgrids that the different CPUs (central processing units), called processing elements (PEs), calculate independently. For every time step each PE updates, the wave field in its subgrid and parameter exchange needs to be carried out with the grid points of neighbouring subgrids. This exchange can slow down the calculation so much that the use of all CPUs and a respective number of subgrids not necessarily delivers the fastest result. Here, the minimum computational time for one test model was achieved with 48 processors and a corresponding number of subgrids.

3 Test site Föhr

The test site is the North Sea island of Föhr (Fig. 1) that was investigated as a pilot area in the Interreg project CLIWAT (Harbo et al., 2011), which was co-financed by the European Union. The aim of CLIWAT was to analyse the influence of climate change to groundwater systems, and one major outcome was a hydrogeological model to forecast the groundwater evolution (Burschil et al., 2012a).

3.1 Geological evolution

The investigation area is located on the German North Sea island of Föhr, which is part of the UNESCO world heritage Wadden Sea (Fig. 1). In general, the deeper subsurface was sedimented as part of the North German Basin while the landscape of Föhr was formed during the glacial and post glacial epoch (Scheer, 2012). The older sediments were deposited under marine conditions since the Cretaceous age until the youngest Tertiary. For this reason, marine clay (mica clay) is found up to Miocene age. Sedimentation changed during the Pliocene, and sandy material (kaolin sand) was deposited until a climate change marked the beginning of the Pleistocene age. Glaciations from the Baltic–Scandinavian area reworked the shallow underground by alternating processes of glacial advance followed by erosion and sedimentation. In the region of Föhr, push moraines as well as a system of tunnel valleys were formed and refilled with glacial deposits (Scheer, 2012). The great outwash plains were increasingly flooded during the Holocene so that tidal mud deposits were accumulated and formed large marshland areas. Finally, heavy floods in historical times eroded large parts and formed the present North Frisian Islands and Wadden Sea.

3.2 Geophysical and geological framework

In the project CLIWAT we accomplished a multidisciplinary geophysical acquisition (Burschil et al., 2012a). Between 2009 and 2011, we acquired seven seismic reflection profiles with P waves (in total 8 km) and three profiles with SH waves (in total 2.4 km). Five vertical seismic profiles (VSPs) were recorded with a 3C borehole geophone and the small electrodynamical vibrator system ELVIS (Polom et al., 2011) with excitation in the vertical and horizontal–transversal directions relative to the borehole location (Fig. 1). Maximum depths of five VSPs were in the range of 39–102 m, depending on the borehole. Additionally, in 2008 the island was mapped within the airborne geophysics mapping project of LIAG (Leibniz Institute for Applied Geophysics; Wiederhold et al., 2010) with the airborne electromagnetic system SkyTEM (Sørensen and Auken, 2004). The result of the P-wave seismic survey was used as a priori information for the electromagnetic data inversion (Burschil et al., 2012b). The information transfer between the different geophysical methods improved the electric resistivity model of the island. Borehole logging data were evaluated statistically regarding electric resistivity as well as seismic velocity. This allowed for a petrophysical classification of sand, till, and clay (Burschil et al., 2012a). These lithological units form structures such as push-moraines and buried valleys, which are consistent with the geological evolution of the region. A simplified hydrogeological model was compiled from all geophysical and geological results for groundwater flow modelling (Burschil et al., 2012a).

4 Seismic reflection field data

4.1 Seismic acquisition

Seismic equipment varied with surveys due to different wave types used for exploration. For two P-wave surveys we used the hydraulic vibrator systems of the LIAG, the MHV2.7 in 2009 and the new HVP-30 in 2010. As source signal we chose a linear sweep in the range of 30–240 Hz with 10 s duration. The source excited the seismic signal on a paved street in the western part of the profile and on grassland in the eastern part. The receivers were vertical geophones (20 Hz), planted in the green strip next to the line of the source locations (cross-line offset ca. 1 m). We chose a receiver spacing of 2 m to enhance the near-surface resolution and fold. A combination of split-spread/roll along geometry was used (Fig. 2a) with a source spacing of twice the receiver spacing (4 m). High-quality data had been acquired with this geometry before. We operated up to 268 active channels with Geometrics Geode seismographs with a maximum in-line source–receiver offset of 360 m. This resulted in a fold between 47 and 72 in the main parts of the profiles. To avoid large data volumes, we recorded the seismic traces after vibro-seismic correlation with the Pelton Vib Pro input signal.

For the shear-wave survey in 2011, an established system comprising LIAG's hydraulic shear-wave vibrator MVP-4S in combination with a 120 m land-streamer system was used (Polom et al., 2010). Horizontal 10 Hz geophones (SH-wave polarization) are mounted every metre on the land-streamer. The whole system was towed by the recording vehicle. The acquisition geometry (Fig. 2b) was similar to the P-wave surveys, except for a shorter spread and that we vibrated within the middle third of the streamer and sequentially pulled thirds of its length along the profile. Shot-point spacing was 4 m. We used a linear sweep between 30 and 160 Hz for 10 s. In contrast to the P-wave surveys, we recorded the uncorrelated traces to allow for precorrelation processing. The shear-wave seismic profile only covers the western half of the P-wave

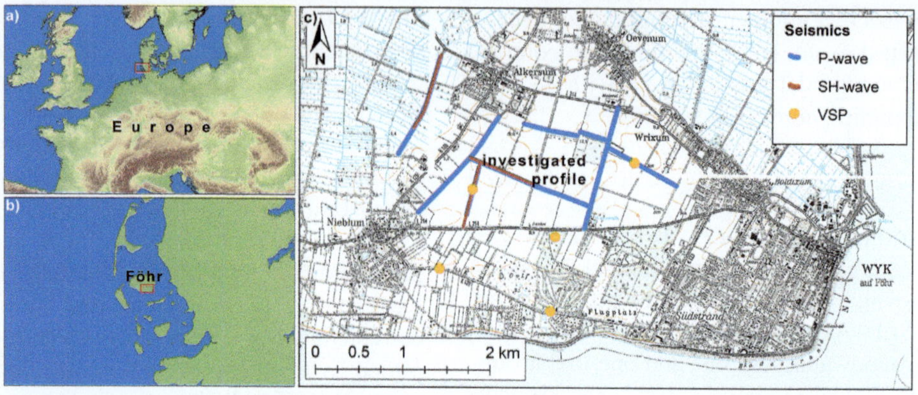

Figure 1. Overview maps with location of the island of Föhr **(a, b)**. Detail map of the seismic locations on the island of Föhr **(c)**. The profile discussed in this paper is labelled.

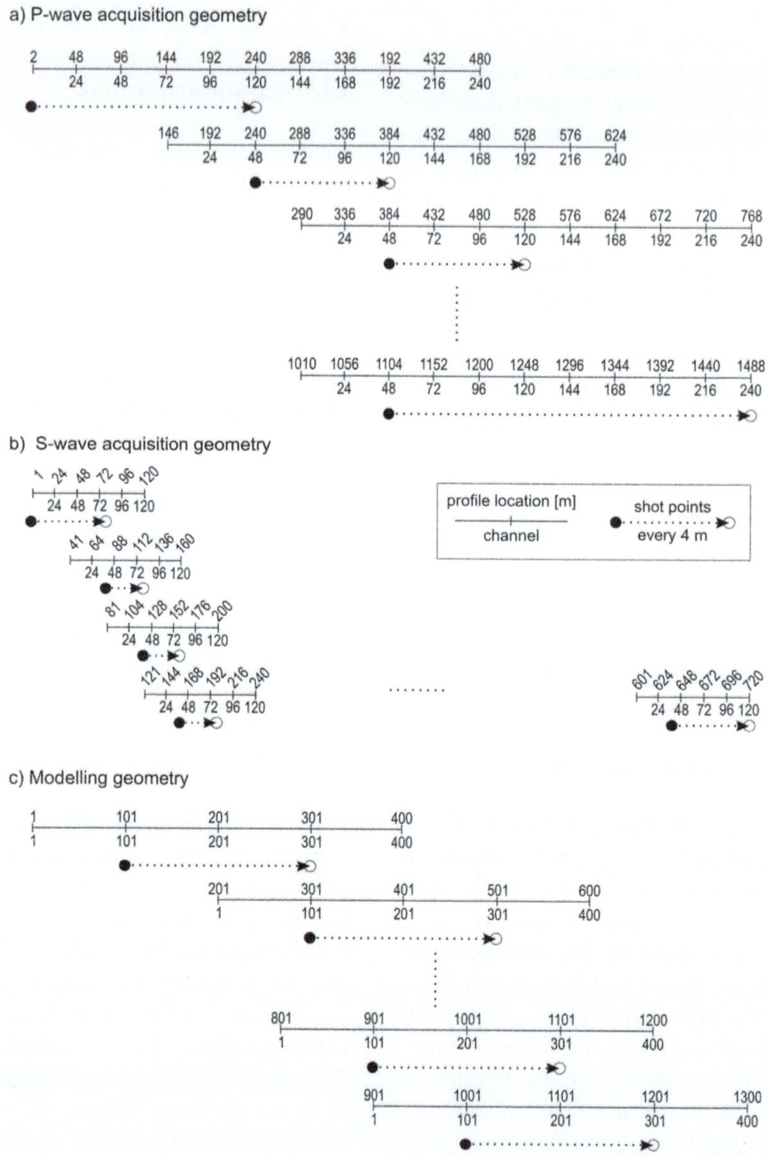

Figure 2. Acquisition geometries for **(a)** the P-wave surveys, **(b)** the SH-wave surveys, and **(c)** FD-modelling geometry simplification.

profile. Here, the surface is paved, which helps avoiding the generation of surface waves.

All P-wave profiles show a good signal-to-noise ratio (Fig. 3). Seismic reflections can be detected down to 1 s two-way-travel time (TWT). In contrast, the shear waves offer a smaller signal-to-noise ratio (Fig. 4). Reflection hyperbola signals are faint and within a reverberating background. Chevron patterns, also called herring-bone pattern, appear irregularly among shot gathers as part of the reverberations. The reflection signal bandwidth decreases with time. Below 0.7 s the signal vanishes, and further analysis in combination with seismic FD modelling is suggested.

4.2 Seismic processing

P-wave and shear-wave data processing differ due to different signal locations within the wave field. For the P-wave data, we set up a processing scheme (Table 1) focusing on the enhancement of the reflections (e.g. Yilmaz, 2001). Processing was carried out with Landmark's ProMAX 2D. The most important processing steps turned out to be muting of the surface waves, spectral whitening, and dynamic corrections, including dip move-out corrections. The detailed velocity analysis left a certain tolerance within the semblance calculations. We therefore included geological a priori knowledge to reduce the uncertainty so that the resulting interval velocity distribution better corresponds to the reflectors (Burschil et al., 2012a). At the end of the processing sequence, we applied a time-migration to the CDP (common-depth-point)-stacked section. The migrated section was time-to-depth converted, using a single velocity function from one borehole, yielding a seismic depth section (Fig. 5).

To test the robustness of the result, a post-stack depth migration of the non-migrated CDP-stacked section was also carried out (not shown here). It provided a similar seismic image as time-migration and time-to-depth conversion. Only small differences in the location of some reflectors of up to 3 m can be identified, but the reflector continuity remains. Therefore, we continued with the time-migrated, depth-converted section.

The shear-wave data contain surface wave interferences related to the specific shear-wave reflection move-out. This is the reason why we cannot simply mute the surface wave noise and purely focus on reflection signals, as we could for P waves. To enhance the reflection signals, we applied several techniques, e.g. fk filtering, spectral whitening, and deconvolution. Finally, automatic gain control (AGC, 300 ms) and fk-filtering with low-cutoff velocities below 350 m s^{-1} provided the best results (Table 1). This filter also removes a wide range of the chevron patterns that are present in the lower part of the seismograms around the shot location, depicted in Fig. 4. In contrast to Polom (1997), who investigated a chevron pattern that originated in ghost sweeps, we cannot identify a comparable increase of the frequency with time in the data. Here, the chevron pattern is rather monofrequent and no ghost sweeps can be detected within the pattern.

Velocity analysis was very difficult because reflections can only be identified sporadically and can hardly be traced through to neighbouring shots. This restrains velocity analysis in CMP (common-mid-point)-sorted data. After normal move-out correction and common midpoint stacking, we converted the section to depth with a velocity function derived from VSP measurements. The result is rather monofrequent, but the till layer as well as deeper reflectors can be identified (Fig. 6). However, the quality of this shear-wave survey is less compared to the P-wave seismic results.

5 Synthetic data from finite-difference modelling

To understand data quality differences between shear-wave measurements and compressional wave measurements on the island of Föhr, seismic wave-field modelling is introduced here for further data analysis. We chose the 2-D P/SV-version of SOFI for P-wave simulations and the 2-D SH version of SOFI for SH-wave modelling.

To create the input velocity and density models, we interpreted the major structures of the P-wave seismic field data depth section (Fig. 5) and assigned P-wave velocities and densities according to geological and geophysical a priori information (Fig. 7; Table 2). To supplement detailed shear-wave velocity information, we generated a cross-plot of velocity data (v_P and v_S) from the VSP surveys on the island of Föhr. This allowed us to calculate mean and median values and the linear regression for different lithological classes. For statistical confidence, the number of classes was limited to those for which more than 20 velocity samples were available. We then picked shear-wave velocities from the linear regression line of this cross-plot and thereby defined 14 lithological units with known v_P and v_S values that constitute the input model (Fig. 8; Burschil et al., 2012a).

For comparison of the synthetic data with the field data, we calculated a number of single shots with geometries only slightly different from the field geometries (Fig. 2). Differences are related to modelling requirements. Although receivers were simulated 1 m apart and shots 4 m apart (Fig. 2c), only every second receiver was used for later P-wave processing. The maximum offset of shear wave channels was restricted in later S-wave processing to respect the field geometry. Additionally, the efficient use of computational capacities required the splitting of the whole model into a number of model segments of which the westernmost is depicted in Fig. 9. Our FD-modelling approach required the use of absorbing frames at the left, right, and lower boundaries of each model segment. Therefore, the model segments were 600 m long and contained an absorbing frame width of 45 m (Table 3). Instead of a free surface, we also implemented an absorbing frame at the surface (Fig. 9). For this purpose, we expanded the model top layer by 50 m that

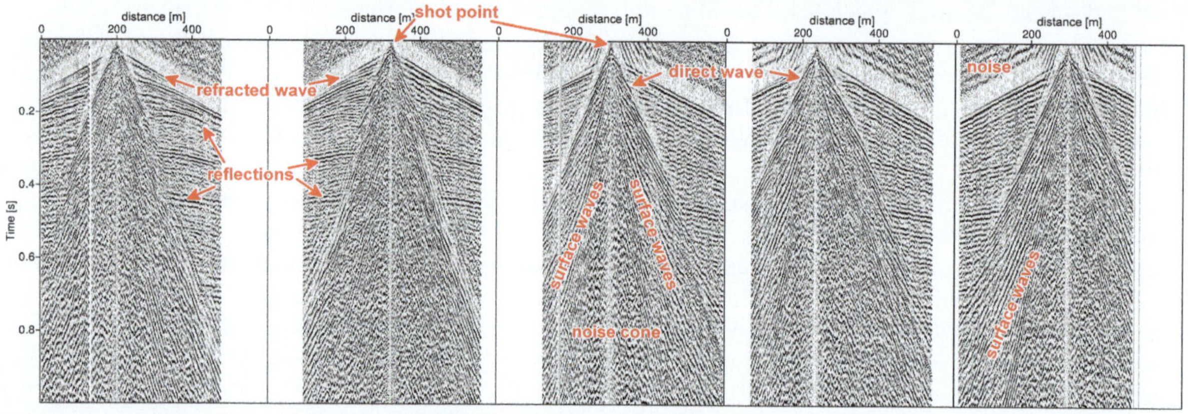

Figure 3. Seismic recordings of five single P-wave shots at different locations along the profile. The amplitude is displayed with an automatic gain control (AGC) of 150 ms.

Table 1. Overview of processing of P-wave and SH-wave seismic field data including normal move-out (NMO) correction, dip move-out (DMO) correction, and common midpoint (CMP) stacking (rms – root mean square).

Processing step	P-wave application	SH-wave application
Data load	SEG2-data load to ProMAX	SEG2-data load to ProMAX
Geometry installation	1 m bin interval	0.5 m bin interval
Vibroseis correlation	applied in the field	using individual sweep
Vertical stacking of records	applied in the field	twofold
Quality control	kill bad traces	kill bad traces
Refraction statics	calculated from first breaks; $v_{replace} = 1600\,\mathrm{m\,s^{-1}}$	not applied
Amplitude scaling	automatic gain control (300 ms length)	automatic gain control (200 ms length)
Fan filtering	not applied	low cut $350\,\mathrm{m\,s^{-1}}$
Deconvolution	zero-phase spiking deconvolution (80 ms length)	not applied
Time-variant filter	bandpass filter (36–220 Hz)	bandpass filter (30–160 Hz)
Trace muting	removal of noise cone	not applied
Interactive velocity analysis	100 m node spacing, iteratively	100 m node spacing, iteratively
Residual statics correction	correlation auto statics	not applied
NMO correction	rms velocity function, 30 % stretch mute	rms velocity function, 300 % stretch mute
DMO preparation	8 m DMO bin spacing, 24 bins	not applied
DMO correction	average single velocity function	not applied
Trace muting	top mute	not applied
CMP stacking		
Amplitude scaling	automatic gain control (300 ms operator length)	not applied
Time-variant filter	bandpass filter (passage window: 30–220 Hz)	not applied
Finite-difference time migration	smoothed velocity function	not applied
Time-to-depth conversion	single velocity function from VSP	single velocity function from VSP

hosts the absorbing frame of 45 m. We did this to suppress surface multiples, which is the same approach that was used by Jones (2013). The effect of this absorbing frame is comparable to a weathering layer with high parameter gradients. Additionally, we introduced a low-velocity layer with a free surface to test the simulation of a weathering layer. We used a zero-phase Ricker wavelet for the simulations instead of a vibro-seismic correlated sweep signal, i.e. a Klauder wavelet. This is a practical compromise which takes into account that the field data signals are absorbed to a certain degree and do not show a white frequency spectrum. In contrast to the field data source signals, we used the same central frequency of the Ricker wavelet for P-wave simulation as well as for SH-wave simulation.

We used the same input models (Fig. 7; Table 2) and settings (Table 3) for P-wave modelling and SH-wave modelling, and only varied the modelled trace length. Due to lower shear-wave velocities, we had to consider longer travel times for the shear-wave signals and simulated data up to 2.5 s TWT.

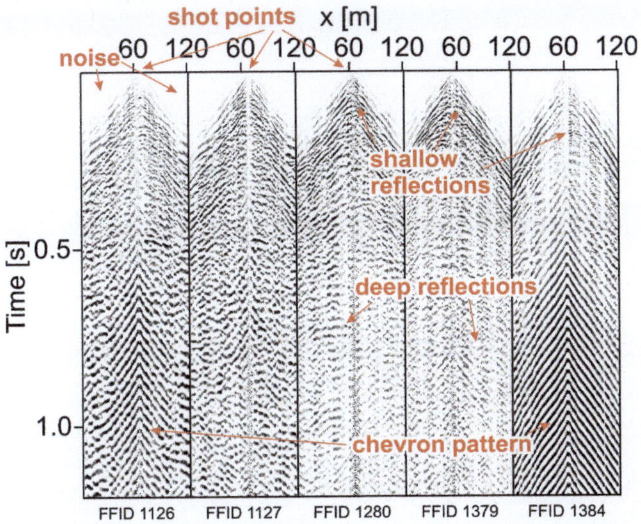

Figure 4. Seismic recordings of five single shear wave shots with spatial divergence correction and AGC of 300 ms applied.

Table 2. Seismic velocities and densities for refined input model (cf. Fig. 7b). Shear-wave velocity was calculated according to the cross-plot relation (Fig. 8).

Unit	v_P (m s^{-1})	v_S (m s^{-1})	ρ (kg m^{-3})
1	1600	330.74	2200
2	1700	366.37	2300
3	1900	437.62	2600
4	1600	330.74	2200
5	1700	366.37	2400
6	1750	384.18	2350
7	1800	401.99	2350
8	1700	366.37	2300
9	1750	384.18	2300
10	1800	401.99	2300
11	1850	419.81	2600
12	1900	437.62	2700
13	2100	508.88	2700
14	2300	580.13	2700

5.1 Computational facility

The computer we are currently using for the simulations is a DELL Poweredge 910. It accommodates four CPUs Intel Xeon E7-4870 with 2.4 GHz, 10 cores each, DDR3 memory, and hyper threading. That results in 80 usable single CPUs with SuSE 11.4 Linux. Our computer hosts 512 GB of RAM, which is by far larger than needed for our 2-D models but will be necessary for larger 3-D simulations.

The computational time for modelling strongly depends on the number of modelled time steps and the size of the model. For P-wave modelling, we simulated the propagation for 1 s and 20 000 time steps (Table 3). The computational time for one single shot with the settings described was ca. 30 min on 2 CPUs using the 2-D P–SV version of SOFI. 2-D shear-wave modelling with the SH version of SOFI was much faster per computational step. Simulating 2.5 s, i.e. 50 000 time steps, of wave propagation took ca. 60 min per shot.

5.2 P-wave modelling

In the simulated single shot data (Fig. 10), we detect direct P and SV waves as well as reflections of P and SV waves. The surface wave pattern of the field data is not present in the synthetic data, (cf. noise cone in Fig. 3). P-wave reflections can be tracked in the synthetic data, even in those regions where they are usually covered by surface waves. Also, the simulated data have an apparent higher frequency content, not showing the typical subsurface low-pass filter effect.

We applied a simple processing scheme to all 300 modelled single shots comprising amplitude control, frequency filtering, normal move-out correction, common midpoint stacking, time-migration, and time-to-depth conversion. Stacking velocities were picked via an interactive velocity analysis from CMP gathers.

In the migrated depth section (Fig. 11a), the uppermost reflector at 30 m is faint. Below, two strong reflectors mark the upper and lower boundaries of the till unit. The upper reflector can be traced through the entire section with varying amplitude. Also, a large number of reflectors with different amplitudes is imaged. In the central part of the seismogram at 150 m depth, two close reflectors mark the lower end of the complex geological units. Below, two nearly horizontal reflectors at 250 and 380 m depths are present. At the western and eastern edges of the section the migration process generated minor artefacts. Because this P-wave depth section was basically able to explain the P-wave field data, we continued with the simulation of the shear-wave field data instead of a further study of additional features present in the P-wave field data, like surface-wave ground roll.

5.3 Shear-wave modelling

Shear-wave propagation was modelled using the same input model and settings. We restricted the modelled data to a maximum offset of 80 m that corresponds to the maximum offset in the field data. Resulting shot gathers show the direct S-wave as well as several shear-wave reflections (Fig. 12). The sharp reflection signal does not change its shape with travel time. Because we used the same frequency band for the P-wave and S-wave source simulations, the stacked section (Fig. 13a) shows an even more detailed image of the subsurface due to the shorter wavelengths of the shear waves compared with the P waves. All structures of the input model are imaged with excellent resolution.

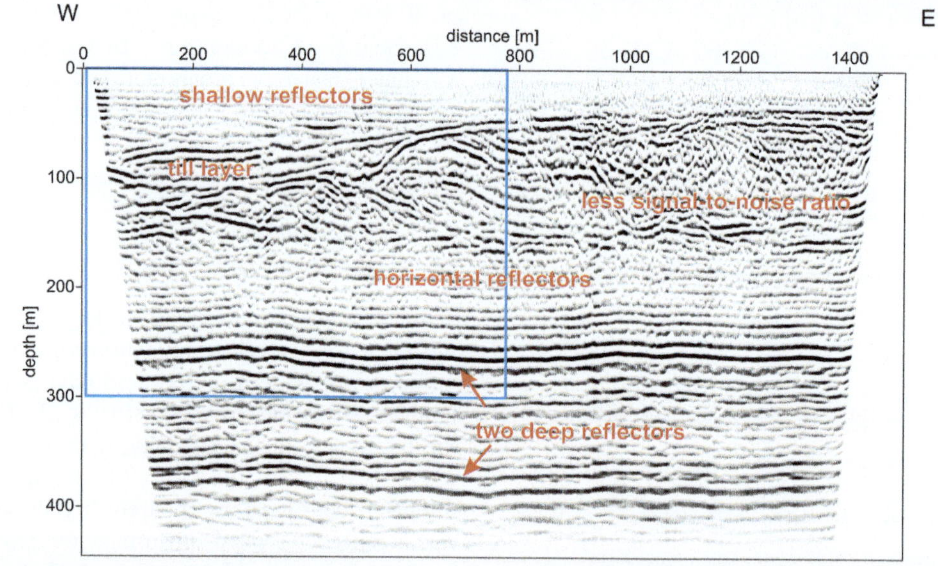

Figure 5. Final time migrated depth section of the P-wave seismic survey with AGC of 200 ms applied before time-to-depth conversion. The blue box marks the position of the SH-wave section (Fig. 6).

Table 3. Input parameters for FD modelling. Grid size is specified by node spacing dh, and absorbing frame width FW characterizes the width of the absorbing frame. dt specifies the calculated time stepping.

	P wave	SH wave
Model size	1200 × 1100 nodes, dh = 0.5 m	1200 × 1100 nodes, dh = 0.5 m
Propagation time	time = 1 s, time steps dt = 5×10^{-5} s	time = 2.5 s, time steps dt = 5×10^{-5} s
Source wavelet	Ricker, 100 Hz central frequency, vertical force	Ricker, 100 Hz central frequency, SH-wave force
Boundary	FW = 90 nodes, 6 % damping per node	FW = 90 nodes, 6 % damping per node
Receiver	vertical geophone, spacing = 2 nodes	horizontal (SH) geophone, spacing = 2 nodes
Seismograms sampling rate	every 10th time step	every 20th time step
Simulation time per shot	ca. 30 min	ca. 60 min

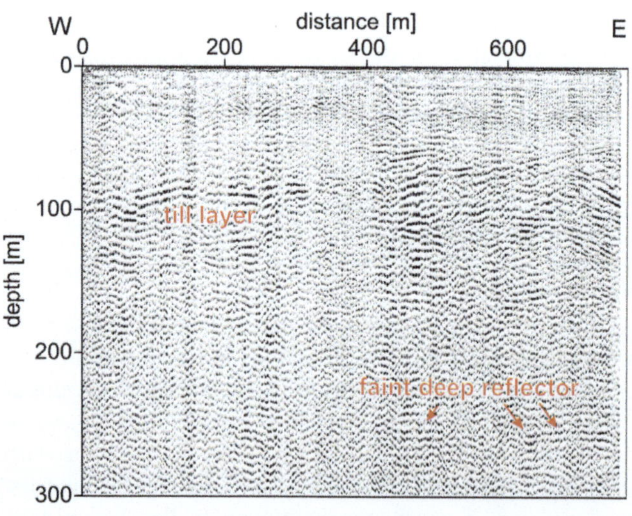

Figure 6. Depth-converted stack of the shear-wave seismic survey with AGC of 300 m applied after time-to-depth conversion.

6 Discussion

In reflection seismic surveys normally only one type of wave is utilized, for instance P waves, shear waves, or converted waves. The use of more than one wave type or even the integration of multicomponent technology (Hardage et al., 2011) is rare but growing. This is even more pronounced for near-surface applications. Recently, a few studies reported the combination of shallow P-wave and shear-wave seismics, comparable to the field work presented in this article (e.g. Pugin et al., 2004, 2009b; Malehmir et al., 2013; Sauvin et al., 2014). Even 9C reflection seismics was tested in near-surface exploration (Pugin et al., 2009a). All of these studies successfully recorded high-resolution data of mainly good quality. The signal-to-noise ratio of the P-wave data is similar to the P-wave data presented in this study, except in Pugin et al. (2009a). In contrast, the shear-wave data of these studies are of good quality and constitute prominent, coherent reflections. The equipment and acquisition geometry slightly vary among the reported surveys, but the combination of vibrator

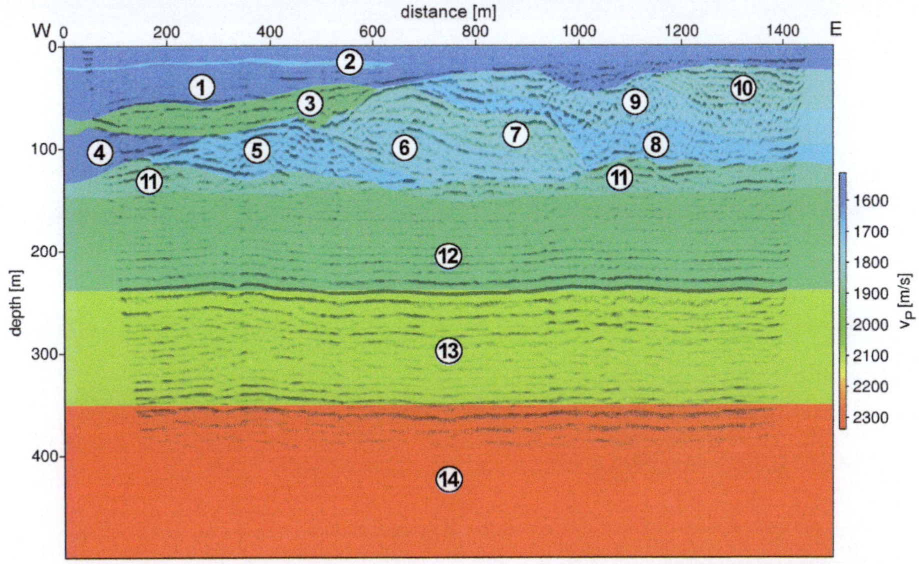

Figure 7. Input model of P-wave velocity. The P-wave field depth section is superimposed. Numbers 1–14 mark the units listed in Table 2.

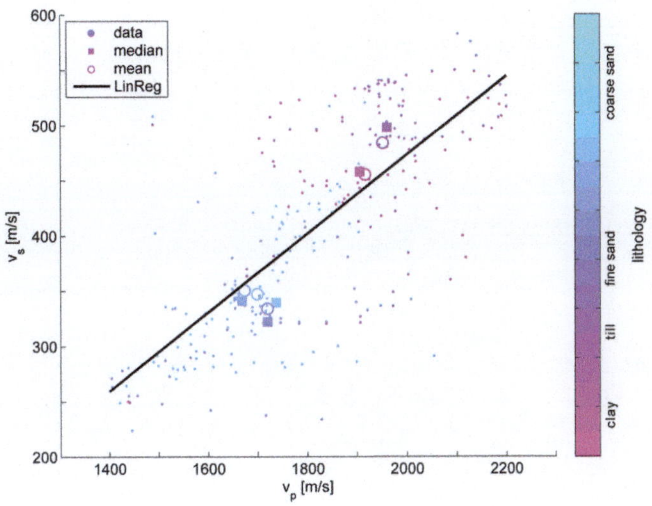

Figure 8. v_P/v_S cross-plot from VSP data colour-coded for different lithologies. Additionally, median values, mean values and the linear regression (LinReg) are indicated. The shear-wave velocity was calculated for each P-wave velocity with the relation resulting from the linear regression.

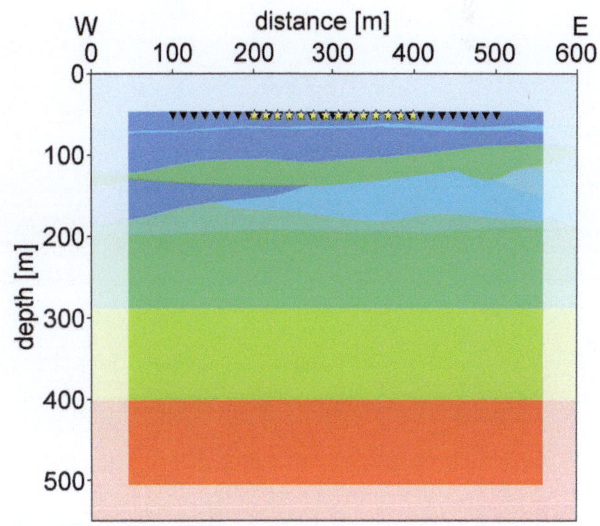

Figure 9. Westernmost model segment of the input model. Individual single shots were modelled and the yellow stars mark the locations of the shot points (every 4 m). Black triangles mark the receiver position (each 1 m). The region of the absorbing frame is faded out.

and land streamer is the favourite choice for shear-wave seismics. General data processing reported in some of the studies is similar to the processing we finally applied (Table 1). Sauvin et al. (2014) reported the application of elevation statics for only one of their shear-wave profiles. None of the authors reported refraction statics for shear waves. In some cases, differences in shear-wave processing are related to deconvolution and spectral whitening, which was applied by Pugin et al. (2009b) and Sauvin et al. (2014). Here, deconvolution and spectral whitening did not provide success.

So far, we have not been able to reproduce the shear-wave field data with seismic modelling. A similar observation is reported by Bellefleur et al. (2012). They calculated synthetic P-wave data of four different input models from the surface to a mineralized zone and compared these data with field data. Their data excellently correlate with data from a VSP, but they consider the correlation with a surface 3-D reflection seismics data set as rather poor. They explain the poor surface data quality with the appearance of surface waves and a higher amount of scattering at inhomogeneities. Surface

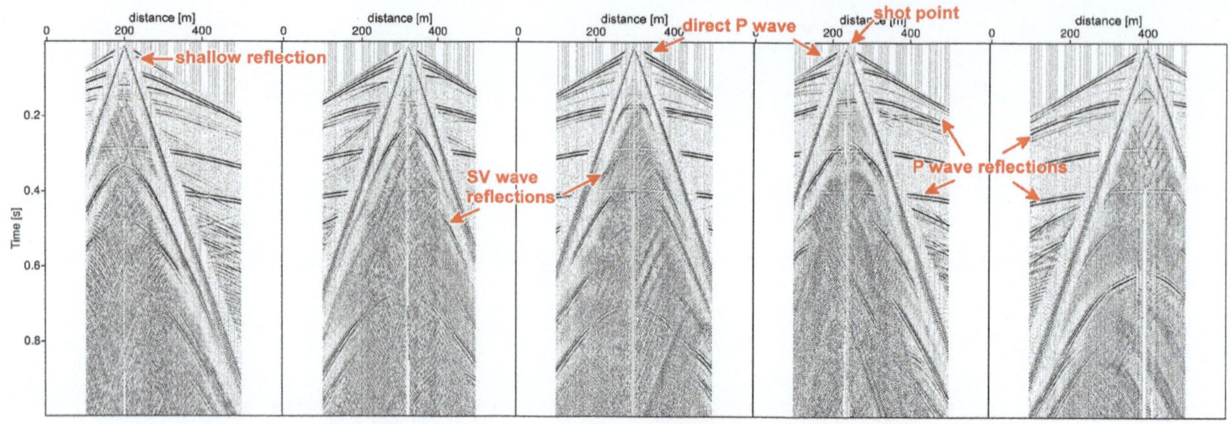

Figure 10. Five different synthetic P-wave single shots. Shot gathers are displayed with 150 ms AGC.

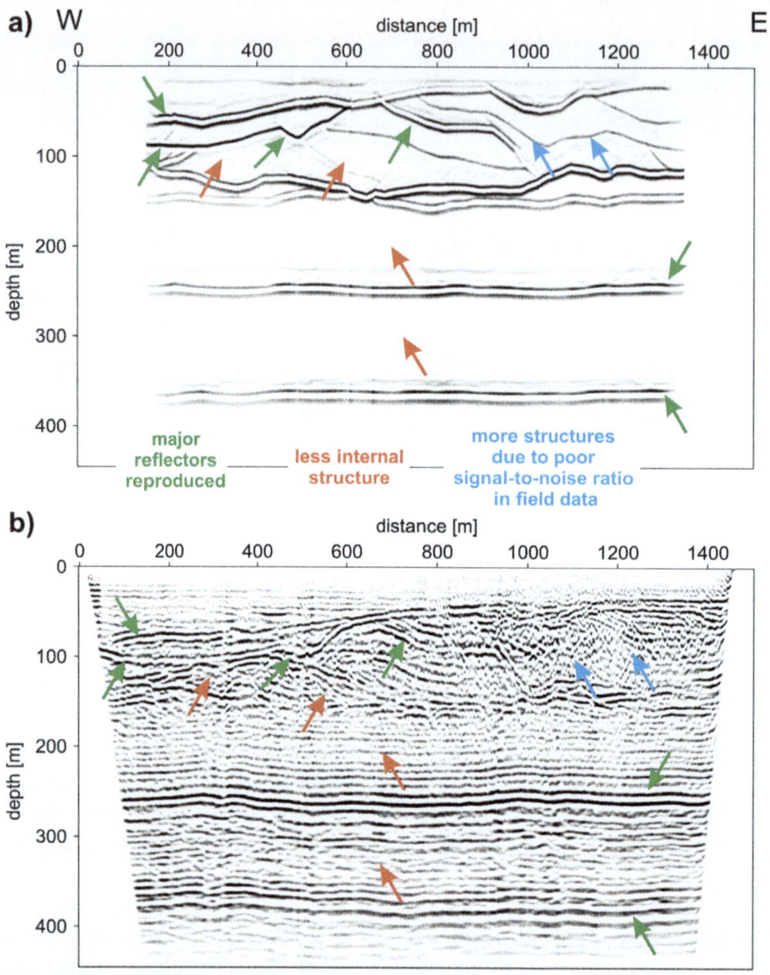

Figure 11. FD modelling of the P-wave section (**a**) and section from field data (**b**). Both sections are displayed with AGC of 100 ms.

waves are not present in VSP data and do not affect data quality significantly. Scattering also affects the surface data more than VSP data because the travel path of seismic body waves from the source to the scattering point and to the receiver is

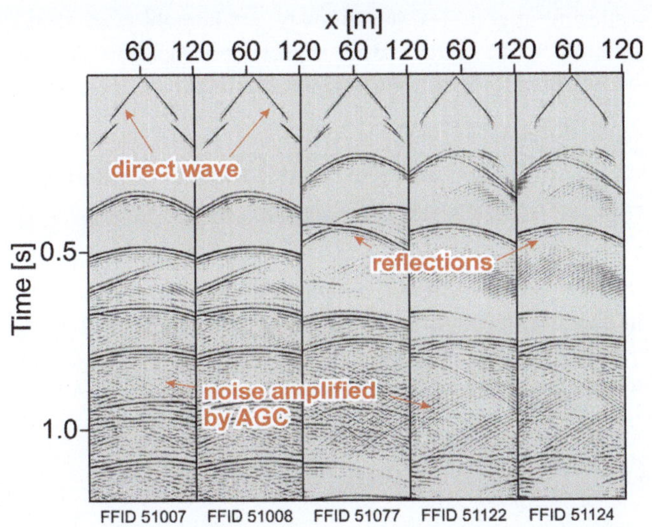

Figure 12. Five different shear wave FD-modelling single shots. Shot gathers with 120 m spread, amplified by 300 ms AGC, are displayed.

longer for surface data than for VSP data. If this explanation is true, we will have to take small-scale inhomogeneities and a weathering layer as origin of surface waves into account for future studies.

6.1 Influence on data quality in land seismic surveys

In general, a number of factors can influence the quality of land seismics data, listed and illustrated in Sheriff (1975). Typical factors, one would assume to be the most likely in our case, are source strength, inappropriate coupling of sources and receivers, superimposed surface waves, multiples, scattering, and intrinsic absorption (Q).

The vertical hydraulic vibrator sources MHV2.7 and HVP-30 have proven their ability to reach reflectors at least 2 km deep (Buness et al., 2009). The LIAG shear-wave vibrator source MHV-4S has been able to generate clear reflections at least as deep as 200 m in fluviatile and marine deposits (Polom et al., 2010). Sauvin et al. (2014) used a wheelbarrow-mounted microvibrator source for the analysis of quick clays. Their data show clear reflections from 40 m below ground level within fluviatile and marine sediments. In the underlying bedrock, reflections can be traced down to 120 m depth. Malehmir et al. (2013) report clear reflections in a similar environment from at least 40 m below ground level with the same source. This means that even small sources are able to create reflections from at least 40 m below the source. Pugin et al. (2009b) used an IVI Minivib on a minibuggy carrier, which is comparable to the MHV-4S. They show SH-wave reflections down to about 50 m in glacial deposits before reaching the bedrock. In the light of these studies, the MHV-4S source can be expected to be strong enough to image the glacially overworked deposits

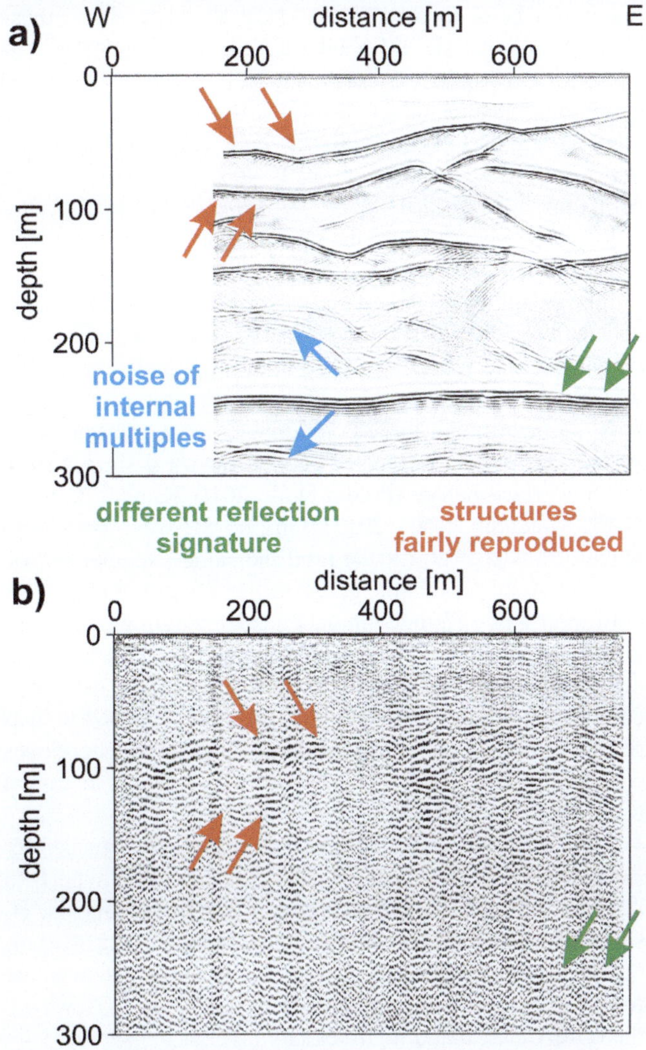

Figure 13. Shear wave FD-modelling section **(a)** and section from field data **(b)**. Both sections are displayed with AGC of 300 ms applied.

down to 150 m. In fact, the SH-wave image shows faint reflections even from ca. 270 m depth (Fig. 6).

The ground coupling of the sources is mainly affected by the vehicle mass and the driving peak force of the system. For P-wave sources this mass acts as a gravitational hold-down force that compensates the vertical force of the vibrator unit (Sheriff and Geldart, 1995). For S-wave sources, this mass increases horizontal friction of the base plate and thus prevents sliding of the shear-wave vibrator unit on the ground. We use the rule of thumb that the gravitational hold-down force of the vibrator mass on loose ground should at least be twice the peak shear force. Experience shows that under favourable conditions, the peak force of a vibrator with a rubber base plate on paved roads can reach 70 % of the hold-down force without sliding. The P-wave hydraulic vibrator systems MHV2.7/HVP-30 have masses of 2.6 t/4 t and maximum peak forces of 27.6 kN/30 kN. Therefore, these vibrator systems should not be affected by bouncing. The shear-wave vibrator system MHV-4S has a mass of 4 t and a maximum peak force of 30 kN. During the survey, we used a peak force of 17 kN on a paved road. Therefore, the gravitational hold-down force of nearly 40 kN should have been sufficient.

Receiver coupling with the ground is another factor. Carefully planted geophones typically offer proper receiver–ground coupling for P-wave surveys (Krohn, 1984). The P-wave field data we acquired support this expectation. For the shear-wave survey, we used the LIAG land streamer on which geophones are directly fixed to aluminium plates that have a gravitational three-point contact to the ground. This system has proven to receive good shear-wave signals from the subsurface before (Polom et al., 2010; Krawczyk et al., 2013; Malehmir et al., 2013). On one SH-wave profile on gravel, Sauvin et al. (2014) used the same streamer but on grassland they planted the geophones in the ground. They do not report any difference in data quality. Pugin et al. (2004, 2009a, b) use a different land streamer system that works successfully as well. In the field, we took care that the geophone coupling was good. Every time a new streamer position was reached, a noise test was carried out and noisy geophones were coupled to the ground by hand. We therefore expect coupling to be sufficient.

Surface waves are an important degradation factor for shear-wave reflection surveys. Here, the reflection hyperbola often interferes with surface wave (Love wave) signals; cf. surface waves (Fig. 3) and reflection hyperbolas (Fig. 4). This factor has complicated the application of shear waves in the past and still limits the application of this method. However, the observation that a high-velocity layer at the surface suppresses the generation of surface waves in shear-wave exploration (e.g. Inazaki, 2004) was an important step to overcome this limitation in many urban applications and even in rural environments if paved or consolidated roads are present. Even though surface waves are often excited during an SH-wave surveys on unconsolidated surfaces, SH-wave reflection seismics surveys have been successful in these conditions (e.g. Malehmir et al., 2013; Sauvin et al., 2014). Polom et al. (2013) identify partly dispersive Love waves that show a similar signature as the chevron pattern depicted in Fig. 4. In their case and in our case, these waves seem to be linked to the shot-point location. Even neighbouring shot points show strong variations in this respect. For instance, in Fig. 4 we show shot gathers with shot-point locations of FFID's 1126 and 1127 that are just 4 m apart. To cancel surface waves with a linear move out or mild dispersive character fk filtering is often successfully used (e.g. Polom et al., 2013; Sauvin et al., 2014). We successfully applied an fk filter as well (Table 1), but this did not improve the coherency of reflectors in the final depth section (Fig. 6).

Further issues for data quality of land seismics data are intrinsic damping (e.g. Kang and McMechan, 1994) and scattering at complex structures (e.g. Cheraghi et al., 2013; Malehmir et al., 2013). Both aspects are crucial to simulate the influence of a near-surface weathering layer.

6.2 Comparison of P-wave and S-wave simulations

We calculated synthetic seismograms for P-wave propagation through two different input models as well as shear-wave propagation through one of these models. In the following, we will compare these modelled data with the corresponding field data and discuss the observations.

6.2.1 P-wave shot gathers

The simulated P-wave and the corresponding field data of single shot seismograms (Figs. 3, 10) contain reflection events that show a similar basic waveform. However, the field data (Fig. 3) are much noisier, in particular before the first break as well as inside the surface-wave cone. No reflections are detectable inside this field data cone whereas in the synthetic data (Fig. 10) reflections can be traced through all parts of the seismogram. This can be explained to a large degree by the lack of a weathering layer in the models. The absence of this layer prevents the build-up of simulated surface waves and thus, a surface wave cone is missing in simulated shot records. Therefore, the refracted wave in the field data corresponds to the direct wave in the synthetic data.

Another noticeable observation is the signature of the reflections. In general, the reflection signals of the synthetic data seem to be more focused whereas many of the reflections within the field data seem to be made up of a number of oscillations/reverberations instead of single reflections (compare reflection signal at 0.3 s in Figs. 3 and 10). We expect an additional fine-layering within the 14 units of the input model to be able to reproduce this observation (Fig. 10). Perhaps, the simulation of a multilayer surface unit, related to the weathering zone might also help explaining the reverberation observation. We also simulated a low-velocity top layer (thickness = 10 m, $v_P = 500\,\mathrm{m\,s^{-1}}$, $v_S = 200\,\mathrm{m\,s^{-1}}$, and $\rho = 2000\,\mathrm{kg\,m^{-3}}$) with a free surface, which corresponds to the unsaturated zone. In this case, the synthetic seismograms contain a large number of multiples between the first impedance contrast and the free surface. The multiples superimpose all reflection signals and cannot be identified as such in the field data. It turned out that it needs additional systematic tests to figure out a surface layer configuration that may be able to simulate this aspect of the field data. For this reason, we used an absorbing frame at the top of the model. Even though we cannot reproduce an exact copy of the reflections of the field data, the major reflections occur in good quality and quantity to use this model for additional shear-wave simulations.

6.2.2 SH-wave shot gathers

Shear-wave shot simulations show larger differences compared with their corresponding field data (Figs. 4 and 12). The clear and continuous reflections in the synthetic data are not present in the field data. Some of the field records show the mentioned chevron pattern parallel to the first break with varying amplitude (Fig. 4) that does not show up in the synthetic data. In the model, we did not consider very shallow structures that could create Love waves.

The spectrum of the synthetic data does not change with time in the seismogram (Fig. 12). The Ricker wavelet has the same spectral shape for shallow reflections as for deep reflections. This is no surprise since up to this point we have not included any kind of signal damping in the model like intrinsic damping or sources for attenuation through scattering or interference. In contrast, the field data signals for longer travel times lose some of the higher-frequency components (Fig. 4). This indicates some kind of intrinsic or scattering attenuation. However, this cannot explain the lateral appearance variation among shot gathers of the shear-wave data.

6.2.3 P-wave depth section

The post-stack time-migrated P-wave depth section of synthetic data shows less noise and less amplitude variability while the corresponding field data show natural levels of different signals and thus contain more information about small-scale and internal structures (Fig. 11). The modelled depth section reproduces the major features of the field data. The till top reflector between 50 and 80 m depth appears continuous and two deep reflectors show as well. Some of the dipping reflectors in this modelled section add details that similarly appear in the field data.

The field data section (Fig. 11b) can be divided into two parts: excitation on paved street and on grassland. Within the seismic field data section, we detected a lack of resolution in the grassland part (about 900–1400 m, marked with blue arrows in Fig. 11b) that is not present in the modelled data. However, we have not implemented the structural complexity of unconsolidated grassland as a near-surface weathering layer, i.e. large velocity contrasts, inhomogeneities, and intrinsic damping, in the model so far.

In general, the reflection signatures in the field data spatially vary more strongly than in the synthetic data (Fig. 11). Sources for these observations in the field data can be intrinsic damping, scattering attenuation including fine layering, and inhomogeneous lithological units. All of these features were not included in the input model (Fig. 7). In a natural environment, complex structures or fine layers can be sources of multiples and wave conversion that subsequently can pose similar challenges to data processing like random noise signals do. Since the model consisted of comparatively large units, this kind of noise was not simulated.

6.2.4 SH-wave stacked sections

The SH-wave stacked sections of synthetic data and field data (Fig. 13) differ more than the P-wave sections. While no clear interpretation can be carried out for the stacked field data, the stacked section from the modelled data set well reproduces the input model. For example reflector segments occur at 90 m depth, which correspond to the interface of the subhorizontal layer at 90 m (top till layer) and at 250 m depth corresponding to the first deep reflector in the P-wave seismic section. Nonetheless, the reflection signatures in the field data are much noisier than in the synthetic data.

7 Summary and outlook

Shear-wave field data recorded on the island of Föhr showed less quality compared with their compressional wave counterparts. To comprehend the reasons for this quality difference, we used seismic wave-field modelling within simple models of the subsurface, using the seismic field geometry. We chose finite-difference modelling to try to reproduce the field data because of its ability to simulate the entire wave field and to allow for arbitrary input models. We come to the following conclusions.

1. We were able to simulate P waves that show clear first-order similarities compared with the P-wave field data.

2. Simplified FD modelling does not explain the small signal-to-noise ratio of the shear-wave field data.

For future analyses, we therefore suggest to consider additional complexity in the subsurface model that will presumably be able to explain the different quality of compressional and shear-wave field data. The most important additional factors are intrinsic damping, thin layers within the modelled units, a complex near-surface weathering layer structure, and heterogeneous material within the layers. While 2-D calculations gain faster results and allow testing the effect of different features, the full complexity of field acquisition may be understood using 3-D simulations in the future.

Acknowledgements. We acknowledge Karlsruhe Institute of Technology (KIT) for providing an academic licence of the SOFI software package. We thank Y. Yang (referee), the anonymous second referee, and Helga Wiederhold for their helpful comments.

Edited by: K. Liu

References

Alterman, Z. and Karal, F.: Propagation of elastic waves in layered media by finite difference methods, B. Seismol. Soc. Am., 58, 367–398, 1968.

Bellefleur, G., Malehmir, A., and Müller, C.: Elastic finite-difference modeling of volcanic-hosted massive sulfide deposits: A case study from Half Mile Lake, New Brunswick, Canada, Geophysics, 77, WC25–WC36, 2012.

Bohlen, T.: Parallel 3-D viscoelastic finite difference seismic modelling, Comput. Geosci., 28, 887–899, 2002.

Buness, H., Gabriel, G., and Ellwanger, D.: The Heidelberg Basin drilling project: Geophysical pre-site surveys, E&G – Quaternary Science Journal, 57, 338–366, 2009.

Burschil, T., Scheer, W., Kirsch, R., and Wiederhold, H.: Compiling geophysical and geological information into a 3-D model of the glacially-affected island of Föhr, Hydrol. Earth Syst. Sci., 16, 3485–3498, doi:10.5194/hess-16-3485-2012, 2012a.

Burschil, T., Wiederhold, H., and Auken, E.: Seismic results as a-priori knowledge for airborne TEM data inversion – A case study, J. Appl. Geophys., 80, 121–128, doi:10.1016/j.jappgeo.2012.02.003, 2012b.

Carcione, J. M., Herman, G. C., and ten Kroode, A.: Seismic modeling, Geophysics, 67, 1304–1325, 2002.

Cheraghi, S., Malehmir, A., Bellefleur, G., Bongajum, E., and Bastani, M.: Scaling behavior and the effects of heterogeneity on shallow seismic imaging of mineral deposits: A case study from Brunswick No. 6 mining area, Canada, Journal of Applied Geophysics, 90, 1–18, 2013.

Fichtner, A., Bleibinhaus, F., and Capdeville, Y.: Full seismic waveform modelling and inversion, Springer, 2011.

Gold, N., Shapiro, S. A., and Burr, E.: Modeling of high contrasts in elastic media using a modified finite difference scheme, in: 68th Annual International Meeting, Soc. Explor. Geophys., Expanded Abstracts, 1997.

Groos, L., Schäfer, M., Forbriger, T., and Bohlen, T.: On the significance of viscoelasticity in a 2D full waveform inversion of shallow seismic surface waves, in: 74th EAGE Conference & Exhibition, Copenhagen, 2012.

Harbo, M. S., Pedersen, J., Johnsen, R., and Peteren, K.: Groundwater in a Future Climate: The CLIWAT Handbook, The CLIWAT Project Group, 2011.

Hardage, B. A., DeAngelo, M. V., Murray, P. E., and Sava, D.: Multicomponent seismic technology, Society of Exploration Geophysicists, 2011.

Inazaki, T.: High-resolution seismic reflection surveying at paved areas using an S-wave type land streamer, Explor. Geophys., 35, 1–6, 2004.

Jones, I. F.: Tutorial: the seismic response to strong vertical velocity change, First Break, 31 (6), 79–90, doi:10.3997/1365-2397.2013018, 2013.

Jørgensen, F., Scheer, W., Thomsen, S., Sonnenborg, T. O., Hinsby, K., Wiederhold, H., Schamper, C., Burschil, T., Roth, B., Kirsch, R., and Auken, E.: Transboundary geophysical mapping of geological elements and salinity distribution critical for the assessment of future sea water intrusion in response to sea level rise, Hydrol. Earth Syst. Sci., 16, 1845–1862, doi:10.5194/hess-16-1845-2012, 2012.

Kang, I. B. and McMechan, G. A.: Separation of intrinsic and scattering Q based on frequency-dependent amplitude ratios of transmitted waves, J. Geophys. Res., 99, 23875–23885, 1994.

Kirsch, R. (Ed.): Groundwater Geophysics: A Tool for Hydrogeology, Springer, 2008.

Krawczyk, C. M., Polom, U., and Beilecke, T.: Shear-wave reflection seismics as a valuable tool for near-surface urban applications, The Leading Edge, 32, 256–263, 2013.

Krohn, C. E.: Geophone ground coupling, Geophysics, 49, 722–731, 1984.

Levander, A. R.: Fourth-order finite-difference P-SV seismograms, Geophysics, 53, 1425–1436, 1988.

Malehmir, A., Bastani, M., Krawczyk, C. M., Gurk, M., Ismail, N., Polom, U., and Persson, L.: Geophysical assessment and geotechnical investigation of quick-clay landslides – a Swedish case study, Near Surf. Geophys., 11, 341–350, 2013.

Miller, R. D.: Introduction to this special section: Urban geophysics, The Leading Edge, 32, 248–249, 2013.

Plessix, R.-E.: Waveform inversion overview: Where are we? And what are the challenges?, in: 74th EAGE Conference & Exhibition, Copenhagen, 2012.

Polom, U.: Elimination of source-generated noise from correlated vibroseis data (the "ghost-sweep" problem), Geophys. Prospect., 45, 571–591, 1997.

Polom, U., Hansen, L., Sauvin, G., L'Heureux, J.-S., Lecomte, I., Krawczyk, C. M., Vanneste, M., and Longva, O.: High-resolution SH-wave seismic reflection for characterization of onshore ground conditions in the Trondheim harbor, central Norway, in: Advances in Near-Surface Seismology and Ground-Penetrating Radar, SEG, Tulsa, 297–312, 2010.

Polom, U., Druivenga, G., Grossmann, E., Grüneberg, S., and Rode, W.: Transportabler Scherwellenvibrator, Patent application, German Patent and Trade Mark Office, 5 pages, 2011 (in German).

Polom, U., Bagge, M., Wadas, S., Winsemann, J., Brandes, C., Binot, F., and Krawczyk, C.: Surveying near-surface depocentres by means of shear wave seismics, First Break, 31 (8), 67–79, 2013.

Pugin, A. J. M., Larson, T. H., Sargent, S. L., McBride, J. H., and Bexfield, C. E.: Near-surface mapping using SH-wave and P-wave seismic land-streamer data acquisition in Illinois, US, The Leading Edge, 23, 677–682, 2004.

Pugin, A. J.-M., Pullan, S. E., and Hunter, J. A.: Multicomponent high-resolution seismic reflection profiling, The Leading Edge, 28, 1248–1261, 2009a.

Pugin, A. J.-M., Pullan, S. E., Hunter, J. A., and Oldenborger, G. A.: Hydrogeological prospecting using P-and S-wave landstreamer seismic reflection methods, Near Surf. Geophys., 7, 315–327, 2009b.

Robertsson, J. O., Blanch, J. O., and Symes, W. W.: Viscoelastic finite-difference modeling, Geophysics, 59, 1444–1456, 1994.

Robertsson, J. O., Bednar, B., Blanch, J., Kostov, C., and van Manen, D.-J.: Introduction to the supplement on seismic modeling with applications to acquisition, processing, and interpretation, Geophysics, 72, SM1–SM4, 2007.

Romdhane, A., Grandjean, G., Brossier, R., Rejiba, F., Operto, S., and Virieux, J.: Shallow-structure characterization by 2D elastic full-waveform inversion, Geophysics, 76, R81–R93, 2011.

Rumpel, H.-M., Binot, F., Gabriel, G., Siemon, B., Steuer, A., and Wiederhold, H.: The benefit of geophysical data for hydrogeological 3D modelling an example using the Cuxhaven buried valley, Z. Dtsch. Ges. Geowiss., 160, 259–269, 2009.

Saenger, E. H., Gold, N., and Shapiro, S. A.: Modeling the propagation of elastic waves using a modified finite-difference grid, Wave Motion, 31, 77–92, 2000.

Sauvin, G., Lecomte, I., Bazin, S., Hansen, L., Vanneste, M., and L'Heureux, J.-S.: On the integrated use of geophysics for quick-clay mapping: The Hvittingfoss case study, Norway, J. Appl. Geophys., 106, 1–13, doi:10.1016/j.jappgeo.2014.04.001, 2014.

Scheer, W.: Geologie und Landschaftsentwicklung von Schleswig-Holstein, in: Der Untergrund von Föhr: Geologie, Grundwasser und Erdwärme; Ergebnisse des Interreg-Projektes CLIWAT, 11–20, Schriftenreihe LLUR SH – Geologie und Boden, Landesamt für Landwirtschaft, Umwelt und ländliche Räume des Landes Schleswig-Holstein, Flintbek, 2012 (in German).

Sheriff, R. E.: Factors affecting seismic amplitudes, Geophys. Prospect., 23, 125–138, 1975.

Sheriff, R. E. and Geldart, L. P.: Exploration Seismology, Cambridge University Press, 1995.

Sørensen, K. I. and Auken, E.: SkyTEM – a new high-resolution helicopter transient electromagnetic system, Explor. Geophys., 35, 194–202, 2004.

Steeples, D. W. and Miller, R. D.: Seismic reflection methods applied to engineering, environmental, and groundwater problems, Geotechnical and Environmental Geophysics, 1, 1–30, 1990.

Virieux, J.: P-SV wave propagation in heterogeneous media: Velocity-stress finite-difference method, Geophysics, 51, 889–901, 1986.

Virieux, J. and Operto, S.: An overview of full-waveform inversion in exploration geophysics, Geophysics, 74, WCC1–WCC26, 2009.

Wiederhold, H., Siemon, B., Steuer, A., Schaumann, G., Meyer, U., Binot, F., and Kühne, K.: Coastal aquifers and saltwater intrusions in focus of airborne electromagnetic surveys in Northern Germany, in: Proc. 21st Salt Water Intrusion Meeting, Azores, 94–97, 2010.

Yilmaz, Ö.: Seismic data analysis: processing, inversion, and interpretation of seismic data, no. 10, in: Investigations in geophysics, SEG Books, 2001.

3

Receiver function images of the Hellenic subduction zone and comparison to microseismicity

F. Sodoudi[1], A. Brüstle[3], T. Meier[4], R. Kind[1,2], W. Friederich[3], and EGELADOS working group

[1]Helmholtz Centre Potsdam, GFZ German Research Centre for Geosciences, Telegrafenberg, 14473 Potsdam, Germany
[2]Department of Earth Sciences, Freie Universität Berlin, Malteserstr. 74–100, 12249 Berlin, Germany
[3] Faculty of Geosciences, Ruhr-Universität Bochum, Universitätsstr. 150, 44801 Bochum, Germany
[4]Institute of Geosciences, Christian-Albrechts Universität zu Kiel, Otto-Hahn-Platz 1, 24118 Kiel, Germany

Correspondence to: F. Sodoudi (foroug@gfz-potsdam.de)

Abstract. New combined P receiver functions and seismicity data obtained from the EGELADOS network employing 65 seismological stations within the Aegean constrained new information on the geometry of the Hellenic subduction zone. The dense network and large data set enabled us to estimate the Moho depth of the continental Aegean plate across the whole area. Presence of a negative contrast at the Moho boundary indicating the serpentinized mantle wedge above the subducting African plate was seen along the entire forearc. Furthermore, low seismicity was observed within the serpentinized mantle wedge. We found a relatively thick continental crust (30–43 km) with a maximum thickness of about 48 km beneath the Peloponnese Peninsula, whereas a thinner crust of about 27–30 km was observed beneath western Turkey. The crust of the overriding plate is thinning beneath the southern and central Aegean and reaches 23–27 km. Unusual low Vp / Vs ratios were estimated beneath the central Aegean, which most likely represent indications on the pronounced felsic character of the extended continental Aegean crust. Moreover, P receiver functions imaged the subducted African Moho as a strong converted phase down to a depth of about 100 km. However, the converted Moho phase appears to be weak for the deeper parts of the African plate suggesting nearly complete phase transitions of crustal material into denser phases. We show the subducting African crust along eight profiles covering the whole southern and central Aegean. Seismicity of the western Hellenic subduction zone was taken from the relocated EHB-ISC catalogue, whereas for the eastern Hellenic subduction zone, we used the catalogues of manually picked hypocentre locations of temporary networks within the Aegean. Accurate hypocentre locations reveal a significant change in the dip angle of the Wadati–Benioff zone (WBZ) from west ($\sim 25°$) to the eastern part ($\sim 35°$) of the Hellenic subduction zone. Furthermore, a zone of high deformation can be characterized by a vertical offset of about 40 km of the WBZ beneath the eastern Cretan Sea. This deformation zone may separate a shallower N-ward dipping slab in the west from a steeper NW-ward dipping slab in the east. In contrast to hypocentre locations, we found very weak evidence for the presence of the slab at larger depths in the P receiver functions, which may result from the strong appearance of the Moho multiples as well as eclogitization of the oceanic crust. The presence of the top of a strong low-velocity zone at about 60 km depth in the central Aegean may be related to the asthenosphere below the Aegean continental lithosphere and above the subducting slab. Thus, the Aegean mantle lithosphere seems to be 30–40 km thick, which means that its thickness increased again since the removal of the mantle lithosphere about 15 to 35 Ma ago.

1 Introduction

The tectonics of the Hellenic subduction zone has been controlled since late Cretaceous by the convergence between Africa and Eurasia, the subduction of African oceanic lithosphere beneath the Aegean lithosphere and the accretion of Africa-derived terranes to Eurasia (e.g. Dercourt et al., 1986; Gealey, 1988; Stampfli and Borel, 2004). The recent kinematics of the Aegean is characterized by a counterclockwise rotation of the Aegean and the Anatolian plates and internal extension of the plates (e.g. McKenzie, 1972, 1978; LePichon et al., 1995; Cocard et al., 1999; Kahle et al., 1999; McClusky et al., 2000). Subduction of African oceanic lithosphere along the Hellenic arc and its associated rollback is supposed to be the cause of the extension and might contribute to the rotation of the Anatolian–Aegean plate (e.g. LePichon et al., 1995; McClusky et al., 2000). From late Miocene time onward, extensional basins developed in the entire Aegean (e.g. Mascle and Martin, 1990; Brun and Sokoutis, 2010), combined with the propagation of the North Anatolian fault into the Aegean region since about ca. 5 Ma (Armijo et al., 1996; Faccenna et al., 2006). Recent studies showed that stretching of the Aegean lithosphere took place in a distinctly episodic fashion, not in a manner that can be described as continuous or progressive (e.g. Forster and Lister, 2009). However, the style of the deformation that accommodates the extension is disputed. Recent seismic anisotropy analysis found evidence for a highly 3-D deformation pattern, strong Miocene extension of the lower crust in the southern Aegean associated with the formation of metamorphic core complexes and significant recent deformation of the entire lithosphere in the northern Aegean (Endrun et al., 2011).

High seismic activity occurs along the Hellenic arc (e.g. Papazachos and Comnikakis, 1971; Hatzfeld and Martin, 1992; Knapmeyer, 1999; Papazachos et al., 2000). The Benioff zone steepens from the western towards the eastern part of the Hellenic subduction zone and reaches depths of about 100–180 km beneath the volcanic arc, where it roughly terminates. Thus, imaging the subducting African plate towards greater depths is challenging and may be restricted to seismic tomography or studies of converted or scattered waves. In general, seismic tomography provided relatively unsharp images of the Hellenic subduction zone (e.g. Spakman et al., 1988, 1993; Papazachos and Nolet, 1997; Bijwaard et al., 1998; Piromallo and Morelli, 2003; Schmid et al., 2004; Chang et al., 2010; Biryol et al., 2011). Studies of converted or scattered waves in the Aegean were mainly obtained from small networks mostly covering the western part of the Hellenic arc (e.g. Suckale et al., 2009; Gesret et al., 2011; Pearce et al., 2012).

Temporal and spatial variations in convergence rate are exhibited along the Hellenic subduction zone (e.g. McClusky et al., 2000). Based on GPS measurements (McClusky et al., 2000; Reilinger et al., 2010; DeMets et al., 1990), subduction becomes increasingly oblique from west to east. In the western part, the vectors of motion are oriented normal to the continental margin, whereas towards the east, in the area of Rhodes (stations RHON, RHOS, Fig. 1), the angle between the motion vector and the active margin decreases causing left-lateral movement between forearc slivers. Özbakir et al. (2013) conducted a detailed analysis on the correspondence of model shear zone structures to observations and

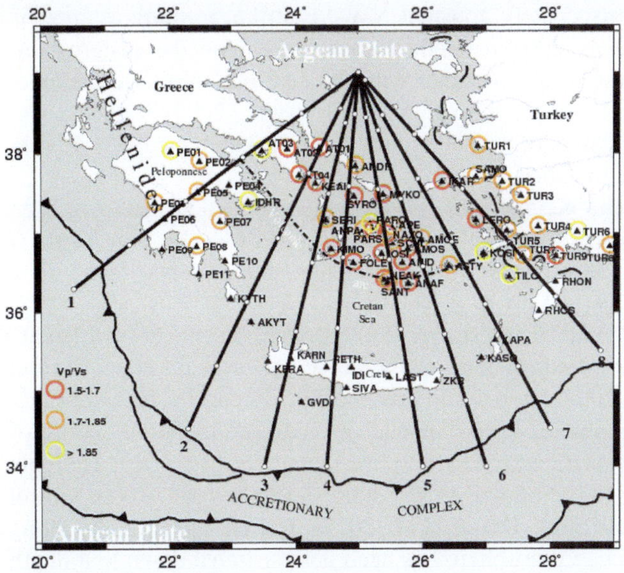

Figure 1. Distribution of the EGELADOS stations used for receiver function analysis (black triangles). The southern deformation front of the Hellenic subduction zone and the border between the central and the inner units of the Mediterranean Ridge are shown with black solid lines with triangles (simplified after Chamot-Rooke et al., 2005). Dash–dotted line denotes location of the volcanic arc. Black thick lines indicate the positions of eight S–N trending profiles used for P receiver functions. Small white circles on the profiles lines mark 100 km distance interval. Estimated values of Vp / Vs ratio are marked with coloured circles.

proposed that the trench perpendicular convergence in the eastern Hellenic arc is taken up by a shear zone and a fold and thrust belt in the Rhodes Basin. Due to the lack of clear geophysical images of the slab in the eastern Hellenic subduction zone Özbakir et al. (2013) question the existence of a NW dipping slab in the region of Rhodes.

The crustal structure of the Hellenic subduction zone has been intensively studied by different reflection, refraction and receiver function studies (e.g. Makris and Vees, 1977; Makris, 1978; Hirn et al., 1996; Mascle and Chaumillon, 1998; Knapmeyer and Harjes, 2000; Bohnhoff et al., 2001; Li et al., 2003; Endrun et al., 2004; Sodoudi et al., 2006). At greater depths, global and regional P-wave tomographies show that the Hellenic subduction zone coincides with high seismic velocity anomalies that extend through the upper mantle into the lower mantle (e.g. Spakman et al., 1988, 1993; Papazachos and Nolet, 1997; Bijwaard et al., 1998; Piromallo and Morelli, 2003; Schmid et al., 2004; Chang et al., 2010; Biryol et al., 2011). Furthermore, they imaged a not completely amphitheatrical geometry for the subducted African slab and showed a gradual increase in the slope of the Hellenic slab from west to east and toward the larger depths of the slab. Based on the model proposed by Papazachos and Nolet (1997), the western part of the slab is much larger than its eastern part. Moreover, the western part has an approximately constant NW–SE strike, forming at approximately 200 km depth a right angle with the eastern part. These results together with the distribution of intermediate-depth seismicity (e.g. Papazachos et al., 2000) implied a segmentation of the subducting African slab in the Hellenic subduction zone.

The geometry of the Hellenic subduction zone has been also studied by other geophysical methods, but none of them has been resolved the subducting slab at depths larger than about 160 km (e.g. Makris and Vees, 1977; Makris, 1978; Knapmeyer and Harjes, 2000; Bohnhoff et al., 2001; Li et al., 2003; Endrun et al., 2004, 2008; Meier et al., 2004; Tirel et al., 2004; Snopek and Casten, 2006). Recently, receiver function studies have obtained new constraints on the nature of the subducting African plate mainly in the western Hellenic arc (e.g. Sodoudi et al., 2006; Suckale et al., 2009; Gesret et al., 2011; Pearce et al., 2012). The base of the subducting African lithosphere has been demonstrated down to a depth of about 225 km beneath the volcanic arc using converted S-to-P waves (e.g. Sodoudi et al., 2006).

The establishment of the temporary, broadband seismic network of EGELADOS (Exploring the Geodynamics of Subducted Lithosphere using an Amphibian Deployment Of Seismographs) within the CRC 526 "Rheology of the Earth" founded by the Deutsche Forschungsgemeinschaft covering the entire Hellenic subduction zone from the Peloponnese in the west to the western Turkish coast in the east, was carried out by the Ruhr-University Bochum in collaboration with partners at the Aristoteles University of Thessaloniki, the National Observatory Athens, the Technical University of Crete, the Istanbul Technical University, the GFZ Potsdam and the University of Hamburg (Fig. 1). Deploying more than 80 temporary on-shore and off-shore stations provides a unique opportunity to obtain high-resolution seismic images of the structure and seismicity of the subducted plate and the overriding Aegean plate.

In the present study, we employ the P receiver function technique to the temporary land stations and the permanent GEOFON stations. Dense seismic sampling obtained from these stations enables us to create a high-resolution image of the Aegean plate and the descending African plate in the southern and central Aegean. We combine our results with hypocentre locations of the EHB-ISC catalogue (Engdahl et al., 1998) for the western Hellenic subduction zone and with new accurate hypocentre locations recorded by the temporary CYCNET and EGELADOS networks (Brüstle, 2012; Brüstle et al., 2014) as well as hypocentre locations by the temporary LIBNET network (Becker et al., 2009) for the eastern Hellenic subduction zone and particularly discuss the correlation of P receiver function images and the hypocentre distribution.

2 Data and methodology

The EGELADOS network consists of 56 on-shore stations deployed on the Greek mainland, on most of the southern Aegean islands and in western Turkey from October 2005 to April 2007. In addition, 24 off-shore stations were deployed in the region. Forty-five of the temporary on-shore stations and all off-shore stations were equipped by the German amphibian seismograph pool (DEPAS). The remaining temporary stations were provided by Ruhr-University Bochum. Additionally, 22 permanent on-shore stations of the Global Seismic Network GEOFON (Geoforschungszentrum Potsdam, Germany), the Hellenic Broadband Seismic Network HL (National Observatory Athens NOA, Greece), and the Mediterranean Network MedNet (Instituto Nazionale di Geofisica e Vulcanologia INGV, Italy) were also included to complete the temporary network. We used at least 1.5 years of teleseismic data recorded at 65 on-shore stations (56 temporary and 9 permanent GEOFON stations) to calculate the P receiver functions (Fig. 1). All available events with a magnitude (mb) larger than 5.5 and at a distance of 30°–95° were analysed. They also provided sufficient azimuthal coverage (0°–360°). More than 11 years of data were available for the permanent GEOFON stations for the P receiver function analysis. We computed P receiver functions (e.g. Langston, 1977; Owens et al., 1984; Kind and Vinnik, 1988; Yuan et al., 1997; Kosarev et al., 1999) for all stations. The processing steps used in this paper are similar to those described by Sodoudi et al. (2006).

To image the seismicity of the Hellenic subduction zone, about 8200 hypocentre locations with a location uncertainty of less than 20 km were selected from various catalogues. Seismicity of the western Hellenic subduction zone (lat < 25°, profiles 1–4 in Fig. 1) was taken from the relocated EHB-ISC catalogue 1960–2007 (Engdahl et al., 1998). Seismicity of the eastern Hellenic subduction zone (lat > 25°, profiles 4–8 in Fig. 1) was taken from catalogues of manually picked hypocentre locations of temporary networks. This data set consists of about 3000 hypocentre locations (Brüstle, 2012; Brüstle et al., 2014) observed by the temporary EGELADOS network, of about 4000 relocated hypocentre locations (Brüstle et al., 2014) obtained by the temporary CYCNET network covering the central subduction zone from 2002–2004 (Bohnhoff et al., 2004, 2006), and of about 2000 hypocentre locations of the temporary LIBNET network (Becker et al., 2009) covering the forearc Southeast of Crete from 2003–2004.

3 Observations

Earthquake catalogues used in this work show that shallow seismicity within the Aegean plate is observed down to about 20 km, predominantly along extensional structures in the thinned brittle Aegean crust. In the regions of transtensional structures in the forearc that have been identified as nearly vertical margins of forearc slivers the seismicity extends down to the subduction interface (Meier et al., 2007; Becker et al., 2009; Brüstle et al., 2014). The seismicity of the plate contact is only incompletely recorded due to the limited station coverage by permanent and temporary networks. In this study, only south of eastern Crete is the plate contact resolved in detail by the dense temporary on-shore/off-shore LIBNET network (Becker et al., 2009). Other temporary networks covering western and central Crete located the seismogenic zone of the plate interface south of Crete with a lateral extent of 40–100 km and at 20–40 km depth (Meier et al., 2004). Another dense temporary network deployed in the region of the Ionian islands at the northwestern margin of the subduction zone located the seismogenic plate interface west and beneath of Cephalonia, with an lateral extension of about 60–70 km, dipping from about 10 km depth in the west to about 20 km in the east (Sachpazi et al., 2000).

Stacked traces of P receiver functions (PRFs) of 65 EGELADOS on-shore stations are shown in Fig. 2. They are filtered with a low-pass filter of 1 s. At all stations, except those located in the forearc of the subduction zone (the area trapped between the Hellenic trench and the volcanic arc, see Fig. 1), we clearly observed a positive P-to-S conversion in the time interval from 2.5 to 6 s delay time, which stem from the Moho of the continental Aegean plate. Multiple converted phases from the Moho boundary can be also observed beneath some stations (Fig. 2). The converted Moho phase is strongly imaged beneath the stations located relatively far from the Hellenic trench (see Fig. 1). At the stations located in the forearc of the subduction zone (marked Forearc in Fig. 2), the converted Moho phase with positive polarity disappeared and instead a negative phase is seen. Such a reversal of sign for the Moho conversion has been previously observed beneath the Cascadia, Hellenic subduction zone and Central Andes (e.g. Knapmeyer and Harjes, 2000; Bostock et al., 2002; Li et al., 2003; Endrun et al., 2004; Sodoudi et al., 2006, 2011).

To have a clear image of the subducted African Moho (hereafter referred to as slab), we separately show the stacked PRFs obtained from the western and eastern parts of the Aegean. For the western part we took stations along profiles 1–3 and for the eastern part those along profiles 4–7 (see Fig. 1). Figure 3 illustrates the stacked PRF traces obtained from these two parts. The PRFs are filtered with a low pass of 3 s and sorted by the relative distance of the stations from the Hellenic trench. Moreover, we show them in a larger time window compared to that in Fig. 2. As Fig. 3 shows, the most pronounced conversions correspond to the slab. This phase is clearly shown at times ranging from 6 s beneath the forearc region to about 11 s beneath Peloponnese (station PE04) in the western Aegean (Fig. 3a), whereas it can be followed from 5 to about 8 s beneath the forearc area (station KAPA) in the eastern Aegean (Fig. 3b). Further north, the continuation of the slab phase is not more clear. Instead, significant con-

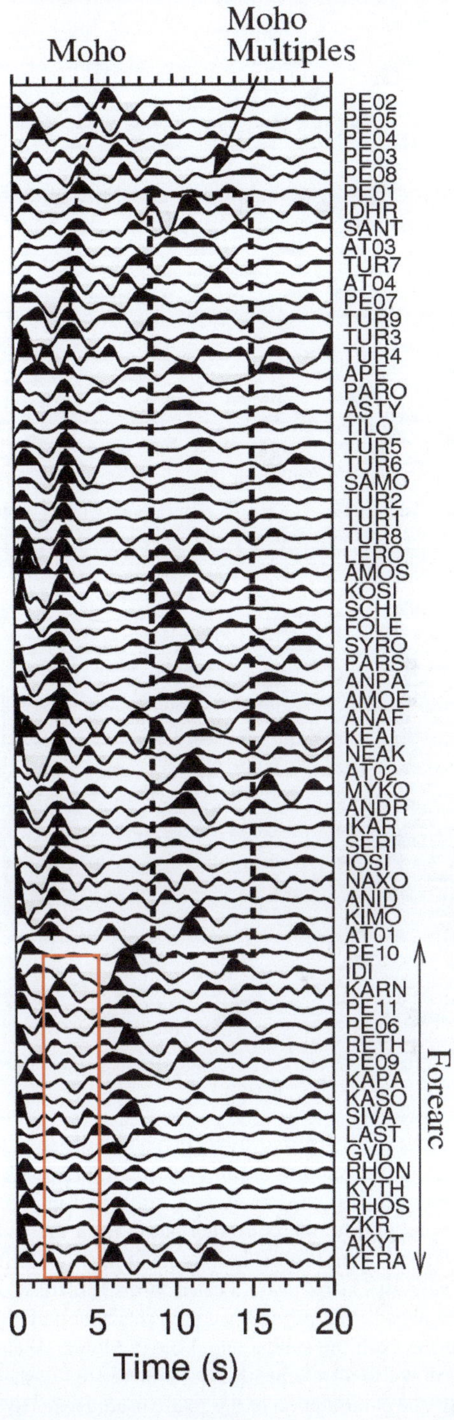

Figure 2. Stacked PRFs of 65 EGELADOS stations. Seismic traces are moveout corrected for a reference distance of 67° before summation. Stations are sorted (except stations in the forearc) according to the arrival time of the Aegean Moho phase from 2.3 to 6 s (black dashed line). Data are filtered using a 1 s low-pass filter. Few positive P-to-S conversions from the Aegean Moho can be seen beneath the forearc stations (red box). Arrival time window of the Moho multiples is shown with black dashed box.

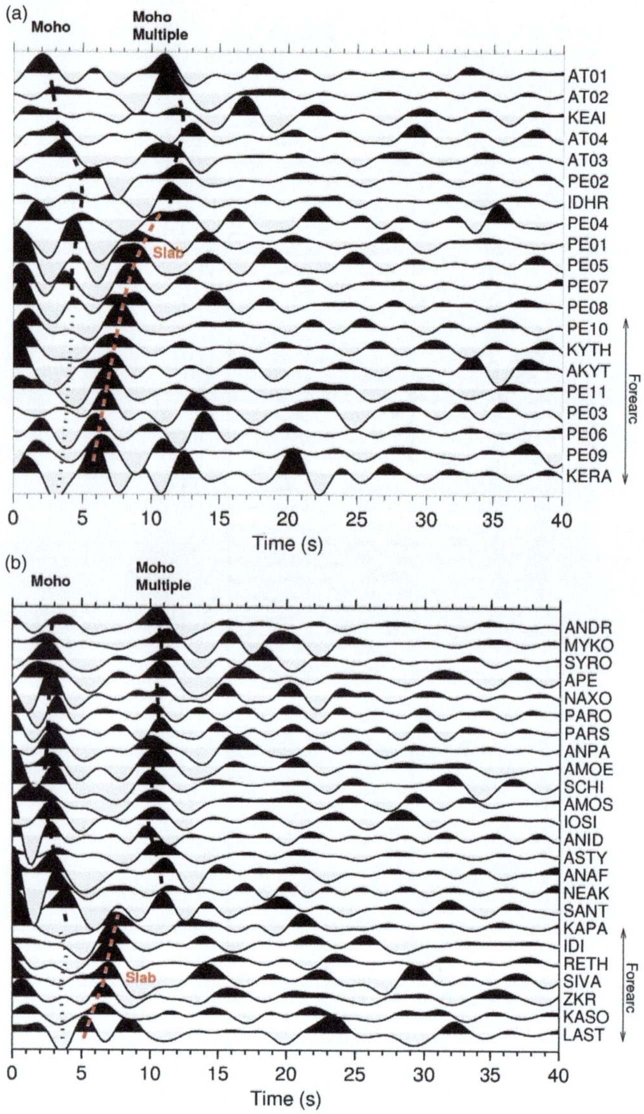

Figure 3. Stacked PRFs sorted by their relative distance from the Hellenic trench from south to north. They are filtered with a bandpass of 3–20 s. **(a)** The western Aegean profiles 1–3 (see Fig. 1) stacked into one profile: the converted phase from the continental Aegean Moho and its multiple (black dashed lines) can be observed beneath all stations except those located in the forearc where conversions with negative amplitudes are seen (black dotted lines). The converted phase from the subducting African Moho (labelled slab) is well seen down to 11 s beneath Peloponnese (red dashed lines). Further north, the continuation of this phase is not clear. **(b)** Same as **(a)** for the eastern Aegean (profiles 4–7, see Fig. 1): the subducting African Moho phase can be followed up to 8 s beneath the forearc region (station KAPA).

verted phases associated with the continental Aegean Moho (hereafter referred to as Moho) and its multiple (Ppps) can be observed.

4 Results

4.1 Aegean Moho depth

4.1.1 Zhu and Kanamori approach

Clear arrivals of the Moho conversions (Fig. 2) allowed a first-order estimation of the crustal thickness beneath the whole region except the forearc area due to the absence of positive Moho conversions. For stations with relatively clear multiples we estimate the crustal thickness and Vp / Vs ratio from the H-k stacking technique (Zhu and Kanamori, 2000), which assumes a flat homogeneous layered structure beneath the station. The receiver functions are stacked at each station. We considered different weights (0.6, 0.3 and 0.2) for the three converted phases (Ps, Ppps and Psps + Ppss). The maximum amplitude of stacked traces appear where the three phases add constructively. We used an average crustal velocity (Vp) of 6.2 km s^{-1} for the depth estimation and performed this analysis for all stations. Our results are listed in Table 1 and are shown in Fig. 1. Figure 4 shows the estimates of crustal thickness and Vp / Vs ratio for eight selected stations. The theoretical arrival times of Ps, Ppps and Psps + Ppss predicted by the estimated crustal thickness and Vp / Vs ratio (thin red lines) mostly coincide with those of the observed converted phases.

Our analysis shows that, given a range of probable values for average crustal velocity (Vp), crustal thickness can change by nearly 4 km, whereas the Vp / Vs ratio can vary by 0.06–0.16. Large uncertainties are probably due to the absence of multiple energy, significant component of slab interference or the deviation of the structure beneath the station of a locally flat and homogeneous layered structure. In general, we found Vp / Vs ratios varying between 1.7 and 1.8 for the stations located in the Peloponnese Peninsula and western Turkey (see Fig. 1), which are in good agreement with those reported by the previous PRF studies (e.g. Karabulut et al., 2013; Vanacore et al., 2013). In contrast, unusually low Vp / Vs ratios in the range of 1.5–1.7 were found for the majority of stations located in the central Aegean Sea and close to the volcanic arc. Anomalous Vp / Vs ratios are frequently shown in subduction zones (e.g. Zandt et al., 1994; Wagner et al., 2005; Zheng and Lay, 2006) and provided useful information on the composition of crustal rocks. Although presence of high values of Vp / Vs ratio in subduction zones can be associated with rocks of more mafic composition, low values of Vp / Vs ratio may reveal felsic crustal composition and high upper-crustal temperatures, which correspond to rocks which contain large amounts of silica.

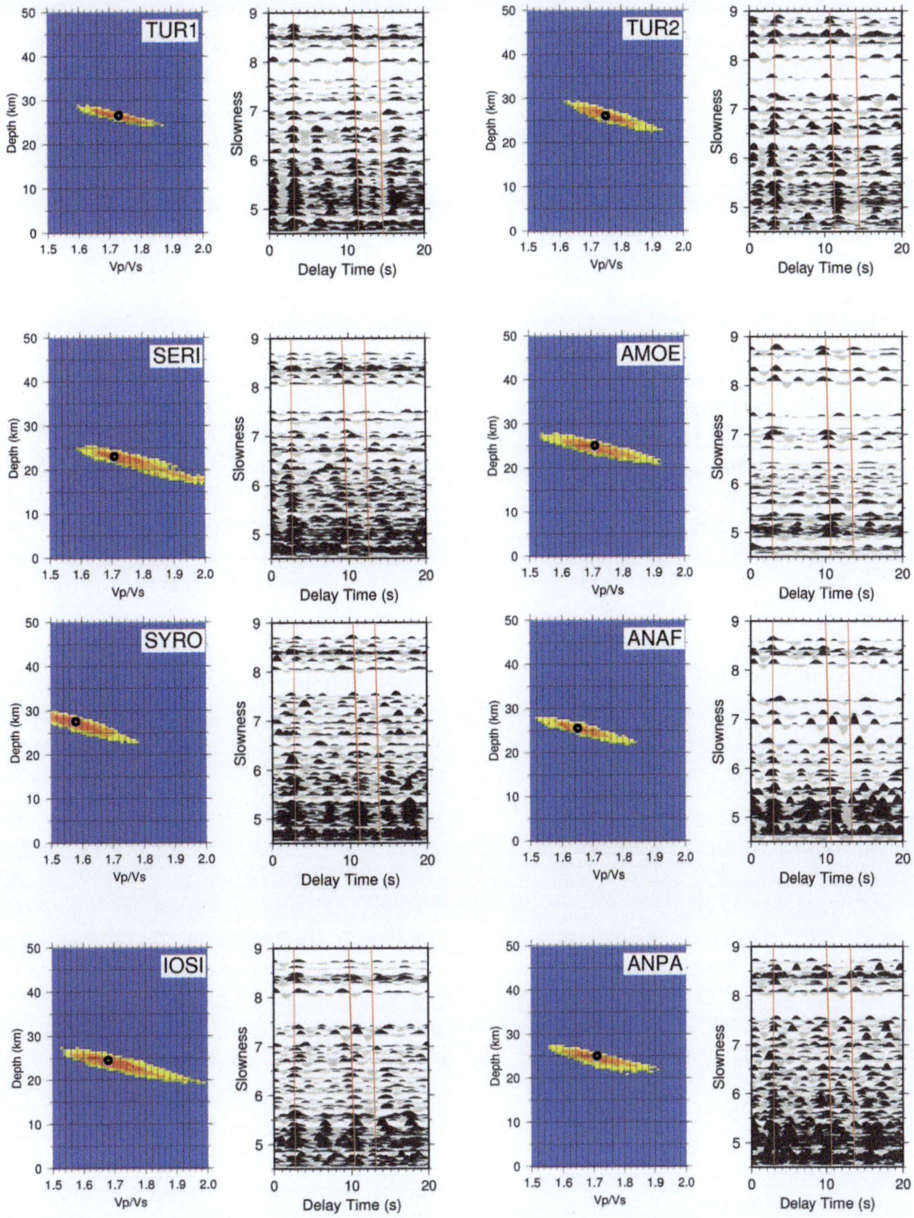

Figure 4. Examples of the Zhu and Kanamori (2000) method applied to PRFs. Black circle shows where the amplitudes of Ps, Ppps and Ppss + Psps phases are stacked most coherently and indicates the best Moho depth and Vp / Vs ratio beneath a station. Red thin lines on slowness sections (right) show theoretical arrival times for the direct Ps conversion and its first two multiples chosen for the best Moho depth and Vp / Vs ratio by the Zhu and Kanamori stacking technique.

Due to the unclear appearance of the Moho multiples at some stations, we also calculated the Moho depth regarding the arrival times of the Moho conversions. For this estimation, we used the modified IASP91 reference model (Kennett and Engdahl, 1991), in which the upper 180–200 km of the global IASP91 model was replaced by a regional minimum 1-D model obtained from the joint inversion of 1130 well-located earthquakes of the EGELADOS catalogue (Brüstle et al, 2014). The calculated Moho depths are listed in Table 1. These values seem mostly to be consistent with those obtained from the Zhu and Kanamori method. However, for stations showing high or low Vp / Vs ratios the calculated crustal thickness may largely differ from that of the H-k stacking technique.

4.1.2 Migration

More detailed images can be provided by our migrated P receiver functions (Kind et al., 2002), in which the PRFs are back-projected along their ray paths and stacked within

Table 1. Stations, their coordinates, observed arrival times of the converted Moho phase (Ps) and Moho multiple (Ppps) of the Aegean plate, corresponding Moho depths and Vp / Vs ratios obtained from the Zhu and Kanamori method, corresponding Moho depths obtained from the modified IASP91 reference model (using a reference distance of 67°).

Station	Lat.	Long.	Moho time (s)	Moho Multiple time (s)	Moho depth Z&K (km)	Vp / Vs	Moho depth modified IASP91 (km)
AMOE	36.9150	25.9788	2.9	10.8	25	1.71	22
AMOS	36.7956	25.7690	3.1	11	26.5	1.51	24
ANAF	36.3581	25.7783	2.9	11.4	25.5	1.65	22
ANDR	37.8361	24.9482	2.7	10.8	24.5	1.78	20
ANID	36.6251	25.6847	2.4	9.5	22.5	1.67	18
ANPA	37.0323	25.0763	2.9	10	25	1.71	22
APE	37.0689	25.5306	3.4	10.9	24	1.83	27
ASTY	36.5795	26.4114	3.4	10.5	23.5	1.83	27
AT01	38.06	24.3784	2.3	9	27.5	1.51	17
AT02	38.0473	23.8638	2.8	11	26	1.67	21
AT03	38.0268	23.4677	3.8	12	21	1.99	30.5
AT04	37.7252	24.0496	3.5	13	30.5	1.67	28
FOLE	36.6216	24.9197	3	10	23	1.69	23
IDHR	37.3908	23.2591	4.1	13.3	23.5	1.99	33.5
IKAR	37.6435	26.3050	2.7	10	23	1.68	20
IOSI	36.7347	25.3618	2.7	10.5	24.5	1.68	20
KEAI	37.6232	24.3191	2.9	11.6	28	1.65	22
KIMO	36.7949	24.5680	2.3	7.5	29.5	1.5	17
KOSI	36.7449	26.9517	3	11.4	20	1.9	23
LERO	37.1634	26.8353	3.1	11.5	26.5	1.7	24
MYKO	37.4822	25.3844	2.8	11.2	27.5	1.57	21
NAXO	36.9800	25.4400	2.5	9.2	26.5	1.63	19
NEAK	36.4087	25.4014	2.85	11.4	28	1.61	21.5
PARO	37.1150	25.1825	3.3	10.4	22.5	1.87	29
PARS	37.0285	25.2253	2.9	10.6	25	1.68	22
PE01	38.0170	22.0283	4.3	13.5	31	1.86	35
PE02	37.8958	22.4907	6	21	45	1.72	51.5
PE03	37.3784	21.7722	4.5	15.5	40	1.83	37
PE04	37.6007	22.9593	4.8	16.2	–	–	39.5
PE05	37.5128	22.4553	5	18.2	43	1.83	41
PE07	37.1478	22.8195	3.5	12.1	27.5	1.8	28
PE08	36.8311	22.4420	4.3	14.8	33	1.81	35
SAMO	37.7043	26.8377	3.4	12	26	1.79	27
SANT	36.371	25.459	4	14.1	32	1.71	32.5
SCHI	36.8744	25.5180	3	10.3	22.5	1.78	23
SERI	37.1610	24.4853	2.7	10	23	1.71	20
SYRO	37.4569	24.9266	2.9	10.5	27.5	1.58	22
TILO	36.4485	27.3535	3.4	11	23	1.88	27
TUR1	38.0865	26.8676	3.2	11.4	26.5	1.73	25
TUR2	37.6420	27.2418	3.3	11.8	26	1.75	26
TUR3	37.4663	27.5380	3.5	11.1	28	1.78	28
TUR4	37.0800	27.8077	3.5	12	28	1.79	28
TUR5	37.0302	27.3167	3.3	11.7	27	1.75	26
TUR6	37.0159	28.4262	3.3	11.4	26.5	1.98	26
TUR7	36.7017	27.5697	3.8	13.5	28	1.81	30.5
TUR8	36.8273	28.9390	3.2	11.5	25	1.79	25
TUR9	36.7015	28.0887	3.5	12	30.5	1.68	28

small bins (2 km × 2 km). The conversion points are calculated by ray tracing using the modified IASP91 reference model. Figure 5 shows depth migrations of PRFs down to 80 km depth along the eight S–N profiles shown in Fig. 1. Accurate hypocentre locations of temporary network catalogues are also shown in Fig. 5b (profiles 5–8). High-frequency P receiver functions reliably indicate the presence of the Aegean and African plates. The continental Aegean Moho can be well seen beneath the northern part of all profiles up to the forearc area (labelled Moho, see Fig. 5). The thickest crust of about 48–50 km is observed beneath the Peloponnese Peninsula, where the Hellenides mountains are located (see profile 1, Fig. 5a). We found relatively thinner crust of approximately 27–30 km beneath the stations located in western Turkey (see profile 8, Fig. 5b). The Aegean crust becomes thinner and is approximately 23–28 km thick beneath the central Aegean Sea. These results are in good agreement with those obtained from the previous PRF studies in western Anatolia, Peloponnese and in central Aegean (e.g. Sodoudi et al., 2006; Zhu et al., 2006; Karabulut et al., 2013; Vanacore et al., 2013). Relatively thin crust beneath the Aegean and adjacent areas likely shows the extensional tectonics that affected the whole Aegean since the Oligocene time (∼ 30 Ma) (e.g. Mascle and Martin, 1990).

Further south, beneath the forearc area, the Moho phase is no longer visible and is continued by negative converted phases. This is a common feature, which can be reliably observed beneath the forearc on all profiles. However, we could not clearly image the transition from Moho to the negative signal (except beneath profile 8 in western Turkey) due to the gap of stations in the Cretan Sea. The negative phase indicates a low-velocity zone (LVZ), which was previously reported and interpreted as serpentinized mantle wedge above the subducting plate (e.g. Endrun et al., 2004; Sodoudi et al., 2006). Very low shear velocities in the forearc mantle were first observed and interpreted by Bostock et al. (2002) in Cascadia. They called their observation "inverted Moho" since the sign of the converted signal has the opposite polarity of the usual Moho conversion. Interestingly, this low-velocity zone shows also low seismicity (see Fig. 5b), whereas the high seismicity occurs just above or below it. From profile 5 – where accurate locations of seismic events at the seismogenic interface are available (Becker et al., 2009) – it becomes evident that the plate contact between the serpentinized mantle and the subducting plate is mainly aseismic. Thus, the serpentinized mantle wedge is deforming mainly aseismically. On the other hand the seismogenic zone seems to terminate where the serpentinized mantle wedge comes in contact with the plate interface (see Fig. 5b). Meier et al. (2007) noted that the seismogenic zone terminates at the southern coastline of Crete and postulate that the recent uplift of Crete is related to the return flow above the seismically decoupled plate interface. Here we show that the serpentinized mantle wedge is present along the entire forearc of the Hellenic subduction zone and its location coincides with the location of the

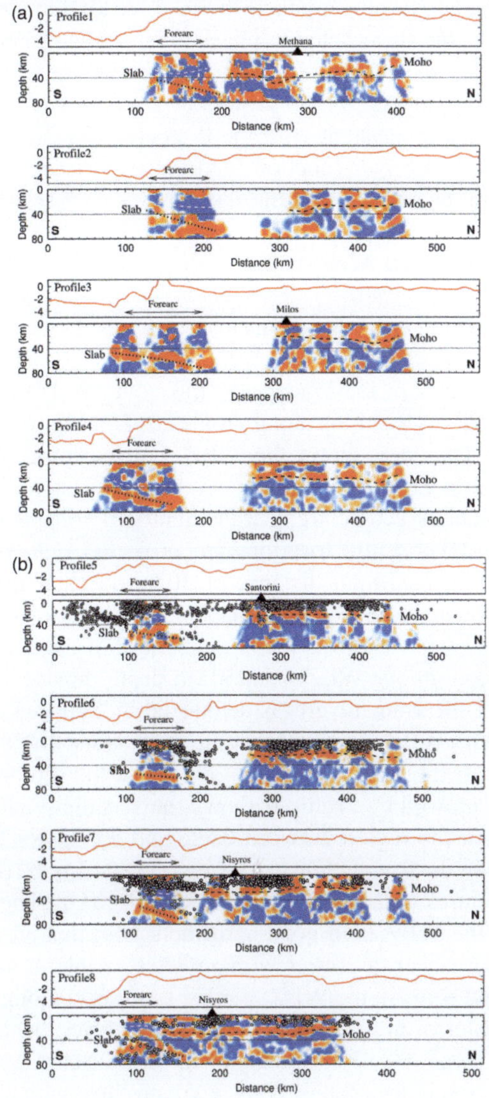

Figure 5. Migrated PRFs along S–N trending profiles shown in Fig. 1. Topography is shown at the top. Positive (negative) phases are shown in red (blue). Positions of the volcanoes are shown with black triangles. **(a)** For the western profiles (1–4): the Moho of the continental Aegean plate (black dashed lines, labelled Moho) is well observed beneath the profiles up to the forearc region. This phase is not visible beneath the forearc area and seems to be continued as negatively converted phases (white dashed lines). The subducted African Moho (black dotted lines, labelled Slab) is well observed beneath the forearc area at depths ranging from 40 to 80 km. Note that the thickest crust of about 48–50 km is seen beneath the Peloponnese (profile 1). **(b)** As **(a)** for the eastern profiles (5–8). The black circles show the seismicity of the temporary network catalogues (Becker et al., 2009; Brüstle et al., 2014). Note that the most seismicity occurs above or below the serpentinized mantle wedge (negative phases in blue) beneath the forearc area. Note that the slab is also observed in the distance range between 100 and 170 km where no seismicity is observed (on profiles 5–7).

forearc high (see upper panels in Fig. 5 for topography). Furthermore, the slab is clearly imaged beneath the forearc area throughout the whole region at depths ranging from 40 to 80 km (Fig. 5). As Fig. 5b shows, the seismicity (for profiles 5–8) is also consistent with the P receiver function results. Both the P receiver functions and the hypocentre locations constrain the geometry of the slab.

4.2 Subducted African slab

4.2.1 Hypocentre locations

The geometry of the Wadati–Benioff zone (WBZ) is shown by three cross-sections along profiles 5–7 in Fig. 6 (see also Fig. 1). Profile 5 shows the WBZ of the central Hellenic Subduction Zone from the forearc high in eastern Crete to the Hellenic volcanic arc near the Santorini volcano (station SANT). Hypocentre locations are projected onto the profile with a maximum distance of 30 km to each side. Further to the north (~ 100 km distance), a nearly aseismic zone of about 50 km width is identified beneath eastern Crete. A nearly continuous WBZ from 50 km depth down to 190 km (beneath the volcanic arc) is delineated by hypocentre locations. The WBZ dips with an angle of about 26° down to 100 km to the north. Further to the north the dip angle increases up to 45°. Profile 6 shows the seismicity along the slab from the region between Crete and Karpathos (station KAPA) in the forearc high to the region between the two volcanic centres with a total profile width of 40 km. The geometry of the WBZ is in good agreement with that along profile 5. However, a cluster with an offset of about 40 km is identified beneath the WBZ. Further to the east along profile 7, a cluster of high seismic activity is observed beneath the volcano region. The seismicity of the cluster corresponds to the position of the eastern WBZ along profile 6 with the same orientation and dip angle (~ 35°) to NNW. The lack of seismicity beneath eastern Crete and further to the east, which is observed on profiles 5–7 (distance 100–150 km) coincides with the negative Moho velocity contrast observed by PRFs and points to a serpentinized mantle wedge (see Fig. 5).

4.2.2 P receiver functions

To follow the continuation of the slab at larger depths, we firstly calculated P-to-S conversion points at 200 km depth using the modified IASP91 reference model. This depth was chosen based on the previous S receiver function study, which exhibited the continuation of the slab down to 180–200 km beneath the central Aegean (Sodoudi et al., 2006). For each profile, we considered the PRFs, whose piercing points are located within the data band of 60 km from each side of the profile. Then we stacked them in bins of 4 km and sorted them according to their distance from the starting point of the profile (see Fig. 1). Figure 7 indicates the binned PRFs along the eight profiles shown in Fig. 1. Besides the Moho conversions (black boxes) with different polarities changing from negative in the forearc to positive in the back arc and the Moho multiples (white dashed lines), there are also significant conversions from the subducting African Moho (black dashed lines). It is important to note that Fig. 7 is more appropriate for deeper phases such as converted phases from the slab. Distribution of P-to-S conversions at 200 km cannot provide accurate images of the Moho variations (for more accurate Moho estimations see Fig. 5).

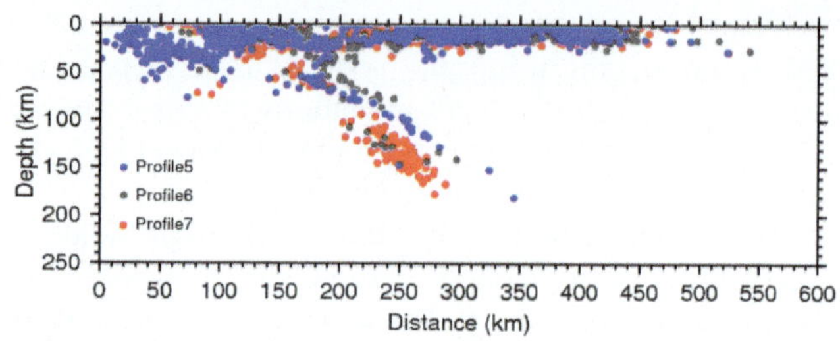

Figure 6. Hypocentre locations along the WBZ. Additionally to the hypocentre locations of the CYYNET and EGELADOS networks, hypocentre locations of the temporary LIBNET (Becker et al., 2009) are also projected. Hypocentre locations along profiles 5–7 are plotted in different colours. A cluster with an offset of about 40 km is identified beneath the WBZ along profile 6. The position, orientation and dip angle of this cluster corresponds with those of a cluster of high seismic activity along profile 7. The lack of seismicity at distances between 100 and 150 km coincides with the negative Moho velocity contrast observed by PRFs (see Fig. 5).

For each profile, local seismicity (Engdahl et al., 1998; Becker et al., 2009; Brüstle et al., 2014) is overlaid on the PRF data. The WBZ of the subducting African plate can continuously be traced down to about 100 km in the western and to about 180 km in the eastern subduction zone, and its dip direction changes from E in the western to N in the central to NW in the eastern part with an increasing dip angle from W (~ 25°) to E (~ 35°) (Brüstle et al., 2014). Clusters of high microseismicity have been observed beneath the volcanic arc (see profile 7, Fig. 7). Generally, there is a good correlation between the geometry of the slab shown by seismicity and that obtained from the PRFs. Both of them can satisfactorily resolve the shallower part of the slab down to at least 90 km depth (~ 10 s).

Our results together with seismicity enabled us to image the slab more accurately, especially for the eastern Hellenic subduction zone, where the resolution of the seismicity of the temporary network catalogues (profiles 5–8) is better than 20 km with an average location uncertainty of less than 10 km as estimated by the location routine NonLinLoc (Becker et al., 2009; Brüstle et al., 2014). Regarding the dominant wave periods of the P wave (approximately 2 s), a maximum depth resolution of about 2 km can be theoretically estimated for P-to-S conversions at discontinuities. Considering some more errors obtained from the velocity model (Brüstle et al., 2014) as well as those from lateral heterogeneities and noise, we

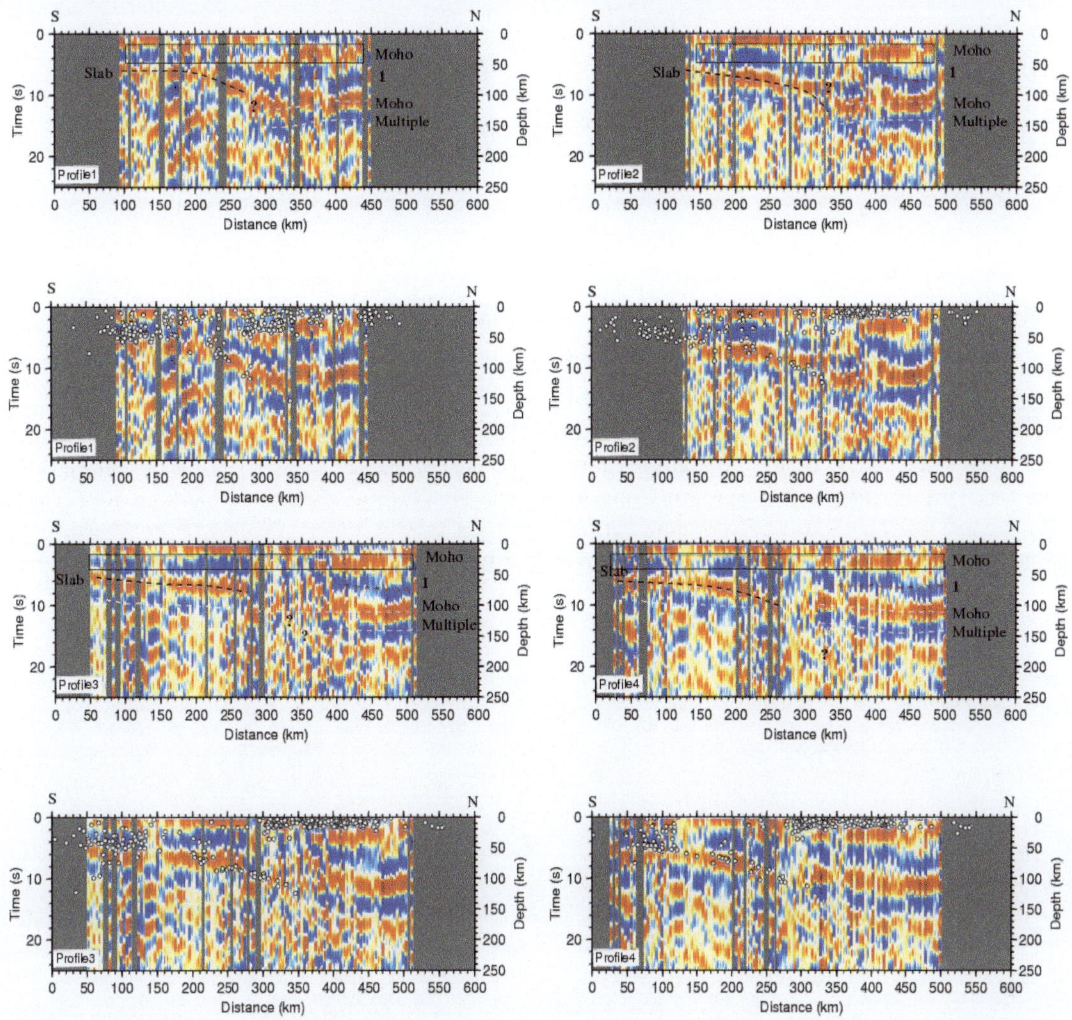

Figure 7. Sections obtained from the binned PRF data in 4 km bins (upper panels) along profiles shown in Fig. 1. Seismicity is also overlaid on the PRF sections (lower panels, small white circles). Positive (negative) conversions are shown in red (blue). The data are filtered with a band-pass filter of 4–30 s. A black box represents the rough location of the Moho boundary, which shows a significant polarity change from the forearc (negative) to the back arc (positive). White dashed lines indicate the Moho multiples, whose polarities are also related to those of the Moho conversions. The shallow part of the downgoing African slab (black dashed lines) is well imaged by PRFs and correlates with the seismicity. Question marks show the same arrival time of the first Moho multiple (shown in red) and the slab phase or weak indications of the continuation of the slab. Note that the strong negative conversion at 4–7 s (dark red dashed lines, marked as 1 in all profiles) may show a LVZ just below the Aegean Moho phase (see also Fig. 8).

expect to have an uncertainty of less than 5 km in depth estimation for the Moho and less than 10 km for the slab. In contrast, resolution of tomographic surface wave and body wave studies is in general lower. Furthermore, previous P receiver function studies could not image the Hellenic slab deeper than 120 km depth (Knapmeyer and Harjes, 2000; Li et al., 2003; Endrun et al., 2004; Sodoudi et al., 2006). In this work, in particular the comparison of P receiver functions with hypocentres of the WBZ allows us to distinguish between primary conversions from the slab and multiples of the Moho. Multiples could otherwise be misinterpreted as flat slabs.

Our results mainly show a relatively shallow dipping slab segment ($\sim 25°$) beneath the western part of the Aegean (profiles 1–4, Fig. 7). Along profile1 located in the westernmost part of the Aegean, the slab is consistently imaged at times ranging from 5 s to about 11 s (black dashed line). Further north, the continuation of the slab is not clear. The arrivals of the Moho multiples (white dashed line) with positive amplitudes make the identification of the converted slab phase difficult (black question mark). Beneath profile 2, the converted slab phase is visible until 10–12 s, where the seismicity stops. The slab phase in profile 3 is resolved at depths up to 80 km (or ~ 8 s).

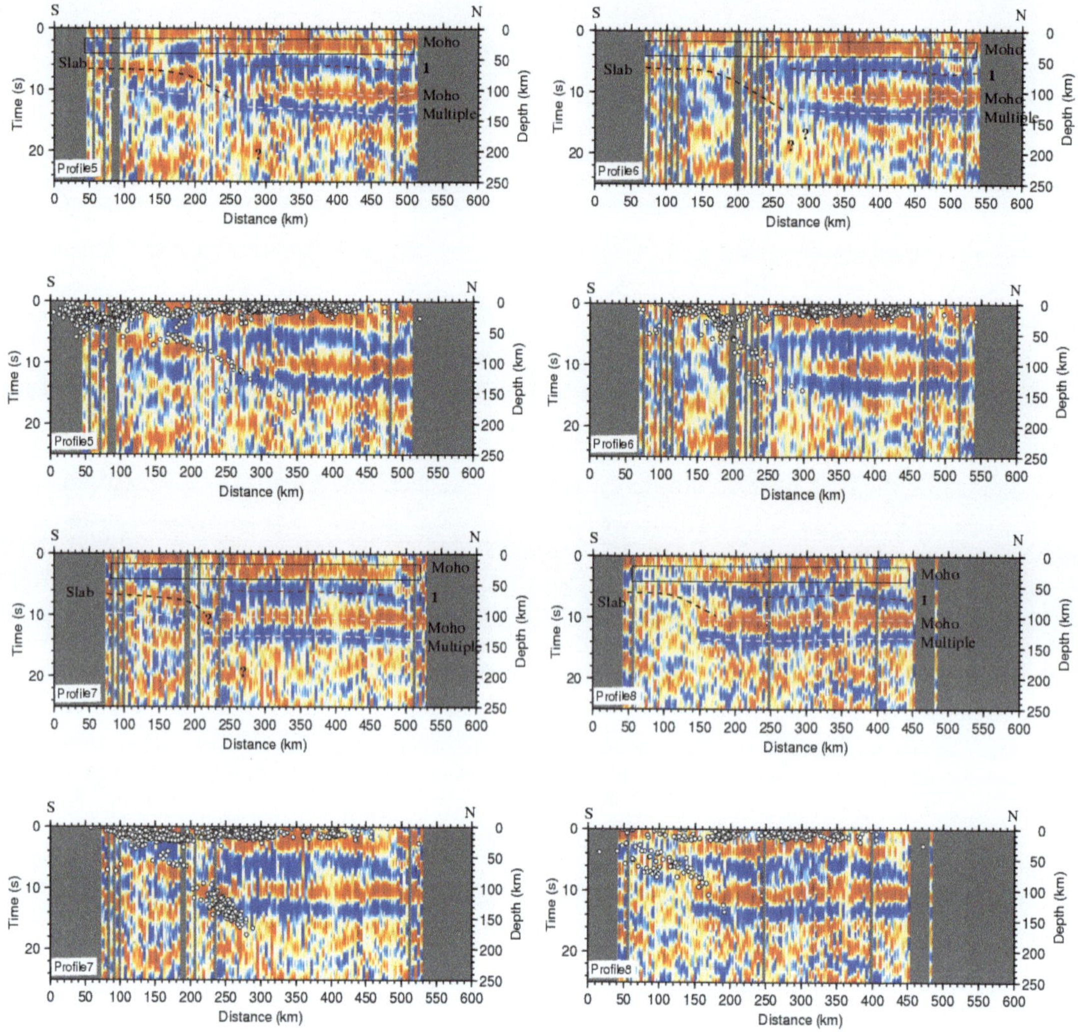

Figure 7. Continued.

The PRFs along profile 4 are similar to those observed along profiles 1–3. The slab down to ~ 10 s is well resolved by PRFs along profile 4. Strong appearance of the Moho multiples with negative amplitudes at the arrival time of the slab phase at about 10 s do not permit clear signals in larger depths.

Along profile 7, the converted slab phases are visible down to 10 s in contrast to seismicity, which shows a cluster down to 180 km depths beneath profile 7. PRFs along profile 8 cannot confirm the presence of the slab at depths larger than 70 km. At depths larger than 80–100 km the slab may be formed by continental rather than oceanic lithosphere because according to tectonic reconstructions not more than about 500 to 700 km of oceanic lithosphere has been subducted at the recent Hellenic arc (Meier et al., 2004; van Hinsbergen, 2005; Brun and Sokoudis, 2010). However, the available data do not permit a clear result.

5 Discussion

Based on hypocentre locations shown in Fig. 6, the eastern Hellenic subduction zone is divided into two portions by the intense seismic deformation of the Aegean plate in the region between eastern Crete and Karpathos (station KAPA, see Fig. 1). On the subducting slab this zone of high deformation is characterized by a vertical offset of about 40 km of the WBZ beneath the eastern Cretan Sea (Fig. 6). This deformation zone separates a shallower N-ward dipping slab in the west from a steeper NW-ward dipping slab in the east (Brüstle, 2012). The change of the dipping is also expressed by the observed seismicity of the overriding Aegean plate. The seismogenic zone between about 20 and 40 km depth of the plate interface is currently seismically active in the southwestern Aegean. This is in contrast to the seismogenic zone in the southeastern Aegean in the region of Rhodes. These differences support a segmentation of the Hellenic subduction zone in the western and eastern part. The low seismicity in the seismogenic zone in the southeastern Aegean may indicate either a low seismic coupling or a locked interface. The different influence of the slab kinematics onto the Aegean plate can also be associated with local variations of GPS velocities and their orientation relative to the central Aegean plate. Velocity vectors of the western region of the eastern Hellenic Subduction Zone are relatively small and preferably S- or SSE-oriented. Relative GPS velocity vectors of the eastern part of the eastern Hellenic Subduction Zone are larger and clearly SE-oriented (McClusky et al., 2000; Reilinger et al., 2010), which may indicate low seismic coupling and a more aseismic deformation along the plate interface in the seismogenic zone in the southeastern Aegean.

The lower dip angle of the western WBZ ($\sim 25°$) and low GPS velocities may point to reduced rollback rates of the African plate. In contrast, in the Rhodes region of the eastern Hellenic subduction zone, where the slab is dipping to NW under larger dip angle ($\sim 35°$), the WBZ is observed to greater depths (~ 180 km) and high GPS velocities may reveal a still active rollback of the African lithosphere to the SE.

These findings (Brüstle, 2012) provided excellent constraints on the geometry of the WBZ, where the resolution of the slab geometry by other studies (e.g. seismicity and tomographic studies) was limited (e.g. Spakman et al., 1993; Knapmeyer, 1999; Papazachos et al., 2000; Meier et al., 2007).

We found evidence for a shallow dipping segment ($\sim 25°$) down to ~ 100 km depth beneath the convergent western part (profile 1–4), whereas the presence of the steeper slab beneath the eastern part is not resolved in our PRFs (profiles 6 and 7, Fig. 7). Very weak conversions observed at larger depths may be related to reduced velocity contrasts implying eclogitization of crustal materials.

Evidence for west–east segmentation of the Hellenic slab was previously suggested by analysing hypocentres of the relocated EHB-ISC catalogue (Engdahl et al., 1998). Meier et al. (2007) identified two slab segments and proposed that they are associated with laterally varying properties of the African slab. Furthermore, in the Mediterranean southeast of Crete, Rhodes and south of western Anatolia, high velocities in the mantle lithosphere – typical for cold oceanic lithosphere – are found by tomographic studies (Bijward et al., 1998; Marone et al., 2004; Endrun et al., 2008; Legendre et al., 2012), whereas south of Peloponnese and western Crete velocities in the mantle lithosphere are lower (Meier et al., 2004; Legendre et al., 2012).

Regional body-wave tomographic study (Papazachos and Nolet, 1997) clearly showed a strong NW–SE delineation of the shallow dipping ($\sim 25°$) and larger western part of the slab. However, the geometry of the smaller eastern part was not resolved probably due to the poor ray geometry. Based on these results, a steeper non-continuous high-velocity anomaly (dip of $\sim 35°$) was seen beneath the eastern part of the subduction forming a more or less right angle at about 200 km depth with the shallower and larger western part of the slab.

The dip of the western segment ($\sim 25°$) obtained from our data is in good agreement with that ($\sim 21°$) shown along the western Hellenic subduction zone using a recently developed imaging technique (2-D GRT), which inverts the teleseismic scattered P-wave coda for discontinuities in elastic properties (e.g. Suckale et al., 2009; Pearce et al., 2012). Our results are also consistent with those derived from regional body-wave travel-time tomography for the western Hellenic subduction zone (e.g. Papazachos and Nolet, 1997), which found evidence for a western (dip of 25°) and eastern (dip of 35°) slab segment. We do not find evidence for a horizontally propagating tear beneath Peloponnese in the region of the western segment of the Hellenic subduction zone as reported by Wortel and Spakman (2000). If present, the horizontal tear is confined to the northern Hellenides beneath mainland Greece north of the Peloponnese and the Gulf of Corinth.

Regarding our PRF images, we cannot confirm the presence of two different segments of the African slab. This is in contrast to hypocentre locations, which infer that the transition between the western and eastern segments may occur beneath profiles 5 and 6 crossing the eastern Cretan Sea (see Fig. 6). In general, seismicity and receiver functions are compatible at shallower depths (~ 100 km depth) as the location uncertainty of microseismicity by temporary networks and the spatial resolution of receiver function imaging are comparable. However, strong multiple arrivals as well as complete phase transitions into denser phases, which results in very weak conversions, may be considered as responsible reasons for the non-coincidence of the seismicity data with PRFs at larger depths.

Furthermore, PRF data may resolve another dominant negative phase at 4–7 s (Fig. 7, dark red dashed lines, marked as 1). This phase is clearly seen beneath the northern part of all profiles just below the Aegean Moho phase. To evaluate the coherency of this phase, we stacked around 785 PRFs beneath the northern part of profile 5 (distance of 360–520 km on the profile line), where this phase is best imaged. The stacked PRF includes PRFs with different epicentral distances ranging between 30° and 95°. We filtered the stacked PRF with different low-pass filters and show them in Fig. 8. Presence of two negative phases in the first 8 s of the stacked PRF can be clearly observed using different filters. The large positive and negative signals between 10 and 20 s are crustal multiples.

Low velocities below 60 km depth beneath the southern Aegean and the volcanic arc have been earlier observed by P-wave tomography (Drakatos et al., 1997; Papazachos and Nolet, 1997). In addition, surface wave dispersion analysis revealed a low-velocity layer with minimum velocities of 4.15 $\mathrm{km\,s^{-1}}$ at depths of 50–70 km between the Aegean Moho and the slab in the Cretan Sea and beneath the central Aegean (Karagianni et al., 2002; Endrun et al., 2008). An even more pronounced low-velocity layer was found by

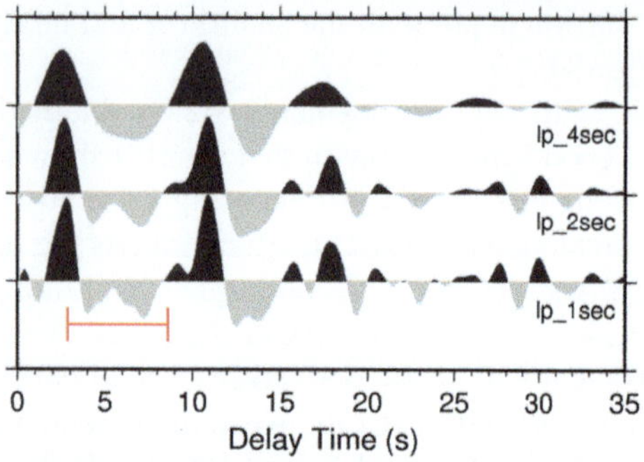

Figure 8. Stacked PRF obtained from the northern part of profile 5 (distance of 360–520 km on the profile line, see Fig. 7). Different low-pass filters were applied to the stacked trace. Independently of the corner frequency of the filter, significant negative signals are visible in the first 8 s of the seismogram (shown with thin red line). The signals between 10 and 20 s are crustal multiples.

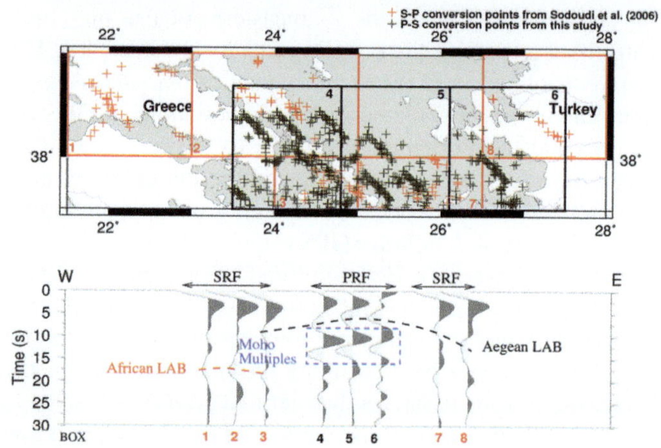

Figure 9. Aegean LAB as seen by PRFs from this study compared with the results obtained from SRFs by Sodoudi et al. (2006). **(a)** Location of P-to-S (black crosses) and S-to-P (red crosses) piercing points. Red boxes (black boxes) show where the SRFs (PRFs) are stacked. **(b)** Summation traces of P and S receiver functions obtained from the boxes shown in **(a)**. The Aegean LAB may lie at about 6–8 s beneath the central Aegean (boxes 4–6) with a significant deepening to about 10 s towards Turkey (box 7). The African LAB as presented by Sodoudi et al. (2006) is shown with red dashed line beneath mainland Greece and western Aegean (boxes 1–3). Blue dotted rectangle shows the arrival time window of the Moho multiples.

Bourova et al. (2005). They observed a large low-velocity anomaly at a depth of 50–100 km beneath the prolongation of the North Anatolian Trough in the northern Aegean, where strong deformation is currently observed at the surface. However, evidence for a LVZ beneath the Aegean plate has not been reported by previous PRF studies.

We have observed the most pronounced and sharp low-velocity zone (6 % velocity reduction) at about 60 km beneath the back arc area (after volcanic arc, see Fig. 1), where no estimation of the lithospheric thickness was made before. The observed LVZ has been constrained consistently between the Moho of the Aegean plate and the slab (see Fig. 7) along all profiles (dark red dashed lines). The depth of our LVZ is consistent with that derived from P-wave tomography and surface wave analysis in the northern and central Aegean (Drakatos et al., 1997; Papazachos and Nolet, 1997; Karagianni et al., 2002; Endrun et al., 2008; Bourova et al., 2005).

Previous S receiver function analysis (Sodoudi et al., 2006) provided earlier constraints on the LAB depth in the Aegean. The S receiver function data resolved LAB signals at about 170 km beneath mainland Greece and at 130 km beneath western Turkey (Fig. 9). Due to the locations of the EGELADOS stations we could not provide any estimation of the Aegean LAB beneath mainland Greece in this work. Furthermore, for the central Aegean, no S receiver function data are available yet. We show in Fig. 9b stacked PRFs obtained from the boxes located beneath the northern parts of the study area, where the negative phase at 4–7 s was significantly observed in our PRFs (see Fig. 7). We combined our results with those previously shown by Sodoudi et al. (2006) using S receiver functions. Stacked PRFs computed for each box (see Fig. 9a) are illustrated along a W–E trending profile covering the central Aegean. We also included the stacked SRFs obtained from five boxes mostly located beneath the mainland Greece and western Turkey from Sodoudi et al. (2006). As Fig. 9b shows, the P and S receiver functions yield consistent results for the LAB signal if multiples in the P receiver functions are carefully identified (compare box 3 with box 4 and box 6 with box 7). Beneath mainland Greece the LAB is found at about 17 s (~ 170 km). This can be seen in boxes 1–3. This signal indicates very likely the bottom of the subducting lithosphere in the region of boxes 1 and 2. The negative signal resolved at 6–8 s beneath the central Aegean (boxes 4–6) is very clear in our PRFs and may represent the Aegean LAB. Furthermore, the signal at about 10 s (~ 100 km) in boxes 7 and 8 showing the Aegean LAB in western Anatolia would imply a thickening of the Aegean lithosphere towards Anatolia.

If the negative discontinuity at about 60 km depth represents the lithosphere–asthenosphere boundary (LAB) of the continental Aegean plate (in good agreement with Endrun et al., 2008), this will be a remarkable result as it indicates a very thin mantle lithosphere in the central and northern Aegean. Taking the error bounds of P receiver functions (±5 km) into account, we found a relatively thicker lithosphere in the central Aegean (60 km) rather than that seen by surface waves analysis beneath the southern Aegean between forearc and volcanic arc (40–50 km) (Endrun et al., 2008). Given a crustal thickness of about 23–27 km in the central Aegean, the mantle lithosphere appears to be about 30–40 km thick. Although there is evidence for lithospheric stretching during the extensional processes governing the more recent (last 30 Ma) evolution of the Aegean, the interpretation of a thinned mantle lithosphere would imply a greater thickness of the mantle lithosphere before 30 Ma. Based on tectonic reconstructions (Dercourt et al., 1986; Gealey, 1988; Stampfli and Borel, 2004), the continuous slab (down to a depth of approximately 1200 km) imaged by seismic tomography would be composed of different sections of oceanic and continental mantle lithosphere (Meier et al., 2004; van Hinsbergen et al., 2005), corresponding to the oceanic basins and continental terranes. Delamination of the continental crust of the terranes from the underlying continental mantle lithosphere has been proposed in order to explain the rather continuous slab (Meier et al., 2004), nappe stacking and metamorphism in the Aegean (van Hinsbergen et al., 2005, Jolivet and Brun, 2010) based on the geological record of Crete (e.g. Thomson et al., 1998) and has been suggested as the reason for the missing deep seismicity in the Hellenic subduction zone (Meier et al., 2004).

Similarly, delamination of the Anatolian mantle lithosphere has been proposed by Keskin (2003). The removal of the mantle lithosphere of accreted continental terranes would mean that despite the lithospheric extension the mantle lithosphere has gained again a thickness of about 30–40 km in the central Aegean since its removal about 35 to 15 Ma ago. Processes involved in the evolution of a younger mantle lithosphere may be cooling of the asthenospheric mantle, magmatic underplating, dehydration of serpentinized mantle, or chemical differentiation of crustal material (Artemieva and Meissner, 2012).

6 Conclusions

We analysed more than 1.5 years of teleseismic data recorded by 65 temporary and permanent stations of EGELADOS network. By combining P receiver function observations and new accurate locations of seismic events, we obtained a detailed image of the subducting African plate beneath the continental Aegean plate in the southern and central Aegean. Observation of the negative Moho velocity contrast of the Aegean plate along the entire forearc clearly showed the serpentinized mantle wedge over the subducted African plate. Furthermore, the plate contact between the serpentinized mantle and the subducting plate seems mainly to be aseismic. We also constrained accurate crustal thicknesses and Vp / Vs ratios for the continental Aegean plate, since it could be sampled at the spacing of the EGELADOS stations. Low Vp / Vs ratios were observed beneath the central Aegean, which most likely reveal felsic crustal composition of the continental Aegean plate. Moreover, we identified the thickest crust of about 48–50 km beneath the Peloponnese Peninsula, whereas a relatively thinner crust of about 27–30 km was observed beneath the western Turkey. The crust of the Aegean plate was estimated to be 23–27 km thick beneath the southern and central Aegean. PRFs significantly imaged the Moho of the subducting African plate as a strong converted phase at shallower depths (down to 100 km) in excellent agreement with the spatial distribution of the earthquake hypocentres. We compared PRFs to hypocentres of the relocated EHB-ISC catalogue (Engdahl et al., 1998) beneath the western Aegean. In the southeastern Aegean, PRFs were compared with microseismicity obtained from the accurately located EGELADOS and relocated CYCNET catalogues (Brüstle et al., 2014) as well as accurate locations from the LIBNET network (Becker et al., 2009). The geometry of the subducted African Moho was obtained along eight different profiles covering the southern and central Aegean. Our results mainly resolved a low dip angle slab segment ($\sim 25°$) down to 100 km depth beneath the western Aegean. In contrast to the accurate hypocentre locations, which show a deformation zone separating a shallower N-ward dipping slab in the west from a steeper ($\sim 35°$) NW-ward dipping slab in the east, we could not find any evidence for the presence of a steeper and deeper slab segment beneath the southeastern Aegean in PRF data. The absence of the converted slab phase at larger depths may support phase transitions and eclogitization of the oceanic crust towards larger depths. Based on our PRFs, a LVL at about 60 km depth may represent the LAB of the Aegean plate beneath the central Aegean. If only crustal material has been accreted to Eurasia due to the collision of Gondwana derived terranes, this would imply a growth of the Aegean mantle lithosphere of about 30–40 km in the last about 15 to 35 Ma.

Acknowledgements. Mobile seismic stations for the temporary networks were provided by the DEPAS and GIPP pools (GFZ Potsdam, AWI Bremerhaven). Furthermore, we thank NOA, Greece, GFZ Potsdam, and INGV, Rome, for data of permanent seismic networks. We also thank Hans Thybo, Wolfram Geissler, Xiaohui Yuan and three anonymous reviewers for their constructive comments. Funding was provided by the German Research Foundation in the framework of the Collaborative Research Centre 526 "Rheology of the Earth " and by GFZ Potsdam.

Edited by: H. Thybo

References

Armijo, R., Meyer, B., King, G., Rigo, A., and Papanastassiou, D.: Quaternary evolution of the Corinth Rift and its applications for the late Cenozoic evolution of the Aegean, Geophys. J. Int., 126, 11–53, 1996.

Artemieva, I. M. and Meissner, R.: Crustal thickness controlled by plate tectonics: a review of crust-mantle interaction processes illustrated by European examples, Tectonophysics, 530–531, 18–49, doi:10.1016/j.tecto.2011.12.037, 2012.

Becker, D., Meier, T., Bohnhoff, M., and Harjes, H.-P.: Seismicity at the convergent plate boundary offshore Crete, Greece, observed by an amphibian network, J. Seismol., 14, 369–392, doi:10.1007/s10950-009-9170-2, 2009.

Bijwaard, H., Spakman, W., and Engdahl, E.R.: Closing the gap between regional and global travel time tomography, J. Geophys. Res., 103, 30055–30078, 1998.

Biryol, C. B., Beck, S. L., Zandt, G., and Özacar, A. A.: Segmented African lithosphere beneath the Anatolian region inferred from teleseismic P-wave tomography, Geophys. J. Int., 184, 1037–1057, 2011.

Bohnhoff, M., Makris, J., Papanikolaou, D., and Stavrakakis, G.: Crustal investigation of the Hellenic subduction zone using wide aperture seismic data, Tectonophysics, 343, 239–262, 2001.

Bohnhoff, M., Rische, M., Meier, T., Endrun, B., Harjes, H. P., and Stavrakakis, G.: CYC-NET: A temporary seismic network on the Cyclades (Aegean Sea, Greece), Seismol. Res. Lett., 75, 352–357, 2004.

Bohnhoff, M., Rische, M., Meier, T., Becker, D., Stavrakakis, G., and Harjes, H. P.: Microseismic activity in the Hellenic Volcanic Arc, Greece, with emphasis on the seismotectonic setting of the Santorini-Amorgos zone, Tectonophysics, 423, 17–33, 2006.

Bostock, M. G., Hyndman, D., Rondenay, S., and Peacock, S. M.: An inverted continental Moho and serpentinization of the forearc mantle, Nature, 417, 536–538, 2002.

Bourova, E., Kassara, I., Pedersen, H. A., Yanovskaya, T., Hatzfeld, D., and Kirtazi, A.: Constraints on absolute S-wave velocities beneath the Aegean Sea from surface wave analysis, Geophys. J. Int., 160, 1006–1019, 2005.

Brun, J.-P. and Sokoutis, D.: 45 m.y. of Aegean crust and mantle flow driven by trench retreat, Geology, 38, 815–818, doi:10.1130/G30950.1, 2010.

Brüstle, A. : Seismicity of the eastern Hellenic Subduction Zone, PhD Thesis, Ruhr University Bochum, 2012, available at: http://www-brs.ub.ruhr-uni-bochum.de/netahtml/HSS/Diss/BruestleAndrea/diss.pdf (last access: 02 November 2012), 2012.

Brüstle, A., Meier, T., Rische, M., Küperkoch, L., Friederich, W., and the EGELADOS working group: Seismicity of the eastern Hellenic Subduction Zone between 2002 and 2007 observed by temporary networks, in preparation, 2014.

Chamot-Rooke N., Rangin, C., Le Pichon, X., and Dotmed Working Group: DOTMED – Deep Offshore Tectonics of the Mediterranean: A synthesis of deep marine data in eastern Mediterranean, Mémoire de la Société géologique de France & American Association of Petroleum Geologists, numéro spécial, 177, 64 pp., 9 maps with cd-rom, 2005.

Chang, S.-J., van der Lee, S., Flanagan, M. P., Bedle, H., Marone, F., Matzel, E. M., Pasyanos, M. E., Rodgers, A. J., Romanowicz, B., and Schmid, C.: Joint inversion for three-dimensional S velocity mantle structure along the Tethyan margin, J. Geophys. Res., 115, B08309, doi:10.1029/2009JB007204, 2010.

Cocard, A., Kahle, H.-G., Peter, Y., Geiger, A., Veis, G., Felekis, S., Paradissis, D., and Billiris, H: New constraints on the rapid crustal motion of the Aegean region: recent results inferred from GPS measurements (1993–1998) across the West Hellenic Arc, Greece, Earth Planet. Sci. Lett., 172, 39–47, 1999.

DeMets, C., Gordon, R. G., Argus, D. F., and Stein, S.: Current plate motions, Geophys. J. Int., 101, 425–478, 1990.

Dercourt, J., Zonenshain, L. P., Ricou, L.-E., Kazmin, V. G., Le Pichon, X., Knipper, A. L., Grandjacquet, C., Sbortshikov, I. M., Geyssant, J., Lepvrier, C., Pechersky, D. H., Boulin, J., Sibuet, J.-C., Savostin, L. A., Sorokhtin, O., Westphal, M., Bazhenov, M. L., Lauer, J. P., Biju-Duval, B.: Geological evolution of the Thetys belt from the Atlantic to the Pamirs since the Lias, Tectonophysics, 123, 241–315, 1986.

Drakatos, G., Karantonis, G., and Stavrakakis, G. N.: P-wave crustal tomography of Greece with use of an accurate two-point ray tracer, Ann. Geofis., 1, 25–36, 1997.

Endrun, B., Meier, T., Bischoff, M., and Harjes, H. P.: Lithospheric structure in the area of Crete constrained by receiver functions and dispersion analysis of Rayleigh phase velocities, Geophys. J. Int., 158, 592–608, doi:10.1111/j.1365-246X.2004.02332.x, 2004.

Endrun, B., Meier, T., Lebedev, S., Bohnhoff, M., Stavrakakis, G., and Harjes, H. P.: S velocity structure and radial anisotropy in the Aegean region from surface wave dispersion, Geophys. J. Int., 174, 593–616, 2008.

Endrun, B., Lebedev, S., Meier, T., Tirel, C., and Friederich, W.: Complex layered deformation within the Aegean crust and mantle revealed by seismic anisotropy, Nat. Geosci., 4, 203–207, doi:10.1038/ngeo1065, 2011.

Engdahl, E. R., Van Der Hilst, R. D., and Buland, R.: Global teleseismic earthquake relocation with improved travel times and procedures for depth relocation, B. Seismol. Soc. Am., 88, 722–743, 1998.

Faccenna, C., Bellier, O., Martinod, J., Piromallo, C., and Regard, V.: Slab detachment beneath eastern Anatolia: A possible cause for the formation of the North Anatolian fault, Earth Planet. Sci. Lett., 242, 85–97, 2006.

Forster, M. and Lister, G.: Core-complex-related extension of the Aegean lithosphere initiated at the Eocene-Oligocene transition, J. Geophys. Res., 114, B02401, doi:10.1029/2007JB005382, 2009.

Gealey, W. K.: Plate tectonic evolution of the Mediterranean-Middle East region, Tectonophysics, 155, 285–306, 1988.

Gesret, A., Laigle, M., Diaz, J., Sachpazi, M., Charalampakis, M., and Hirn, A.: Slab top dips resolved by teleseismic converted waves in the Hellenic subduction zone, Geophys. Res. Lett., 38, L20304, doi:10.1029/2011GL048996, 2011.

Hatzfeld, D. and Martin, C.: The Aegean intermediate seismicity defined by ISC data, Earth Planet. Sci. Lett., 113, 267–275, 1992.

Hirn, A., Sachpazi, M., Siliqi, R., McBride, J., Marnelis, F., and Cernobori, L., and the STREAMERS-PROFILES group: A traverse of the Ionian islands front with coincident normal incidence and wide-angle seismics, Tectonophysics, 264, 35–49, doi:10.1016/S0040-1951(96)00116-3, 1996.

Jolivet, L. and Brun, J.-P.: Cenozoic geodynamic evolution of the Aegean, Int. J. Earth Sci., 99, 109–138, doi:10.1007/s00531-008-0366-4, 2010.

Kahle, H. G., Cocard, M., Peter, Y., Geiger, A., Reilinger, R., McClusky, S., King, R., Barka, A., and Veis, G.: The GPS strain rate field in the Aegean Sea and western Anatolia, Geophys. Res. Lett., 26, 2513–16, 1999.

Karabulut, H., Paul, A., Ergün, T. A., Hatzfeld, S., Childs, D. M., and Aktar, M.: Long-wavelength undulations of the seismic Moho beneath the strongly stretched Western Anatolia, Geophys. J. Int., 194, 450–464, doi:10.1093/gji/ggt100, 2013.

Karagianni, E. E., Pabagiotopoulos, D. G., Panza, G. F., Suhadolc, P., Papazachos, C. B., Papazachos, B. C., Kirtazi, A., Hatzfeld, D., Makropoulos, K., Priestley, K., and Vuan, A.: Rayleigh wave group velocity tomography in the Aegean area, Tectonophysics, 358, 187–209, 2002.

Kennett, B. L. N. and Engdahl, E. R.: Travel times for global earthquake location and phase identification, Geophys. J. Int., 105, 429–465, doi:10.1111/j.1365-246X.1991.tb06724.x, 1991.

Keskin, M.: Magma generation by slab steepening and breakoff beneath a subduction-accretion complex: an alternative model for collision-related volcanism in eastern Anatolia, Turkey, Geophys. Res. Lett., 30, 8046, doi:10.1029/2003GL018019, 2003.

Kind, R. and Vinnik, L. P.: The upper mantle discontinuities underneath the GRF array from P-to-S converted phases, J. Geophys., 62, 138–147, 1988.

Kind, R., Yuan, X., Saul, J., Nelson, D., Sobolev, S.V., Mechie, J., Zhao, W., Kosarev, G., Ni, J., Achauer, U., and Jiang, M.: Seismic images of crust and upper mantle beneath Tibet: evidence for Eurasian plate subduction, Science, 298, 1219–1221, 2002.

Knapmeyer, M.: Geometry of the Aegean Benioff zones, Annali di Geofisica, 42, 27–38, 1999.

Knapmeyer, M. and Harjes, H. -P.: Imaging crustal discontinuities and the down-going slab beneath western Crete, Geophys. J. Int., 143, 1–21, 2000.

Kosarev, G., Kind, R., Sobolev, S. V., Yuan, X., Hanka, W., and Oreshin, S.: Seismic evidence for a detached Indian lithosphere mantle beneath Tibet, Science, 283, 1306–1309, 1999.

Langston, C. A.: The effect of planar dipping structure on source and receiver responses for constant ray parameter, B. Seismol. Soc. Am., 67, 1029–1050, 1977.

Legendre, C., Meier, T., Lebedev, S., Friederich, W., and Viereck-Götte, L.: A shear-wave velocity model for the European upper mantle from automated inversion of seismic shear and surface waveforms, Geophys. J. Int., 191, 282–304, 2012.

Le Pichon, X., Chamot-Rooke, N., and Lallemant, S.: Geodetic determination of the kinematics of central Greece with respect to Europe: Implications for eastern Mediterranean tectonics, J. Geophys. Res., 100, 12675–12690, 1995.

Li, X., Bock, G., Vafidis, A., Kind, R., Harjes, H. -P., Hanka, W., Wylegalla, K., Van der Meijde, M., and Yuan, X.: Receiver function study of the Hellenic subduction zone: imaging crustal thickness variations and the oceanic Moho of the descending African lithosphere, Geophys. J. Int., 155, 733–748, 2003.

Makris, J.: The crust and upper mantle of the Aegean region from deep seismic soundings, Tectonophysics, 46, 269–284, 1978.

Makris, J. and Vees, R.: Crustal structure of the central Aegean Sea and the islands of Evia and Crete, Greece, obtained by refractional seismic experiments, J. Geophys., 42, 330–341, 1977.

Marone, F., Van der Lee, S., and Giardini, D.: Three-dimensional upper mantle S-velocity model for the Eurasia-Africa plate boundary region, Geophys. J. Int., 158, 109–130, doi:10.1111/j.1365-246X.2004.02305.x, 2004.

Mascle, J. and Martin, L.: Shallow structure and recent evolution of the Aegean Sea: A synthesis based on continuous reflection profiles, Mar. Geol., 94, 271–299, 1990.

Mascle, J. and Chaumillon, E.: An overview of Mediterranean Ridge collisional accretionary complex as deduced from multichannel seismic data, Geo Mar. Lett., 18, 81–89, doi:10.1007/s003670050056, 1998.

McClusky, S., Balassanian, S., Barka, A., Demir, C., Ergintav, S., Georgiev, I., Gurkan, O., Hamburger, M., Hurst, K., Kahle, H., Kastens, K., Kekelidze G., King, R., Kotzev, V., Lenk, O., Mahmoud, S., Mishin, A., Nadariya, M., Ouzounis, A., Paradissis, D., Peter, Y., Prilepin, M., Reilinger, R., Sanli, I., Seeger, H., Tealeb, A., Toksöz, M. N., Veis, G.: Global Positioning System constrains on plate kinematics and dynamics in the eastern Mediterranean and Caucasus, J. Geophys. Res., 105, 5695–5719, 2000.

McKenzie, D.: Active tectonics of the Mediterranean region, Geophys. J. Roy. Astr. S., 30, 109–185, 1972.

McKenzie, D.: Active tectonics of the Alpine–Himalayan belt: the Aegean Sea and surrounding regions, Geophys. J. Roy. Astr. S., 55, 217–254, 1978.

Meier, T., Dietrich, K., Stöckhert, B., and Harjes, H. P.: 1-dimensional models of the shear-wave velocity for the eastern Mediterranean obtained from the inversion of Rayleigh wave phase velocities and tectonic implications, Geophys. J. Int., 156, 45–58, 2004.

Meier, T., Becker, D., Endrun, B., Rische, M., Bohnhoff, M., Stöckhert, B., and Harjes, H. P.: A model for the Hellenic subduction zone in the area of Crete based on seismological investigations, in: The Geodynamics of the Aegean and Anatolia, edited by: Taymaz, T., Yilmaz, Y., and Dilek, Y., Geol. Soc. Spec. Publ., 291, 183–199, 2007.

Owens, T. J., Zandt, G., and Taylor, S. R.: Seismic evidence for an ancient rift beneath the Cumberland Plateau, Tennessee: A detailed analysis of broadband teleseismic P waveforms, J. Geophys. Res., 89, 7783–7795, 1984.

Özbakir, A. D., Sengör, A. M. C., Wortel, M. J. R., and Govers, R.: The Pliny–Strabo trench region: A large shear zone resulting from slab tearing, Earth Planet. Sci. Lett., 375, 188–195, 2013.

Papazachos, B. C. and Comninakis, P. E.: Geophysical and tectonic features of the Aegean arc, J. Geophys. Res., 76, 8517–8533, 1971.

Papazachos, B. C., Karakostas, V. G., Papazachos, C. B., and Scordilis, E. M.: The geometry of the Wadati-Benioff zone and lithospheric kinematics in the Hellenic arc, Tectonophysics, 319, 275–300, 2000.

Papazachos, C. and Nolet, G.: P and S deep velocity structure of the Hellenic area obtained by robust nonlinear inversion of travel times, J. Geophys. Res., 102, 8349–8367, 1997.

Pearce, F. D., Rondenay, S., Schapazi, M., Charalampakis, M., and Royden, L. H.: Seismic investigation of the transition from continental to oceanic subduction along the western Hellenic Subduction Zone, J. Geophys. Res., 117, B07306, doi:10.1029/2011JB009023, 2012.

Piromallo, C. and Morelli, A.: P wave tomography of the mantle under the Alpine-Mediterranean area, J. Geophys. Res., 108, 2065, doi:10.1029/2002JB001757, 2003.

Reilinger, R., McClusky, S., Paradissis, D., Ergintav, S., and Vernant, P.: Geodetic constraints on the tectonic evolution of the Aegean region and strain accumulation along the Hellenic subduction zone, Tectonophysics, 488, 22–30, doi:10.1016/j.tecto.2009.05.027, 2010.

Sachpazi, M., Hirn, A., Clément, C., Haslinger, F., Laigle, M., Kissling, E., Charvis, P., Hello, Y., Lépine, J.-C., Sapin, M., and Ansorge, J.: Western Hellenic subduction and Cephalonia Transform: local earthquakes and plate transport and strain, Tectonophysics, 319, 301–319, 2000.

Schmid, C., Van Der Lee, S., and Giardini, D.: Delay times and shear wave splitting in the Mediterranean region, Geophys. J. Int., 159, 275–290, 2004.

Snopek, K. and Casten, U.: 3GRAINS: 3-D Gravity Interpretation Software and its application to density modeling of the Hellenic subduction zone, Comput. Geosci., 32, 592–603, doi:10.1016/j.cageo.2005.08.008, 2006.

Sodoudi, F., Kind, R., Hatzfeld, D., Priestley, K., Hanka, W., Wylegalla, K., Stavrakakis, G., Vafidis, A., Harjes, H. P., and Bohnhoff, M.: Lithospheric structure of the Aegean obtained from P and S receiver functions, J. Geophys. Res., 111, B12307, doi:10.1029/2005JB003932, 2006.

Sodoudi, F., Yuan, X., Asch, G., and Kind, R.: High-resolution image of the geometry and thickness of the subducting Nazca lithosphere beneath northern Chile, J. Geophys. Res., 116, B04302, doi:10.1029/2010JB007829, 2011.

Spakman, W., Wortel, M. J. R., and Vlaar N. S.: The Hellenic subduction zone: A tomographic image and its geodynamical implications, Geophys. Res. Lett., 15, 60–63, 1988.

Spakman, W., Van der Lee, S., and Van der Hilst R. D.: Travel time tomography of the European-Mediterranean mantle down to 1400 km, Phys. Earth Planet. In., 79, 3–73, 1993.

Stampfli, G. M. and Borel, G. D.: The transmed transect in space and time: Constraints on the paleotectonic evolution of the mediterranean domain, in: The TRANSMED Atlas: the Mediterranean Region from Crust to Mantle, edited by: Cavazza, W., Roure, F., Spakman, W., Stampfli, G. M., and Ziegler, 53–80, Springer Verlag, Berlin, 2004.

Suckale, J., Rondenay, S., Sachpazi, M., Charalampakis, M., Hosa, A., and Royden, L. H.: High-resolution seismic imaging of the western Hellenic subduction zone using teleseismic scattered waves, Geophys. J. Int., 178, 775–791, 2009.

Thomson, S.N., Stöckhert, N., and Brix, M.R.: Thermochronology of the high-pressure metamorphic rocks of Crete, Greece: implications for the speed of tectonic processes, Geology, 26, 259–262, 1998.

Tirel, C., Gueydan, F., Tiberi, C., and Brun, J. -P.: Aegean crustal thickness inferred from gravity inversion, Geodynamical implications, Earth Planet. Sci. Lett., 228, 267–280, doi:10.1016/j.epsl.2004.10.023, 2004.

Vanacore, E. A., Taymaz, T., and Saygin, E.: Moho structure of the Anatolian Plate from receiver function analysis, Geophys. J. Int., 193, 329–337, doi:10.1093/gji/ggs107, 2013.

Van Hinsbergen, D. J. J., Hafkenscheid, E., Spakman, W., Meulenkamp, J. E., and Wortel, R.: Nappe stacking resulting from subduction of oceanic and continental lithosphere below Greece, Geology, 33, 325–328, 2005.

Wagner, L. S., Beck, S., and Zandt, G.: Upper mantle structure in the south central Chilean subduction zone (30° to 36° S), J. Geophys. Res., 110, B01308, doi:10.1029/2004JB003238, 2005.

Wortel, R. and Spakman, W.: Subduction and slab detachment in the Mediterranean-Carpathian region, Science, 290, 1910–1917, 2000.

Yuan X., Ni, J., Kind, R., Mechie, J., and Sandvol, E.: Lithospheric and upper mantle structure of southern Tibet from a seismological passive source experiment, J. Geophys. Res., 102, 491–500, 1997.

Zandt, G., Velasco, A. A., and Beck, S. L.: Composition and thickness of the southern Altiplano crust, Bolivia, Geology, 22, 1003–1006, 1994.

Zheng, Y. and Lay, T.: Low Vp / Vs ratios in the crust and upper mantle beneath the Sea of Okhotsk inferred from teleseismic pMP, sMP, and sMS underside reflections from the Moho, J. Geophys. Res., 111, B01305, doi:10.1029/2005JB003724, 2006.

Zhu, L. and Kanamori, H.: Moho depth variation in southern California from teleseismic receiver functions, J. Geophys. Res., 105, 2890–2969, 2000.

Zhu, L., Mitchell, B. J., Akyol, N., Cemen, I., and Kekovali, K.: Crustal thickness variations in the Aegean region and implications for the extension of continental crust, J. Geophys. Res., 111, B01301, doi:10.1029/2005JB003770, 2006.

Variations of the crustal thickness in Nepal Himalayas based on tomographic inversion of regional earthquake data

I. Koulakov[1,2], **G. Maksotova**[1,2], **S. Mukhopadhyay**[3], **J. Raoof**[3], **J. R. Kayal**[4], **A. Jakovlev**[1,2], **and A. Vasilevsky**[1,2]

[1]Trofimuk Institute of Petroleum Geology and Geophysics SB RAS, Prospekt Koptyuga, 3, 630090, Novosibirsk, Russia
[2]Novosibirsk State University, Novosibirsk, Russia, Pirogova 2, 630090, Novosibirsk, Russia
[3]Department of Earth Sciences, IIT Roorkee, Roorkee, India
[4]School of Oceanographic Studies, Jadavpur University, Kolkata 700032, India

Correspondence to: I. Koulakov (koulakoviy@ipgg.sbras.ru)

Abstract. We estimate variations of the crustal thickness beneath the Nepal Himalayas based on tomographic inversion of regional earthquake data. We have obtained a low-velocity anomaly in the upper part of the model down to depths of 40 to 80 km and proposed that the lower limit of this anomaly represents variations of the Moho depth. This statement was supported by results of synthetic modeling. The obtained variations of crustal thickness match fairly well with the free-air gravity anomalies: thinner crust patterns correspond to lower gravity values and vice versa. There is also some correlation with magnetic field: higher magnetic values correspond to the major areas of thicker crust. We propose that elevated magnetic values can be associated with more rigid segments of the incoming Indian crust which cause more compression in the thrust zone and lead to stronger crustal thickening.

1 Introduction

Collision processes are related to the convergence of continental blocks and lead to significant shortening and thickening of the crust. The collision zones with strong seismic activity often coincide with highly populated areas, leading to damage and destruction of human habitation and suffering of population. The Himalayas, which are the highest mountain chain on the Earth, has formed due to the collision of the Indian and Asian plates. The mechanisms of mountain building in Himalayas and Tibet have been extensively discussed by many authors for decades (e.g., Dewey and Bird, 1970; Seeber et al., 1981; Molnar and Tapponier, 1975; Allegre et al. 1984). As the Indian landmass moved northwards the sedimentary piles with their older crystalline foundation complexly folded, faulted and thrusted, which caused varied crustal structure all along the 2500 km long Himalayan arc from west to east. According to the most popular tectonic model of Himalayan collision (Seeber et al., 1981), the Indian plate underthrusts the Asian plate along a gentle north-dipping (4–10° N) detachment plane, called the Main Himalayan Thrust (MHT) (Fig. 1). Most of the Himalayan earthquakes are shallow and occur at 15 to 20 km depth on the MHT. Several recent seismological studies, however, suggest that the tectonic model varies from west to east. For example, earthquakes in the eastern Himalaya tend to be much deeper than in the western part (Kayal, 2001, 2010; Mukhopadhyay and Sharma, 2010). More definitive geodynamic concepts can only be constructed based on reliable information on the deep structures in the crust and the mantle. However, due to many political and natural reasons the Himalayas is a difficult region to make a detailed study with most of geophysical methods.

The Mohorovičić (Moho) discontinuity depth is one of the key types of information which is directly linked with the major geodynamical processes. For the Himalayas, the existing Moho depth models are either too generalized or too fragmentary. For example, an existing global model CRUST2.0 (Bassin et al., 2000) provides an over-smoothed solution in Himalayas with extrapolation in some parts. The gravity modeling in the Himalayas also provides fairly smooth variations of the Moho depth (e.g., Tenzer and Chen, 2014).

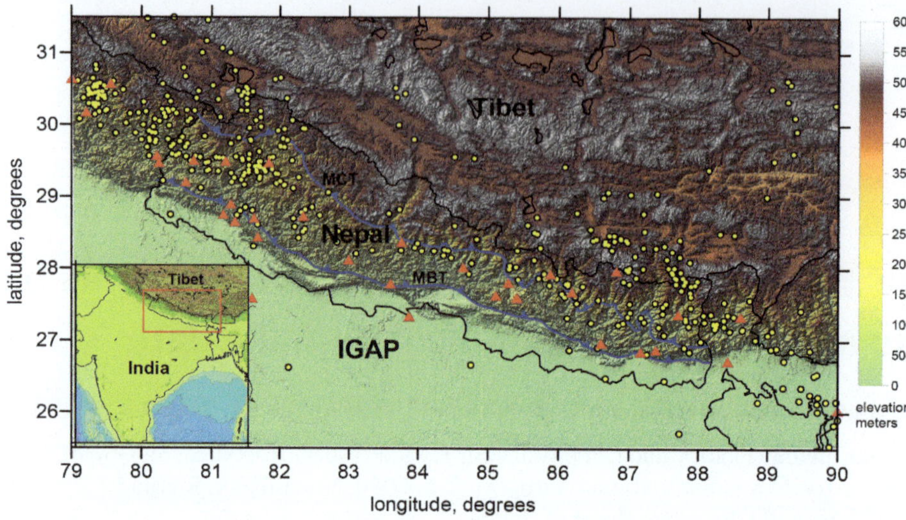

Figure 1. Map of the study area and data distributions. Background is topography. Yellow dots are the earthquakes, and red triangles are the stations used in this study. Locations of three profiles used for visualization of the results are shown. Blue lines indicate the Main Central Thrust (MCT) and the Main Boundary Thrust (MBT). IGAP is the Indo-Gangetic Alluvial Plains. Inset shows the location of the region.

Another gravity study by Jin et al. (1996) reported that the Moho depth varies from 38 km below Indo-Gangetic Alluvial Plains (IGAP) to about 75 km below high Himalayas. The regional tomography models though depict reliable images of the lithospheric behavior beneath Himalayas and Tibet (e.g., Li et al., 2008; Koulakov, 2011), but they cannot provide many constraints on the crustal structures. On local scale, the existing receiver function sites, deep sounding profiles and local earthquake tomography results from east to west of the Himalayas and Tibet (e.g., Kind et al., 2002; Kumar et al., 2005; Galve et al., 2002; Hauck et al., 1998; Mitra et al., 2005; Ramesh et al., 2005; Schulte-Pelkum et al., 2005; Rai et al., 2006, Mukhopadhyay and Sharma, 2010) provide reliable but too local and sparse information which is hard to use to build a generalized crustal model for the entire Himalayas.

In this paper we make an attempt to estimate the variations of the crustal thickness based on results of tomography inversion using the travel time data recorded by the networks of Nepal and northern India. In most cases, seismic tomography is used to derive smooth velocity distributions and appears to be not sensitive to sharp first-order interfaces. However, in some cases it can provide useful information to estimate the variations of the main interfaces. For example, Koulakov and Sobolev (2006) provided the map of the Moho depth beneath the Middle East area based on the inversion of the regional travel time data from the ISC catalogue. This model is fairly corroborated by later studies based on receiver functions and active seismic profiles (Mechie et al., 2013). Koulakov and Sobolev (2006), however, put forward some conditions which would make possible studying the Moho depth using travel time tomography: (1) stations in the study area should be distributed densely and uniformly as much as possible; (2) sufficient number of sources should be located inside the study region; (3) size of the area should be in the range of 150–500 km; (4) both travel times of crustal (Pg, Sg) and mantle (Pn, Sn) rays should be presented in the data set. To some extent, all these conditions are fulfilled in the Nepal Himalayas region. Thus, we claim that the tomographic results in this study provide new information on the variations of the Moho depth beneath the Nepal Himalayas.

2 Data analysis and tomography model

We have combined the data of regional networks in northern India (run by India Meteorological Department, IMD) and Nepal (run by the Department of Mines and Geology, Nepal, DMN) along with the global ISC catalogue for the years of 2004–2014. In total, we used the information from 78 seismic stations installed in India and Nepal. The data selection was based on three criteria: (1) the residuals for the P data and S data after location of sources in the 1-D model should not exceed 2 and 3 s, respectively; (2) the number of picks per event should not be less than 8; (3) the distance from an event to the nearest recording station should not be more than 250 km. In total, 10 864 P arrival times and 5293 S arrival times from 821 events in the study region were selected for this study (on average, almost 20 picks per event). The distributions of stations and selected events used for computations are shown in Fig. 1. Note that only in Nepal we have fairly dense distributions of both stations and earthquakes. In China to the north, there were many events, but no stations were available; in India, there were some stations available, but very few events were reported.

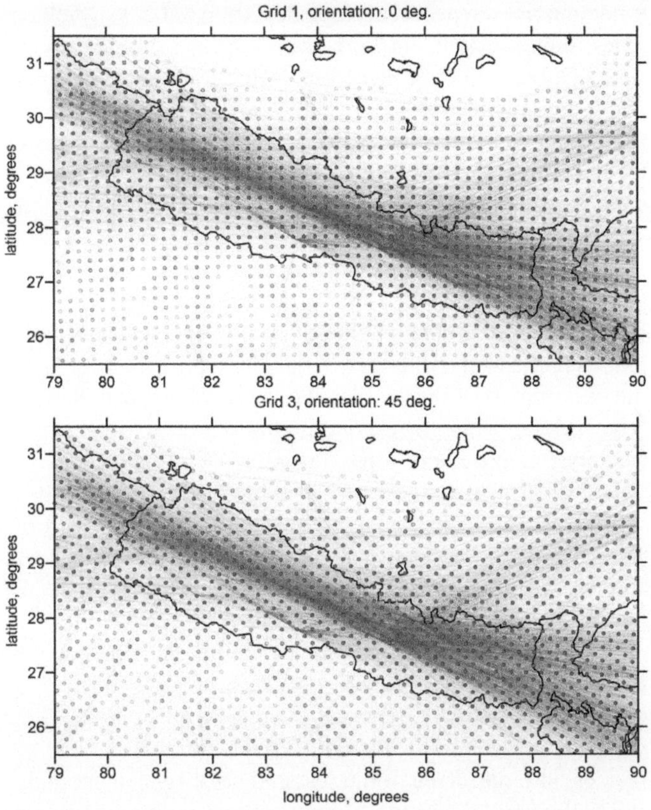

Figure 2. Distribution of P rays and two parameterization grids corresponding to basic orientations of 0 and 45 degrees. Intensity of the points represents the number of nodes in z direction corresponding to the current x–y coordinates.

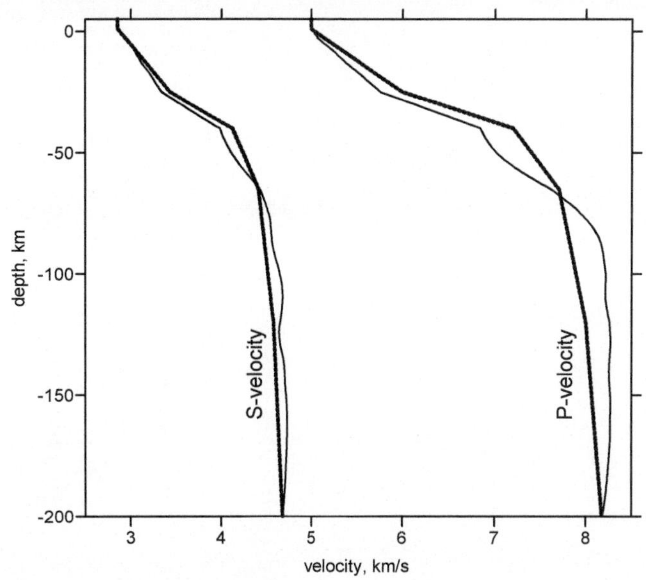

Figure 3. P velocity and S velocity versus depth. Dotted lines depict starting 1-D velocity models; solid line represents average velocities along section 1 for the main result shown in Figs. 6–7.

The analysis of data is performed using the iterative tomographic algorithm LOTOS (Koulakov, 2009). Because of the large size of the area, we have modified the code by taking into account the sphericity of the Earth. All the calculations are performed in the Cartesian coordinates. However, the reference model is kept radially symmetric, and Z coordinates for the events and stations are corrected according to the spherical shape of the Earth. In other aspects, the workflow of the analysis was similar to that used in other studies based on this algorithm (e.g., Koulakov et al., 2010). The processing starts with preliminary source locations with the use of reference table containing travel times in the 1-D model. In the next step, the sources are re-located using 3-D algorithm of ray tracing based on bending method. The velocity distributions are parameterized with nodes distributed inside the study area according to the ray density. To avoid any bias of the model due to predefined parameters of the grid, we performed the inversions for four different grids with different basic orientations. Examples of node distributions for two grids with basic orientations of 0° and 45° are shown in Fig. 2. Note that, in map view, the node projections look regularly spaced. However, along the vertical lines, the number of nodes and spacing depend on the data distribution (cases with denser node distributions are depicted with darker points in Fig. 2). The inversion was performed simultaneously for the 3-D P velocity distributions and S velocity distributions, source parameters and station correction. The matrix was inverted using the LSQR method (Paige and Saunders, 1982; Nolet, 1987). The inversion results obtained using differently oriented grids are averaged into one model which was then used to update the 3-D model for the next iteration. In total, for the analysis of both synthetic and observed data, we used three iterations, each of which included the steps of source locations in the updated velocity models, matrix calculations and inversions.

To avoid any predefinition for the Moho depth, we set the reference model without any sharp interfaces and even without high gradient levels. We defined a constant V_P/V_S ratio equal to 1.75 and set the P-velocity values at different depth levels: 5 km s−1 at −1 km, 6 km s−1 at 25 km, 7.2 km s−1 at 40 km, 7.7 km s−1 at 65 km, 8 km s−1 at 120 km and 8.2 km s−1 at 210 km depth. Between these levels, the velocity was linearly interpolated. Starting velocities for P and S models are shown in Fig. 3 with dotted lines.

Unlike the tomography algorithms used by Koulakov and Sobolev (2006) for studying the Moho depth in the Middle East, here we do not parameterize the Moho as a sharp first-order interface with variable depth. Instead, we derive the geometry of Moho based on consideration of velocity anomalies and absolute velocities. In the starting 1-D model the velocity around the Moho depths was faster than expected crustal velocities and slower than mantle velocities. As a result, the crust was revealed as low-velocity anomaly, whereas the uppermost mantle is associated with

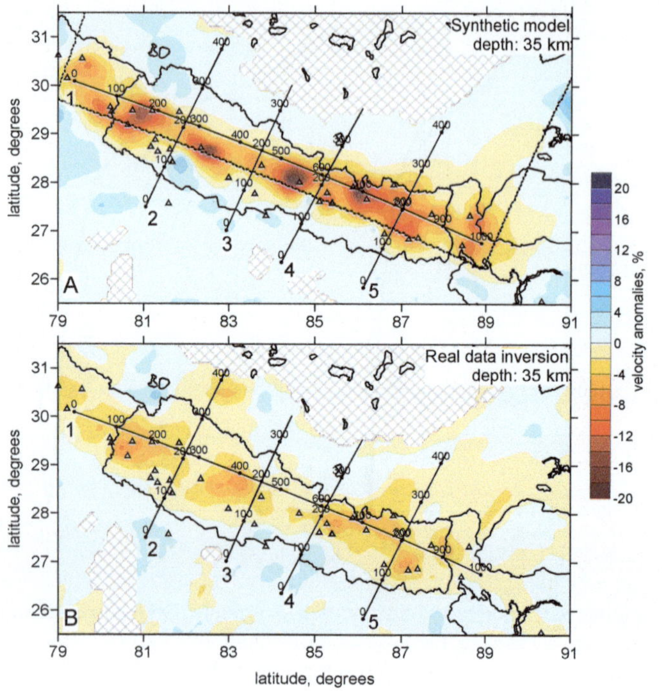

Figure 4. Velocity anomalies at 35 km depth after inversion of synthetic **(a)** and observed data **(b)**. For the case of synthetic modeling, the limits of "thick crust" are marked with a dotted line. Triangles denote stations. Locations of profiles presented in Figs. 3 to 5 are shown.

high-velocity anomaly. The variation of thickness of the crust-related low-velocity anomaly may represent the perturbations of the Moho depth. After inversion, absolute velocity forms a zone of higher gradient around the presumed location of the Moho interface, as seen in representation of average absolute velocities in Fig. 3. Note that in this case we present an average for a long profile, for which the crustal thickness may vary significantly. For local points, the Moho-related high-gradient zone is seen more prominently.

To examine the adequacy of the detection of crustal thickness, we have performed a series of synthetic tests. In Figs. 4a and 5 we present result of one of the tests. The synthetic model was defined as a superposition of a reference 1-D velocity model and a low-velocity anomaly with the amplitude of -15 % of variable thickness representing the crust. The lower limit of this anomaly is indicated in vertical sections in Fig. 5 with solid lines. In map view (Fig. 4a), a zone of variable thicker crust is highlighted with a dotted line. Outside this zone, the Moho is set at 30 km depth. Note that starting 1-D model for the synthetic reconstruction was different for the "true" reference models. This represents the realistic situation in the case of observed data analysis when the true reference model is unknown.

To compute the synthetic data, we have used the same source–receiver pairs as in the case of the real data analysis. The computed synthetic travel times were perturbed with random noise having an average deviation of 0.1 s, which enables approximately the same variance reduction as in the case of observed data inversion. After computing synthetic travel times using the 3-D ray tracer, we "forgot" all information on the velocity distributions and source locations. Then we performed the full data processing including the steps of source locations. The restored anomalies at 35 km depth are shown in Fig. 4a. In vertical sections in Fig. 5, we present the restored relative anomalies and absolute velocities. It is seen that, in section 1 along Himalayas, the thickness of the derived low P-velocity anomalies correctly represents the undulations of the Moho interface in the input model, especially in the eastern part of the profile. We see that the location of the "true" Moho better corresponds to the velocity of 7.4 km $s-1$ (yellow zone) in the restored absolute velocity model.

For section 2 to section 5, the low-velocity anomaly is visible only beneath Nepal. Neither in the Indian nor in the Tibetan side is the crust-related anomaly restored. This test shows that the robust reconstruction of the Moho depth using the tomographic reconstruction can be achieved only in case of coexistence of stations and events in a sufficiently large area. Just availability of only stations (like in India) or only seismicity (in Tibet) is not enough for this purpose. The resolved area can be estimated from this test that roughly coincides with the shape of the restored crust-related anomaly.

The horizontal resolution is examined with another synthetic test, which is presented in Fig. S1 of the Supplement.

As mentioned earlier, for the inversion of the observed data, we used three iterations. During the inversion procedure, the residuals reduced from 0.807 to 0.500 s for the P data (37.99 % of reduction) and from 1.57 s to 0.83 for the S data (47.14 % of variance reduction). Note that the similar remnant residuals were obtained in the final synthetic model discussed above.

The results of tomographic inversions for P velocity are presented in five vertical sections: one along and four sections across Himalayas. We present both relative deviations with respect to the starting model (Fig. 6) and absolute velocities (Fig. 7). In addition, one horizontal section of P-velocity anomalies at 35 km depth is presented in Fig. 4b. More horizontal sections, as well as S-velocity anomalies, are presented in the supplementary material in Fig. S2. We have defined the resolved area according to the results of various synthetic tests, mainly based on the capacity to retrieve crust-related anomalies in the test shown in Fig. 4. Areas with lower resolution in sections 2 to 5 are shaded in the resulting plots in Figs. 6 and 7.

As we see from the results of the synthetic test, thickness of the low-velocity anomaly beneath Nepal may represent the depth variations of Moho. We can also identify Moho in a zone of generally higher gradient of the absolute velocity, which is observed in plots of absolute velocities in Fig. 7. Based on the results of synthetic modeling, we have identified the Moho depth approximately at the contour line of 7.4 km $s-1$ (yellow zones in Fig. 7).

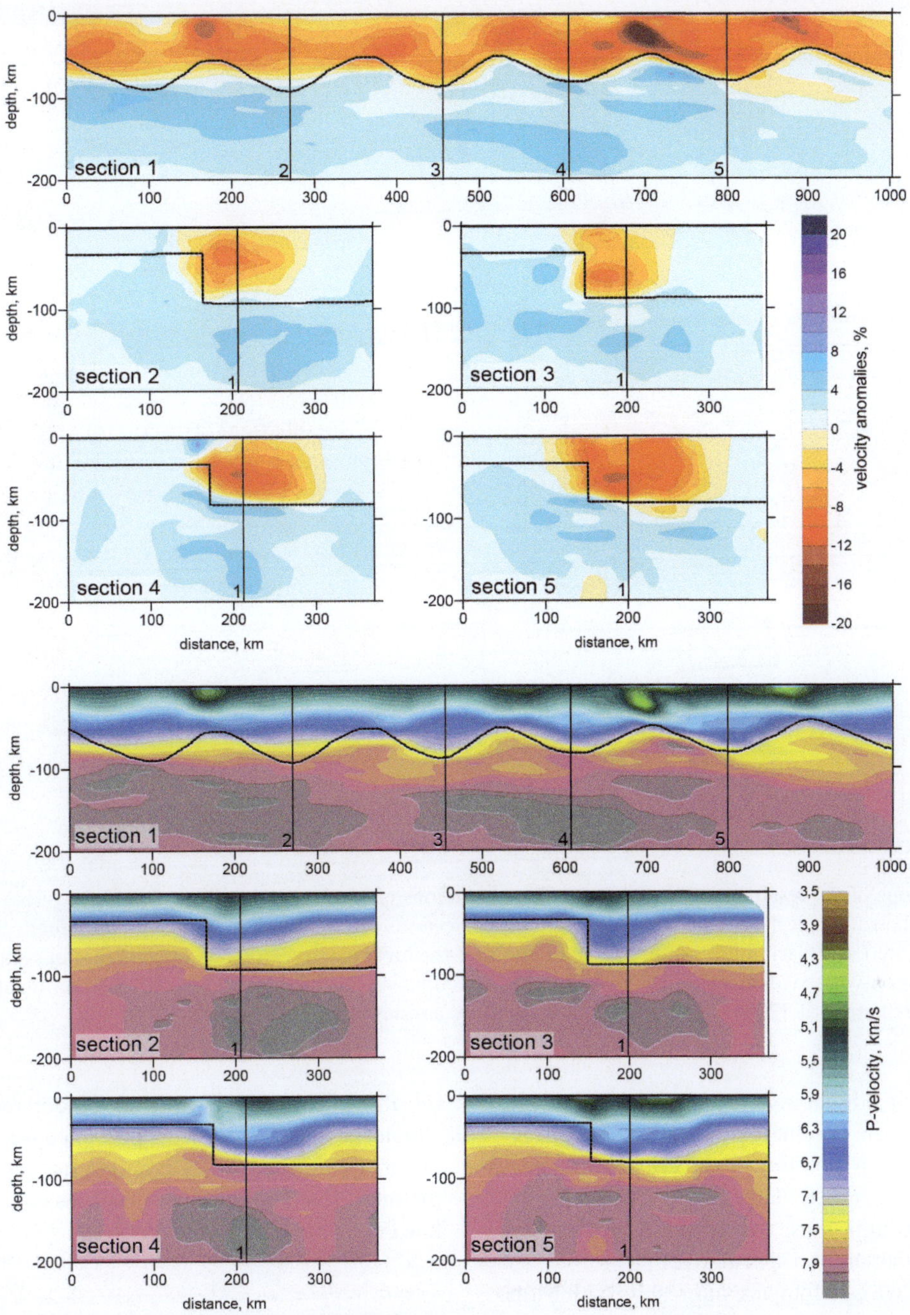

Figure 5. Synthetic test with reconstruction of the "variable Moho" model in relative anomalies (upper part) and absolute P velocities (lower part). Locations of the profiles are shown in Fig. 2. The configuration of the synthetic "Moho" is indicated with a dotted line. Vertical lines with numbers mark locations where sections cross each other.

We have manually traced the lower limit of this low-velocity anomaly in 21 vertical sections passing across the Himalayan chain (see Fig. S3) and created the 2-D surface of this limit beneath Nepal (Fig. 8a). Projections of this surface to the vertical sections below Nepal, where a satisfactory resolution is achieved, are shown in Figs. 6 and 7 with solid lines. It should be noted that the unambiguous tracing of the Moho is not possible everywhere. For example in section 5 to 7 (see Fig. S3) below the main low-velocity anomaly, there is another low-velocity pattern which appears to be weaker and separated from the upper one. This transitional anomaly may or may not be included to the crust. In case it is included, the total thickness of the crust in the frontal zone beneath Himalayas may reach 80 km (see Figs. S3 and S4), which seems to be less plausible. Thus the latter model may be accepted as more realistic model; the former one is shown in

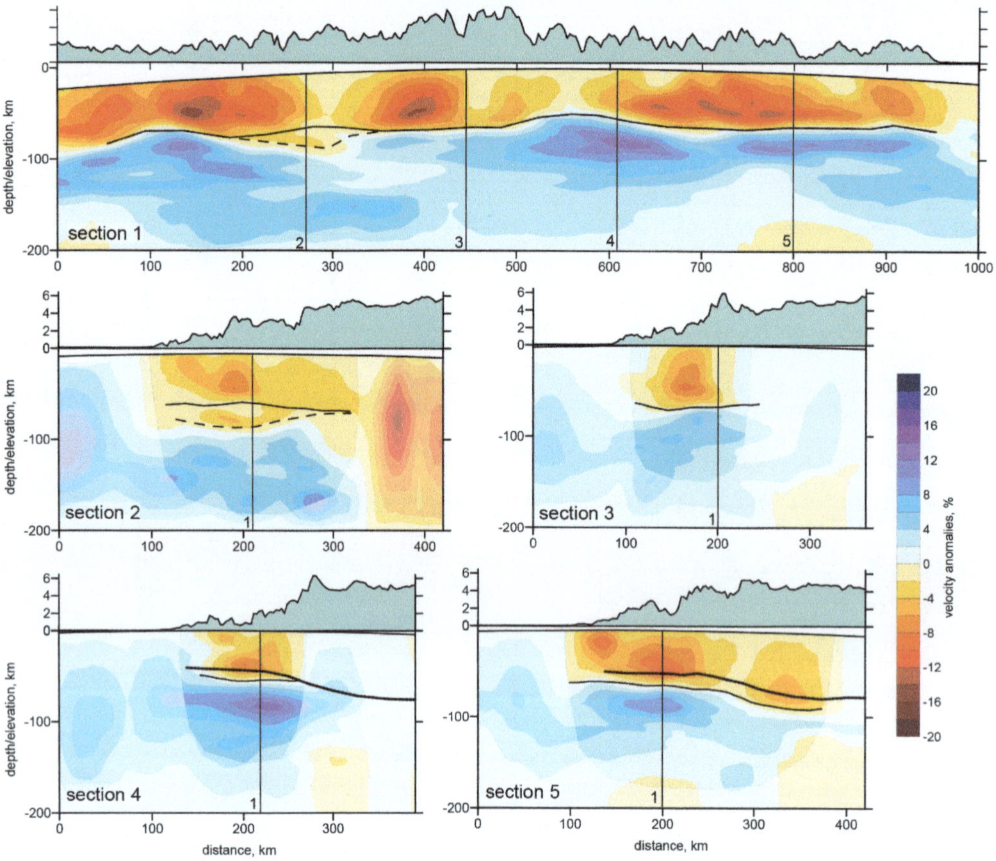

Figure 6. Vertical sections of the resulting P-velocity anomalies. Locations of sections are indicated in Fig. 4. Above each section, exaggerated topography is shown. Vertical lines indicate locations where sections cross each other. Areas with poorer resolution in sections 2 to 5 are shaded. Moho interface (black line) is traced on the bottom of the low-velocity anomaly. Dashed line in sections 1 and 2 indicates an alternative interpretation which is less plausible. Dotted lines in sections 4 and 5 indicate Moho depth determinations from Hetényi et al. (2006) and Schulte-Pelkum et al. (2005), respectively. More sections are shown in Supplement.

Supplement. Thick transitional zone in this area may represent underthrusting of one continental block underneath another one, which results in doubling of the crustal thickness. However, as shown by synthetic tests, the resolution in this part of the area is not high; thus, we should be prudent and avoid interpretations that are too speculative.

In sections 4 and 5, we can compare our results with Moho depth determinations obtained in other studies. In section 4, dotted line depicts the Moho depth derived from combined interpretation of receiver function, gravity and thermomechanical modeling by Hetényi et al. (2006). Same as in our results, they depict flat Moho in the area where our model is resolved. To the north, their model shows gradual thickening of the crust; however, this trend is observed outside the resolved area of our model. In section 5, we can compare with the Moho depth determined from receiver functions by Schulte-Pelkum et al. (2005). This model shows generally same dipping trend as in our result. In both profiles, Moho interface in our results appears to be deeper to 5–7 km than the one determined by other authors. We should emphasize that the absolute values of Moho depth derived from tomography should be considered with prudence because of unambiguity of the conversion of continuous seismic anomalies into the interface. The relative variations of the crustal thickness, however, appear to be correct.

3 Discussion

The variations of the crustal thickness in the frontal zone of the Himalayan thrust belt, as seen in our tomographic model (Fig. 8a), may be attributed to variable mechanical properties of the collided plates. The existence of weaker or more rigid segments in the underlying Indian plate may cause weaker or stronger folding in the Himalayan thrust zone. However, due to several reasons, it is not easy to quantify this correlation because the Indian plate is mostly covered by thick sediments of the Gangetic alluvium, which hide major tectonic features.

To identify hidden crustal structures, the observations of potential fields might be useful. In Fig. 8b we show the free-air gravity anomalies for the Nepal Himalayas region ex-

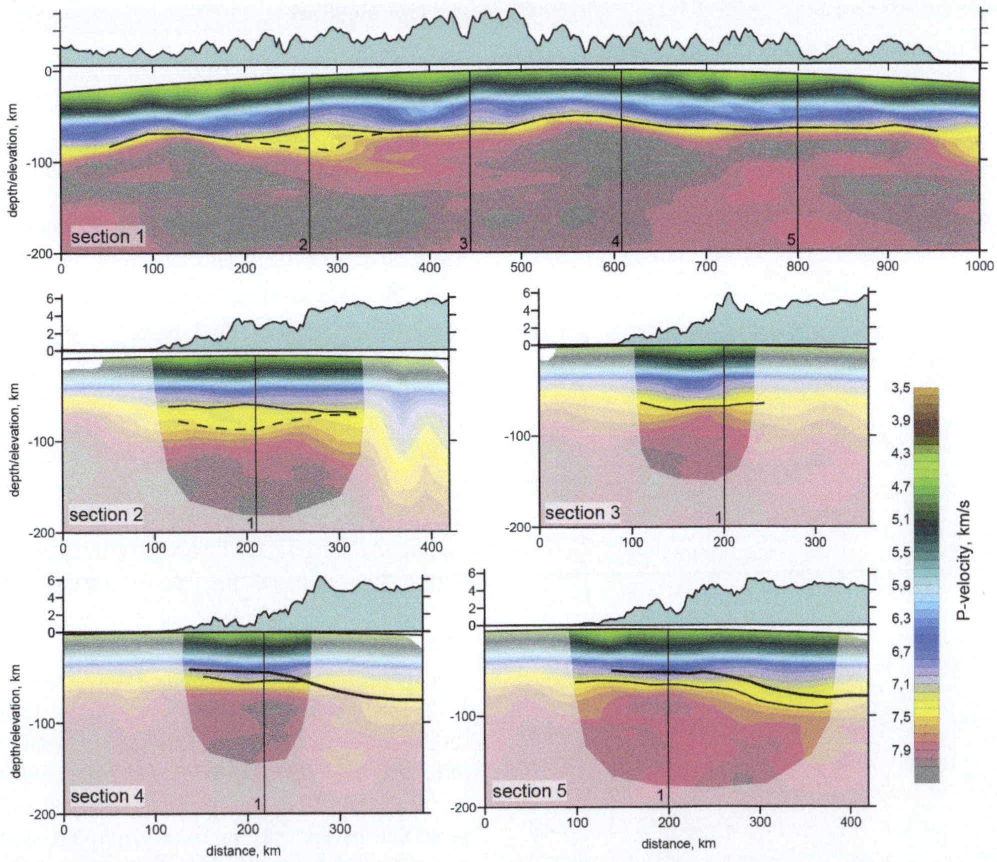

Figure 7. Same as Fig. 6, but for absolute P velocity.

tracted from the global model by Andersen et al. (2010); the smoothed anomaly is obtained using a pseudo-Gaussian weight function with the characteristic radius of 10 km. In the Nepal Himalayas, the gravity field demonstrates very strong variations. To the south of the Himalayas, there is a strong negative anomaly, which is partly caused by isostatic compensation related to the mountain growth, and it might also reflect thick sediments of the Gangetic alluvium brought from the Himalayas due to very fast erosion. The maximum value of the free-air gravity field is observed in the higher Himalayas along the Nepal–China border. It is clear that these strong variations across the Himalayan thrust zone are mostly due to abrupt Moho dipping from relatively thinner crust in the Indian Plate to almost doubled crust beneath the Himalayas and Tibet. Along the Himalayas we also observe strong variations of the gravity anomalies which might be associated with laterally inhomogeneous thickness of the crust. The lateral variations in gravity anomalies in Nepal correlate rather well with our estimates of the crustal thickness. For example, areas of thinner crust indicated with “1”, “3”, “6” and “8” correspond to lower-gravity anomaly patterns. On the contrary, thicker crust segments numbered with “2”, “4”, “5”, “7” and “9” are associated with higher values of gravity anomalies.

Here we also examine the magnetic anomalies extracted from the global compilation by Maus et al. (2009). Besides the map for the Nepal Himalayas and adjacent areas in Fig. 8c, we present the map of magnetic anomalies for a much larger area in Fig. S5. In the Nepal Himalayas, the correlation of crustal thickness with the magnetic anomalies is not as clear as that found with the gravity map (partly due to non-availability of high-resolution magnetic data in Nepal). However, it is worth noting that the largest pattern of thinner crust “6” corresponds to the negative magnetic anomaly. On the contrary, the positive anomaly “4” in the frontal thrust zone is located close to the positive magnetic anomaly to the south.

Based on the observed correlation of the seismic model with gravity and magnetic anomalies, we propose a mechanism which may explain the variability of crustal thickness along the Nepal Himalayas. As observed in regional magnetic map of India (Fig. S5), the crust of the Indian plate appears to be very heterogeneous. One of the reasons for strong magnetic anomalies may be the presence of large provinces affected by relict igneous processes. Although these occurred in the geological past, the large magmatic intrusions might considerably strengthen the crust. The negative magnetic anomalies, on the other hand, may be explained by thicker

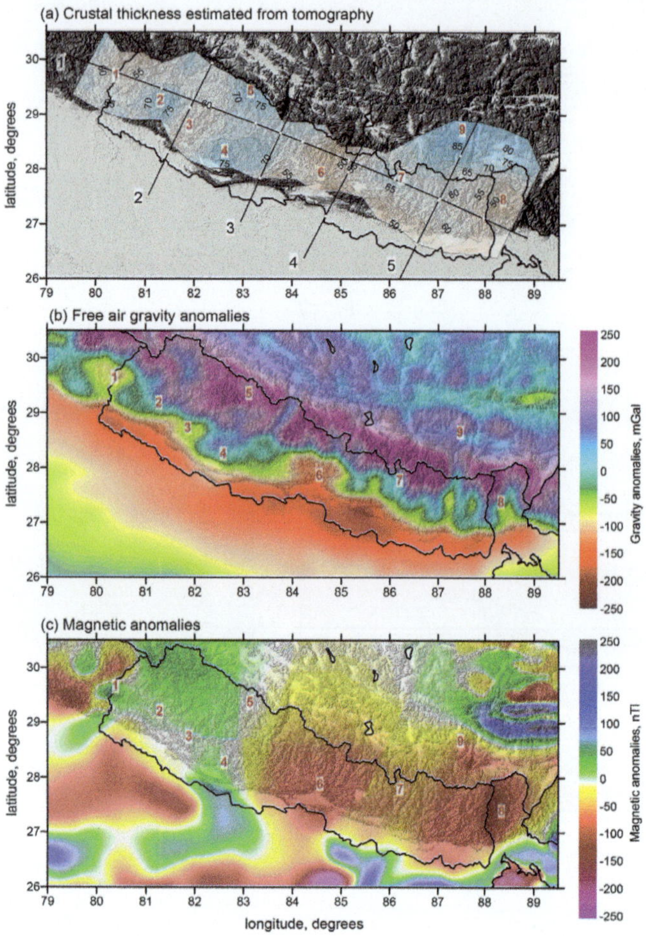

Figure 8. Map of estimated Moho depth beneath the Nepal Himalayas **(a)** together with free-air gravity anomalies (Andersen et al., 2010) **(b)** and magnetic anomalies (Maus et al., 2009) **(c)**. Red numbers indicate the locations discussed in the text. In **(a)**, the locations of profiles used for presenting the main results are shown.

sediments. In case of collision, these two types of the crust behave differently and cause different mechanical effects. The crust affected by igneous processes is stronger, and thus the compression of the overlying crust in the collision zone would be more prominent. This would explain the thicker crust in anomaly "4" close to the contact area, which can be explained by a stronger pushing effect of more rigid partition of the Indian plate. The presence of southward curve of the mountain limit line in front of the thicker crustal pattern "4" may represent a broader shortening zone produced by the more rigid incoming block. The segment of the anomalously thinner crust "6" may be explained by a smaller compression rate of the crust because of weaker incoming crust and lubricating effect of thicker sediments having a lower magnetic effect. In this case, the mountain front line is curved northward, indicating less intensive shortening.

4 Conclusions

In most tomographic studies, the main target is the smooth distribution of seismic properties which is not sensitive to the geometry of first-order interfaces. In this study we made an attempt to reconstruct variations of the Moho interface beneath the Nepal Himalayas where more or less uniform distribution of stations and sources takes place and travel times of both crustal (Pg, Sg) and mantle rays (Pn, Sn) are available. Based on synthetic modeling, we found that, for most of the Nepal Himalayas area, the crustal thickness variations can be robustly retrieved. For the surrounding areas, like northern India and Tibet, crustal structures cannot be resolved with the available data.

The obtained crustal thickness varies from 40 to 75 km along the Nepal Himalayas. There is a fair correlation of the derived crustal structures with the observed gravity and magnetic anomalies. The areas of thicker crust are associated with higher values of the free-air gravity field and vice versa. This correlation is a good argument to prove the reliability of our findings. The magnetic anomalies may provide important information on the mechanical properties of the crust. We see that different segments of the Indian crust behave differently leading to various collision rates. We expect that thicker crust in the frontal thrust zone can be associated with the more rigid incoming crust. Weaker crust segments may penetrate underneath an overlying plate with less resistance, and thus the weaker compression rate leads to thinner crust in the frontal thrust zone. The presence of thick sediments may have a lubricating effect and thus may also reduce the shortening of the crust.

This study gives us a fair understanding of the Moho configuration beneath the central Himalayas. However, western and eastern parts of the Himalayas have not yet been well studied. Such a comprehensive study based on joint consideration of seismic, gravity and magnetic data for the entire Himalayas will make a better understanding of the mechanisms of the India–Asia collision possible.

Author contributions. J. Raoof and S. Mukhopadhyay provided seismic data and performed the preliminary analysis and preparation. G. Maksotova with the help of A. Jakovlev performed all tomographic calculations. A. Vasilevsky provided necessary data and information on magnetic and gravity fields used in the paper. J. R. Kayal, S. Mukhopadhyay and I. Koulakov provided geodynamical interpretation of presented results. A. Jakovlev prepared presented graphic materials. I. Koulakov prepared the manuscript with contributions of all co-authors.

Acknowledgements. This study was performed in the framework of the joint Russia–India research Project INT/RUS/RFBR/P-156 (from DST, India) and RFBR #13-05-92691-ind. This paper is partly supported by the SB RAS IP 76.

Edited by: T. Taira

References

Allegre, C. O., Courtillot, V., Tapponnier, P., Hirn, A., Mattauer, M., Coulon, C., and Xu, R.: Structure and evolution of the Himalaya – Tibet Organic Belt, Nature, 30, 17–22, 1984.

Andersen, O. B., Knudsen, P., and Berry, P.: The DNSC08GRA global marine gravity field from double retracked satellite altimetry, J. Geod., 84, 191–199, doi:10.1007/s00190-009-0355-9, 2010.

Baruah, S., Saikia, S., Baruah, S., Bora, P. K., Tatevossian, R., and Kayal, J. R.: The September 2011 Sikkim Himalaya earthquake Mw 6.9: is it a plane of detachment earthquake? Geomatics, Nat. Haz. Risk, 1–17, 2014

Bassin, C., Laske, G., and Masters, G.: The current limits of resolution for surface wave tomography in North America, EOS, Trans. AGU, 81, F897, 2000.

Dewey, J. F. and Bird, J. M.: Mountain belts and the new global tectonics, J. Geophys. Res., 75, 2625–2647, 1970.

Galvé, A., Sapin, M., Hirn, A., Diaz, J., Lépine, J. C., Laigle, M., Gallart, J., and Jiang, M.: Complex images of Moho and variation of Vp/Vs across the Himalaya and South Tibet, from a joint receiver-function and wide-angle-reflection approach, Geophys. Res. Lett., 29, L15308, doi:10.1029/2006GL026076, 2002.

Gansser, A.: The Geology of the Himalayas, Wiley Interscience, London, 289 pp., 1964.

Hauck, M. L., Nelson, K. D., Brown, L. D., Zhao, W., and Ross, A. R.: Crustal structure of the Himalayan orogen at 90°east longitude from Project INDEPTH deep reflection profiles, Tectonics, 17, 481–500, 1998.

Hetényi, G., Cattin, R., Vergne, J., and Nábělek, J. L.: The effective elastic thickness of the India Plate from receiver function imaging, gravity anomalies and thermomechanical modelling, Geophys. J. Internat., 167, 1106–1118, 2006

Jin, Y., Mcnutt, M. K., and Zhu, Y.-S.: Mapping the descent of Indian and Eurasian plates beneath the Tibetan plateau from gravity anomalies, J. Geophys. Res., 101, 275–11,290, 1996.

Kayal, J. R.: Microearthquake activity in some parts of the Himalaya and the tectonic model, Tectonophysics, 339, 331–351, 2001.

Kind, R., Yuan, X., Saul, J., Nelson, D., Sobolev, S.V., Mechie, J., Zhao, W., Kosarev, G., Ni, J., Achauer, U., and Jiang, M.: Seismic images of crust and upper mantle beneath Tibet: Evidence for Eurasian plate subduction, Science, 298, 1219–1221, 2002.

Koulakov, I.: LOTOS code for local earthquake tomographic inversion. Benchmarks for testing tomographic algorithms, Bulletin of the Seismological Society of America, 99, 194–214, 2009.

Koulakov, I.: High-frequency P and S velocity anomalies in the upper mantle beneath Asia from inversion of worldwide traveltime data, J. Geophys. Res., 116, B04301, doi:10.1029/2010JB007938, 2011.

Koulakov, I. and Sobolev, S. V.: Moho depth and three-dimensional P and S structure of the crust and uppermost mantle in the Eastern Mediterranean and Middle East derived from tomograpic inversion of local ISC data, Geophys. J. Internat., 164, 1, 218–235, 2006.

Koulakov, I., Zaharia, B., Enescu, B., Radulian, M., Popa, M., Parolai, S., and Zschau, J.: Delamination or slab detachment beneath Vrancea? New arguments from local earthquake tomography, G-cubed, 10, Q03002, doi:10.1029/2009GC002811, 2010.

Kumar, P., Yuan, X., Kind, R., and Kosarev, G.: The lithosphere-asthenosphere boundary in the Tien Shan-Karakoram region from S receiver functions: Evidence for continental subduction, Geoph. Res. Lett., 32, L07305, doi:10.1029/2004GL022291, 2005.

Li., C., van der Hilst, R. D, Meltzer, A. S., and Engdahl, E. R.: Subduction of the Indian lithosphere beneath the Tibetan Plateau and Burma, Earth Planet Sci Lett., 274, 157–168, 2008.

Maus, S., Barckhausen, U., Berkenbosch, H., Bournas, N., Brozena, J., Childers, V., Dostaler, F., Fairhead, J. D., Finn, C., von Frese, R. R. B., Gaina, C., Golynsky, S., Kucks, R., Lühr, H., Milligan, P., Mogren, S., Müller, R. D., Olesen, O., Pilkington, M., Saltus, R., Schreckenberger, B., Thébault, E. and Caratori Tontini, F.: EMAG2: A 2–arc min resolution Earth Magnetic Anomaly Grid compiled from satellite, airborne, and marine magnetic measurements, Geochem. Geophys. Geosys, 10, Q08005, doi:10.1029/2009GC002471, 2009.

Mechie, J., Ben-Avraham, Z., Weber, M., Götze, H.-J., Koulakov, I., Mohsen, A., and Stiller, M.: The distribution of Moho depths beneath the Arabian plate and margins, Tectonophysics, 609, 234–249, 2013.

Mitra, S., Priestley, K., Bhattacharyya, A. K., and Gaur, V. K.: Crustal structure and earthquake focal depths beneath northeastern India and southern Tibet, Geophys. J. Internat., 160, 227–248, 2005.

Molnar, P. and Tapponnier, P.: Cenozoic tectonics of Asia: Effects of a continental collision, Science, 189, 419–426, 1975.

Mukhopadhyay, S. and Sharma, J.: Crustal scale detachment in the Himalayas: a reappraisal, Geoph. J. Int., 183, 850–860, 2010

Nakata, T.: Active fault of the Himalayas of India and Nepal, In: Tectonics of Western Himalayas, Eds. Malinconico Jr., L.L. and Lillie R., Geol. Soc. Am. Sp. Pap., 232, 243–264, 1989.

Ni, J. F.: Active tectonics of the Himalaya, Proc. Indian Acad. Sci. (Earth Planet Sci), 98, 71–89, 1989.

Nolet, G.: Seismic wave propagation and seismic tomography, in: Seismic Tomography, edited by: Nolet, G., Reidel, Dordrecht, 1–23, 1987.

Paige, C. C. and Saunders, M. A.: LSQR: An algorithm for sparse linear equations and sparse least squares, ACM Trans. Math. Soft, 8, 43–71, 1982.

Rai, S. S., Priestley, K., Gaur, V. K., Mitra, S., Singh, M. P., and Searle, M.: Configuration of the Indian Moho beneath the NW Himalaya and Ladakh, Geophys. Res. Lett., 33, L15308, doi:10.1029/2006GL026076, 2006.

Ramesh, D. S., Kumar, M. R., Devi, E.U., Raju, P.S., and Yuan, X.: Moho geometry and upper mantle images of northeast India, Geophys. Res. Lett., 32, 14301–14304, 2005.

Schelling, D.: The tectonostratigraphy and structure of the eastern Nepal Himalaya, Tectonics, 11, 925–943, 1992.

Schulte-Pelkum, V., Sheehan, A., Wu, F., and Billham, R.: Imaging the Indian subcontinent beneath the Himalaya, Nature, 435, 1222–1225, 2005.

Seeber, L., Armbruster, J. G., and Quittmeyer, R.: Seismicity and continental subduction in the Himalayan Arc, in: Zagros, Hindu Kush, Himalaya, Geodynamic Evolution, edited bg: Gupta, H. K. and Delany, F. M., Geodyn. Ser., 3, 215–242, 1981.

Tenzer, R. and Chen, W.: Regional gravity inversion of crustal thickness beneath the Tibetan plateau, Earth Science Informatics, 1–12, 2014.

Valdiya, K. S.: Himalayan transverse faults and their parallelism with subsurface structures of north Indian plains, Tectonophysics, 32, 352–386, 1976.

5

Volcanological aspects of the northwest region of Paraná continental flood basalts (Brazil)

F. Braz Machado[1], **E. Reis Viana Rocha-Júnior**[2], **L. Soares Marques**[3], **and A. J. Ranalli Nardy**[4]

[1]Departamento de Ciências Exatas e da Terra, Universidade Federal de São Paulo, Diadema, SP09920-540, Brazil
[2]Departamento de Física da Terra e do Meio Ambiente, Instituto de Física, Universidade Federal da Bahia, Salvador, BA40170-115, Brazil
[3]Instituto de Astronomia, Geofísica e Ciências Atmosféricas, Universidade de São Paulo, Sao Paulo, SP05508-090, Brazil
[4]Departamento de Petrologia e Metalogenia, Instituto de Geociências e Ciências Exatas, Universidade Estadual Paulista, Rio Claro, SP13506-900, Brazil

Correspondence to: F. Braz Machado (fabiobrazmachado@gmail.com)

Abstract. There has been little research on volcanological aspects of Paraná continental flood basalts (PCFBs), and all investigations have mainly been concentrated on the internal portions of the lava flows. Thus, this study describes for the first time morphological aspects of lava flows and structural characteristics caused by lava–sediment interaction in the northwestern PCFB province (NW-PCFB). Early Cretaceous (134 to 132 Ma) tholeiitic rocks of the PCFB were emplaced on a large intracratonic Paleozoic sedimentary basin (Paraná Basin), mainly covering dry eolian sandstones (Botucatu Formation). As this sedimentary unit is overlain by the basic lava flows of the PCFB, the interaction of lavas and unconsolidated sediments resulted in the generation of fluidal peperites. This aspect is significant because it shows that restricted wet environments should have existed in the Botucatu desert. The peperite zones of the NW-PCFB are associated with compound pahoehoe-type (P-type) flows and are always related to the first volcanic pulses. These flows have dispersed vesicles and sand-filled cracks in their base and top borders, as well as the presence of interlayered sandstones with irregular contacts and varied thicknesses. It is remarkable that, to the best of current knowledge, only in this area of the whole PCFB did the volcanic activity start with low-Ti basalt flows of Ribeira type (TiO_2 < 2.3 wt %), which are scarce in the province.

1 Introduction

Continental flood basalts (CFBs) are the result of brief igneous events that may last only a few million years or less, producing vast volumes of lavas and intrusive rocks (Coffin and Eldholm, 1994). As emphasized by Self et al. (1997) and Skilling et al. (2002), the physical volcanology of CFBs has received relatively little attention, since many previous studies have tried to understand their petrological, geophysical, geodynamical, and geochemical features, identifying possible mantle sources, crustal or mantle contaminants, and low-pressure magma evolution of the basalts generated in these magmatic processes.

Initial studies on physical volcanology of CFBs have been devoted to mainly describe the flow architectures as well as the mechanisms involved in their emplacement. One of the first provinces where this kind of investigation was accomplished is the Columbia River basalts, whose flows are mainly compound pahoehoe (Riedel and Tolan, 1992), presenting also peperites as a result of the interaction of lava flows with unconsolidated sedimentary deposits (Riedel, 1998). For the Deccan volcanic province, Duriaiswami et al. (2008) described rubbly pahohoe flows in upper stratigraphic unities, whereas Boundre et al. (2004) emphasized the occurrence of compound pahohoe in the lower and middle sequences, as well as peperites (only just at the edges of the province). Regarding the Ferrar volcanic province, Antarctica, the peperites formed by mingling of basalt flows with coal, as well as where these lavas encountered continental sedimentary successions of the Mawson Formation (McClintock and White, 2002).

The lava flows of the Paraná continental flood basalt (PCFB) province were emplaced in the Early Cretaceous on a large intracratonic sedimentary basin (Paraná Basin) that began subsiding in the early Paleozoic, covering sandstones of the Botucatu Formation (dry eolian system in extensive fields of dunes). Although considerable progress on the geochemical, geochronological, geophysical, and geodynamical aspects of the PCFB, as well as about the characteristics of the mantle sources involved in the magmatism, has been made, very little is known about the physical volcanology of this event. The geochemical and petrological studies have mainly been concentrated in the internal portions of the lava flows, where supposedly the rocks have no significant changes from their original geochemical composition. This fact partially explains the restricted number of studies that typify the flows and their internal structures, although PCFB constitutes one of the largest volcanic provinces of the world. Another important aspect that does not encourage investigations on this subject is the lack of well-preserved exposures due to weathering under the tropical–subtropical humid climate and the discontinuity of the outcrops.

The early studies that have attempted to understand the lava–sediment interaction in the PCFB were concentrated on exposures from its southern and central areas, and mainly focused on the peperite occurrences. The investigations were carried out by Petry et al. (2007) and Waichel et al. (2007, 2008) for basalt flows and Luchetti et al. (2014) for basalt and rhyolite flows or ignimbrites. According to these workers, the textures of the mafic peperite breccia vary from globular to amoeboid (anhedral form), while the rhyolitic ones range from blocky (also called jigsaw-fit) to irregular form (subhedral form). Regarding the peperite origin of the PCFB flows there is some controversy. For Waichel et al. (2007, 2008) these structures are the result of mixing and mingling processes involving lavas and unconsolidated or poorly consolidated wet sediments, agreeing with the usually proposed models for other provinces (White et al., 2000; Cas et al., 2001; Dadd and Van Wagoner, 2002; Skilling et al., 2002). However, Petry et al. (2007), following the reasoning of Jerram and Stollhofen (2002), explain that some peperites of the PCFB (southern region) were not originated in a wet environment, requiring only lava overspreading the lee side of the dunes (slopes $\sim 30°$).

In this context, the main goal of this study is to describe the morphological aspects of lava flows and the structural characteristics caused by lava–sediment interaction in the northwestern PCFB (NW-PCFB; Fig. 1), as well as to discuss the mechanisms involved during the extrusion processes. In addition, we present new geochemical data in order to establish the accurate stratigraphy of the flows in the area. It is important to stress that the investigated region has a distinctive geomorphology, with flat topography, encompassing some geological faults, rare outcrops, and significant erosive processes due to the tropical humid climate. These characteristics make it much more difficult to recognize the volcanological features, in comparison to other areas of the PCFB, as well as in locations where the eruptions occurred more recently (e.g., Columbia River) and/or the rocks are very well preserved (e.g., Etendeka province – Namibia).

2 Geological setting

The tholeiites of the PCFB province are of Early Cretaceous age and were emplaced on a large intracratonic Paleozoic sedimentary basin (Paraná Basin) covering 1500 000 km^2 that developed between the Ordovician and the Cretaceous (Fig. 1). This basin is located in terrains of the South American platform, which were extensively affected by the tectonic, metamorphic, and magmatic events of the Brasiliano–Pan-African orogenic cycle, whose most important accretion episodes are the ones between 650 and 500 Ma (e.g., Cordani et al., 2003).

The PCFB occupies an area of about 917 000 km^2 (encompassing $\sim 75\,\%$ of the Paraná Basin), covering large areas in south Brazil, Argentina, Uruguay, and Paraguay, with an estimated original volume in excess of 600 000 km^3 (Frank et al., 2009). Associated with the volcanic activity there was intrusive magmatism, represented by sills, which outcrop mainly in the eastern and northeastern Paraná Basin, and dyke swarms (Ponta Grossa, Serra do Mar, and Florianópolis; Piccirillo et al., 1990; Raposo et al., 1998; Deckart et al., 1998; Almeida et al., 2012, and references therein).

Geochronological data (based on $^{40}Ar/^{39}Ar$ method), combined with paleomagnetic studies, show that the main volcanic activity of the PCFB occurred within a short time interval between 133 and 132 Ma (Renne et al., 1992; Turner et al., 1994; Ernesto et al., 2002). Most recently, Thiede and Vasconcelos (2010) placed PCFB extrusion at 134.6 ± 0.6 Ma, concluding that the duration of the PCFB volcanism was less than 1.2 Ma. Baddeleyite/zircon LA-ICP-MS (laser ablation-inductively coupled plasma-mass spectrometry) and U-Pb age dating on felsic volcanic rocks from the PCFB yielded an age of 134.3 ± 0.8 Ma (Janasi et al., 2011), confirming the idea that this magmatic event was emplaced over a short period of about 1–2 Ma.

Previous geological and geochemical studies (e.g., Bellieni et al.,1984; Piccirillo and Melfi, 1988; Peate et al., 1992; Marques et al.,1999) allowed the division of the PCFB into two main regions: (1) the southern PCFB is characterized by dominant tholeiitic basalts with dominantly low Ti (LTi; $TiO_2 \leq 2$ wt %), low concentrations of incompatible elements – such as P, Ba, Sr, Zr, Hf, Ta, and Y – and light REEs; (2) the northern PCFB is characterized by prevalent tholeiites with relatively high concentrations of TiO_2 (HTi; $TiO_2 > 2$ wt %) and incompatible elements. These studies also showed that rare LTi basalts are found in the northern

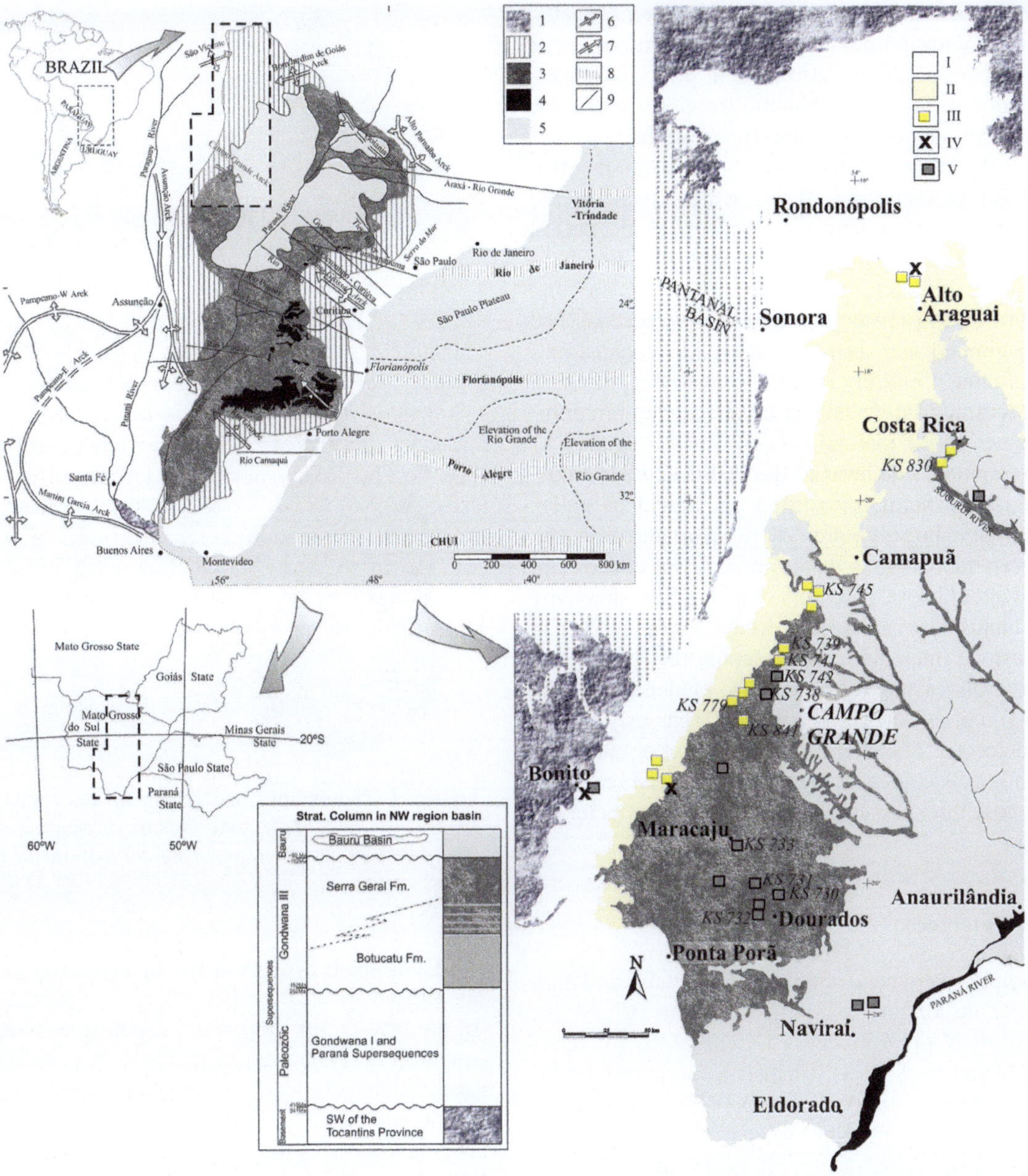

Figure 1. Simplified geological map (CPRM, 2006) of the investigated area indicating sample locations and the stratigraphic column of the area. The geological map of the Paraná Basin (Nardy et al., 2002) is also shown. 1 – Tocantins province (basement); 2 – pre-volcanic sedimentary rocks; 3 – basalts (PCFB); 4 – acid volcanics (PCFB); 5 – post-volcanic sedimentary rocks; 6 – anticlinal structure; 7 – synclinal structure; 8 – oceanic lineaments; 9 – continental lineaments; I – Paleozoic sequences of the Paraná Basin; II – Botucatu Formation; III – Ribeira magma-type samples; IV – Paranapanema magma-type samples; V – Pitanga magma-type samples.

PCFB and that scarce HTi basalts are encountered in the southern PCFB.

According to Peate et al. (1992), the basalt rocks from PCFB can be grouped into six distinct magma types based on titanium, incompatible trace elements (Sr, Y, and Zr) and their ratios, which were also used in order to minimize fractional crystallization effects. The HTi basalts were organized into three groups: Urubici ($TiO_2 > 3.3$ %; Sr > 550 ppm; Ti / Y > 500), Pitanga ($TiO_2 > 2.8$ wt %; Sr > 350 ppm; Ti / Y > 350), and Paranapanema ($1.7 < TiO_2 < 3.2$ wt %; 200 < Sr < 450 ppm; Ti / Y > 330). The LTi basalts were also organized into three groups: Gramado ($0.7\,\% < TiO_2 < 2$ wt %; 140 < Sr < 400 ppm; Ti / Y < 300), Esmeralda ($1.1 < TiO_2 < 2.3$ wt %; Sr < 250 ppm; Ti / Y < 330), and Ribeira ($1.5 < TiO_2 < 2.3$ wt %; 200 < Sr < 375 ppm; Ti / Y > 300). It is important to note that in comparison to the detailed geochemical analyses carried out on other magma types of the PCFB there have been few prior geochemical studies done on Ribeira basalts, especially those outcropping at the northwestern border of the province. In this study, we also present new whole-rock analysis of major, minor, and lithophile trace elements for a significant number of Ribeira magma-type samples.

Throughout the province, the lavas often cover the Botucatu Formation of the Paraná Basin, which is interpreted as a typical dry eolian system (e.g., Scherer et al., 2000), reaching an area of approximately 1300 000 km^2. The sedimentation process initiated at the Mid-Jurassic (Leonardi and Oliveira, 1990) and lasted until the Early Cretaceous. The fine–medium-grained and well-selected reddish sandstones are composed almost entirely of quartz, presenting textural bimodality. Medium to large tabular cross bedding is a common feature.

In that environment, the lavas of the PCFB first buried the areas between dune fields; then filled the inter-dune areas; and finally covered largest eolic desert accumulations. The contact between the two geological formations is concordant and abrupt, with a lack of paleosols, which might show any depositional hiatus between the two events. Furthermore, the frequent sandstone intercalations, present at the bottom and intermediate levels of the volcanic piles, evidence that the two geological unities developed contemporaneously during the Early Cretaceous. The intermittent character of both geological formations is evidenced by the interlayered sedimentary rocks, whose thicknesses vary from a few centimeters to several meters.

3 Analytical methods

Thirty-six samples were collected in the NW-PCFB and analyzed for major, minor, and some trace elements (Cr, Ni, Sr, Y, Zr, and Nb) by X-ray florescence (XRF), at the Universidade Estadual Paulista (UNESP, Brazil). A subset of 12 samples was selected for rare earth elements (REEs; La, Ce, Nb, Sm, Eu, Gd, Dy, Ho, Er, Yb, and Lu) analysis by inductively coupled plasma mass spectrometry (ICP-MS) at the Universidade Estadual de Campinas (UNICAMP, Brazil). Sample locations are shown in Fig. 1.

Prior to analysis, unaltered parts of every rock were covered with a cloth and crushed to smaller pieces with a rock hammer. Selected portions were cleaned with distilled water in ultrasonic bath and ground to thin powder with a tungsten carbide mill for XRF analysis and with an agate mill for ICP-MS analysis. The technique used is that described by Nardy et al. (1997), Machado et al. (2009), Cotta and Enzweiler (2012), and Rocha-Júnior et al. (2013). In general, XRF relative accuracy for major and minor elements is ± 1 %, whereas for trace elements it is within ± 3 %. For the REEs determined by ICP-MS the analytical uncertainties is typically better than 5 %.

Figure 2. Photographs and schematic illustrations of large and small clasts in peperite zone showing vesicles and stretched vesicles (filled by calcite in **a** and fluidal margins in **b**). Coin diameter = 27 mm.

4 Geological aspects of the investigated area

In the NW-PCFB the Botucatu sandstones are widely overspread (Fig. 1), presenting medium cross bedding and no distinctive characteristics in comparison to the reported ones for the other areas of the Paraná Basin (e.g., Scherer et al., 2000). The sediment flux measurements performed in three outcrops indicate sand movements to the north-northeast.

Since the first stages of basalt lava emplacement were concurrent with those related to the end of Botucatu sand deposition and diagenesis, several features of lava–sediment interaction were generated in the investigated area. The peperite zones are always associated with the first basalt flow and reach up to 2 m thick, whose host is composed of very fine grained sandstone, with low-angle stratification (5°). This reddish and silicified sedimentary rock contains irregular clasts of volcanic rock, varying from submillimeter-to-millimeter size to 30 cm. The larger ones (centimeter-to-decimeter size) are vesicular and in general characterized by irregular contacts and fluidal margins (Fig. 2), whereas the smaller ones (submillimeter-to-millimeter size) are subrounded, and some of them exhibit quenched rims.

The peperites are amoeboid-shaped and may extend up to 80 m in a unique outcrop. Based on the relative proportions

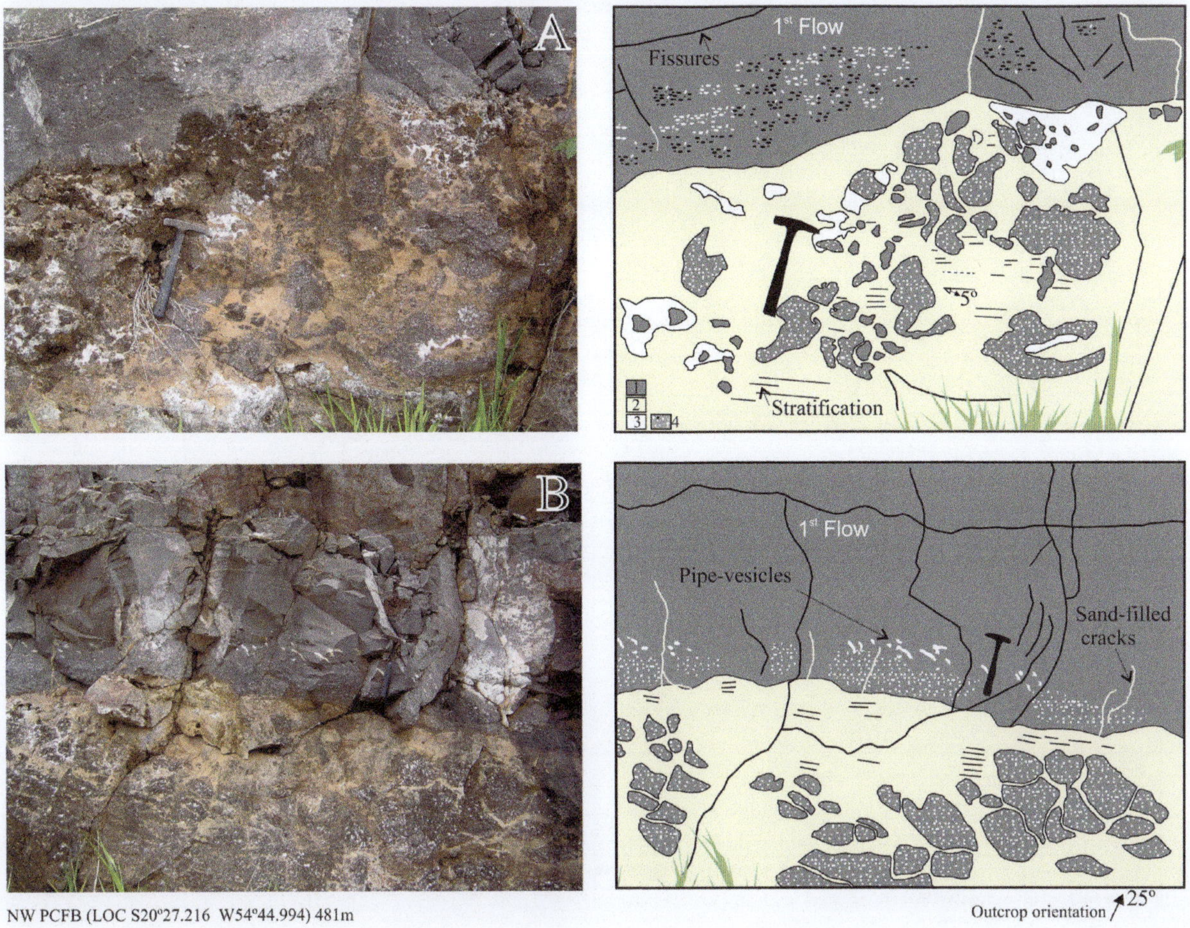

Figure 3. Peperite layers showing basalt volcanic clasts hosted in eolian sandstone. **(a)** Peperite texture in very fine grained sandstone with cross-bedding, low-angle stratification (5°) with juvenile volcanic clasts of amoeboid shape. **(b)** Pipe vesicles and sand-filled cracks in the base of the sequences.

Figure 4. Rope surface identified on the top of the compound pahoehoe flows outcropping near Nivaraí village.

of host sediment and juvenile clasts, these peperites are classified in the literature as dispersed (e.g., Skilling et al., 2002).

It is also important to emphasize the presence of aligned vesicle zones (up to 1 m thick) parallel to the contact with sedimentary host rock or even dispersed in it. Pipe vesicles are also found (Fig. 3), indicating an apparent direction of 25° NE for the lava flow. The vesicles may achieve 2 cm in length, and there is no significant variation in their sizes if they are in the center or near the borders of the largest volcanic clasts, which may also present fractures. In general, the vesicles are filled with sediments of the host rock (Fig. 2b) or even by secondary minerals, such as calcite (Fig. 2a), natrolite or quartz, although some of them are empty. Stretched vesicles are often found, indicating plastic deformation, before complete cooling (Fig. 2).

The peperites are associated with pahoehoe flows and are only found in the early volcanic pulses of the NW-PCFB. The thicknesses vary from 1.5 to 3.0 m, occurring as compound lava flows spreading over gently sloping sandy surfaces (5°), preserving the Botucatu paleo-erg or causing a slight deformation of it. Rope surfaces are rare, and some were identified on the top of the flows outcropping near Nivaraí vil-

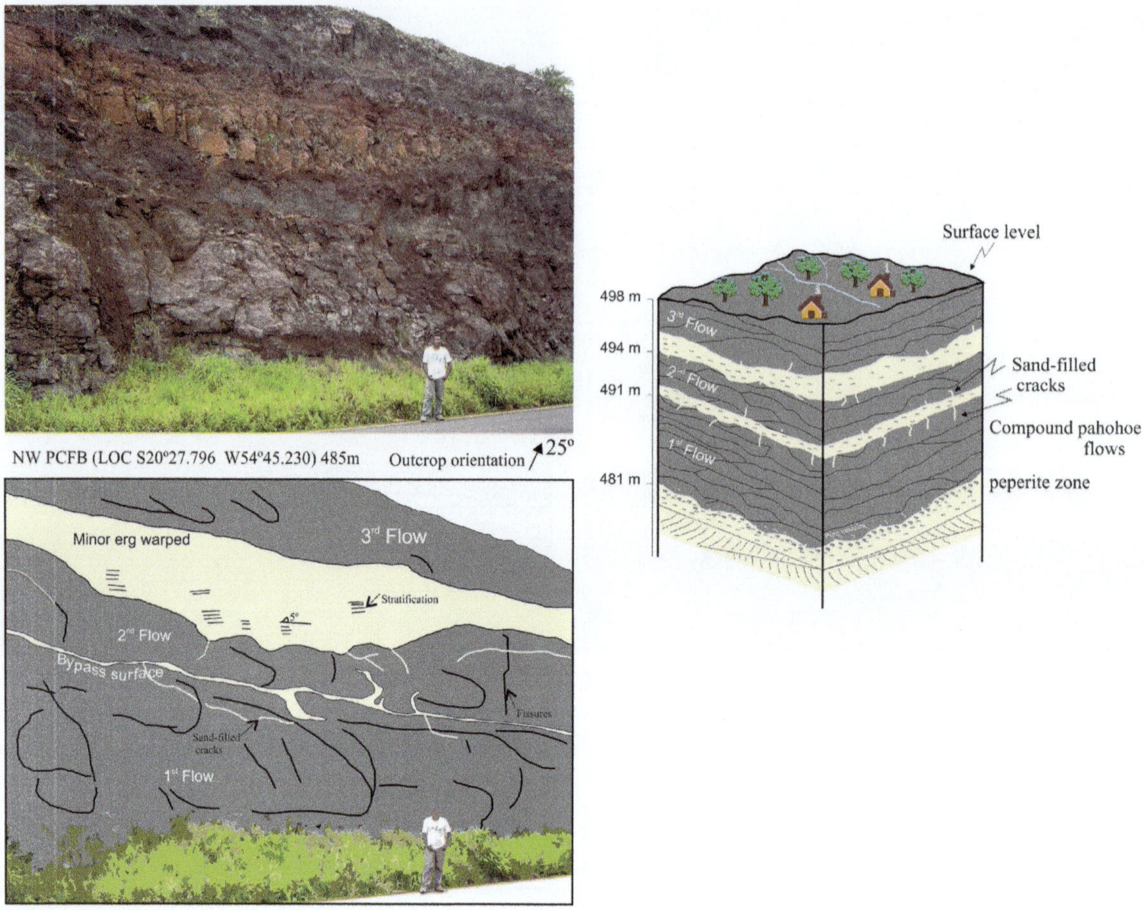

Figure 5. Compound pahoehoe lava flows with interlayered sandstones of the Botucatu Formation, showing bypass sediment features and minor erg deformation.

Figure 6. Outcroup with ascending and descending sand-filled cracks in two compound pahoehoe lava flows separated by Botucatu sandstone.

lage (Fig. 4). The individual lobes are small, attaining up to 2.5 m thick. The flows exhibit dispersed vesicles and sand-filled cracks in their base and top borders, where in contact with sediments (Figs. 5, 6, 7), suggesting a pahoehoe-type (P-type) lava flow (Wilmoth and Walker, 1993). The presence of interlayered sandstones is common, showing irregular contacts and varied thicknesses, which are caused by the filling of empty spaces left by the movement of lobes (Figs. 6, 7).

NW PCFB (LOC S22°06.660 W54°47.700) 399 m

Figure 7. Massive simple pahoehoe flows, without sandstone intercalations, exposed in a quarry near the town of Dourados (Fig. 1). The outcrop location is shown in the inset (image from Google Earth).

What is remarkable is the presence of sand-filled cracks, which are associated with the compound pahoehoe lava lobes, occurring either near the peperite zones, in the case of the first volcanic pulse over the sediments, or at the bottom and top of the subsequent flows, when they are separated from each other by a sediment layer. The sand penetration into the base of the first flow achieves a length of up to 30 m, whereas the youngest ones reach 1 m length. The thicknesses of the sand-filled cracks vary from a few millimeters to 10 cm (Figs. 5, 6).

The peperites are observed only in the bottom of the volcanic pile or in the contact of the first layer of basalt flow with Botucatu sandstone, despite breccias of sedimentary rock clasts and sand-filled cracks being observed in different levels of the volcanic pile.

Typical features of the sandstone emplacement just located below the pahoehoe flows are rare. In some locations, the paleodune slip face was identified, as well as striation and chevron marks. The same characteristics were also noticed by Holz et al. (2008) in a study conducted at the northern border of the study area.

The thicknesses of the flows, as well as of all the below sedimentary units, increase towards the interior of the basin (to the E and SE). Near the of Dourados (Fig. 1) the flows are simple pahoehoe (Fig. 6), massive, without sandstone intercalations, and stratigraphically located above those of compound pahoehoe.

Intrusive rocks are rare and have been found in only two mafic dykes 3 m thick, located in the southern area, near Bonito village (Fig. 1), and cutting the Botucatu Formation in a N–S trend parallel to the Assunção Arch located far west. Although some publications have reported a small dyke swarm in the area (Bellieni et al., 1986; Peate et al., 1992), it was not found. Still regarding the intrusive rocks, one sill of about 25 km long and at least 15 m thick occurs in the northern border of the NW-PCFB, intruding into the lower portion of Paraná Supersequence (Fig. 1).

Finally, besides the Quaternary alluvial and colluvial deposits, continental sedimentary rocks of Bauru Basin (Fig. 1) covered the early cretaceous PCFB, in discordant contact, during the Paleogene. The occurrence of basalt clast-supported conglomerates suggests considerable erosion processes in the upper portions of the volcanic piles before the Bauru Basin sediment deposition.

5 Petrography of the basalts

The mineralogy of the investigated rocks is the same for all identified flow types, being generally composed of plagioclase, pyroxene (augite and pigeonite), and opaque minerals (magnetite, ilmenite, and sulphides), with subordinated olivine (most often as pseudomorphs) and apatite. The groundmass is commonly microgranular and rarely micrographic (granophyric texture). There have also been identified alteration minerals like celadonite (quite often), kaolinite, serpentine, biotite, chlorite, calcite, and quartz. The last one is also found as a primary mineral.

The average mode consists of 48 % plagioclase, 33.7 % augite, 0.9 % pigeonite, 7.3 % opaque minerals (magnetite, ilmenite), 1.2 % olivine, and 7.5 % microgranular groundmass. Among the secondary minerals, it is worth highlighting celadonite (1.6 %) and goethite (1.5 %).

Figure 8. Different textures identified in basalt samples of the NW-PCFB. **(a)** Intergranular, with plagioclase subhedral crystals and interstitial augite crystals (parallel nicols). **(b)** Subophitic, with plagioclase crystals partially included in crystals of augite (parallel nicols). **(c)** Intersertal, with plagioclase crystals corroded by the groundmass (parallel nicols). **(d)** Graphic, with fuzzy aspect micrographic matrix containing quartz (crossed nicols). Legend: plag – plagioclase; aug – augite; op – opaque minerals (magnetite/ilmenite); mat - matrix; oli – olivine.

The presence of vesicles is more common in compound pahoehoe flows than in the massive ones. The segregation structures in the central portion of the lobes, along with microvesicles, located between plagioclase and pyroxene crystals in diktytaxitic texture, are indicative of highly volatile contents in magmas feeding those flows (Jerram and Petford, 2011).

The mineralogical diversity, along with the rheological characteristics of lavas (cooling speed, temperature, viscosity, volatile content, among others), generates a significant variety of textures (Fig. 8). The most common intergranular ones and their subophitic and ophitic variations are associated with rocks that cooled slowly, as evidenced by the size and shape of the mineral phases. Intersertal (with plagioclase crystals corroded by microgranular matrix) and graphic (characterized by an intergrowth of feldspar and microgranular quartz with vermicular or dendritic forms) textures are also noticed, indicating magmatic differentiation processes, with separation of rich-volatile residual liquid forming anomalous mineral phases (quartz, potassic feldspar) for a basic lava.

6 Major, minor, and trace element geochemistry

The NW-PCFB is still lacking in detailed geochemical studies; thus, additional major, minor, and lithophile trace element abundance data were obtained for whole-rock basalt samples (Table 1). All the samples analyzed for major and trace elements are fresh and have loss on ignition (LOI) values lower than 2.5 wt %. Bulk-rock compositions, recalculated on an anhydrous basis, range from sub-alkaline basalt to basaltic andesite according to the total alkali versus silica (TAS) classification (Le Bas and Streckeisen, 1991; Fig. 9). According to the classification for the PCFB flows proposed by Peate et al. (1992), 20 samples belong to Pitanga, 12 to Ribeira, and 4 to Paranapanema magma type.

As a whole, the NW-PCFB rocks have moderately evolved compositions, with MgO contents varying between 6.5 and 3.3 wt %. Compositionally, the analyzed rocks exhibit a tholeiitic affinity, presenting high Fe / Mg ratios (Fe_2O_{3t} / MgO ~ 2.1–4.6) for relatively high silica contents (SiO_2 = 50–53 wt %). The tholeiitic nature is reinforced by the modal mineralogy of the rocks, with the presence of two pyroxenes (augite and pigeonite).

Table 1. Chemical analyses, magma types, lava flow characteristics, and locations of representative basalt samples of the NW-PCFB.

Magma type Pahoehoe type Sample	Pitanga Simple KS 730	Pitanga Simple KS 731	Pitanga Simple KS 732	Pitanga Simple KS 733	Pitanga Simple KS 738	Pitanga Simple KS 742	Ribeira Compound KS 739	Ribeira Compound KS 741	Ribeira Compound KS 745	Ribeira Compound KS 779	Ribeira Compound KS 830	Ribeira Compound KS 841
SiO_2	51.44	50.30	51.15	50.70	50.45	50.37	51.10	50.80	50.79	50.79	49.98	49.91
TiO_2	3.58	4.18	3.68	3.77	3.78	3.98	1.79	1.82	2.20	1.82	1.74	1.80
Al_2O_3	12.64	12.92	12.53	13.34	13.08	12.69	13.39	13.75	13.37	13.34	13.31	13.13
$Fe_2O_{3\,(t)}$	15.86	15.58	15.54	15.22	15.11	15.34	14.12	12.99	15.03	13.89	13.87	13.88
MnO	0.20	0.21	0.21	0.17	0.20	0.22	0.23	0.21	0.22	0.22	0.21	0.23
MgO	3.44	3.98	3.51	3.59	4.22	4.02	5.69	6.00	5.54	6.32	6.31	5.95
CaO	7.72	8.24	7.82	8.33	8.50	8.07	10.12	10.36	9.34	10.16	10.13	9.88
Na_2O	2.73	2.53	2.70	2.58	2.62	2.60	2.17	2.21	2.31	2.24	2.16	2.12
K_2O	1.81	1.56	1.76	1.45	1.41	1.64	0.85	0.81	0.99	0.84	0.73	0.94
P_2O_5	0.68	0.53	0.68	0.42	0.41	0.58	0.21	0.21	0.25	0.21	0.19	0.20
LOI	0.81	0.85	1.20	1.18	1.08	1.09	0.68	1.23	0.95	0.83	0.81	1.23
Sum	100.91	100.88	100.78	100.75	100.86	100.60	100.35	100.39	100.99	100.66	99.45	99.26
Latitude (S)	22°6.66′	22°6.66′	22°8.7′	21°38.58′	20°26.46′	20°25.14′	20°23.58′	20°25.2′	19°52.68′	20°26.1′	18°32.4′	20°33.36′
Longitude (W)	54°47.7′	54°47.7′	54°49.44′	55°7.38′	54°46.5′	54°51.48′	54°40.26′	54°44.28′	54°26.52′	54°44.4′	53°11.94′	54°39.9′
Height (m)	399	399	382	375	394	342	478	505	440	440	568	496
$Ba^{(1)}$	645	645	571	531	521	578	256	274	336	302	261	279
$Rb^{(1)}$	34	36	34	35	27	27	19	13	25	25	16	22
$Sr^{(1)}$	417	413	405	413	417	398	246	277	280	262	281	276
$Zr^{(1)}$	243	228	253	236	233	246	114	138	158	134	123	127
$Y^{(1)}$	36	43	48	35	41	43	26	31	35	25	29	29
$Nb^{(1)}$	23	30	24	20	22	22	8	13	12	13	9	9
$Cr^{(1)}$	49	75	78	43	38	47	135	94	117	113	177	113
$Ni^{(1)}$	18	18	29	20	21	26	55	57	46	53	68	54
$La^{(2)}$	192.8	170.0	153.2	143.0	134.6	156.5	71.3	74.3	95.4	72.2	69.6	72.6
$Ce^{(2)}$	164.4	145.7	130.3	118.1	115.2	135.2	60.8	63.3	82.1	61.2	58.6	61.2
$Nd^{(2)}$	117.3	108.5	95.8	89.7	84.7	96.1	44.6	44.2	57.3	45.1	43.8	45.1
$Sm^{(2)}$	80.5	71.5	62.4	60.7	56.8	65.5	31.8	32.0	42.2	32.5	31.5	32.4
$Eu^{(2)}$	65.4	57.5	52.0	52.2	48.8	27.2	28.2	54.9	34.8	27.2	27.2	27.7
$Tb^{(2)}$	46.0	42.7	37.7	37.7	35.2	23.5	22.7	37.1	29.4	23.8	23.5	23.8
$Yb^{(2)}$	26.5	24.1	21.3	21.4	20.1	18.1	17.8	21.9	22.7	17.8	17.6	18.0
$Lu^{(2)}$	25.6	22.8	20.3	20.3	18.7	17.5	17.5	21.1	23.2	17.5	16.7	17.5
Eu / Eu*	1.04	1.07	1.04	1.01	1.01	1.01	1.06	1.02	1.07	1.08	1.05	1.06
Ti / Zr	88	110	88	96	97	97	94	80	83	82	86	86
Ti / Y	595	582	461	648	554	557	414	355	377	437	370	378
Zr / Y	7	5	5	7	6	6	4	4	5	5	4	4
Sr / Y	12	10	8	12	10	9	9	9	8	10	10	9
Ba / Y	18	15	12	15	13	13	10	9	10	12	9	10
$(La/Lu)_n$	7.5	7.5	7.5	7.0	7.2	7.4	4.1	4.2	4.1	4.1	4.2	4.2
$(Sm/Yb)_n$	3.0	3.0	2.9	2.8	2.8	3.0	1.8	1.8	1.9	1.8	1.8	1.8
$(La/Yb)_n$	7.3	7.1	7.2	6.7	6.7	7.1	3.9	4.2	4.2	4.0	3.9	4.0
$(La/Sm)_n$	2.4	2.4	2.5	2.4	2.4	2.4	2.2	2.3	2.3	2.2	2.2	2.2

* Major and minor element oxides (wt %) were determined by X-ray fluorescence, whereas trace element concentrations (ppm) were obtained by X-ray fluorescence (1) and ICP-MS (2); $Fe_2O_{3(t)}$ corresponds to the total Fe present in each sample; LOI stands for loss on ignition; REE abundance ratios normalized to chondrites (McDonough and Sun, 1995).

Major and minor elements, against MgO, considered as representative of the magmatic evolution of the rocks, show how the concentrations of these elements vary in the samples (Fig. 10). The results provide evidence that the majority of the NW-PCFB tholeiites have high-TiO_2 contents (2.6–4.2 wt %), as do most of the basalts from the northern PCFB reported in previous studies (e.g., Piccirillo and Melfi, 1988; Peate et al., 1992; Marques et al., 1999; Rocha-Júnior et al., 2012, 2013). Of this high-Ti group, we can distinguish two magma types which emerge from the major, minor, and trace element compositions, referred to as Pitanga (predominant) and Paranapanema (subordinate) magma types, according to the parameters proposed by Peate et al. (1992). In addition, low-Ti tholeiites (namely Ribeira type; $TiO_2 < 2.3$ wt %) have been found further west of Pitanga and Paranapanema outcrops at the northwestern margins of the Paraná Basin. It is important to note that Ribeira samples are less evolved ($MgO = 6.5$–5.3 wt %) than Pitanga and Paranapanema ones, which have MgO contents of 5.8–3.3 wt %.

As can be seen in Fig. 10, in general, SiO_2, TiO_2, Fe_2O_{3t}, Na_2O, K_2O, and P_2O_5 increase, whereas CaO and Al_2O_3 decrease, with decreasing MgO. Despite some scattering, the trends seem to be compatible with fractional crystallization, displaying in general a continuous evolutionary line, which is consistent with literature data (Bellieni et al., 1984; Piccirillo and Melfi, 1988; Peate et al., 1992; Rocha-Júnior et al., 2013). However, at similar MgO contents, Ribeira rocks contain higher SiO_2 and lower P_2O_5 and K_2O than Pitanga and Paranapanema tholeiites. These geochemical characteristics, along with major and minor element mass balance calculations, rule out differentiation of Ribeira magmas by this petrogenetic process to generate Pitanga basalts (e.g., Piccirillo

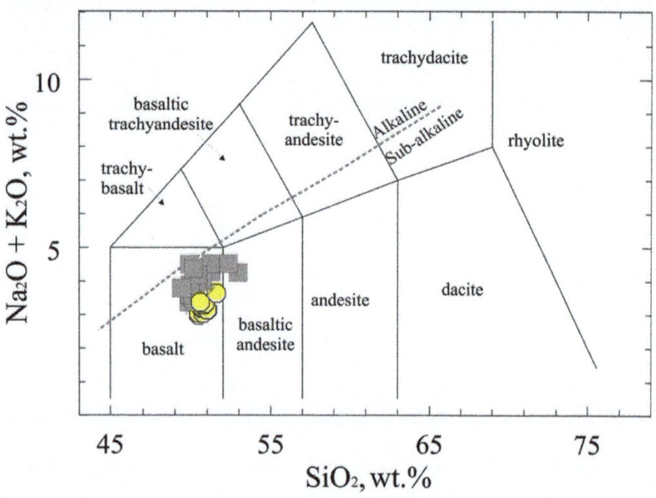

Figure 9. Total alkali versus silica (TAS) diagram (Le Bas et al., 1986; Le Bas and Streckeisen, 1991) used for the nomenclature of basic volcanic rocks of Paraná continental flood basalt province. The dividing dotted line between alkali and sub-alkaline magma series is from Irvine and Baragar (1971). Symbols as in Fig. 1.

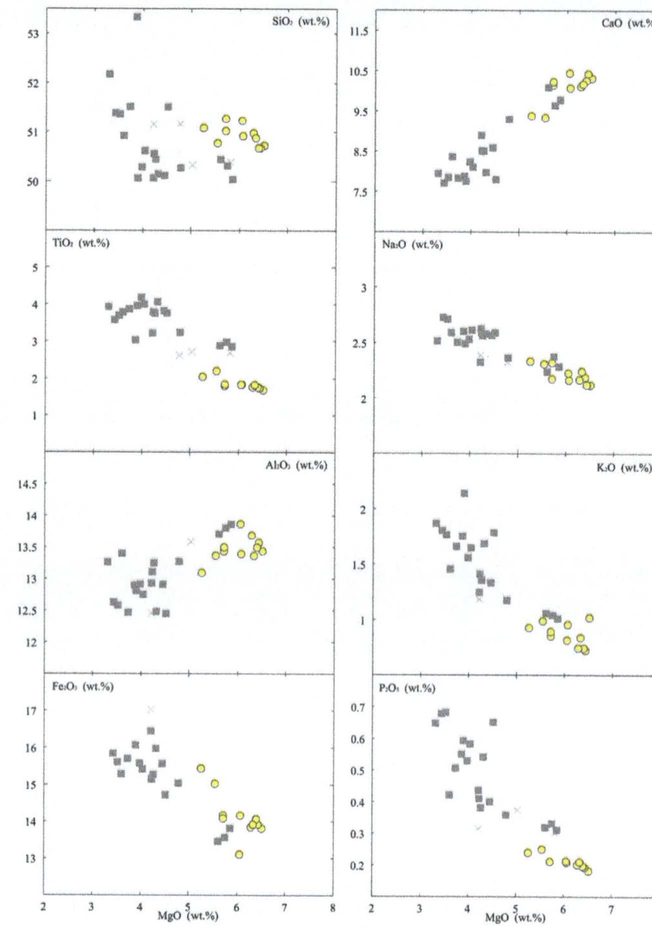

Figure 10. MgO (wt %) versus major (wt %) and minor (wt %) variation diagrams of the investigated tholeiites. Symbols as in Fig. 1.

and Melfi, 1988; Marques et al., 1989, 1999). On the other hand, fractional crystallization mainly controlled by clinopyroxene, plagioclase, and titanomagnetite may account for the geochemical variability inside each basalt group.

The NW-PCFB tholeiites show positive correlations between compatible trace elements (i.e., Ni, Cr) and MgO, although the most evolved samples (Pitanga and Paranapanema) have variable Ni contents. The concentrations of incompatible trace elements such as Nb, Rb, Ba, Zr, and REEs generally increase with the degree of differentiation (MgO; Fig. 11) inside each group of basalts (Ribeira and Pitanga), suggesting evolution by fractional crystallization, reinforcing the behavior of major and minor element oxides. The difference between the two magma types is very clear for the Sr concentrations, with a considerable gap dividing both groups.

REE abundances of the NW-PCFB, normalized to chondrites (McDonough and Sun, 1995), are displayed in Fig. 12, being very similar to those reported in previous studies by Marques et al. (1989) and Rocha-Júnior et al. (2013). The patterns are strongly enriched in light REEs relative to heavy REEs with $(La / Yb)_N = 3.9–7.3$. It should be noted that the HTi rocks investigated in this study have distinctive higher $(La / Yb)_N$ ratios, varying from 6.7 to 7.3, than the Ribeira volcanics, which present $(La / Yb)_N$ from 3.9 to 4.2. The europium anomalies are very slightly positive for both groups of basalts, with Eu / Eu* varying from 1.01 to 1.07.

7 Lava flow morphology and lava–sediment interaction relationships

As emphasized by Dadd and Van Wagoner (2002), peperites are found both in marine and continental settings, and involve variable rock chemistry, from basalt to rhyolite. In their study of lava flows of Southeastern Canada (Passamaquoddy Bay), they concluded that the main determining factors for the generation of different volcanic clast morphologies are lava viscosity and wet unconsolidated sediment properties. According to these authors, basic lavas tend to generate peperites with more variable clast shapes in comparison to acid ones, which originate blocky clast peperites. In fact, these observations are in agreement with the occurrence of PCFB peperites. In the NW-PCFB, the basic lava interaction with wet, fine-grained sandstone resulted in fluidal peperites, whereas those containing irregular juvenile clasts are found in the central PCFB (Waichel et al., 2007, 2008). In addition, the jigsaw-fit texture peperites associated with rhyolites are reported for the southern PCFB by Luchetti et al. (2014).

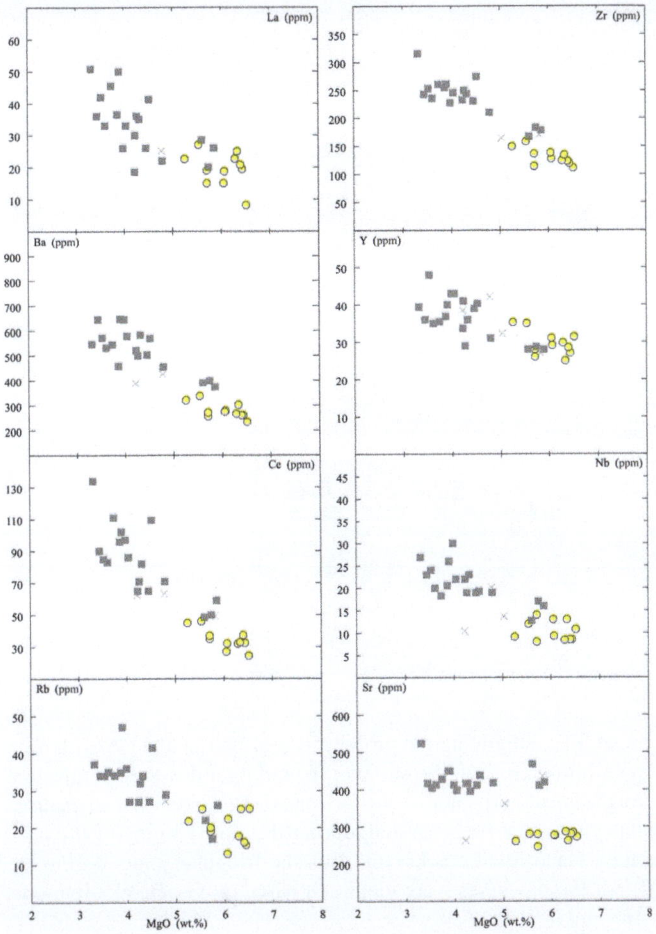

Figure 11. MgO (wt %) versus trace element (ppm) variation diagrams of the investigated tholeiites. Symbols as in Fig. 1.

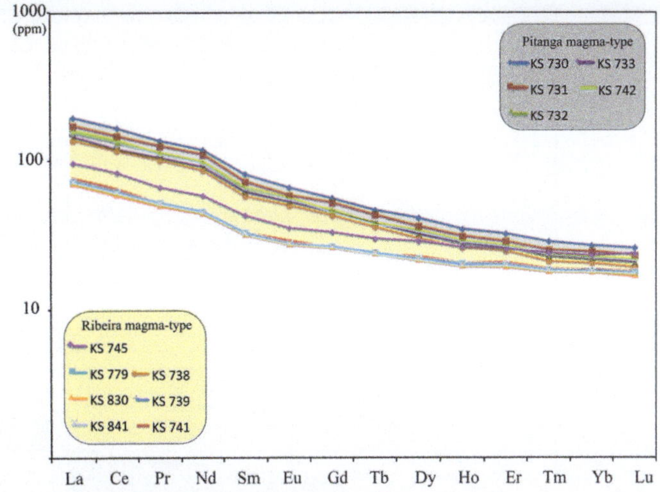

Figure 12. Chondrite-normalized (McDonough and Sun, 1995) rare earth element distribution patterns of Pitanga and Ribeira basalts investigated in this study. Sample location in Fig. 1.

The fluidal peperites associated with pahoehoe lavas of the NW-PCFB seem to be very similar to those of Hawaii. Therefore, it is most likely that the effusive materials have such low viscosities (10^1–10^3 Pa s; McBirney and Murase, 1984) and eruption temperatures ranging near liquidus (~ 1150 °C; Helz and Thornber, 1987; Cashman et al., 1999). Instead, the rhyolite lavas that generated the peperites reported by Luchetti et al. (2014) have the same characteristics as those found in Passamaquoddy Bay (Dadd and Van Wagoner, 2002), with low temperatures (~ 800 °C) and high viscosity (10^8–10^{10} Pa s).

As mentioned above, the NW-PCFB peperites are only associated with the first compound pahoehoe flow, although there are flow intercalations with the Botucatu sandstone all over the area. Thus, if the peperite generation were not related to wet unconsolidated sediments, this same feature would also appear in the upper contacts of the interlayered sediments with the compound pahoehoe flows, and not just in the first contact with the sandstones. In addition, the low-angle stratification (5°) of the sandstone corroborates the inter-dune field environment, with no significant variation of the paleotopography in the region.

The presence of water in the Botucatu desert is a controversial issue, since it is generally recognized as a huge, hot, and high-pressure desert system (e.g., Kocurek and Haveholm, 1993). However, its limits were probably subjected to larger variations in weather conditions during the Jurassic and Early Cretaceous periods. The sedimentation environment would be locally modified by three factors: (a) atmosphere, which would be changed by the PCFB's own volcanism; (b) geographical position that allowed the presence of climatic transitional areas in the desert boundaries; and (c) nearby mountains, allowing for possible riverine channels at the borders of the desert. Therefore, the existence of a humid environment at the edges of the desert cannot be ruled out, permitting peperite generation due to lava and wet unconsolidated sandy sediment interactions.

Contrasting to the peperites, which are observed only to be associated with the first flow, the sand-filled cracks take place in all compound pahoehoe flows intercalated with sandstones of the Botucatu Formation. Therefore, the sand-filled crack generation has no direct relationship with the sedimentation environment, occurring in either wet (peperite zone) or dry sediments.

According to the field observations, the sand-filled cracks originated in two stages. Firstly, fissures opened during the inflation and subsequent expansion of pahoehoe lava feeder tubes, whose tips were in direct contact with the sediments (Fig. 13). Thereafter, these cracks were filled with fine sand, still unconsolidated, and may be ascending (subordinate) or descending (predominant). Contrarily to the pipe vesicles, the sand-filled cracks do not necessarily occur in the same direction as the lobe movement, being vertical to subvertical. Note that sand-filled cracks were also described by Jerram

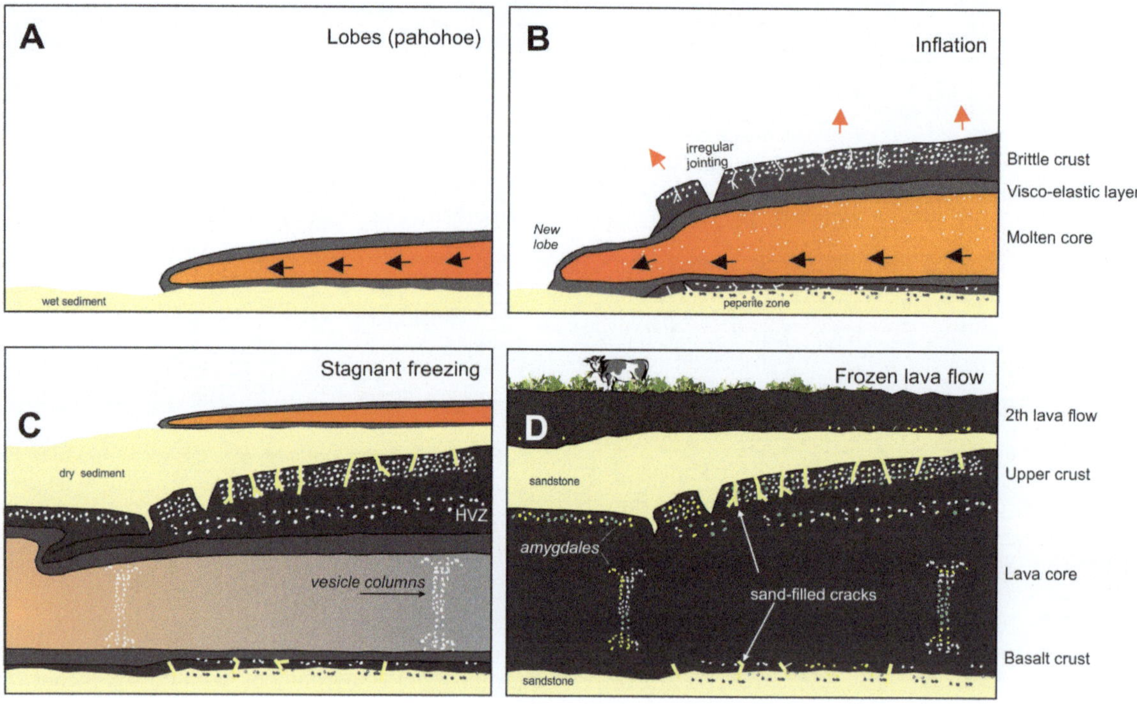

Figure 13. Sketches of sand-filled crack generation observed in flows of the PCFB, integrating the propositions of Self et al. (1996), Jerram and Stolfens (2002), and Jerram and Petford (2011). **(a)** Lobe formation with brittle crust on the surface; **(b)** inflation due to lava injection into the lobe, causing irregular jointing in the basalt upper crust and generating trapped horizontal vesicular zones (HVZ); **(c)** after stagnation, columns of vesicles may form vertical cylinders within the lobe and regular joints may be generated, during the slow cooling of the lava. Before diagenesis, by lithostatic weight sand invades the bottom of the flow (sand-filled cracks) and fills the irregular joints during the sedimentation process, which was synchronous to the lava outpouring. **(d)** All the processes were identified based on vesicle distributions, jointing patterns, and sand-filled cracks. Vertical scale varies from 1 to 8 m.

and Stollhofen (2002) in the Awahab Formation volcanics in the Huab Basin (African counterpart of the PCFB).

On the other hand, the simple pahoehoe flows do not have intercalations with sandstones, corresponding to the top of the volcanic sequence in the region, which is located east of the compound pahoehoe flows. However, the thickness of both volcanic facies architectures could not be determined due to tectonic faulting, which took place during the Pleistocene (Fulfaro and Perinoto, 1994). The tectonism caused the uplift of the region, leading to the Maracaju Mountain formation (Fig. 1; Serra do Maracaju), which is 300 m higher than the general topography. This explains the fact that the samples collected from compound pahoehoe flows of the NW-PCFB border have higher altitudes than those from the simple pahoehoe ones (Table 1). Another consequence of the faulting was the generation of the Ponta Porã Formation (e.g., Fúlfaro and Perinoto, 1994), composed of clastic unconsolidated rocks, rich in basalt clasts, which outcrop to the south of the investigated area.

8 Concluding remarks

This paper reports the main volcanic aspects of the northwestern PCFB, where until now no study has been already, showing not only the flow architectures and the different interaction features with contemporaneous sediments but also the geochemical characteristics of the basalts and their relationship with the magma types of the province. The main conclusions of this investigation are the following:

- In the investigated area, the basalt lavas are in contact and cover the eolic sandstones of the Botucatu Formation of the Paraná Basin, whose sedimentary unit in the African counterpart during the Pangea supercontinent is the Twyfelfontein Formation of the Huab Basin.
- The occurrence of a restricted humid environment in the investigated area is evidenced by peperite occurrences. These wetlands were possibly interdune fields, wherein three main factors might take place in the hydrological cycle: atmosphere change, geographical position, and relief (nearby mountains). These factors could cause the water table to rise and consequently cause its exposition on surface. The most favorable places for the existence of this humid environment were the edges of the Paraná

Basin. Another alternative would be the existence of some small lakes in the Botucatu desert.

- The fluidal peperites of the NW-PCBF are only associated with the first compound pahoehoe flow and were found outcropping by an extent of approximately 80 m long. Juvenile clasts range from submillimeter up to 30 cm long, containing vesicles of variable and fluidal margin sizes. Sometimes the clasts have fine-grained sandstone injections. The wet sedimentary host is a fine-grained sandstone with a low angle of cross-bedding stratification.
- The sand-filled cracks observed in the compound pahoehoe flows are not related to the peperites, but instead to the morphology of the lavas (pahoehoe). The sand-filled cracks were generated by fissures that opened during the lava inflation into the lobes and subsequent expansion of the pahoehoe flow.
- The early basalt P-type pahoehoe flows (containing pipe vesicles), which are in contact with Botucatu sandstones, have the same geochemical signatures of Ribeira magma type. To the east, as well as toward the interior of the Paraná Basin, the pahoehoe flows are overlaid by simple pahoehoe basalt flows, massive and without sand intercalations, presenting identical geochemical characteristics of the Pitanga (dominant) and Paranapanema (very subordinate) magma types.
- In contrast to the central and southern portion of the PCFB, the northwestern area underwent intense tectonic uplift during the Pleistocene, which mainly affected the area of Ribeira basalts, making it difficult to infer their thicknesses. However, the flow stratigraphy was completely defined by the contact relationship with the sandstones of the Botucatu Formation.

Acknowledgements. This study was supported by São Paulo Research Foundation (FAPESP – 2004/10081-9 and 2012/06082-60 numbers) and National Council for Scientific and Technological Development (CNPq – 482611/2011-5 number). We thank Breno Leitão Waichel, Mario Assine, and Alexandre Perinotto for the stimulating discussions about flow and sediment interaction. The paper benefitted greatly from comments and suggestions of two anonymous reviewers. In particular, we would like to thank Scarlato for his indispensable help and encouragement. This is a GEO-SEDex – USP contribution (no. 07).

Edited by: P. Scarlato

References

Almeida F. F. M., Carneiro, C. S. R., and Bartorelli, A.: Magmatismo pós-paleozoico no Brasil, in: Geologia do Brasil, edited by: Hasui, Y., Carneiro, C. D. R., Almeida F. F. M., and Bartorelli, A., Beca Ed., 430–452, 2012.

Bellieni, G., Comin-chiaramonti, P., Marques, L. S., Melfi, A. J., Piccirilo, E. M., Nardy, A. J. R., and Roisenberg, A.: High- and low-Ti flood basalts from the Paraná plateou (Brazil): petrology end geochemical aspects bearing on their mantle origin, Neues Jahr. Miner. Abh., 150, 272–306, 1984.

Bellieni, G., Comin-Chiaramonti, P., Marques, L. S., Melfi, A. J., Nardy, A. J. R., Papatrechas, C., Piccirillo, E. M., and Roisenberg, A.: Petrogenetic aspects of acid and basaltic lavas from the Paraná plateau (Brazil): geological, mineralogical and petrochemical relationships, J. Petrol., 27, 915–944, 1986.

Bondre, N. R., Duraiswani A., and Dole, G.: Morphology and emplacement of flows from the Deccan Volcanic Province, India, Bull Volcanol., 66, 29–45, 2004.

Cas, R. A. F., Edgar, C., Allen, R. L., Bull, S., Clifford, B. A., Giordano, G., and Wright, J. V.: Influence of magmatismo and tectonics on sedimentation in a extensional lake basin: the Upper Devonian Bunga Beds, Boyd Volcanic Complex, southeastern Australia, in: Volcanoclastic Sedimentation in Lacustrine Settings, edited by: White, J. D. L. and Riggs, N. R., Int. Assoc. Sediment. Spec. Publi. 30, 83–108, 2001

Cashman, K. V. and Thornber, C., Kauahikau, J. P.: Cooling and cristalization of lava in open channels and the transition of pahohoe lava to aa, Bull. Volcanol., 61, 306–323. 1999.

Coffin, M. F. and Eldholm, O.: Large Igneous Provinces: Crustal structure, dimensions, and external consequences, Rev. Geophys., 32, 1–36, 1994.

Cordani, U. G., Brito-Neves, B. B., and D'Agrella-Filho, M. S.: From Rodinia to Gondwana: a review of the available evidence from South America, Gondwana Res., 6, 275–283, 2003.

Cotta, A. J. B. and Enzweiler, J.: Classical and New Procedures of Whole Rock Dissolution for Trace Element Determination by ICP-MS, Geostand. Geoanal. Res., 36, 27–50, 2012.

CPRM – Companhia de pesquisa e Recursos Minerais, Mapa Geológico de Mato Grosso do Sul, escala 1 : 1000 000, Brasflia (Brazil), 2006.

Dadd, K. A. and Van Wagoner, N. A.: Magma composition and viscosity as controls on peperite texture: an example from Passamaquoddy Bay, southeastern Canada, J. Volcanol. Geotherm. Res., 114, 63–80, 2002.

Deckart, K., Féraud, G., Marques, L. S., and Bertrand, H.: New time constraints on dyke swarms related to the Paraná – Etendeka magmatic province, and subsequent South Atlantic opening, southeastern Brazil, J. Volcan. Geotherm. Res., 80, 67–83, 1998.

Duriaiswami, R. A., Bondre, N. R., and Managave, S.: Morphology of rubbly pahoehoe (simple) flows from the Deccan Volcanic Province: Implications for style of emplacement, J. Volcanol. Geotherm. Res., 177, 822–836, 2008.

Ernesto, M., Marques, L. S., Piccirillo, E. M., Molina, N., Ussami, P., Comin-chiaramonti, G. P., and Bellieni, P.: Paraná Magmatic Province-Tristan da Cunha plume system: fixed versus mobile plume, petrogenic considerations and alternative heat sources, J. Volcanol. Geotherm. Res., 118, 15–36, 2002.

Frank, H. T., Gomes, M. E. B., and Formoso, M. L. L.: Review of the extent and the volume of the Serra Geral Formation, Paraná Basin, South America, Pesquisas em Geociências, 36, 49–57, 2009.

Fulfaro, V. J. and Perinotto, J. A. J.: A tectônica fanerozóica no Mato Grosso do Sul: o Graben de Corguinho e a Formação Ponta Porã, in: Anais do IV Simpósio de Geologia do Centro-Oeste, 61–63, 1994.

Helz, R. T. and Thornber, C. R.: Geothermometry of Kilauea Iki lava lake, Hawaii, Bull. Volcanol., 49, 651–668, 1987.

Holz, M., Soares, A. P., and Soares, P. C.: Preservation of aeolian dunes by pahoehoe lava: An example from the Botucatu Formation (Early Cretaceous) in Mato Grosso do Sul state (Brazil), western margin of the Paraná Basin in South America, J. South Am. Earth Sci., 25, 398–404, 2008.

Irvine, T. N. and Baraguar, W. R. A.: A guide to the chemical classification of the common volcanic rocks, Canad. J. Earth Sci., 8, 523–548, 1971.

Janasi, V. A., Freitas, V. A., and Heaman, L. H.: The onset of flood basalt volcanism, Northern Paraná Basin, Brazil: A precise U Pb baddeleyite/zircon age for a Chapecó-type dacite, Earth Planet. Sci. Lett., 302, 147–153, 2011.

Jerram, D. A. and Petford, N.: The Field Description of Igneous Rocks (Geological Field Guide), (Wiley-Blackwell), 2nd Edn., 264 pp., 2011.

Jerram, D. A. and Stollhofen, H.: Lava-sediment interaction in desert settings; are all peperite-like textures the result of magma-water interaction?, J. Volcanol. Geotherm. Res., 114, 231–249, 2002.

Kocurek, G. and Haveholm, K.G.: Eolian sequence stratigraphy – a conceptual framework, in: Siliciclastic sequence stratigraphy – recent developments and applications, edited by: Weimer, P. and Posamentier, H., Tulsa, American Association of Petroleum Geologists, AAPG Memoir, 58, 393–409, 1993

Le Bas, M. J. and Streckeisen, A.: The IUGS systematics of igneous rocks, J. Geol. Soc. London, 148, 825–833, 1991.

Leonardi, G. E. and Oliveira, F. H.: A revision of the Triassic and Jurassic tetrapod footprints of Argentina and a new approach on the age and meaning of Botucatu Formation footprints (Brazil), Revista Brasileira de Geociências, 20, 216–229, 1990.

Luchetti, A. C. F., Nardy, A. J. R., Machado, F. B., Madeira, J. E. O., and Arnosio, J. M.: New insights on the occurrence of peperites and sedimentary deposits within the silicic volcanic sequences of the Paraná Magmatic Province, Brazil, Solid Earth, 5, 121–130, doi:10.5194/se-5-121-2014, 2014.

Machado, F. B., Nardy, A. J. R., Rocha-Júnior, E. R. V., Marques, L. S., and Oliveira, M. A. F.: Geologia e Litogeoquímica da Formação Serra Geral nos Estados de Mato Grosso e Mato Grosso do Sul, Geociências (São Paulo. Impresso), 28, 523–540, 2009.

Marques, L. S., Piccirilo, E. M., Melfi, A. J., Comin-Chiaramonti, P., and Bellieni, G.: Distribuição de terras raras e outros elementos traços em basaltos da Bacia do Paraná, Geochim. Brasil., 3, 33–50, 1989.

Marques, L. S., Dupre, B., and Piccirillo, E. M.: Mantle source compositions of the Paraná Magmatic Province (southern Brazil): evidence from trace elements and Sr-Nd-Pb isotope geochemistry, J. Geodynam., 28, 439–458, 1999.

McBirney, A. and Murase, T.: Rheological properties of magmas, Annual Rev. of Earth and Planet. Sci., 12, 337–357, 1984.

McClintock, M. C. and White, J. D. L.: Granulation of weak rock as a precursor to peperite formation: coal peperite, Coombs Hills, Antarctica, J. Volcanol. Geotherm. Res., 114, 205–217, 2002.

McDonough, W. F. and Sun, S. S.: The composition of the Earth, Chem. Geol., 120, 223–254, 1995.

Nardy, A. J. R., Enzweiler, J., Bahia F°, O., Oliveira, M. A. F., and Peneiro, M. A. V.: Determinação de Elementos Maiores e Menores em Rochas Silicáticas por Espectometria de Fluorescência de Raios-X: Resultados Preliminares. In: VI Congresso Brasileiro De Geoquímica, 1997, Salvador (BA), 346–348, 1997.

Nardy, A. J. R., Oliveira, M. A. F., and Machado, F. B.: Geologia e estratigrafia da Formação Serra Geral, Geociências (São Paulo. Impresso), São Paulo-SP, 21, 13–30, 2002.

Peate, D. W., Hawkeswort, C. J., and Mantovani, M. S. M.: Chemical stratigraphy of the Paraná lavas (South America): classification of magma types and their spatial distribution, Bull. Volcanol., 55, 119–139, 1992.

Petry, K., Jerram, D. A., Almeida, D. del P. M. de, and Zerfazz, H.: Volcanic-sedimentary features in the Serra Geral Fm., Paraná Basin, southern Brazil: Examples of dynamic lava-sediment interactions in an arid setting, J. Volcanol. Geotherm. Res., 159, 313–325, 2007.

Piccirillo, E. M., Comin-Chiaramonti, P., Melfi, A. J.; Stolfa, D., Bellieni, G.; Marques, L. S., Giaretta, A., Nardy, A. J. R., Pinese, J. P. P., Raposo, M. I. B., and Roisenberg, A.: Petrichemistry of Continental Flood Basalt-Rhyolite Suites and Related Intrusives from the Paraná Basin (Brazil), in: The Mesozoic Flood Volcanism of the Paraná Basin: Petrogenetic and Geophysical Aspects, edited by: Piccirillo, E. M. and Melfi, A. J., Universidade de São Paulo, 107–156, 1988.

Piccirillo, E. M., G. Bellieni, H. Cavazzini, P. Comin- Chiaramonti, R. Petrini, A. I. Melfi, J. P. P., and Pinese, P.: Zantadeschi, A. de Min, Lower Cretaceous tholeiitic dyke swarms in the Ponta Grossa Arch (South East Brazil): petrology, Sr-Nd isotopes, and genetic relationships from ParanAflood volcanics, Chem. Geol., 89, 19–48, 1990.

Raposo, M. I. B., Ernesto, M., and Renne, P. R.: Paleomagnetism and 40Ar/39Ar dating of the Eraly Cretaceous Florianópolis dike swarm (Santa Catarina Island), Southern Brazil, Physics of the Earth and Planetary Interiors, 108, 257–290, 1998.

Reidel, S. P. and Tolan, T. L.: Eruption and emplacement of River flood basalt. An example from the large-volume Teepee Butte member, Columbia River basalt Group, Geol. Soc. Am. Bull., 104, 1650–1671, 1992.

Renne, P. R., Ernesto, M., Pacca, I. G., Coe, R. S., Glen, J. M., Prévot, M., and Perrin, M.: The Age of Paraná Flood Volcanism, Rifting of Gondwanaland, and the Jurassic-Cretaceous Boundary, Science, 258, 975–979, 1992.

Riedel, S. P.: Emplacement of Columbia River basalt, J. Geophy. Res., 103, 27393–27410, 1998.

Rocha-Júnior, E. R. V., Puchetel, I. S., Marques, L. S., Walker, R. J., Machado, F. B., Nardy, A. J. R., Babinski, M., and Figueiredo, A. M. G.: Re Os isotope and highly siderophile element systematics of the Paraná continental flood basalts (Brazil), Earth Planet. Sci. Lett., 337/338, 164–173, 2012.

Rocha-Júnior, E. R. V., Marques, L. S., Babinski, M., Nardy, A. J. R., Figueiredo, A. M.G., and Machado, F. B.: Sr-Nd-Pb isotopic constraints on the nature of the mantle sources involved in the genesis of the high-Ti tholeiites from northern Paraná Continental Flood Basalts (Brazil), J. South Am. Earth Sci., 46, 9–25, 2013.

Scherer, C. M. S.: Eolian dunes of the Botucatu Formation (Cretaceous) in Southernmost Brazil: morphology and origin, Sediment. Geol., 137, 63–84, 2000.

Self, S., Thordarson, T., Keszthelyi, L., Walker, K. H., Hon, K., Murphy, M. T., Lon, P., and Finnemore, S.: A new model for the emplacement of Columbia River basalts as large, inflated pahoehoe lava flow fields, Geophys. Res. Lett., 23, 2689–2692, 1996.

Self, S., Thordarson, T., and Keszthelyi, L.: Emplacement of continental flood basalts flows, in: Large Igneous Provinces, edited by: Mahoney, J. J. and Coffin, M., Am. Geophys. Union Geophys. Monogr., 100, 381–410, 1997.

Skilling, I. P., White, J. D. L., and McPhie, J.: Peperite: a review of magma-sediment mingling, in: Peperite: processes and products of magma-sediment mingling, edited by: Skilling, I. P., White, J. D. J., and McPhie, J., J. Volcanol. Geotherm. Res., 114, 1–17, 2002.

Stewart, K., Turner, S., Kelley, S., Hawkesworth, C., Kirstein, L., and Mantovani, M.: 3-D, $^{40}Ar/^{39}Ar$ geochronology in the Paraná continental flood basalt province, Earth Planet. Sci. Lett., 143, 95–109, 1996.

Turner, S., Regelous, M., Kelley, S. Hawkeswort, C. J., and Mantovani, M. S. M.: Magmatism and continental break-up in the South atlantic: high precision $^{40}Ar/^{39}Ar$ geochronology, Earth Planet. Sci. Lett, 121, 333–348, 1994.

Waichel, B. L., Lima, E. F., Sommer, C. A., and Lubachesky, R.: Peperite formed by lava flows over sediments: An example from the central Paraná Continental Flood basalt, Brazil, J. Volcanol. Geotherm. Res., 159, 343–354, 2007.

Waichel, B. L., Scherer, C. M. S., and Frank, H. T.: Basaltic lava flows covering active aeolian dunes in the Paraná Basin in southern Brazil: Features and emplacement aspects, J. Volcanol. Geotherm. Res., 171, 59–72, 2008.

White, J. D. L., McPhie, J., and Skilling, I. P.: Peperite: a useful genetic term, Bull. Volcanol., 62, 65–66, 2000.

6

Short-lived tectonic switch mechanism for long-term pulses of volcanic activity after mega-thrust earthquakes

M. Lupi[1] **and S. A. Miller**[2]

[1]ETH Zurich, Department of Earth Sciences, Zurich, Switzerland
[2]Steinmann Institue, Geodynamics/Geophysics, University of Bonn, Bonn, Germany

Correspondence to: M. Lupi (matteo.lupi@erdw.ethz.ch)

Abstract. Eruptive rates in volcanic arcs increase significantly after subduction mega-thrust earthquakes. Over short to intermediate time periods the link between mega-thrust earthquakes and arc response can be attributed to dynamic triggering processes or static stress changes, but a fundamental mechanism that controls long-term pulses of volcanic activity after mega-thrust earthquakes has not been proposed yet.

Using geomechanical, geological, and geophysical arguments, we propose that increased eruption rates over longer timescales are due to the relaxation of the compressional regime that accompanies mega-thrust subduction zone earthquakes. More specifically, the reduction of the horizontal stress σ_h promotes the occurrence of short-lived strike-slip kinematics rather than reverse faulting in the volcanic arc. The relaxation of the pre-earthquake compressional regime facilitates magma mobilisation by providing a short-circuit pathway to shallow depths by significantly increasing the hydraulic properties of the system. The timescale for the onset of strike-slip faulting depends on the degree of shear stress accumulated in the arc during inter-seismic periods, which in turn is connected to the degree of strain-partitioning at convergent margins.

We performed Coulomb stress transfer analysis to determine the order of magnitude of the stress perturbations in present-day volcanic arcs in response to five recent mega-thrust earthquakes; the 2005 M8.6, 2007 M8.5, and 2007 M7.9 Sumatra earthquakes; the 2010 M8.8 Maule, Chile earthquake; and the 2011 M9.0 Tohoku, Japan earthquake. We find that all but one the shallow earthquakes that occurred in the arcs of Sumatra, Chile and Japan show a marked lateral component. We suggests that the long-term response of volcanic arcs to subduction zone mega-thrust earthquakes will be manifested as predominantly strike-slip seismic events, and that these future earthquakes may be followed closely by indications of rising magma to shallower depths, e.g. surface inflation and seismic swarms.

1 Introduction

One of the fundamental uncertainties in earth science is the mechanism controlling volcanism in compressional environments. Horizontally oriented stresses during compression produce high confining pressures on incipient flow paths, suppressing vertical flow. Instead, this favours sill development when the pressure of magma at depth exceeds the minimum principal stress. Reviews of the structural and petrological characteristics of volcanic systems associated with reverse and strike-slip faults (Tibaldi et al., 2010) show how evolutionary effects of the local stress field affect eruption rates and volcanic behaviours. Over non-geological time periods, the complexity of these systems is partially due to variations in the loading environment throughout mega-thrust earthquake cycles, from sub-horizontal compression during convergence, to short-lived viscoelastic relaxation during post-seismic periods. Indeed, volcanic belts in convergent margins undergo different tectonic regimes that either restrict or promote vertical migration of deep fluids and magmas by the cyclic loading and unloading of the effective normal stress caused by the recurrence of mega-thrust slips at the subduction interface.

A link between mega-thrust earthquakes and volcanic activity was first recognised by Charles Darwin when he noted increased volcanism in Chile following the M8.5 Concepción 1835 earthquake (Darwin, 1840). Seismically induced volcanic activity with varying time- and space-scales is now recognised around the globe (Eggert and Walter, 2009), with examples from the USA (Lipman and Mullineaux, 1981; Sanchez and McNutt, 2004), Kamchatka (Walter, 2007), Central America (Williams and Self, 1983; White and Harlow, 1993), Japan (Koyama, 2002), and Italy (Sharp et al., 1981; Nercessian et al., 1991; Marzocchi et al., 1993; Nostro et al., 1998).

Triggered volcanic activity spans a range of timescales and distances (Hill et al., 2002; Delle Donne et al., 2010). In the immediate or short term associated with passing seismic waves, mechanical perturbation of the magma reservoir may result in processes such as rectified diffusion, gas exsolution, and/or unclogging of migration pathways (Hill et al., 2002; Brodsky and Prejean, 2005; Manga and Brodsky, 2006). Such processes may also be enhanced by the overall geological structure at depth (Lupi et al., 2013). Examples include the Kamchatka M9.0 earthquake in 1952 (Johnson and Satake, 1999), which was followed by eruptions of the Karpinsky and Tao-Rusyr volcanoes one and seven days, respectively, after the main shock (Walter and Amelung, 2007), and the Minchinmavida and Cerro Yanteles volcanoes that erupted less than 24 hours after the Concepción, Chile, earthquake (Darwin, 1840; Cembrano and Lara, 2009).

Static stress changes induced by slip on the subduction interface develop at short to intermediate timescales, principally by relaxing the normal stress that acts on incipient flow pathways (Cembrano and Lara, 2009). The delay between earthquakes and volcanic unrest can extend to years, with some relationship of the hypo-central distance to the volcanic system (Walter and Amelung, 2007). The rising speed of the magma is principally driven by the content of volatiles in the melt (e.g. Sigurdsson et al., 1999). Hence, it is not possible to address any relationship between the depth of the magmatic reservoir and the delay of the eruption. However, few examples may provide some insights about the timescale of these processes. For instance, a M8.3 subduction earthquake in 1923 in Kamchatka triggered many subsequent volcanic eruptions in the region, with two in 1923, one in 1925, two in 1926, and one each in 1927, 1928, 1929 and 1930 (Eggert and Walter, 2009). More recent examples include the Planchón-Peteroa, Puyehue-Cordón Caulle, and Copahue volcanoes that erupted approximately seven, sixteen, and thirty-four months after the M8.8 February 2010 Maule earthquake, respectively. The mechanism driving static stress triggering at intermediate to long timescales remains unclear, but recent studies (Walter and Amelung, 2007; Bonali et al., 2012) suggest that reductions of the effective stress in response to slip along mega-thrust subduction interfaces may contribute to magma mobilisation.

In this paper we expand on this concept to explore long-term and evolving changes in the stress state in the arc following mega-thrust subduction zone earthquakes. We propose a mechanism that operates at timescales greater than the immediate effects of dynamic-stress triggering and long after static stress changes from mega-thrust earthquakes have perturbed the system. We explore the mechanical consequences of a decadal change in the kinematics associated with the post-seismic relaxation and suggest that the tectonic switch from pre-earthquake compression to post-earthquake relaxation results in a dominantly strike-slip stress state emplaced during the "less compressive" time window that follows mega-thrust slips. In this context, increases of crustal permeability via normal stress reduction and strike-slip faulting become the most efficient mechanism for vertical migration of melts and deep slab-derived trapped fluids.

2 Short-lived tectonic switch: proposed mechanism

We hypothesise that long-term increase in arc volcanism is kinematically controlled whereby post-seismic relaxation of the compressional stress regime initiates short-lived strike-slip kinematics in the volcanic arc. The onset of strike-slip faulting is controlled by (i) the degree of strain partitioning at convergent margins, (ii) the inter-seismic duration that controls the shear stress amplitudes, and (iii) the relative values of the vertical σ_v and horizontal σ_h principal stresses before the mega-thrust earthquake. The onset of strike-slip faulting promotes an increase of volcanic activity because it provides a short-circuit pathway for upwelling fluids and magmas. Strike-slip faulting dramatically increases the vertical permeability of the crustal transport network and may induce high strain rates in the magmatic reservoir if the magmatic chamber is affected by the fault plane (Chaussard et al., 2013).

The degree of strain-partitioning strongly affects the switch from reverse to strike-slip environments (Chemenda et al., 2000). In addition, the closer the absolute values of σ_h and σ_v are before the mega-thrust earthquake, the more the switch from reverse to strike-slip faulting is favoured. The mechanical argument is illustrated in the 3-D Mohr diagram in Fig. 1. In compressive subduction systems, thrust faulting is favoured prior to mega-thrust earthquakes, with the maximum principal stress ($\sigma_1 = \sigma_H$) assumed to be horizontal in the direction of convergence, the minimum principal stress ($\sigma_3 = \sigma_v$) vertical (weight of the overburden), and the intermediate principal stress ($\sigma_2 = \sigma_h$) also horizontal and orthogonal to σ_H. This stress state favours sill development and thrust faulting but hinders vertical dike formation and propagation (Tibaldi et al., 2010). The maximum shear stress developed during compression is $\tau_{max} = (\sigma_H - \sigma_v)/2$ and it is proportional to the angle of incidence of the ongoing subduction (Chemenda et al., 2000). In this stress state, optimally oriented faults are thrust faults dipping at around 30°

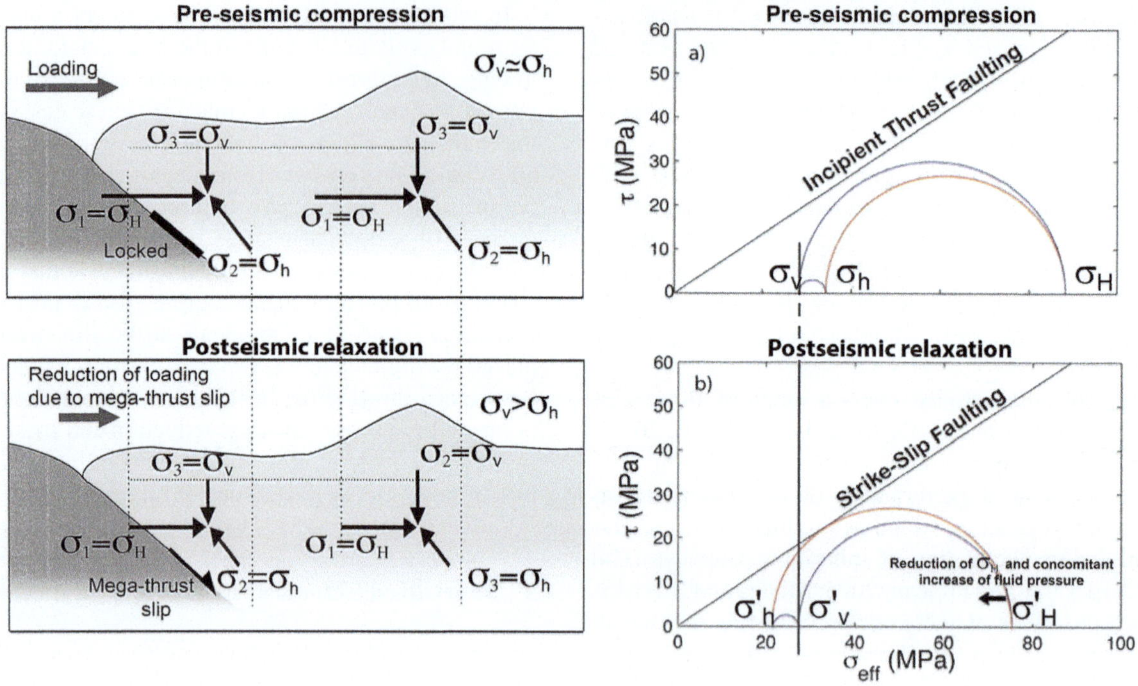

Fig. 1. Mechanical argument for a long-term increase in volcanism following mega-thrust earthquakes. The stress state during compression (loading) encourages thrust faulting and sill development, but limits vertical fluid flow. Beneath the arc, the value of σ_v is higher (and close to σ_h) than at same depths in the fore-arc region. Post-seismic unloading reduces the maximum horizontal stress, while simultaneously increasing vertical permeability (increasing fluid pressure) due to the reduction in fault normal stress. Due to the reduction of σ_H and moreover σ_h, σ_v becomes the intermediate principal stress initiating strike-slip faulting in the arc.

assuming a friction coefficient of 0.6. Slip along the mega-thrust interface predominantly relaxes σ_H. Ignoring poro-elastic effects, σ_H and σ_h are coupled. Both σ_H and σ_h are continuously reduced during the relaxation process even though σ_h is reduced at a lower rate. The overburden σ_v remains unchanged, so if σ_h reduces below σ_v the stress state in the arc emerges as a strike-slip regime. The time required to achieve strike-slip faulting depends on the stress difference between σ_h and σ_v prior to the mega-thrust earthquake. For this reason, the switch from reverse to strike-slip regimes is favoured in the arc where σ_v and σ_h are closer (Fig. 1). This implies that, compared to fore- or back-arc regions, smaller reductions of σ_h are needed beneath the volcanic arc for σ_v to become the intermediate principal stress.

2.1 Supporting arguments

Geochemical studies in arc lavas suggest transport of slab-derived fluids originated at depth and mobilised over short timescales (e.g. Sigmarsson et al., 1990, 2002). A textbook example that illustrates the mechanism proposed in this paper is provided by the recent evolution of the Laguna del Maule system, which is a volcanic complex located in the fold and thrust belt of Central Chile, facing the main slip of the Maule earthquake. InSAR studies highlight an inflation phase that began in 2007 and reached an average value of 227 mm yr^{-1} in early 2012 (Feigl, 2013). The inflation strongly and suddenly accelerated to 388 mm yr^{-1} after the occurrence of a strike-slip M6.0 earthquake on 17 June 2012, which occurred beneath the volcanic system (Fig. 2). An example of a magmatic province that was likely developed during a period of extension or strike-slip faulting are the Sierra Nevada Batholiths, California, which are thought to have formed as soon as compression ceased (Tikoff and Greene, 1997). In addition, Mazzini et al. (2009) show that strike-slip faulting may favour the development of piercement structures and faulting as a mechanism for seismic pumping of deeply trapped fluids has been proposed by Sibson et al. (1975), and contributed to explain aftershock sequences in the Italian Apennines (Miller et al., 2004), or the emplacement of ore deposits (Cox and Ruming, 2004).

Our conceptual model is also consistent with other speculative but yet intriguing examples where magma mobilisation is restricted during thrusting and accelerates significantly once large part of the compressive stress is removed via large-scale slip on the subduction zone interface. For instance, studies of crystal growth (Druitt et al., 2012) show that Santorini volcano, Greece, lay dormant for 18 kyr before evolving into a caldera-forming eruption (ca. 1600 BC) in a mere 100 yr. Driessen and Macdonald (1997) highlight the occurrence of seismically induced ruins of the Minoan civilisation on the isle of Thera pre-dating the eruption of

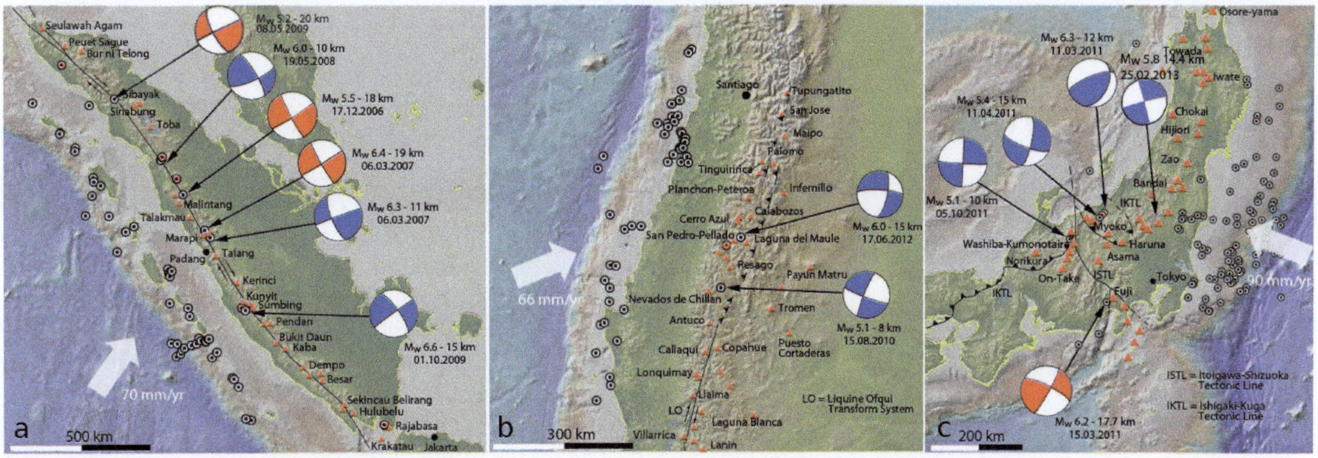

Fig. 2. Maps of Sumatra, Chile, and Japan showing locations of shallow aftershock events that followed the mega-thrust events of Sumatra (M8.7, M8.4, M7.9), Maule (M8.8), and Tohoku (M9.0). Blue focal mechanisms identify seismic events with focus depths shallower than 15 km whereas red focal mechanisms show events deeper than 15 km but shallower than 20 km. The depth range was selected to identify crustal events only (i.e. relevant to brittle deformation of the arc). Focal mechanisms of seismic events with magnitudes smaller than M5 were not always available from USGS, ESMC, and GMT online catalogues and were therefore not plotted to avoid biased estimates of ongoing crustal deformation. White and black-cored circles indicate aftershock activity M > M5 at the subduction interface while red and black-cored circles indicate seismic events (M > M5) in the arc for which focal mechanisms are not available. All the focal mechanism plotted here can be retrieved from the CMT seismic catalogue.

Santorini. Given that the earthquake occurred on the Hellenic subduction zone, this event is consistent with our hypothesis where a mega-thrust earthquake transformed the stress state in the arc hosting Santorini into extension or transtension, and possibly initiated the observed rapid magma mobilisation. Another example is the most recent eruption of the Chaitén volcano, Chile, that entered a new eruptive phase in May 2008 after more than 9 kyr of dormancy. Wicks et al., 2011 suggest that mobilisation of the viscous rhyolitic magma that fed the eruption was accelerated by the introduction of (mechanically created) permeability and pressurisation of the magmatic chamber due to the 1960 Valdivia mega-thrust earthquake. Interestingly, a temporary network deployed in 2004 and 2005 around the Chaitén volcano showed clusters of seismicity at the source of that eruption with strike-slip focal mechanisms (Wicks et al., 2011). Finally, Mount Sinabung in Sumatra, after 400 yr of quiescence, erupted in August and September 2010, 5 yr after the 2005 M8.6 Nias earthquake. The Sinabung volcano, which resides upon the Great Sumatran strike-slip fault, is located about 180 km east of the region of maximum slip during the Nias earthquake (Hendrasto, 2011).

3 Strain partitioning and observed focal mechanisms in the arc

If strike-slip faulting contributes to magma mobilisation, then it is important to determine to what extent strike-slip faults in the arc were perturbed from slip on the subduction interface. Figure 3 shows the tectonic settings, the slip distributions, and the direction of the volcanic arcs for Sumatra, Chile, and Japan. In oblique subduction zones the shear component is accommodated in the overriding plate by trench-parallel, regional-scale, strike-slip faults that are localised within the volcanic belt because of the higher geothermal gradients. Several authors investigated the dynamics of strain partitioning using mathematical (Platt, 1993; Braun and Beaumont, 1995), experimental (Pinet and Cobbold, 1992; Burbidge and Braun, 1998), and field (Rosenau et al., 2006) studies. Fitch (1972) suggests that the strike-slip component relates linearly to the incidence angle of the subduction, with some relation between normal and shear stresses. Chemenda et al. (2000) highlight that the degree of stress partitioning is proportional to the rate of along-trench translation and inversely proportional to the convergence rate and incidence angle. These observations are consistent with the behaviour in the oblique-subducting margins of Chile and Sumatra where strain is partitioned and where strike-slip faulting occurs in the arc (Fig. 2). Little partitioning is expected in Japan because of the frontal collision. However, some strike-slip faulting following the Tohoku earthquake was observed along the periphery of the perturbed regions, and close to the Itoigawa–Shizuoka and Ishigaki–Kuga tectonic lines, in proximity of mapped reverse faults. This is in agreement with a progressive reduction of σ_h, which promotes σ_v to become the intermediate principal stress. However, fault reactivation and magma mobilisation are not always coupled and when this happens (i.e. Laguna del Maule) many (unknown) factors (such as crustal permeability, magma viscosity, current stress-state of the volcanic system) may affect the time delay between faulting and magma mobilisation.

4 Coulomb stress transfer to the arc

Coulomb stress transfer is a commonly used tool to determine the change in normal and shear stress on defined planes (receiver faults) due to slip on elastic dislocations in a three-dimensional half-space (King et al., 1994; Stein, 1999; Toda et al., 2011). We performed Coulomb stress analysis using published slip distributions (USGS, 2010; Ji, 2013) as the source faults for the 2005 M8.6, 2007 M8.5, and 2007 M7.9 Sumatra earthquakes; the 2010 M8.8 Maule, Chile earthquake; and the 2011 M9.0 Tohoku, Japan earthquake. As receiver faults we used the strike, dip, and rake angle provided by moment tensor solutions of shallow (less than 20 km deep) strike-slip earthquakes listed in the CMT catalogs that occurred in each of the volcanic arcs following the mega-thrust earthquake. Moment tensor solutions for seismic events weaker than M5 are not always available. We only account for the focal mechanism of events larger than M5. We also acknowledge the sensitivity of Coulomb analyses to the selected parameters. Here we are only concerned about the order of magnitude of the stress variations imparted in the arc by the mega-thrust earthquake.

4.1 Sumatra

Oblique convergence in Sumatra is accommodated by trench-normal subduction and trench-parallel transform motion along the Sumatra transform fault. Three mega-thrust subduction zone earthquakes occurred in 2005 and 2007, producing trench-normal slip along most of the island. We do not consider the 2004 M9.2 earthquake because slip from that earthquake occurred primarily to the north of Sumatra where there is no arc. Substantial shear stress along the Sumatran fault is expressed by the occurrence of six $M > 5.0$ strike-slip faulting events between 2006 and 2009. Figure 4 shows that the regions where these events occurred were mechanically perturbed by the mega-thrust slip either by increase of shear or normal stress. In the first case, seismic or aseismic lateral slip may shear the magmatic reservoir (for instance located upon the Great Sumatran Fault) reducing its effective viscosity by orders of magnitude and thus enhancing its mobility. In the second case, the reduction of normal stress promotes an increase of permeability and the upwelling of deep fluids that further reduce the effective normal stress bringing faults closer to failure. Based on this we would expect additional events with strike-slip focal mechanism to occur within the next years as post-seismic relaxation of the compressional regime will continue to trigger the transform component of strain-partitioning.

4.2 Chile

The region that faces the slip of the M8.8, 2010, Maule earthquake is located north of the Liquiñe–Ofqui transform system, but oblique convergence of the Nazca and South American plates also provides for shear stress development in the thick and old crust north of 37° S. This is shown by two $M > 5$ strike-slip earthquakes that occurred 6 months and 30 months after the Maule earthquake in the fold and thrust belt of Central Chile (Bonali et al., 2012), where lateral motion is not highlighted by mapped strike-slip faults. In the Coulomb analysis in Fig. 5, we choose receiver faults with strike, dip, and rake determined from moment tensor solutions of the M6.0 strike-slip event in the Chilean arc.

The results in Fig. 5 refer to the fault plane subparallel to the direction of the arc; results for the antithetic fault plane striking NW–SE are discussed later. The stress transfer analysis shows effects of the Maule earthquake, and highlights that fault planes subparallel to the direction of the arc were loaded towards failure. Although much of the Coulomb failure stress is due to the reduction of the normal stresses, we also observe a remarkable increase in shear stress. Figure 5 (post-seismic state) shows increased shear and reduced normal stress where observed strike-slip focal mechanisms occurred, indicating that the stress state in the arc is currently accommodating the lateral component.

4.3 Japan

In Japan, strain partitioning is negligible because of the nearly frontal subduction of the Pacific Plate beneath the Eurasian Plate. This is also confirmed by the reverse focal mechanisms of large magnitude seismic events occurring in the Japanese arc before the 11 March 2011 mega-thrust earthquake (i.e. on 13 June 2008 or 20 September 2010). The distribution of the volcanic systems in the Japanese arc may also reflect the absence of a through-going strike-slip fault system (Nakamura et al., 1977). The strike-slip focal mechanisms that immediately followed the Tohoku earthquake are located in complicated regions that accommodate the differential (strike-slip) motion of the Pacific and Philippine plates with respect to the Eurasian Plate. Strike-slip seismic events (blue circles of Fig. 2c) occur in regions characterised by thrust faulting (i.e. close to the Kuroiwayama, Furumi, Juchihara, and Zenkoji faults – see AIS for a complete database of Japanese faults). The reactivation of such systems with strike-slip kinematics can be explained by the short-lived tectonic switch illustrated in Fig. 1. Figure 6 shows a Coulomb stress transfer analysis from the slip of the M9.0 2011 Tohoku earthquake (USGS, 2010). As receiver fault we used the plane subparallel to the direction of the arc identified by the focal solution of a M5.4 event (11 April 2011). The analysis shows that the entire region was mechanically perturbed and faults were brought closer to failure, consistently with static stress transfer as an earthquake triggering mechanism.

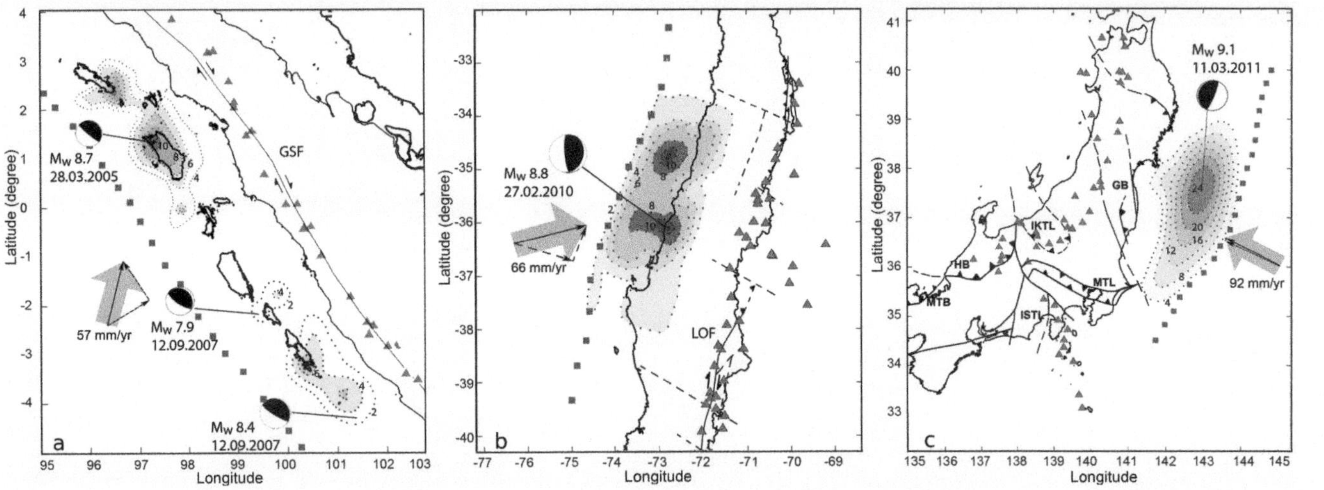

Fig. 3. Tectonic setting and co-seismic slip of the Sumatran, Chilean, and Japanese earthquakes. **(a)** The M8.7 Nias earthquake occurred on 28 March 2005 at the interface between the Sunda and Australian plates. The maximum slip was calculated to be approximately 14 m (30 km deep) with a rupture plane striking N °35 and dipping 15°, which propagated laterally for approximately 400 km NW–SE and for more than 180 km down-dip (USGS, 2010). The M8.4 and M7.9 Sumatran earthquakes occurred on September 12 2007 and caused maximum slips of 4 and 7 m, respectively. **(b)** The M8.8 Maule earthquake struck Chile on 27 February 2010 rupturing the Concepción–Constitucion gap at the interface of the Nazca and South America plates. The slip plane (18° NE, 18° S dip) ruptured for more than 100 km downdip and extended approximately 500 km in the north–south direction (USGS, 2010)) inducing a maximum slip of 14 m at 35 km depth (**?**). **(c)** The M9.0 Tohoku earthquake struck Japan on 11 March 2011, close to the northeast coast of Honshu, Japan, where the Pacific Plate subdues beneath the North America Plate. The slip plane (striking N 340° and dipping is 14° WNW) propagated 400 km along strike and 150 km in the dip direction with a maximum slip of 24 m according to USGS. GSF and LOF stand for Great Sumatran Fault and Liquiñe–Ofqui Fault system, respectively. ISTL stands for Itoigawa–Shizuoka Tectonic Line, MTL for Median Tectonic Line, IKTL for Ishigaki–Kuga Tectonic Line, MTB for Mino–Tanba Belt, GB for Gosaisho Belt, HB for Hida Belt

5 Discussion

Sumatra, Central Chile and Japan constitute a spectrum of strain partitioning, and therefore the behaviour of their respective arcs may record this. The Coulomb stress transfer analysis in Sumatra, Chile, and Japan shows that earthquakes with strike-slip focal mechanisms were all brought closer to failure after each of the mega-thrust earthquakes. We recognise that stress transfer results are sensitive to many of the input parameters, and the results can change with uncertainties in the slip distribution and/or the orientation or slip of the receiver faults. Nevertheless, the general behaviour of volcanic arcs is consistent with the Coulomb stress transfer picture that emerges. Namely, a strike-slip regime is activated by mega-thrust earthquakes with an apparent correlation between the degree of strain partitioning and the number of strike-slip earthquakes. Although the time window is different between the three systems (with Japan being the most recent event), there have been six large strike-slip events in the Sumatran arc, two in Chile (far from the Liquiñe–Ofqui transform system), and one in the arc directly behind the Tohoku rupture zone (four around the tectonically complex Itoigawa–Shizuoka and Ishigaki–Kuga tectonic lines).

Geological and geophysical observations show that strain partitioning is most pronounced in Sumatra where the high obliquity of convergence between the Australian and Sunda plates results in convergence accommodated by subduction and the strike-slip component taken up by the Great Sumatran Fault. The recent large intra-oceanic plate earthquakes off Sumatra (Duputel et al., 2012) now appear to also take up some strike-slip component, but the existence of the through-going Great Sumatran Fault indicates significant shear stress development along the arc. In Central Chile, where a well-developed lateral transport system is not apparent, focal mechanisms of recent events confirm that considerable shear stress has developed in the arc (see also Cembrano and Lara (2009) for additional focal mechanisms).

Focal mechanisms are ambiguous as to which fault plane failed. In Sumatra, the likely true fault plane is subparallel to the strike of the Sumatra Fault, while in Japan there appears to be no visible strike-slip system in the arc behind the Tohoku slip zone. Central Chile, where the Maule earthquake occurred, is interesting because it lies to the north of the through-going Liquiñe–Ofqui Fault, and does not appear to exhibit a regional scale strike slip system. Tectonic settings controlled by a strong compressional component show a general diffuse distribution of volcanic centres (Nakamura et al., 1977), with a direct correlation between convergence and diffusion of the volcanic systems (Tibaldi et al., 2010). This is consistent with the diffuse distribution of volcanoes in Japan. In Sumatra, volcanic centres are aligned along the

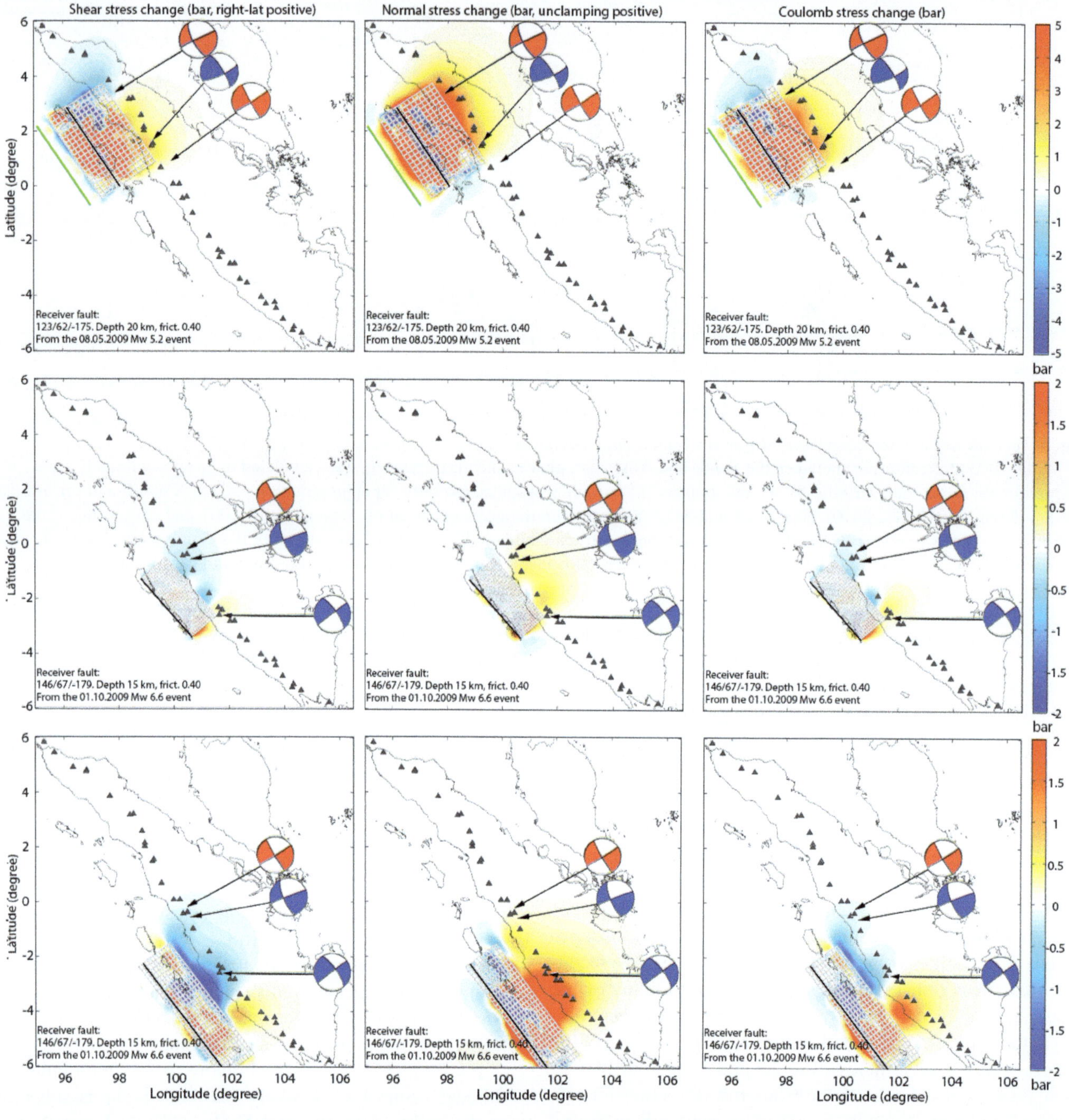

Fig. 4. Coulomb stress analysis for Sumatra. The slip distributions of the events were downloaded from Ji (2013). The first row highlights shear stress, normal stress, and Coulomb stress induced by the 2005 earthquake while the second and third rows indicate effects of the M7.9 and M8.5 2007 earthquakes, respectively. The receiver fault for the northernmost mega-thrust slip (Nias earthquake) is taken from the fault plane solution of the M5.2 event that occurred on 8 May 2009. Depth of the event and strike, dip, and rake angles are shown in each box. For the M7.9 and M8.5 mega-thrust earthquakes we selected the fault plane solution identified by the M6.6 event that occurred on 1 October 2009.

Great Sumatran Fault, while in southern Chile the arc is cut by the Liquiñe–Ofqui transform system. The Liquiñe–Ofqui terminates around 38° S and towards the north the volcanic arc still runs subparallel to the subduction zone. However, an accurate analysis shows that Chilean and Argentinean volcanoes may also align along apparent NW–SE oriented lineaments, antithetic to the direction of the arc. Interestingly, these lineaments are subparallel to basement

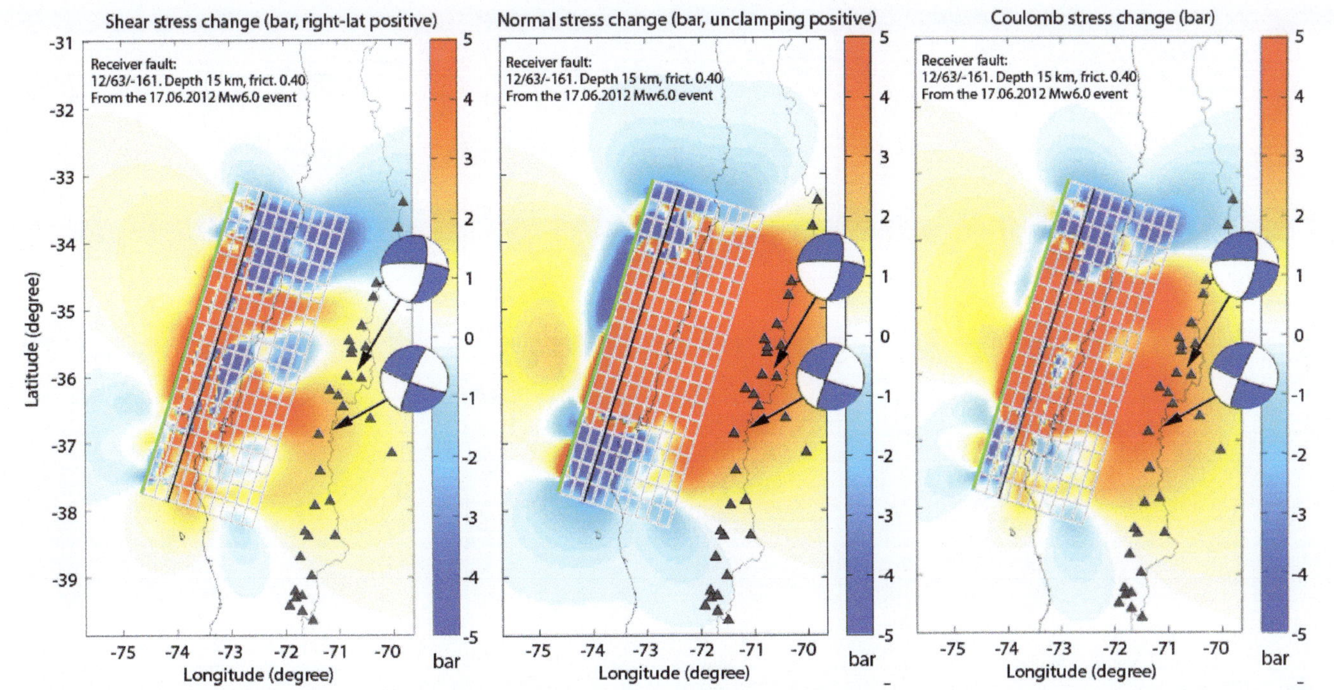

Fig. 5. Shear, normal, and Coulomb stress analysis for Chile. The slip distributions of the Maule earthquake was downloaded from the USGS (2010). The receiver fault, subparallel to the strike of the volcanic arc, is taken from the fault plane solution of the M6.0 event that occurred on 17 June 2012. Depth of the event and strike, dip, and rake angles are shown the boxes.

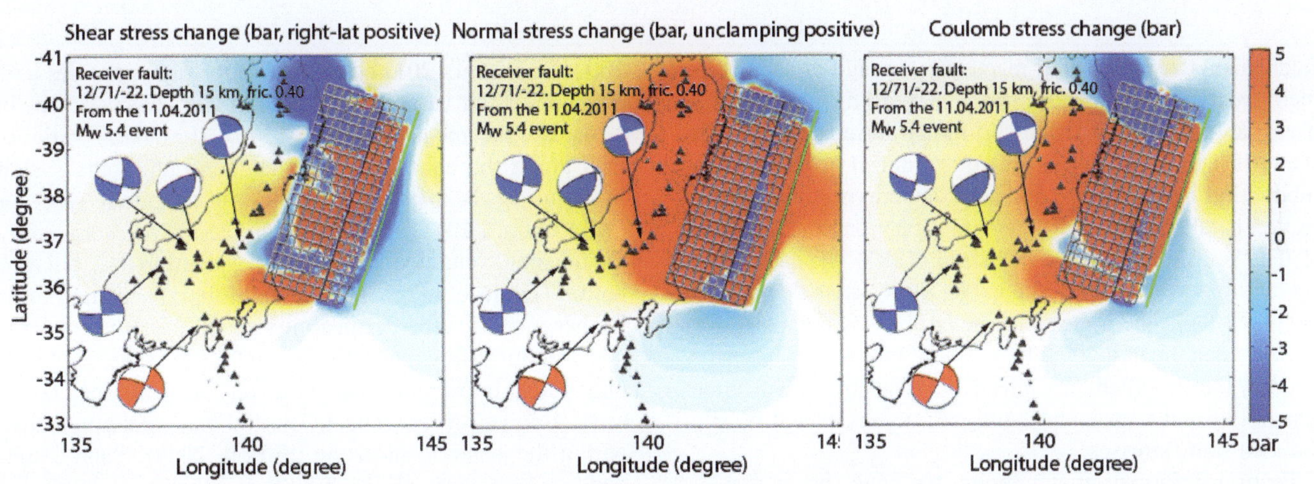

Fig. 6. Coulomb failure calculations of shear, normal, and Coulomb stress induced by the Tohoku–Honshu earthquake on fault plane solutions subparallel to the strike of the volcanic arc. The receiver fault is identified by the focal mechanisms of the M5.4 event in the arc that occurred on 11 April 2011.

NW–SE trending lineaments that have been interpreted as ancient deep structures reactivated as left-lateral strike-slip faults during the development of the volcanic belt (Cembrano and Lara, 2009). Such structures subparallel to the compressional component of the subduction are comparable to the regional-scale lineaments along which Italian volcanic systems align (Rosenbaum et al., 2008). Similar structures have also been identified in northern Fiji, Sulawesi, Indonesia, and at the margins of Calabrian, Caribbean and Scotia arcs (Schellart et al., 2002; Rosenbaum et al., 2008). Cembrano and Lara (2009) highlight that these lineaments can be reactivated by supralithostatic magmatic pressures and our analysis shows that also NW–SE striking faults have been perturbed by mega-thrust earthquakes (Fig. 7).

If the volcanoes in Central Chile are affected by strike-slip motion along NW–SE trending lineaments, then a conjectural

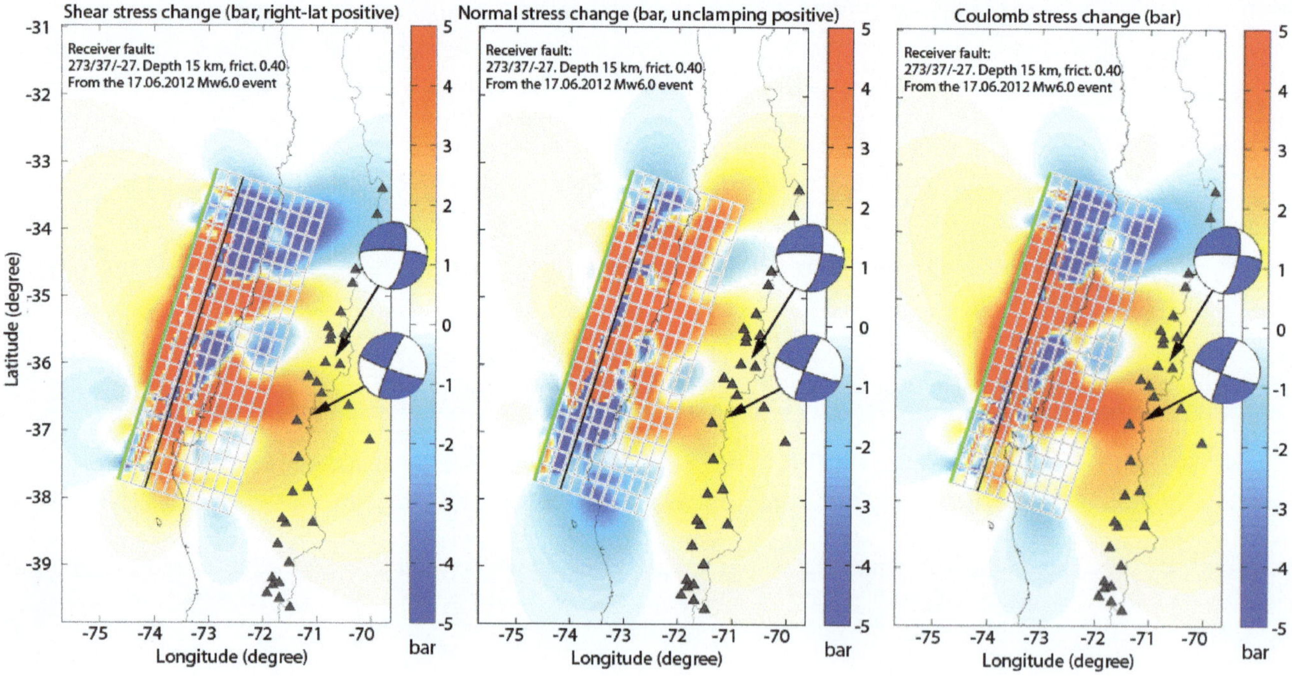

Fig. 7. Shear, normal, and Coulomb stress analysis for Chile using the other fault plane solution of the M6.0 event that occurred on 17 June 2012 (antithetic to the direction of the volcanic arc). The depth of the event and strike, dip, and rake angles are shown the boxes. The Coulomb analysis shows that results similar to the ones in Fig. 5 are obtained when using alternative fault planes. This implies that the Maule earthquake may have perturbed also ancient fault striking at approximately N 270°.

mechanism to explain their reactivation would be the "bookshelf mechanism" (Mandl, 1987). According to this, shear stress develops along strike-slip faults during convergence, but as the compressional stress σ_H is reduced from slip on the subduction zone interface and post-seismic relaxation, deformation is taken up by left-lateral motion along the interfaces of the "books". As stated in Mandl (1987), "[...] when the lateral support is relaxed, the books tilt to one side and shift the plank [...]. The rotation of the books is accompanied by slip between the individual books [...]." Indeed, regions separated by the alignment of Chilean and Argentinean volcanoes resemble tilted books, and faults can be reactivated by slip-induced static stresses.

From a deformation standpoint, we have shown that the reactivation of the strike-slip component in volcanic arcs is a common post-seismic response to mega-thrust subduction zone earthquakes. Strike-slip faulting is initiated by stress transfer from a slip on the subduction interface. Also, the post-seismic increase of crustal hydraulic properties (due to reductions of the normal stress shown in Figs. 4, 5, 6 and 7) promotes the development of this mechanism. The permeability is strongly (e.g. exponentially) dependent on the effective normal stress acting on steeply dipping fractures (e.g. Sibson (1990); Rice (1992); Miller et al. (2004); Sibson (2007)) and vertical diffusion and advection of fast-migrating deep fluids further decrease the effective normal stress at shallower crustal depths (Sibson, 1992; Cox and Ruming, 2004; Lupi et al., 2011). Our Coulomb study provides a first estimate of the variations of normal and shear stresses imposed by the mega-thrust slip. Such (few-bars) variations may affect the equilibrium of magmatic plumbing systems by reducing the confining pressure on magmatic reservoirs (Wicks et al., 2011). This can promote the pressurisation of the magmatic chambers by volatile exsolution and initiate inflation (Druitt et al., 2012). In addition, co-seismic faulting further increases the permeability of the system by creating new fractures (flow paths) inducing fluid pressure cycles that may cause important changes in physical parameters such as seismic velocities, reflectivity, and electrical conductivity within the seismogenic zone (Sibson, 2009). Such coupled and transient permeability fluid-flow behaviours are thought to be relevant in volcanic and hydrothermal systems, as well as in critically stressed geological environments like subduction zones (e.g. Ingebritsen and Sanford, 1999). They may also contribute to the formation of gold deposits (Weatherley and Henley, 2013).

Figure 8 summarises the short-lived switch mechanism that we propose to explain the long-term pulse of increased volcanic activity after mega-thrust earthquakes. Namely, the mega-thrust slip reduces the magnitude of the horizontal stress tensors and σ_h becomes smaller than σ_v.

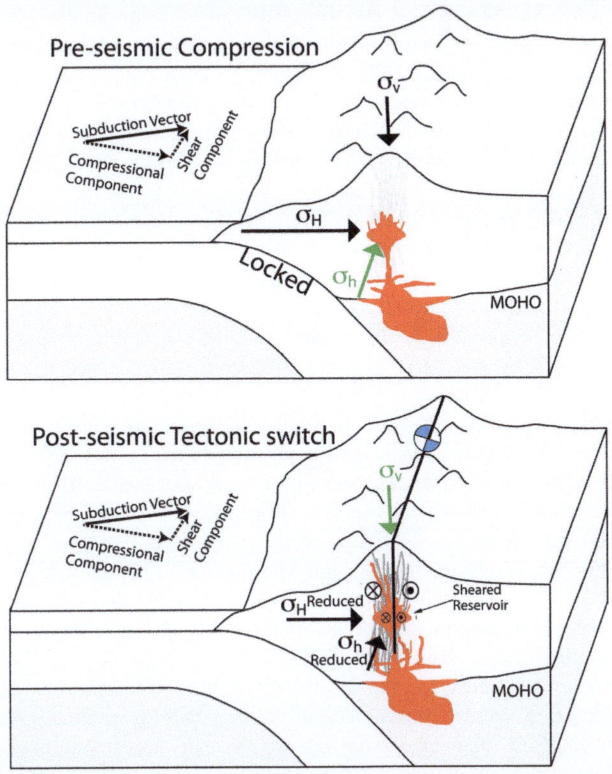

Fig. 8. Simplified representation of the short-lived tectonic switch that takes place in the arc after mega-thrust earthquakes. The green arrow indicates σ_2 and changes from the pre-seismic to the post-seismic state due to the reduction of σ_h as described in Fig. 1. Notice that σ_H and σ_h are reduced, while σ_v remains unchanged. The diagram also highlights the increase of crustal permeability due to lateral motion (darker grey lines).

6 Conclusions

We proposed a mechanism to explain the pulsing eruptive rates in volcanic arcs over decades in response to mega-thrust earthquakes. Our hypothesis is based on the enhancement/establishment of strike-slip regimes in the overriding plate of oblique subduction zones after mega-thrust earthquakes. Beneath the volcanic arc, in regions where pre-slip $\sigma_v = \sigma_3$ is close to $\sigma_h = \sigma_2$, the slip-induced reduction of σ_H and σ_h and post-seismic relaxation of compressional stresses cause in the vertical stress σ_v to adopt the intermediate principal stress. The time delay between mega-thrust earthquakes and subsequent volcanism reflects the time needed to accommodate the new tectonic loading environment, which depends on the degree of strain-partitioning and shear stress development in the arc. Activation of the (steeply dipping) strike-slip regime then promotes vertical magma migration by providing a shorter pathway to shallow depths and through faulting-induced increases in the permeability structure. Upwelling of deep fluids increases pore pressure along the flow path, and thus the effective normal stress acting on steeply dipping fractures and faults reduces even further, initiating a feedback of fracture, flow, and volatile exsolution.

Coulomb stress transfer analyses for Sumatra, Japan, and Chile all show that strike-slip faults that failed after the mega-thrust earthquake were brought closer to failure by the main event. Future studies and observations should focus on strike-slip events expected to occur in theses volcanic arcs, with efforts in Central Chile to determine which of the ambiguous focal mechanism solutions is the true fault plane and which is the auxiliary fault plane. We also expect that strike-slip faulting in the arcs of Chile and Sumatra will be followed closely by indicators of a rising magma source, including swarms, inflation, and eruptions. Our hypothesis can be tested in a way that includes (i) structural studies of volcanic arcs to confirm the coexistence of strike-slip and reverse faulting regimes; and (ii) seismic, geodetic, and geochemical networks to capture the expected behaviour described in this paper.

Acknowledgements. The authors thank the Humanitus Sidoarjo fund and the ETH Zurich postdoctoral fellowship program for supporting this study. Discussions with N. Mancktelow and G. L. Cardello are highly appreciated. J.P. Burg and M. Frehner are thanked for commenting the manuscript. We also thank Boris Kaus for reviewing and improving the quality of this manuscript.

Edited by: T. Gerya

References

AIST, available at: http://riodb02.ibase.aist.go.jp/activefault/index_e.html, Active fault database of Japan; Last Access date: 13th of June 2013; Year of Publication (of the English Version) 1st of December 2005.

Rapid uplift during the 2007–2012 at Laguna del Maule volcanic field, Andean Southern volcanic zone, 2012.

Bonali, F., Tibaldi, A., Corazzato, C., Tormey, D., and Lara, L.: Quantifying the effect of large earthquakes in promoting eruptions due to stress changes on magma pathway: the Chile case, Tectonophysics, 583, 54–67, doi:10.1016/j.tecto.2012.10.025, 2012.

Braun, J. and Beaumont, C.: Three-dimensional numerical experiments of strain partitioning at oblique plate boundaries: Implications for contrasting tectonic styles in the southern Coast Ranges, California, and central South Island, New Zealand, J. Geophys. Res., 100, 18059–18074, 1995.

Brodsky, E. E. and Prejean, S. G.: New constraints on mechanisms of remotely triggered seismicity at Long Valley Caldera, J. Geophys. Res.-Sol. Ea., 110, B04302, doi:10.1029/2004JB003211, 2005.

Burbidge, D. R. and Braun, J.: Analogue models of obliquely convergent continental plate boundaries, J. Geophys. Res., 103, 15221–15237, 1998.

Cembrano, J. and Lara, L.: The link between volcanism and tectonics in the southern volcanic zone of the Chilean Andes: a review, Tectonophysics, 471, 96–113, 2009.

Chaussard, E., Amelung, F., and Aoki, Y.: Characterization of open and closed volcanic systems in Indonesia and Mexico using InSAR time series, J. Geophys. Res.-Sol. Ea., 118, 3957–3969, 2013.

Chemenda, A., Lallemand, S., and Bokun, A.: Strain partitioning and interplate friction in oblique subduction zones – constraints provided by experimental modeling, J. Geophys. Res., 105, 5567–5581, 2000.

Cox, S. and Ruming, K.: The St Ives mesothermal gold system, Western Australia – a case of golden aftershocks?, J. Struct. Geol., 26, 1109–1125, 2004.

Darwin, C.: On the connexion of certain volcanic phenomena in South America; and on the formation of mountain chains and volcanoes, as the effect of the same power by which continents are elevated, Trans. Geol. Soc. Lond. 5, (2nd ser.), 601–631, 1840.

Delle Donne, D., Harris, A. J., Ripepe, M., and Wright, R.: Earthquake-induced thermal anomalies at active volcanoes, Geology, 38, 771–774, 2010.

Driessen, J. and Macdonald, C. F.: The troubled island: Minoan Crete before and after the Santorini eruption (Aegaeum 17), Liège and Austin, Aegaeum, 1997.

Druitt, T., Costa, F., Deloule, E., Dungan, M., and Scaillet, B.: Decadal to monthly timescales of magma transfer and reservoir growth at a caldera volcano, Nature, 482, 77–80, 2012.

Duputel, Z., Kanamori, H., Tsai, V. C., Rivera, L., Meng, L., Ampuero, J.-P., and Stock, J. M.: The 2012 Sumatra great earthquake sequence, Earth Planet. Sc. Lett., 351, 247–257, 2012.

Eggert, S. and Walter, T.: Volcanic activity before and after large tectonic earthquakes: observations and statistical significance, Tectonophysics, 471, 14–26, 2009.

Fitch, T.: Plate convergence, transcurrent faults, and internal deformation adjacent to southeast Asia and the western Pacific, J. Geophys. Res., 77, 4432–4460, 1972.

Feigl, K. L., Le Mével, H., Ali, S. T., Córdova, L., Andersen, N. L., DeMets, C. and Singer, B. S.: Rapid uplift in Laguna del Maule volcanic field of the Andean Southern Volcanic zone (Chile) 2007–2012, Geophys. J. Int., doi:10.1093/gji/ggt438, 2013.

Hendrasto, M.: Learn from 2010 Eruptions at Merapi and Sinabung Volcanoes in Indonesia, Annuals of Disas. Prev. Res., 54, 185–194, 2011.

Hill, D., Pollitz, F., and Newhall, C.: Earthquake–volcano interactions, Phys. Today, 55, 41–47, 2002.

Ingebritsen, S. and Sanford, W.: Groundwater in Geologic Processes, Cambridge University Press, USA, 1999.

Ji, C.: Large Earthquake Database, available at: http://www.geol.ucsb.edu/faculty/ji/big_earthquakes/home.html, last access: 13th of June 2013.

Johnson, J. and Satake, K.: Asperity distribution of the 1952 great Kamchatka earthquake and its relation to future earthquake potential in Kamchatka, Pure Appl. Geophys., 154, 541–553, 1999.

King, G. C., Stein, R. S., and Lin, J.: Static stress changes and the triggering of earthquakes, B. Seismol. Soc. Am., 84, 935–953, 1994.

Koyama, M.: Mechanical coupling between volcanic unrests and large earthquakes: a review of examples and mechanisms, J. Geogr., 111, 222–232, 2002.

Lipman, P. and Mullineaux, D.: The 1980 eruptions of Mount St. Helens, Washington, US Dept. of the Interior, US Geological Survey, United States Government Printing Office, Washington, D.C., 1981.

Lupi, M., Geiger, S., and Graham, C.: Numerical simulations of seismicity-induced fluid flow in the Tjörnes Fracture Zone, Iceland, J. Geophys. Res., 116, B07101, doi:10.1029/2010JB007732, 2011.

Lupi, M., Saenger, E., Fuchs, F., and Miller, S.: Lusi mud eruption triggered by geometric focusing of seismic waves, Nat. Geosci., 6, 642–646, 2013.

Mandl, G.: Tectonic deformation by rotating parallel faults: the "bookshelf" mechanism, Tectonophysics, 141, 277–316, 1987.

Manga, M. and Brodsky, E.: Seismic triggering of eruptions in the far field: volcanoes and geysers, Annu. Rev. Earth Pl. Sc., 34, 263–291, 2006.

Marzocchi, W., Scandone, R., and Mulargia, F.: The tectonic setting of Mount Vesuvius and the correlation between its eruptions and the earthquakes of the Southern Apennines, J. Volcanol. Geotherm. Res., 58, 27–41, 1993.

Mazzini, A., Nermoen, A., Krotkiewski, M., Podladchikov, Y., Planke, S., and Svensen, H.: Strike-slip faulting as a trigger mechanism for overpressure release through piercement structures, implications for the Lusi mud volcano, Indonesia, Mar. Petrol. Geol., 26, 1751–1765, 2009.

Miller, S., Collettini, C., Chiaraluce, L., Cocco, M., Barchi, M., and Kaus, B.: Aftershocks driven by a high-pressure CO_2 source at depth, Nature, 427, 724–727, 2004.

Nakamura, K., Jacob, K. H., and Davies, J. N.: Volcanoes as possible indicators of tectonic stress orientation – Aleutians and Alaska, Pure Appl. Geophys., 115, 87–112, 1977.

Nercessian, A., Hirn, A., and Sapin, M.: A correlation between earthquakes and eruptive phases at Mt Etna: an example and past occurrences, Geophys. J. Int., 105, 131–138, 1991.

Nostro, C., Stein, R., Cocco, M., Belardinelli, M., and Marzocchi, W.: Two-way coupling between Vesuvius eruptions and southern Apennine earthquakes, Italy, by elastic stress transfer, J. Geophys. Res., 103, 24487–24504, 1998.

Pinet, N. and Cobbold, P.: Experimental insights into the partitioning of motion within zones of oblique subduction, Tectonophysics, 206, 371–388, 1992.

Platt, J.: Mechanics of oblique convergence, J. Geophys. Res., 98, 16–239, 1993.

Rice, J.: Fault stress states, pore pressure distributions, and the weakness of the San Andreas fault, International Geophysics Series, 51, 475–475, 1992.

Rosenau, M., Melnick, D., and Echtler, H.: Kinematic constraints on intra-arc shear and strain partitioning in the southern Andes between 38° S and 42° S latitude, Tectonics, 25, TC4013, doi:10.1029/2005TC001943, 2006.

Rosenbaum, G., Gasparon, M., Lucente, F. P., Peccerillo, A., and Miller, M. S.: Kinematics of slab tear faults during subduction segmentation and implications for Italian magmatism, Tectonics, 27, TC2008, doi:10.1029/2007TC002143, 2008.

Sanchez, J. and McNutt, S.: Intermediate-term declines in seismicity at Mt. Wrangell and Mt. Veniaminof volcanoes,

Alaska, following the 3 November 2002 Mw 7.9 Denali Fault earthquake, B. Seismol. Soc. Am., 94, 6B, S370–S383, doi:10.1785/0120040602004, 2004.

Schellart, W., Lister, G., and Jessell, M.: Analogue modeling of arc and backarc deformation in the New Hebrides arc and North Fiji Basin, Geology, 30, 311–314, 2002.

Sharp, A., Lombardo, G., and Davis, P.: Correlation between eruptions of Mount Etna, Sicily, and regional earthquakes as seen in historical records from AD 1582, Geophys. J. Roy. Astr. S., 65, 507–523, 1981.

Sibson, R.: Conditions for fault-valve behaviour, Geol. Soc. Spec. Publ., 54, 15–28, 1990.

Sibson, R.: Implications of fault-valve behaviour for rupture nucleation and recurrence, Tectonophysics, 211, 283–293, 1992.

Sibson, R.: An episode of fault-valve behaviour during compressional inversion? – The 2004 M_J 6.8 Mid-Niigata prefecture, Japan, earthquake sequence, Earth Planet. Sc. Lett., 257, 188–199, 2007.

Sibson, R.: Rupturing in overpressured crust during compressional inversion – the case from NE Honshu, Japan, Tectonophysics, 473, 404–416, 2009.

Sibson, R., Moore, J. M. M., and Rankin, A.: Seismic pumping: a hydrothermal fluid transport mechanism, J. Geol. Soc. London, 131, 653–659, 1975.

Sigmarsson, O., Condomines, M., Morris, J., and Harmon, R.: Uranium and 10Be enrichments by fluids in Andean arc magmas, Nature, 346, 163–165, doi:10.1038/346163a0, 1990.

Sigmarsson, O., Chmeleff, J., Morris, J., and Lopez-Escobar, L.: Origin of 226Ra–230Th disequilibria in arc lavas from southern Chile and implications for magma transfer time, Earth Planet. Sc. Lett., 196, 189–196, 2002.

Sigurdsson, H., Houghton, B., Rymer, H., Stix, J., and McNutt, S.: Encyclopedia of volcanoes, Access Online via Elsevier, 1999.

Stein, R.: The role of stress transfer in earthquake occurrence, Nature, 402, 605–609, 1999.

Tibaldi, A., Pasquarè, F., and Tormey, D.: Volcanism in reverse and strike-slip fault settings, in: New Frontiers in Integrated Solid Earth Sciences, 315–348, doi:10.1007/978-90-481-2737-5, SpringerVerlag, Dordrecht, The Netherlands, 2010.

Tikoff, B. and Greene, D.: Stretching lineations in transpressional shear zones: an example from the Sierra Nevada Batholith, California, J. Struct. Geol., 19, 29–39, 1997.

Toda, S., Stein, R., Sevilgen, V., and Lin, J.: Coulomb 3. 3 Graphic-rich deformation and stress-change software for earthquake, tectonic, and volcano research and teaching-user guide, Tech. rep., USGS, 2011.

USGS: USGS online archive of significant earthquakes, Tech. rep., available at: http://earthquake.usgs.gov/earthquakes/eqinthenews/, USGS, last access: 13th of June 2013, 2010.

Walter, T.: How a tectonic earthquake may wake up volcanoes: stress transfer during the 1996 earthquake-eruption sequence at the Karymsky volcanic group, Kamchatka, Earth Planet. Sc. Lett., 264, 347–359, 2007.

Walter, T. and Amelung, F.: Volcanic eruptions following M 9 megathrust earthquakes: implications for the Sumatra-Andaman volcanoes, Geology, 35, 539–542, doi:10.1130/G23429A.1, 2007.

Weatherley, D. K. and Henley, R. W.: Flash vaporization during earthquakes evidenced by gold deposits, Nat. Geosci., 6, 294–298, doi:10.1038/ngeo1759, 2013.

White, R. and Harlow, D.: Destructive upper-crustal earthquakes of Central America since 1900, B. Seismol. Soc. Am., 83, 1115–1142, 1993.

Wicks, C., de La Llera, J., Lara, L., and Lowenstern, J.: The role of dyking and fault control in the rapid onset of eruption at Chaitén volcano, Chile, Nature, 478, 374–377, 2011.

Williams, S. and Self, S.: The October 1902 plinian eruption of Santa Maria Volcano, Guatemala, J. Volcanol. Geotherm. Res., 16, 33–56, 1983.

7

Petrophysical constraints on the seismic properties of the Kaapvaal craton mantle root

V. Baptiste and A. Tommasi

Géosciences Montpellier, Université Montpellier 2 & CNRS, CC 60, Place E. Bataillon, 34095 Montpellier cedex 5, France

Correspondence to: V. Baptiste (virginie.baptiste@gm.univ-montp2.fr)

Abstract. We calculated the seismic properties of 47 mantle xenoliths from 9 kimberlitic pipes in the Kaapvaal craton based on their modal composition, the crystal-preferred orientations (CPO) of olivine, ortho- and clinopyroxene, and garnet, the Fe content of olivine, and the pressures and temperatures at which the rocks were equilibrated. These data allow constraining the variation of seismic anisotropy and velocities within the cratonic mantle. The fastest P and S_2 wave propagation directions and the polarization of fast split shear waves (S_1) are always subparallel to olivine [100] axes of maximum concentration, which marks the lineation (fossil flow direction). Seismic anisotropy is higher for high olivine contents and stronger CPO. Maximum P wave azimuthal anisotropy (AV_p) ranges between 2.5 and 10.2 % and the maximum S wave polarization anisotropy (AV_s), between 2.7 and 8 %. Changes in olivine CPO symmetry result in minor variations in the seismic anisotropy patterns, mainly in the apparent isotropy directions for shear wave splitting. Seismic properties averaged over 20 km-thick depth sections are, therefore, very homogeneous. Based on these data, we predict the anisotropy that would be measured by SKS, Rayleigh (S_V) and Love (S_H) waves for five endmember orientations of the foliation and lineation. Comparison to seismic anisotropy data from the Kaapvaal shows that the coherent fast directions, but low delay times imaged by SKS studies, and the low azimuthal anisotropy with with the horizontally polarized S waves (S_H) faster than the vertically polarized S wave (S_V) measured using surface waves are best explained by homogeneously dipping (45°) foliations and lineations in the cratonic mantle lithosphere. Laterally or vertically varying foliation and lineation orientations with a dominantly NW–SE trend might also explain the low measured anisotropies, but this model should also result in back-azimuthal variability of the SKS splitting data, not reported in the seismological data. The strong compositional heterogeneity of the Kaapvaal peridotite xenoliths results in up to 3 % variation in density and in up to 2.3 % variation of V_p, V_s, and V_p/V_s ratio. Fe depletion by melt extraction increases V_p and V_s, but decreases the V_p/V_s ratio and density. Orthopyroxene enrichment due to metasomatism decreases the density and V_p, strongly reducing the V_p/V_s ratio. Garnet enrichment, which was also attributed to metasomatism, increases the density, and in a lesser extent V_p and the V_p/V_s ratio. Comparison of density and seismic velocity profiles calculated using the xenoliths' compositions and equilibration conditions to seismological data in the Kaapvaal highlights that (i) the thickness of the craton is underestimated in some seismic studies and reaches at least 180 km, (ii) the deep sheared peridotites represent very local modifications caused and oversampled by kimberlites, and (iii) seismological models probably underestimate the compositional heterogeneity in the Kaapvaal mantle root, which occurs at a scale much smaller than the one that may be sampled seismologically.

1 Introduction

A wealth of seismological studies have investigated the structure of the Kaapvaal craton aiming at unraveling the thermal structure, the composition, and the deformation fabric of the cratonic root and at understanding the causes of its stability since Archean times, as inferred from Re–Os model ages obtained in kimberlite-born mantle xenoliths (Pearson et al., 1995). Tomographic models and receiver function data agree on the presence of a high velocity upper mantle lid on top of a low velocity layer (Jordan, 1978; Priestley, 1999; James et al., 2001; Li and Burke, 2006; Priestley et al., 2006). Most body and surface wave tomography models indicate that this lid is 200–250 km thick beneath the Kaapvaal craton (James et al., 2001; Chevrot and Zhao, 2007; Fishwick, 2010). A thinner high-velocity layer, extending to depths of 150 or 180 km, was, however, imaged by other models (Sebai et al., 2006; Wang et al., 2008). Receiver function studies mapped sharp decreases in seismic velocities at 150 and 200 km depths (Wittlinger and Farra, 2007; Savage and Silver, 2008; Hansen et al., 2009; Adams and Nyblade, 2011).

Discriminating between these models, which point to different cratonic root thicknesses, and interpreting these geophysical data in terms of composition and temperature requires independent observations. The high seismic velocities beneath the Kaapvaal craton have indeed been alternatively attributed to thermal effects (Priestley and Tilmann, 2009) or to changes in composition (Begg et al., 2009). Similarly, the conversion at 150 km imaged by *S* wave receiver functions was interpreted as produced by compositional layering, by anisotropy contrasts within the cratonic mantle (Wittlinger and Farra, 2007; Savage and Silver, 2008) or the lithosphere–asthenosphere boundary beneath the craton (Hansen et al., 2009). Small-scale variations in seismic properties within the cratonic root also may not be imaged or interpreted unambiguously. Beneath the Bushveld complex, some, but not all body wave tomography models imaged velocities lower than the average ones in the craton, which have been attributed to either more fertile compositions associated with the 2 Ga-old Bushveld magmatism (James et al., 2001; Fouch et al., 2004a) or to a hotter present-day geotherm (Fouch et al., 2004a).

The deformation of the Kaapvaal craton has also been extensively investigated by seismic anisotropy studies (Silver et al., 2001; Fouch et al., 2004a; Fouch et al., 2004b; Silver et al., 2004; Adam and Lebedev, 2012; Vinnik et al., 1995, 2012). Silver et al. (2001) measured fast polarization directions that consistently followed the trend of geological structures and small SKS delay times (0.62 s on average), with almost null delays in the central Kaapvaal craton. They attributed this anisotropy to fossil deformation in the mantle lithosphere. Vinnik et al. (1995) attributed the same NNE–SSW fast polarization directions to asthenospheric deformation in response to African plate motion. Significant variations of SKS delay times were measured near Kimberley, which was interpreted as the boundary between a strongly and a weakly anisotropic domain (Fouch et al., 2004b). More recently, a Rayleigh wave azimuthal anisotropy model (Adam and Lebedev, 2012) proposed that the mantle fabric at lithospheric depths parallels the Archean–Paleoproterozoic crustal structures in the Limpopo belt and in the northern Kaapvaal, but is perpendicular to the crustal structures in the western part of the craton. Vinnik et al. (2012), using *P* receiver functions and SKS waveform inversion, also observed belt-parallel, fast directions in the uppermost mantle near the Limpopo belt. They also detected a change with depth in the orientation of the fast direction; the latter aligns progressively with current plate motion at depths > 160 km. In summary, seismic anisotropy data derived from body and surface waves are mostly, but not always, consistent. Their interpretation is also not unique and may benefit from direct observations of the lithospheric mantle deformation and resulting anisotropy.

The Kaapvaal craton was affected by numerous kimberlite eruptions, mostly between the late Jurassic and the Cretaceous, but also during the Meso-Proterozoic and Paleozoic (Kramers and Smith, 1983; Allsopp et al., 1985; Phillips et al., 1998). These magmas carried xenoliths that provide direct sampling of the lithospheric mantle beneath the craton. The seismic properties of these peridotite xenoliths can be calculated from their crystal-preferred orientation (CPO) and composition data (Mainprice and Silver, 1993; Mainprice and Humbert, 1994). Such data constrain the possible seismic velocities and anisotropies within the Kaapvaal mantle root, as well as their dependence on the peridotites composition and crystal-preferred orientations. In the present study, we calculated the anisotropic seismic properties of 47 peridotite xenoliths from 9 kimberlite pipes, which sample different domains, depths, and ages of the Kaapvaal mantle root (Fig. 1). The present study represents a significant enrichment of the database of seismic properties of the Kaapvaal mantle root, as previous studies focused on either the composition- and depth-dependence of the isotropic seismic velocities (James et al., 2004) or the crystal-preferred orientation-induced anisotropy (Ben Ismaïl et al., 2001).

2 The studied data set

The 47 peridotite xenoliths used in this study come from 9 kimberlitic pipes (Kimberley, Jagersfontein, Monastery, Lentseng, De Beers, Finsch, Kamfersdam, Premier, and Mothae) from the Kaapvaal craton, with eruption ages of 1200 Ma (Premier) and 120–87 Ma (all other pipes, see Baptiste et al., 2012, for details). These peridotites are mainly composed of olivine, orthopyroxene, clinopyroxene ±garnet. When spinel is present, it is a minor phase. Thermobarometry data indicates that the studied peridotites were equilibrated at depths ranging between 83 and 187 km on a cold cratonic geotherm (Baptiste et al., 2012). The full description of the microstructures, crystal-preferred orientations (CPOs), and data on the water contents in olivine and pyroxenes in these xenoliths is presented in Baptiste et al. (2012), but their main characteristics are recalled below. The modal content, texture, olivine CPO type and intensity, and equilibration P–T conditions of all studied samples are summarized in Table 1.

The studied mantle xenoliths can be separated into two textural groups: the coarse-grained, which are composed by plurimillimetric to centimetric olivine and pyroxene crystals, and the sheared peridotites, which are characterized by

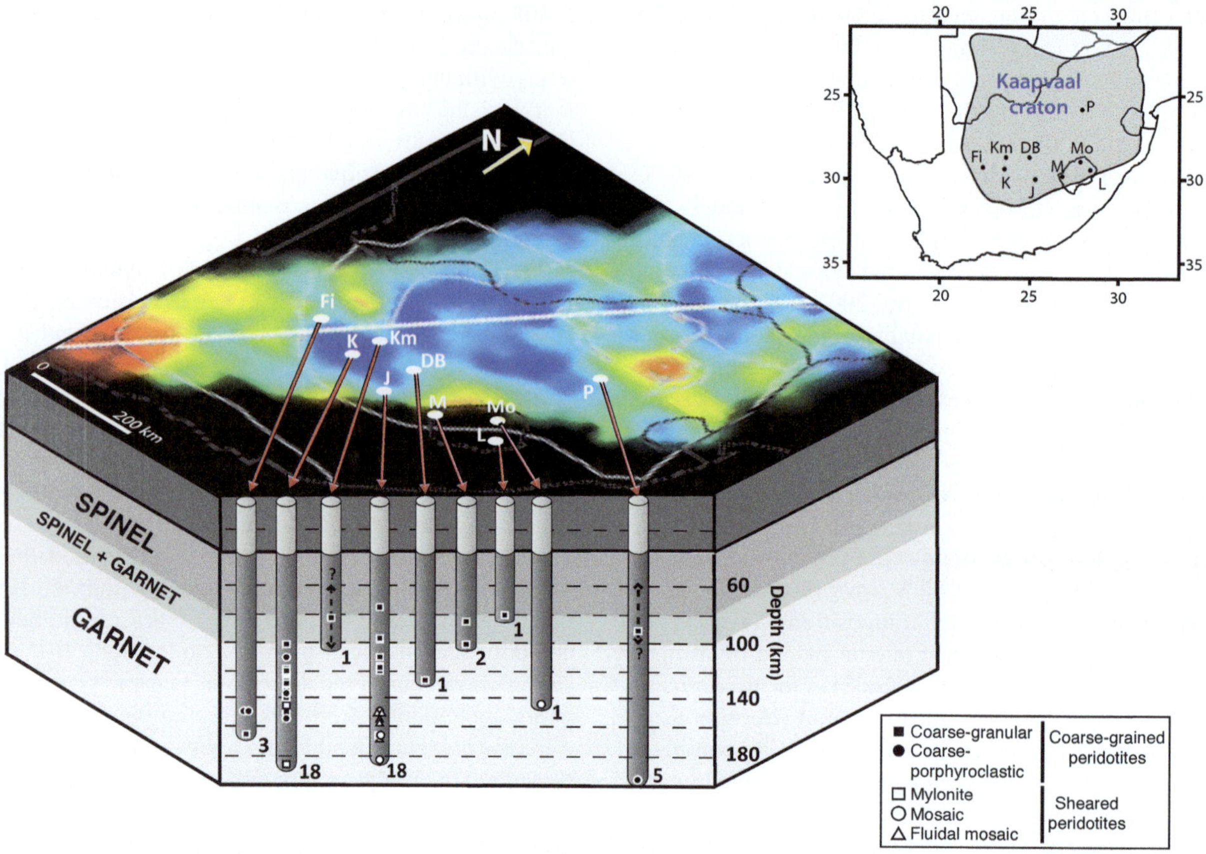

Fig. 1. Sketch displaying the location of the sampled kimberlitic pipes as well as the samples' depths and microstructures (after Baptiste et al., 2012) relatively to the *S* wave tomography model of Fouch et al. (2004a) at 150 km. Fi: Finsch, K: Kimberley, Km: Kamfersdam, J: Jagersfontein, DB: De Beers, M: Monastery, Mo: Mothae, L: Lentseng, P: Premier.

a matrix of fine-grained olivine (a few to 100 μm) and, occasionally, orthopyroxene that encloses millimeter-size garnet, pyroxene, and, sometimes, olivine porphyroclasts. The coarse-grained peridotites are the dominant textural group (72 % of the studied samples), being observed at all depths and in all pipes (Fig. 1). They display most often strongly annealed textures, but coarse-porphyroclastic microstructures are also common. The sheared peridotites are only observed in the deeper part of the sampled columns (equilibration pressures > 4.5 GPa). Their textures range from mylonitic to fluidal or mosaic-like, probably due to variable post-deformational annealing. Based on their occurrence, microstructures and thermal evolution, Baptiste et al. (2012) interpreted the coarse-grained microstructures as the typical cratonic root microstructure, resulting from an early deformation episode followed by a long quiescence time, and the sheared peridotites as recording local modifications of the cratonic microstructure just before the xenolith extraction. Their formation is probably associated with the kimberlitic activity as proposed by Green and Guegen (1974).

Coarse-grained peridotites have in average more refractory compositions than the sheared peridotites (Table 1). They are mainly harzburgites, with high orthopyroxene contents (> 20 %). Sheared peridotites are mainly lherzolites, with highly variable garnet contents (0–18 %). In the studied samples, water contents in olivine and pyroxenes are very variable; maximum water contents in olivine increase up to 150 km depth, reaching up to 150 wt. ppm H_2O (values obtained using the Bell et al. (2003) calibration; Baptiste et al., 2012). In contrast, olivines from samples equilibrated at depths > 150 km are almost dry.

3 Methods

Seismic properties of Kaapvaal xenoliths were calculated based on the olivine, orthopyroxene, clinopyroxene, and garnet CPOs, and on the modal compositions (Mainprice, 1990). The CPOs were measured by electron-backscattered diffraction (EBSD), producing crystal orientation maps, which covered the entire thin section (see Baptiste et al., 2012 for details on the EBSD data acquisition and treatment). Because olivine is the dominant phase in peridotites, largely determining their seismic properties, and because olivine compositions vary strongly among the studied samples (Table 1),

Table 1. Olivine Mg#, modal contents, equilibration conditions, calculated seismic properties and densities of the studied Kaapvaal peridotites.

Locality	Sample	Ol Mg# (%)	Modal compositions (%) Ol	Opx	Cpx	Gt	Temperature (°C)	Pressure (GPa)	Depth (km)	Density ($g\,cm^{-3}$)	$V_P^{(1)}$ ($km\,s^{-1}$)	Max AV_P (%)	$V_S^{(2)}$ ($km\,s^{-1}$)	Max AV_s pol (%)	Max AVs_1 (%)	Max AVs_2 (%)
KIMBERLEY	KBBF6a	90.8	74	13	4	9	926	3.8	121	3.3690	8.27	9.7	4.67	8.0	4.8	6.0
	KBBF6b	91.0	82	8	4	6	908	3.7	118	3.3683	8.26	9.2	4.66	7.0	3.5	5.9
	KBBF8	94.3	81	19	0	0	1029	–	138	3.3500	8.19	10.0	4.64	7.4	6.1	2.6
	KBBF9	89.8	67	16	9	8	1102	4.7	149	3.3623	8.28	5.4	4.65	4.2	2.8	2.1
	KBBF10	92.4	72	27	1	0	948	–	125	3.3398	8.19	7.9	4.65	6.0	3.5	4.9
	KBBF11	93.6	88	11	1	0	1159	5.9	187	3.3618	8.17	4.7	4.55	3.3	4.6	2.9
	KBBF14	93.0	83	16	1	0	996	–	133	3.3518	8.20	5.4	4.64	5.4	3.6	2.8
	KBBF15	92.6	65	31	0	4	1067	4.3	137	3.3442	8.21	5.2	4.65	4.7	1.2	4.5
	KBBF16	93.3	71	26	0	3	–	4.0	127	3.3476	8.21	4.0	4.66	4.1	2.2	3.0
	KBBF18	92.2	75	17	6	0	850	–	110	3.3451	8.22	7.2	4.67	6.3	4.6	3.7
	KBBF20	–	65	35	0	0	800	–	102	*3.3320	*8.17	*7.8	*4.67	*6.4	*3.2	*5.7
	FRB1402	93.2	83	9	0	8	1023	4.6	146	3.3770	8.25	4.2	4.64	3.4	3.2	0.5
	FRB1404	93.5	59	36	1	4	987	4.4	140	3.3380	8.21	3.2	4.67	3.2	2.6	2.4
	FRB1422	93.5	65	31	0	4	994	4.4	140	3.3442	8.21	5.1	4.66	4.1	3.7	2.0
	FRB1423	91.8	68	18	4	10	851	3.8	121	3.3658	8.28	7.2	4.69	5.5	2.8	3.7
	FRB1447	93.4	73	23	1	3	968	4.2	133	3.3501	8.22	3.4	4.66	3.2	1.1	2.6
	FRB348	92.3	58	28	7	7	938	3.7	118	3.3478	8.27	5.6	4.70	4.3	2.4	2.9
JAGERSFONTEIN	KBJ6	90.7	84	9	3	4	1163	4.9	152	3.3673	8.22	6.0	4.61	4.8	3.7	1.8
	KBJ8	90.9	87	13	0	0	1219	–	168	3.3689	8.21	10.2	4.60	6.8	2.3	6.4
	KBJ14	91.3	70	5	12	13	1140	4.7	149	3.3822	8.36	5.7	4.69	4.1	2.0	3.3
	KBJ30	89.4	63	11	11	15	1282	5.8	184	3.3813	8.35	4.5	4.66	3.9	2.5	1.6
	KBJ33	89.3	68	8	6	18	1217	4.8	152	3.3925	8.35	7.6	4.68	5.3	1.9	4.5
	KBJ52	92.7	75	25	0	0	900	–	118	3.3425	8.18	7.3	4.64	5.7	4.1	3.5
	KBJ54	92.3	75	23	2	0	907	3.8	121	3.3431	8.19	8.2	4.65	7.5	5.9	4.4
	KBJ56	92.5	65	28	0	7	895	3.5	111	3.3524	8.24	6.3	4.69	5.4	5.1	2.3
	KBJ60	91.5	86	13	1	0	1190	5.2	165	3.3574	8.17	7.5	4.58	6.3	4.3	2.9
	KBJ62	92.6	64	36	0	0	759	3.1	98	3.3311	8.18	7.3	4.68	7.4	4.4	5.6
	KBJ63	91.3	82	10	3	5	1250	4.9	156	3.3689	8.23	7.5	4.61	7.7	3.9	4.4
	J34	92.0	59	29	5	7	656	2.4	76	3.3457	8.26	9.8	4.75	6.8	1.8	6.1
	J41	92.4	61	39	0	0	668	3.7	118	3.3283	8.17	4.0	4.66	4.9	4.5	2.0
	J47	90.7	85	6	8	1	1227	4.7	149	3.3615	8.22	7.1	4.61	5.7	4.6	1.6
	J57	92.4	79	18	3	0	765	3.1	98	3.3457	8.21	8.6	4.67	6.0	3.4	4.7
	J63	87.7	78	10	10	2	1176	5.2	165	3.3581	8.23	4.5	4.59	3.3	2.2	1.6
MONASTERY	ROM23	92.2	66	28	4	1	910	3.3	105	3.3373	8.21	5.0	4.69	4.9	3.1	3.0
	ROM69	92.2	76	20	3	1	784	2.6	83	3.3438	8.22	2.5	4.70	3.1	2.3	2.5
LENTSENG	PHN4274	93.1	62	33	2	3	1015	4.3	137	3.3387	8.21	3.5	4.66	3.5	3.0	2.5
MOTHAE	PHN1925	89.9	63	22	11	4	1306	4.6	137	3.3462	8.26	3.4	4.67	2.7	0.8	2.5
KAMFERSDAM	PHN5580	92.6	73	27	0	0	–	–	–	**3.2814	**8.34	**5.5	**4.89	**5.2	**4.0	**3.6
DE BEERS	KBB2	93.3	66	33	0	1	917	4.0	127	3.3363	8.19	6.8	4.67	5.7	2.6	4.5
PREMIER	FRB1330	91.3	80	17	3	0	744	–	93	3.3455	8.21	9.4	4.68	6.7	2.3	6.1
	FRB1336	88.9	70	30	0	0	–	–	–	**3.3052	**8.28	**7.1	**4.85	**6.2	**4.9	**2.2
	FRB1339	91.7	71	18	8	3	–	–	–	**3.3118	**8.33	**5.1	**4.86	**4.8	**3.8	**1.9
	PHN5266	92.7	70	26	4	0	–	–	–	**3.3109	**8.29	**8.5	**4.86	**6.7	**5.0	**3.9
	PHN5267	91.3	70	19	6	5	1464	6.5	171	3.3667	8.29	6.9	4.66	4.8	2.5	4.4
FINSCH	FRB1501	93.0	70	22	0	8	1030	4.6	149	3.3620	8.24	5.5	4.65	4.6	3.3	2.1
	FRB1512	92.0	68	22	0	10	1058	4.7	149	3.3657	8.26	5.0	4.66	4.1	3.3	1.9
	FRB1513	92.6	70	27	1	2	1037	5.2	165	3.3443	8.19	4.9	4.62	5.6	4.8	2.3

Ol Mg#, modal content and P, T data from Baptiste et al. (2012).
(1) Mean P wave velocity.
(2) Mean S wave velocity.
AV_P: V_P azimuthal anisotropy; AV_S: V_S azimuthal anisotropy; AVS_1: fast V_S azimuthal anisotropy; AVS_2: slow V_S azimuthal anisotropy; maximum values.
* Seismic properties and density were calculated by only taking into account the sample equilibrium pressure and temperature, since olivine Mg# was unknown.
** Seismic properties and density were calculated by only taking into account the mean olivine Mg# of the sample, since the equilibrium conditions were unknown.

olivine elastic tensors were recalculated to account for the actual olivine mean forsterite content or Mg# (Mg + Fe / Mg) in each sample. The elastic constants variation as a function of the olivine Mg# was calculated by linear interpolation between the single-crystal elastic constant tensors measured in laboratory for fayalite, forsterite, and olivine with Mg# of 90, 91, and 93 (Kumazawa and Anderson, 1969; Suzuki et al., 1983; Webb et al., 1989; Isaak et al., 1993; Abramson et al., 1997). For orthopyroxene, clinopyroxene, and garnet, the single crystal elastic tensors of Chai et al. (1997a), Collins and Brown (1998), and Chai et al. (1997b) at ambient pressure (P) and temperature (T) were used. Due to lack of experimental data, compositional changes could not considered for these phases. The density, the elastic constants and resulting seismic properties of each sample were calculated both at ambient pressure and temperature conditions and at its equilibration pressure and temperature. In both cases, the actual olivine Mg# is considered. To account for the effect of pressure and temperature on the seismic properties, the elastic tensors and density of olivine, pyroxenes, and garnet were recalculated for depths between 40 and 200 km (a point at every 10 km depth) using the AnisPT8 software by D. Mainprice, based on published pressure and temperature derivatives of the elastic constants for all major mineral phases (Abramson et al., 1997; Chai et al., 1997a, b; Collins and Brown, 1998) and the Kaapvaal geotherm of Baptiste et al. (2012). A Voigt–Reuss–Hill averaging was applied in all calculations. Equivalent "isotropic" seismic velocities were also estimated for each sample by averaging *P* and *S* wave velocities over all directions. Calculated seismic properties and elastic constants of all samples are summarized in Tables 1 and 2, respectively. The equilibration temperature and pressure of the samples were estimated using traditional geothermometers and geobarometers for mantle rocks, which have an uncertainty of up to ±60° C and ±0.2 GPa, respectively (see Baptiste et al. 2012 for more details). This uncertainty in the determination of the equilibration temperature and pressure results in an uncertainty in V_p and V_s of ±0.04 and 0.03 km s^{-1}, respectively. In the present study, the seismic properties as a function of depth were not calculated using the elastic tensors of the different minerals calculated at the exact equilibration depth of the samples, but at every 10 km interval along the Kaapvaal geotherm of Baptiste et al. (2012). This simplification corresponds to a maximum pressure variation of 0.15 GPa, which is lower than the geobarometer uncertainty. It does not, by consequence, increase the uncertainty of the estimated seismic velocities. A probable additional source of uncertainty in the seismic velocities estimations at depth is the one related with the uncertainties in temperature and pressure derivatives of the elastic constants. However, the existing experimental data does not allow constraining this uncertainty. The three-dimensional distribution of seismic velocities and birefringence of *S* waves relative to the samples' structural reference frame are displayed on lower hemisphere stereograms.

The mean seismic properties of six 20 km-thick sections between 70 and 190 km depth were calculated by averaging the olivine composition and modal contents and adding up the crystal-preferred orientations of all samples equilibrated within the corresponding depth interval. This approach results in an upper bound for the estimated anisotropy, since the CPO summation is performed considering that the structural reference frames of all samples are parallel, that is, that the orientation of the foliation and lineation is the same for all samples within each depth interval.

To analyze the relations between the olivine CPO symmetry and intensity and the seismic anisotropy, the olivine CPO symmetry was characterized using the dimensionless BA index defined as

$$\mathrm{BAindex} = \frac{1}{2}\left(2-\left(\frac{\mathrm{P010}}{\mathrm{G010+P010}}\right)-\left(\frac{\mathrm{G100}}{\mathrm{G100+P100}}\right)\right), \quad (1)$$

where P and G are indexes used to characterize the shape of the distribution (P for Point and G for Girdle) of the olivine principal axes ([100], [010] and [001]); these indexes were calculated from the eigenvalues of the normalized orientation matrix as $\mathrm{P} = \lambda 1 - \lambda 3$ and $\mathrm{G} = 2(\lambda 2 - \lambda 3)$, calculated using the MTEX texture analysis toolbox (Hielscher and Schaeben, 2008; Mainprice et al., 2011). This index allows the classification of the olivine CPO into 3 types: axial [010] (BA < 0.4), orthorhombic (0.4 < BA < 0.6), and axial [100] (BA > 0.6).

The olivine CPO intensity is characterized by the dimensionless *J* index, which is the volume-averaged integral of the squared orientation densities defined as

$$J = \mathrm{f(g)}^2\,\mathrm{d}g, \quad (2)$$

where $f(g)$ is the orientation distribution function (ODF) and $\mathrm{d}g = \mathrm{d}\varphi_1 \mathrm{d}\varphi \mathrm{d}\varphi_2 \sin\varphi\pi^2 \cdot \varphi_1$, φ, and φ_2 are the Euler angles that define the rotations allowing for coincidence between the crystallographic and external reference frames. In natural peridotites, olivine CPO *J* indexes vary between 2 and 20, with a peak at 8–10 (Ben Ismaïl and Mainprice, 1998; Tommasi et al., 2000). *J* indexes were calculated based on the mean orientation of each grain, using the SuperJctf program by D. Mainprice (ftp://www.gm.univ-montp2.fr/mainprice//CareWare_Unicef_Programs/) with a 10° Gaussian half-width, 1° cells, and truncation of the ODF at 22°.

4 Results

4.1 Seismic anisotropy

Seismic anisotropy patterns for samples representative of the different microstructures and olivine CPO symmetry patterns observed among Kaapvaal xenoliths (Table 1 and Baptiste et al., 2012) are illustrated in Fig. 2. At first order, seismic anisotropy patterns for all samples have similar characteristics, independently of the peridotite microstructure and olivine CPO symmetry. The fastest and slowest *P* wave and slow *S* wave (S_2) propagation directions are aligned with the olivine [100] and [010] axes maxima, respectively. The orientation of the [100] maximum also controls the direction of polarization of the fast *S* wave (S_1). Moreover, when the lineation (crystals elongation marking the structural *X* direction) and foliation (crystals flattening marking the plane normal to the *Z* structural direction) could be identified, the [100] maximum corresponds to the lineation, which marks the fossil flow direction. The fast split shear wave (S_1) is thus polarized in the plane containing the wave propagation direction and the fossil flow direction in the mantle lithosphere. The highest V_p/V_{S_1} ratio is also parallel to this direction and the lowest V_p/VS_1 ratio is observed for waves propagating normal to it.

Changes in olivine CPO symmetry result in second order variations in the seismic anisotropy pattern. Peridotites displaying an axial [100] CPO tend to show slow *P* wave velocities for all directions normal to the lineation, instead of a clear V_p minimum at high angle to the foliation. *S* wave polarization anisotropy patterns are also sensitive to the CPO symmetry. Samples with an axial [100] olivine CPO display high V_s anisotropy (AV_s) for directions of propagation in the foliation plane, but at high angle to the lineation (*Y* structural direction) and a minimum anisotropy for propagation along

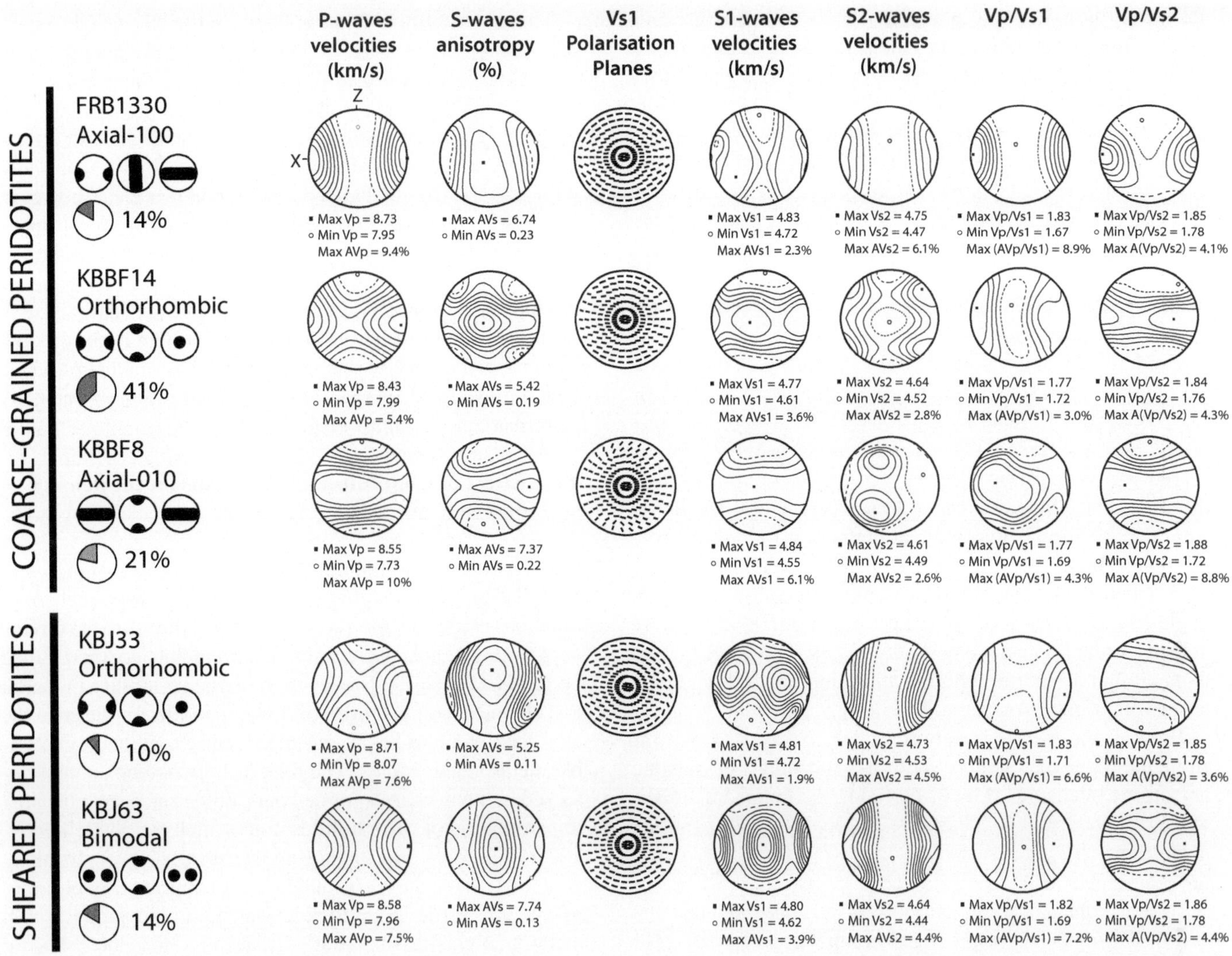

Fig. 2. Calculated seismic properties of the Kaapvaal coarse-grained and sheared peridotites showing different olivine CPO patterns. From left to right, are displayed the 3-D distributions of P wave velocities (V_p), S wave polarization anisotropy (V_s), the orientation of fast shear wave polarization plane, S_1 wave velocities (V_{S_1}), S_2 wave velocities (V_{S_2}), V_p/V_{S_1}, and V_p/V_{S_2}. Black squares and white spots indicate maximum and minimum values, respectively.

the lineation (X direction) and more rarely at 45° to the lineation in the XZ plane. The slowest S_1 wave propagation is always located at high angle to the foliation (parallel to the Z structural direction), while the fastest propagation is in the foliation plane, at 45° to the lineation. S_2 wave velocity is maximum close to the lineation or in the XZ plane, at $\sim 10°$ to the lineation. The V_p/V_{S_1} ratio is highest parallel to the lineation (X structural direction) and lowest at high angle to the foliation (Z structural direction).

In contrast, peridotites with an axial [010] CPO show high S wave polarization anisotropy for propagations along the foliation plane and low anisotropy for S waves traveling at high angle to the foliation. The strongest S wave birefringence is usually observed for waves propagating at low angle to the lineation, but it may also be oblique to it. S_1 wave velocity is higher in the foliation plane. The fastest S_1 waves may be aligned with either the X or the Y structural directions or even at 45° to these directions. This variability in the orientation of the maximum S wave birefringence and of the fastest S_1 waves depends on the actual symmetry of the olivine CPO; samples with lower BA indexes have both maxima at low angle to the X structural direction. S_2 wave velocity is minimum along the XZ plane. The V_p/V_{S_1} ratio tends to show high values in the foliation plane and low values at high angle to it (Z structural direction).

Peridotites with an orthorhombic CPO show patterns intermediate between the two described previously. The maximum S wave polarization anisotropy is at low angle to the Y structural direction, similarly to that observed for samples with an axial [100] CPO, but, the minimum S wave anisotropy is in the XZ plane, at 45° to both the lineation (X structural direction) and to the foliation (Z structural direction). The S_1 wave velocity is most often the highest along the foliation plane, parallel to both X and Y structural directions, or at 45° to these axes. The slowest S_1 wave propagation is at high angle to the foliation (Z structural direction). The S_2 wave velocity is maximum parallel to the lineation or at 45° to the lineation in the XZ plane, as in the example displayed in Fig. 2. It is minimum along the YZ plane. The maximum V_p/V_{S_1} ratio is observed along the foliation plane and the minimum V_p/V_{S_1} ratio is located close to the Z structural direction.

Peridotites with bimodal olivine CPO have also intermediate patterns, with maximum S wave polarization anisotropy for propagation directions close to the Y structural direction. The minimum S wave polarization anisotropy is observed for propagation directions at 45° to the lineation in the XZ plane. In a few cases, however, the S wave polarization anisotropy is minimum at high angle to the foliation. S_1 wave velocity is the highest in the foliation plane, close to both X and Y structural directions and it is the lowest at high angle to the foliation (Z structural direction). The S_2 wave velocity is the highest along the YZ plane. The V_p/V_{S_1} ratio is maximum along the foliation and minimum at high angle to the foliation (Z structural direction).

Despite the weak variation in the patterns, xenoliths from the Kaapvaal craton show a wide range of seismic anisotropy intensities (Fig. 3, Table 1). Maximum anisotropy values range between 2.5 and 10.2 % for P wave azimuthal anisotropy (AV_p), between 2.7 and 8 % for S wave polarization anisotropy (AV_s), between 0.8 and 6.1 % for S_1 wave propagation anisotropy (AVS_1), and between 0.5 and 6.4 % for S_2 wave propagation anisotropy (AV_{S2}). All anisotropies show a positive covariance with the olivine CPO strength (Fig. 3a, b); samples with stronger olivine CPOs are more anisotropic. Yet, this variation is not linear, as observed in previous studies (Ben Ismaïl and Mainprice, 1998): anisotropy increases fast with increasing CPO strength up for low J indexes (≤ 4), but slow afterwards, tending to stabilize for J indexes >8. Coarse-grained peridotites show more variable olivine CPO intensities (J indexes range between 2–11). However, their maximum seismic anisotropies are in the same range as those displayed by the sheared peridotites (Fig. 3a, b). These variations suggest that the modal composition has also an important effect on the seismic anisotropy of these samples.

The presence of pyroxenes and garnet is known to dilute the bulk anisotropy of the rock (Mainprice and Silver, 1993; Mainprice et al., 2000). Samples with high modal olivine, and therefore lower pyroxene and/or garnet contents, tend to have higher maximum AV_p, AVS_1 and AV_{S2}. However, the dependence is weaker than the one on the olivine CPO strength (compare Fig. 3c and a). The dependence on the olivine content is even less marked for the S wave polarization anisotropy (Fig. 3d).

There is no simple relation between the olivine CPO symmetry, characterized by the BA index, and the maximum P, S, S_1, and S_2 wave anisotropies (Fig. 3e, f). Samples with an olivine axial [100] CPO tend to show higher maximum P and S_2 propagation anisotropies than those with an orthorhombic CPO, which are themselves more anisotropic than those with an axial [010] CPO. The relation is less clear, however, for S wave polarization and S_1 propagation anisotropies. Moreover, this relation probably reflects more a variation in the olivine CPO intensity then a strong dependence on the olivine CPO symmetry. In the studied data set, samples with axial [100] tend to have stronger olivine CPO, but this behavior is not a general one in naturally deformed peridotites, which may also display very strong axial [010] olivine CPO patterns (e.g., Tommasi et al., 2008).

4.2 Effect of compositional variations on the density and isotropic seismic velocities

Kaapvaal mantle xenoliths record a long and complex history. Early partial melting and extraction of high melt fractions are evidenced by the refractory compositions (olivine Mg# > 92), which are common among the coarse-grained peridotites (Table 1). However, the large variability in modal and chemical compositions of these xenoliths implies that the craton mantle root has been subsequently modified by multiple metasomatic events. The high orthopyroxene content observed in many coarse-grained peridotites has been attributed to percolation of Si-rich fluids and/or melts (e.g., Bell et al., 2005; Wasch et al., 2009; Baptiste et al., 2012). In many xenoliths, secondary crystallization of phlogopite, clinopyroxene and garnet point to reactions with percolating K-rich melts at high pressure (e.g., Grégoire et al., 2003; Griffin et al., 2003; Bell et al., 2005; Baptiste et al., 2012). These metasomatic events may have also allowed for rehydration of the Kaapvaal mantle root above 150 km depth (Baptiste et al., 2012). Finally, local high-stress deformation, which was probably shortly predated by a late metasomatic event by K-rich fluids and/or melts (Wasch et al., 2009), produced the sheared peridotites (Baptiste et al., 2012).

This compositional variability resulted in significant variations in both density and seismic velocities as illustrated in Fig. 4, where densities and isotropic P and S wave velocities at ambient conditions are displayed as a function of the olivine Mg#, the orthopyroxene, and the garnet content of each sample. The density of Kaapvaal samples varies by up to 3 % (Fig. 4a e, i). It decreases with increasing olivine Mg# and orthopyroxene content, but strongly increases with increasing garnet content. The linear correlation with the garnet content is particularly striking. This later effect is dominant in the sheared peridotites, where the garnet content may attain up to 18 %.

P wave velocities vary by up to 2.2 % with the peridotites composition (Fig. 4b, f, j). There is a weak positive covariance between P wave velocities and garnet content. P wave

Table 2. Elastic constants in megabars (Mbar) calculated using Voigt–Reuss–Hill averaging and the elastic constants of enstatite (Chai et al., 1997a), diopside (Collins and Brown, 1998), garnet (Chai et al., 1997b), and olivine elastic tensors recalculated to account for the actual olivine Mg# in each sample, based on the single-crystal elastic constant tensors for fayalite, forsterite, and olivine with Mg# of 90, 91 and 93 (Kumazawa and Anderson, 1969; Suzuki et al., 1983; Webb et al., 1989; Isaak et al., 1993; Abramson et al., 1997).

Locality	Sample	C11	C22	C33	C44	C55	C66	C12	C13	C14	C15	C16	C23	C24	C25	C26	C34	C35	C36	C45	C46	C56
KIMBERLEY	KBBF6a	2.1600	2.6174	2.2429	0.7689	0.6740	0.7329	0.8049	0.8439	0.0014	0.0056	0.0159	0.8139	0.0077	0.0012	0.0161	0.0223	0.0120	0.0085	0.0093	−0.0051	0.0025
	KBBF6b	2.1614	2.5764	2.2501	0.7556	0.6664	0.7421	0.8135	0.8370	−0.0054	−0.0156	0.0171	0.8165	−0.0169	−0.0058	0.0196	−0.0034	−0.0020	0.0029	0.0014	−0.0034	−0.0060
	KBBF8	2.0052	2.3564	2.4182	0.7834	0.6926	0.6801	0.7947	0.7908	0.0040	0.0156	−0.0021	0.8264	−0.0262	−0.0027	−0.0015	−0.0221	0.0196	−0.0003	0.0005	0.0002	−0.0073
	KBBF9	2.1736	2.4113	2.3372	0.7573	0.7079	0.7222	0.8452	0.8475	0.0018	0.0148	0.0131	0.8649	0.0037	−0.0011	0.0113	0.0142	0.0134	-0.0016	0.0017	0.0014	0.0106
	KBBF10	2.1134	2.4711	2.2052	0.7482	0.6726	0.7237	0.7736	0.7940	−0.0017	−0.0021	0.0095	0.7817	0.0137	−0.0029	0.0098	0.0062	−0.0029	−0.0009	0.0085	−0.0045	0.0041
	KBBF11	2.1584	2.3682	2.2252	0.7137	0.6704	0.7026	0.8370	0.8492	0.0017	0.0022	0.0063	0.8611	0.0108	−0.0002	0.0021	0.0081	−0.0014	0.0031	0.0015	−0.0023	0.0055
	KBBF14	2.1386	2.3764	2.6650	0.7599	0.6847	0.7116	0.8074	0.8050	0.0022	0.0039	0.0031	0.8123	0.0071	0.0015	0.0078	0.0184	0.0086	−0.0011	0.0014	0.0048	0.0007
	KBBF15	2.1818	2.3652	2.2124	0.7426	0.6903	0.7403	0.8056	0.8075	−0.0132	0.0132	0.0107	0.8058	0.0366	−0.0003	0.0216	0.0384	0.0086	−0.0018	0.0082	0.0064	0.0123
	KBBF16	2.2368	2.3379	2.1751	0.7433	0.7055	0.7431	0.8045	0.8074	0.0069	−0.0122	−0.0190	0.8099	0.0041	0.0064	−0.0181	0.0121	−0.0032	−0.0052	−0.0141	−0.0067	0.0088
	KBBF18	2.1390	2.4690	2.2471	0.7535	0.6870	0.7203	0.7915	0.7984	0.0016	0.0000	−0.0016	0.7852	0.0088	−0.0005	0.0082	−0.0106	0.0001	−0.0001	0.0035	−0.0058	−0.0033
	KBBF20	*2.1224	*2.345	*2.1857	*0.7783	*0.705	*0.7222	*0.7523	*0.7805	*−0.0149	*−0.0033	*0.0159	*0.8011	*0.0786	*0.0118	*−0.0034	*0.0558	*0.0001	*−0.0015	*−0.0004	*−0.0047	*0.0383
	FRB1402	2.2061	2.3580	2.3435	0.7481	0.7164	0.7143	0.8438	0.8511	0.0006	−0.0300	0.0054	0.8454	−0.0018	−0.0036	0.0101	−0.0035	−0.0343	0.0014	0.0004	−0.0119	0.0008
	FRB1404	2.1869	2.3028	2.2378	0.7427	0.7145	0.7375	0.8114	0.7987	−0.0035	0.0001	−0.0175	0.7990	−0.0050	−0.0008	−0.0227	0.0118	−0.0066	0.0046	−0.0170	−0.0052	0.0011
	FRB1422	2.3020	2.1990	2.2797	0.7157	0.7279	0.7266	0.8119	0.7965	0.0052	0.0247	0.0476	0.7985	−0.0138	0.0052	0.0436	−0.0100	0.0027	−0.0066	0.0125	0.0113	0.0080
	FRB1423	2.1669	2.4982	2.3049	0.7677	0.6995	0.7445	0.8137	0.8346	−0.0053	−0.0015	0.0073	0.8259	−0.0154	0.0053	0.0070	−0.0169	−0.0026	0.0002	0.0074	−0.0062	−0.0077
	FRB1447	2.2415	2.3276	2.2290	0.7341	0.7024	0.7401	0.8055	0.8075	0.0019	0.0189	0.0064	0.8056	0.0197	0.0034	0.0024	0.0206	0.0117	−0.0047	0.0013	0.0030	0.0096
	FRB348	2.3286	2.3646	2.1892	0.7347	0.7297	0.7578	0.8134	0.8002	−0.0027	0.0033	0.0421	0.8119	0.0291	−0.0083	0.0346	0.0268	−0.0062	0.0003	0.1120	0.0065	0.0195
JAGERSFONTEIN	KBJ6	2.2117	2.3819	2.3190	0.7547	0.6901	0.7053	0.8441	0.8364	−0.0029	−0.0033	0.0068	0.8614	−0.0092	−0.0002	0.0057	−0.0190	−0.0063	0.0057	−0.0046	0.0026	−0.0081
	KBJ8	2.1437	2.5764	2.1707	0.7389	0.6493	0.7355	0.8320	0.8593	0.0008	−0.0163	0.0267	0.8386	0.0163	−0.0106	0.0458	0.0111	−0.0170	−0.0025	0.0134	0.0023	0.0063
	KBJ14	2.2506	2.4998	2.3425	0.7652	0.7101	0.7546	0.8722	0.8754	−0.0026	−0.0147	0.0115	0.8805	0.0071	−0.0025	0.0252	0.0048	−0.0145	−0.0014	0.0097	−0.0041	0.0023
	KBJ30	2.2488	2.4582	2.3794	0.7645	0.7093	0.7313	0.8902	0.8869	−0.0035	−0.0073	0.0065	0.8980	0.0015	−0.0043	0.0035	−0.0037	−0.0091	0.0021	0.0019	−0.0036	−0.0019
	KBJ33	2.2248	2.5585	2.3411	0.7687	0.6996	0.7582	0.8629	0.8864	−0.0043	−0.0155	0.0232	0.8917	0.0233	−0.0077	0.0316	0.0130	−0.0179	−0.0010	0.0107	−0.0034	0.0097
	KBJ52	2.2035	2.4416	2.1234	0.7254	0.6871	0.7387	0.7953	0.7927	−0.0010	−0.0212	0.0079	0.7794	0.0118	−0.0097	−0.0143	0.0006	−0.0161	0.0058	0.0000	−0.0208	0.0015
	KBJ54	2.2153	2.2867	2.4037	0.7625	0.7020	0.6838	0.7860	0.7916	0.0009	−0.0192	−0.0310	0.7927	0.0102	−0.0075	−0.0489	0.0442	−0.0174	−0.0112	−0.0221	−0.0137	0.0137
	KBJ56	2.1418	2.3781	2.3544	0.7536	0.7152	0.7349	0.7922	0.8008	−0.0049	−0.0040	−0.0275	0.7929	−0.0061	0.0041	−0.0509	0.0057	−0.0090	0.0138	−0.0006	0.0017	−0.0021
	KBJ60	2.0658	2.3988	2.2948	0.7523	0.6646	0.6907	0.8171	0.8336	0.0013	0.0039	0.0088	0.8512	0.0038	−0.0005	0.0078	−0.0018	−0.0015	−0.0005	−0.0036	−0.0016	0.0030
	KBJ62	2.1179	2.4316	2.1733	0.7726	0.6807	0.7351	0.7584	0.7640	−0.0004	0.0002	0.0022	0.7763	−0.0364	−0.0035	−0.0210	−0.0027	−0.0119	0.0242	0.0064	−0.0182	0.0064
	KBJ63	2.1334	2.4778	2.2230	0.7773	0.6667	0.7192	0.8461	0.8455	0.0009	−0.0028	−0.0023	0.8786	0.0027	−0.0027	−0.0058	−0.0035	0.0008	−0.0006	−0.0039	0.0019	−0.0003
	J34	2.1540	2.5660	2.1857	0.7884	0.6897	0.7790	0.7848	0.7536	−0.0031	0.0206	−0.0073	0.7843	0.0087	0.0102	−0.0165	−0.0015	0.0214	0.0155	−0.0027	0.0061	0.0042
	J41	2.1366	2.2972	2.2589	0.7585	0.7003	0.7067	0.7728	0.7694	−0.0009	0.0057	0.0082	0.7780	0.0153	−0.0045	0.0012	−0.0179	−0.0023	−0.0053	−0.0083	−0.0013	0.0013
	J47	2.0888	2.4032	2.3554	0.7546	0.6832	0.6981	0.8293	0.8381	0.0013	0.0078	0.0090	0.8698	−0.0066	0.0050	0.0071	0.0151	0.0097	0.0070	0.0045	0.0044	0.0033
	J57	2.2094	2.4552	2.1555	0.7644	0.6801	0.7316	0.7790	0.7931	−0.0047	0.0242	−0.0098	0.7893	0.0281	0.0081	−0.0044	0.0180	0.0210	0.0053	-0.0027	0.0096	−0.0009
	J63	2.1674	2.3572	2.2985	0.7324	0.6919	0.7018	0.8557	0.8508	0.0028	0.0077	0.0073	0.8719	−0.0123	0.0001	0.0047	−0.0180	0.0069	−0.0007	0.0001	0.0048	−0.0088
MONASTERY	ROM23	2.1676	2.3192	2.2399	0.7534	0.7260	0.7418	0.7872	0.7800	0.0017	−0.0391	0.0188	0.7975	−0.0075	−0.0016	0.0030	−0.0303	−0.0236	−0.0004	0.0099	−0.0185	−0.0103
	ROM69	2.2753	2.2929	2.2192	0.7600	0.7223	0.7367	0.7644	0.7835	−0.0031	−0.0066	0.0048	0.7937	0.0035	−0.0124	0.0077	−0.0113	-0.0215	0.0036	0.0067	−0.0027	0.0010
LENTSENG	PHN4274	2.1888	2.3263	2.2424	0.7428	0.7190	0.7144	0.7991	0.8029	0.0159	−0.0095	0.0013	0.8002	−0.0199	0.0046	−0.0078	−0.0126	−0.0111	−0.0047	−0.0093	0.0020	0.0018
MOTHAE	PHN1925	2.2229	2.3552	2.2732	0.7414	0.7096	0.7357	0.8226	0.8247	0.0003	−0.0066	−0.0004	0.8326	0.0183	0.0002	-0.0083	0.0241	−0.0057	0.0025	−0.0035	−0.0045	0.0109
KAMFERSDAM	PHN5580	**2.1736	**2.373	**2.2914	**0.8095	**0.7614	**0.7901	**0.7172	**0.7128	**−0.0114	**−0.0046	**−0.0152	**0.7163	**−0.032	**−0.0143	**−0.0482	**−0.0107	**0.0071	**0.0256	**0.0069	**−0.0025	**−0.0085
DE BEERS	KBB2	2.1587	2.4200	2.1824	0.7440	0.6834	0.7392	0.7736	0.7905	−0.0042	0.0159	−0.0089	0.7725	0.0078	0.0013	−0.0088	−0.0025	0.0332	0.0139	0.0029	0.0004	−0.0014
PREMIER	FRB1330	2.1317	2.5519	2.1425	0.7644	0.6702	0.7469	0.7729	0.7992	0.0000	0.0118	−0.0079	0.7807	−0.0113	0.0042	−0.0089	−0.0004	0.0096	−0.0014	−0.0038	0.0043	−0.0007
	FRB1336	**2.2550	**2.1943	**2.3222	**0.8001	**0.7928	**0.7580	**0.7155	**0.7224	**0.0034	**0.5060	**−0.0450	**0.7226	**0.0491	**−0.0018	**−0.0439	**0.5300	**0.0431	**−0.0083	**−0.0238	**0.0174	**0.0145
	FRB1339	**2.2486	**2.3529	**2.3271	**0.7881	**0.7751	**0.7731	**0.7287	**0.7385	**−0.0001	**−0.0510	**0.0225	**0.7201	**−0.0199	**0.0122	**0.0136	**0.0031	**−0.0546	**0.0074	**0.0074	**−0.021	**−0.0110
	PHN5266	**2.2466	**2.2858	**2.2703	**0.7922	**0.7830	**0.7936	**0.7158	**0.7234	**0.0008	**0.0552	**−0.0103	**0.7152	**−0.0564	**0.0136	**−0.0104	**−0.0565	**0.0743	**−0.0190	**−0.0308	**0.0413	**−0.0435
	PHN5267	2.2229	2.4937	2.2670	0.7534	0.6895	0.7366	0.8408	0.8572	−0.0078	−0.0324	−0.0068	0.8623	0.0021	−0.0138	−0.0110	0.0076	−0.0234	−0.0131	−0.0111	−0.0134	−0.0067
FINSCH	FRB1501	2.1482	2.4002	2.3306	0.7620	0.6975	0.7185	0.8152	0.8274	0.0056	0.0036	0.0002	0.8403	0.0012	0.0025	−0.0028	−0.0076	0.0036	−0.0067	−0.0033	−0.0004	−0.0007
	FRB1512	2.1689	2.3907	2.3505	0.7573	0.7040	0.7194	0.8214	0.8350	−0.0033	−0.0124	0.0012	0.8417	−0.0001	−0.0122	0.0050	0.0042	−0.0250	−0.0024	−0.0016	−0.0056	−0.0014
	FRB1513	2.1300	2.3269	2.3105	0.7450	0.6878	0.6906	0.8008	0.8197	−0.0035	0.0182	0.0107	0.8198	0.0117	0.0038	0.0029	−0.0214	0.0035	0.0083	0.0037	0.0005	−0.0044

* Seismic properties were calculated by only taking into account the sample equilibrium pressure and temperature, since olivine Mg# was unknown.
** Seismic properties were calculated by only taking into account the mean olivine Mg# of the sample, since the equilibrium conditions were unknown.

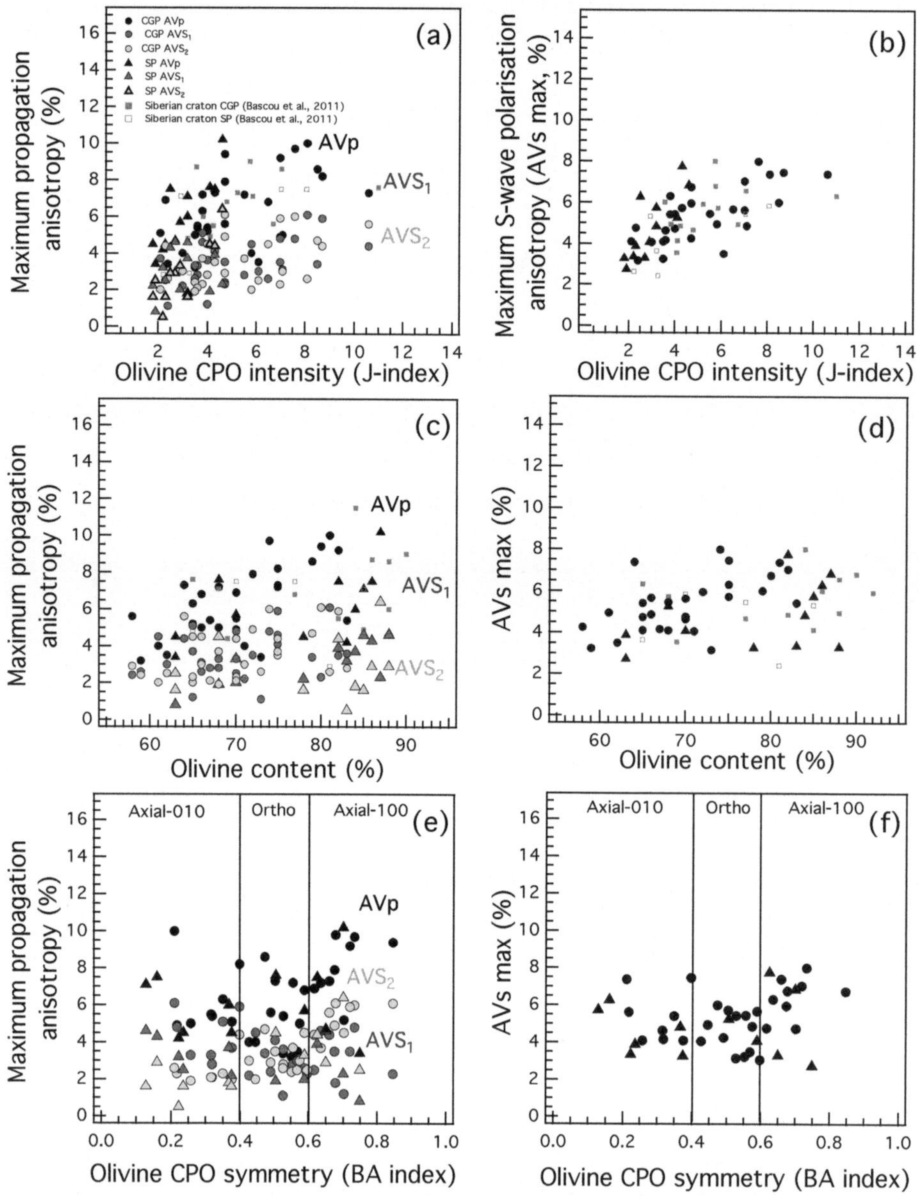

Fig. 3. Dependence of the maximum P, S_1 and S_2 wave anisotropies (AV_p, AVS_1 and AV_{S_2}), and of the maximum S wave polarization anisotropy (AV_s) on the olivine CPO strength (characterized by the J index), the olivine content and the BA index.

velocities in coarse-grained peridotites also tend to increase with increasing olivine Mg# (Fig. 4b) and decreasing orthopyroxene content (Fig. 4f), but the competing effects of olivine Mg# and orthopyroxene content results in dispersion of P wave velocities of the coarse-grained peridotites in all diagrams. In contrast, P wave velocities in sheared peridotites increase strongly with olivine Mg#, probably because the latter have lower orthopyroxene contents.

The isotropic S wave velocities of the Kaapvaal mantle xenoliths show a clear linear correlation with olivine Mg# (Fig. 4c), varying by up to 2.4 %. This variation is consistent with previous data showing that an increase in the olivine Mg# in a dunite results in a linear increase in V_s (Tommasi et al., 2004). No clear co-variation is observed between the orthopyroxene and garnet contents and V_s in neither coarse-grained nor sheared peridotites, although S wave velocities are on average higher in the coarse-grained peridotites, which have higher orthopyroxene contents (Fig. 4g and k).

The V_p/V_s ratio is also sensitive to compositional changes (Fig. 4d, h, l), varying by up to 2.3 %. It decreases with increasing olivine Mg# and orthopyroxene content and increases with garnet content. The V_p/V_s ratio is very sensitive to the orthopyroxene content, as indicated by the strong linear correlation between the two parameters (Fig. 4h).

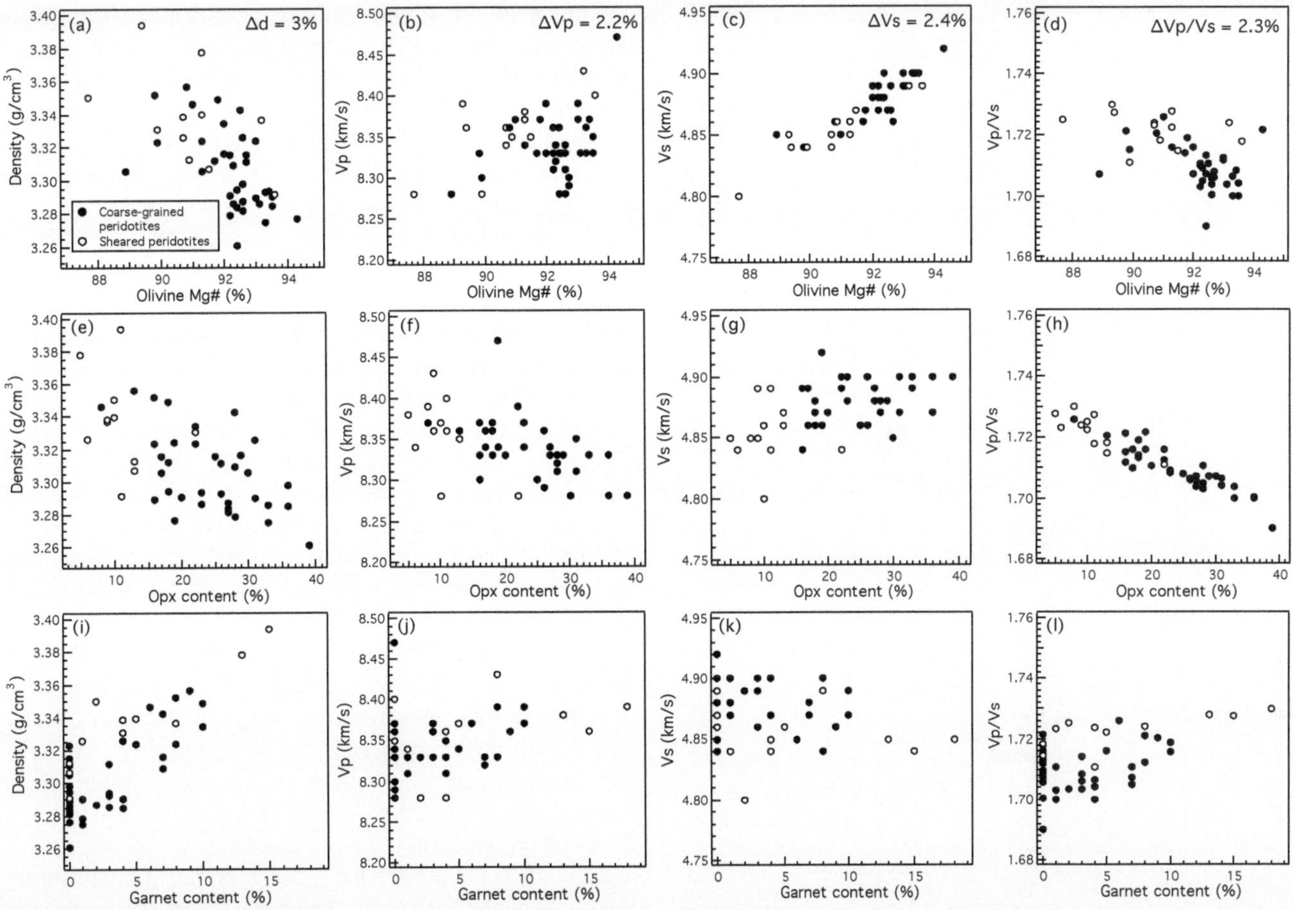

Fig. 4. Evolution of the samples' densities, P wave velocities (V_P), S wave velocities (V_S), and V_P/V_S ratios with olivine Mg#, orthopyroxene (opx) content, and garnet contents.

Sheared peridotites have a higher and less variable V_P/V_S ratio than coarse-grained peridotites, probably due to their lower modal orthopyroxene.

4.3 Evolution of seismic properties and density with depth

The variations of the isotropic P and S wave velocities (V_P, V_S), V_P/V_S ratio, and density with depth in the present Kaapvaal xenolith suite and in the one studied by James et al. (2004) are plotted in Fig. 5. The isotropic P wave velocities calculated for the equilibration pressure and temperature and for the composition of our samples do not show a monotonic increase with depth. They are highly heterogeneous, being comprised between 8.17 and 8.36 km s^{-1} over the entire depth section. This result contrasts with the one from James et al. (2004), who observed a regular increase in V_P up to 180 km depth, though with some dispersion in the depth interval of 120–150 km, and a marked decrease below this depth. One may argue that the present data set does not sample well the deepest part of the root (below 180 km), but at shallower depths the peridotites analyzed in this study tend to display, on average, higher and more variable P wave velocities than those reported by James et al. (2004). In contrast, the isotropic S wave velocities decrease and the V_P/V_S ratio increases almost linearly with depth, in excellent agreement with the results of James et al. (2004) for depths above 180 km. The dispersion of calculated densities and P wave velocities at a given depth tends to be higher than for V_S or the V_P/V_S ratio, indicating a stronger dependence on the composition.

Compositions, olivine CPOs, and anisotropic seismic properties of average samples (obtained by summing the CPO and averaging the compositions of all samples in a 20 km-thick depth interval) for six depths between 70 and 190 km depth are presented in Fig. 6. With increasing depth, olivine and garnet contents increase, while the orthopyroxene content decreases. The variability of olivine Mg# increases as well, with a higher proportion of normal to Fe-rich olivines (Mg#≤90) in the deeper part of the cratonic root, and the olivine CPO patterns change from axial [100] to more orthorhombic. These variations do not result, however, in marked changes in seismic anisotropy with depth.

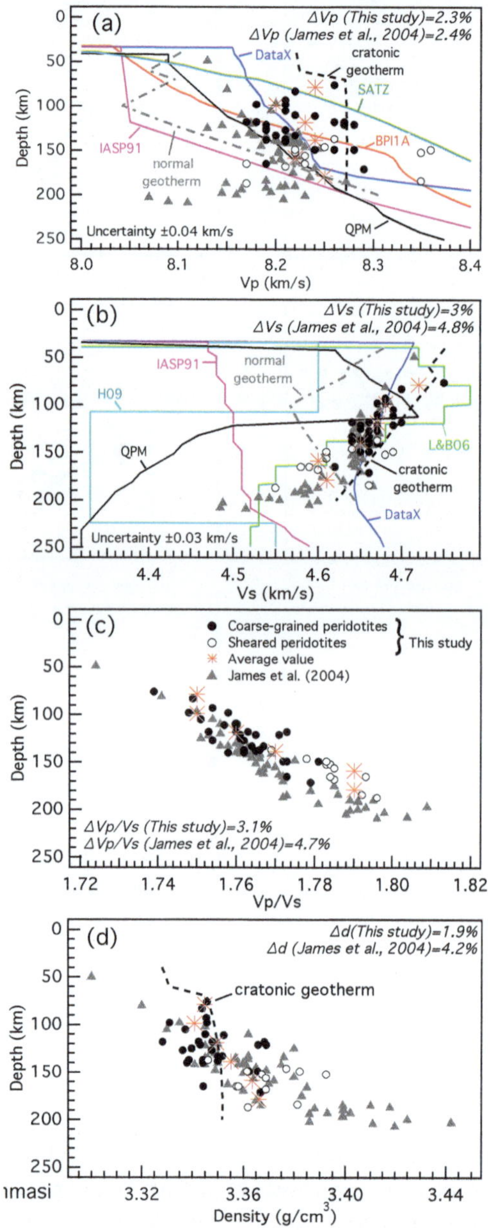

Fig. 5. Variation with depth of the isotropic seismic velocities and density of the studied peridotite xenoliths: **(a)** V_p, **(b)** V_s, **(c)** V_p/V_s ratio, and **(d)** density. For comparison, the density and seismic properties of a peridotite with a standard composition (between 40 and 70 km: 65 % olivine with Mg#90, 23 % orthopyroxene, 12 % clinopyroxene; between 70 and 200 km: 65 % olivine with Mg#90, 20 % orthopyroxene, 10 % clinopyroxene, 5 % garnet) for both a "cratonic geotherm" as estimated by Baptiste et al. (2012) and a "normal geotherm" producing a 100 km-thick lithosphere are shown as black and gray dashed lines, respectively. The IASP91 reference model and different seismic models for southern Africa (BPI1A: Simon et al., 2002; DataX: Larson et al., 2006; H06: Hansen et al., 2009; L&B06: Li and Burke, 2006; QPM: Qiu et al., 1996; SATZ: Zhao et al., 1999) are also shown (colored lines). Uncertainties in the seismic velocity estimates due to the uncertainty in the determination of the equilibration pressure and temperature of the xenoliths are shown in **(a)** and **(b)**.

The maximum P wave anisotropy varies between 4.5 and 7.9 %. The fastest P wave propagation direction is always aligned with the olivine [100] axes maximum, that is, with the lineation. Between 70 and 110 km, P wave propagation is slow in all directions at high angle to the olivine [100] axes maximum. At greater depths, P wave propagation is the slowest parallel to the olivine [010] axes maximum.

The maximum S wave birefringence varies between 3.8 and 5.8 %. The fast split shear wave (S_1) is always polarized in a plane containing the propagation direction and the olivine [100] axes maximum. The maximum S wave polarization anisotropy is observed for waves propagating along the structural direction Y at all depths. The apparent isotropy direction, however, changes slightly with depth. It is always contained in the XZ plane (that is the plane that contains the lineation and the normal to the foliation), but it is very close to the lineation above 100 km depth, where axial [100] olivine CPO patterns dominate, and at 45° to the lineation in the deeper layers, where the olivine CPO has a dominantly pattern.

The propagation of S_1 waves is always fastest along the foliation plane and slowest at a high angle to it, close to the Z structural axis. However, the fastest propagation direction varies with depth. Between 70 and 90 km, it is parallel to the lineation and Y structural axis, while between 90 and 130 km, it is located at 45° to the lineation. At greater depths, S_1 wave propagation is fast in all directions within the foliation plane. In contrast, the S_2 wave propagation anisotropy does not change with depth. The fastest propagation direction of S_2 waves is always located on the XZ plane, at about 30° to the lineation, while the slowest velocities are found along the YZ plane. Finally, the V_p/V_{S_1} ratio is always maximum parallel to the alignment of olivine [100] axes and minimum along the YZ plane.

5 Discussion

5.1 Compositional effects on isotropic seismic properties

In the studied data set, P and S wave velocities, the V_p/V_s ratio, and density vary with sample composition (Fig. 4) by 2.2, 2.4, 2.3, and 3 %, respectively. Melt depletion tends to decrease the density and the V_p/V_s ratio, but to increase V_p and V_s. Metasomatism may result in refertilization, that is, enrichment in pyroxenes and/or garnet, depending on the melt composition and depth. Enrichment in orthopyroxene decreases the whole rock density and P wave velocities, but increases S wave velocities and, hence, strongly reduces the V_p/V_s ratio. Garnet enrichment tends to strongly increase density and slightly P wave velocities and the V_p/V_s ratio. However, it has little effect on S wave velocities.

The V_p/V_s ratio is more sensitive to orthopyroxene content than to the olivine Mg# or garnet content. High V_s

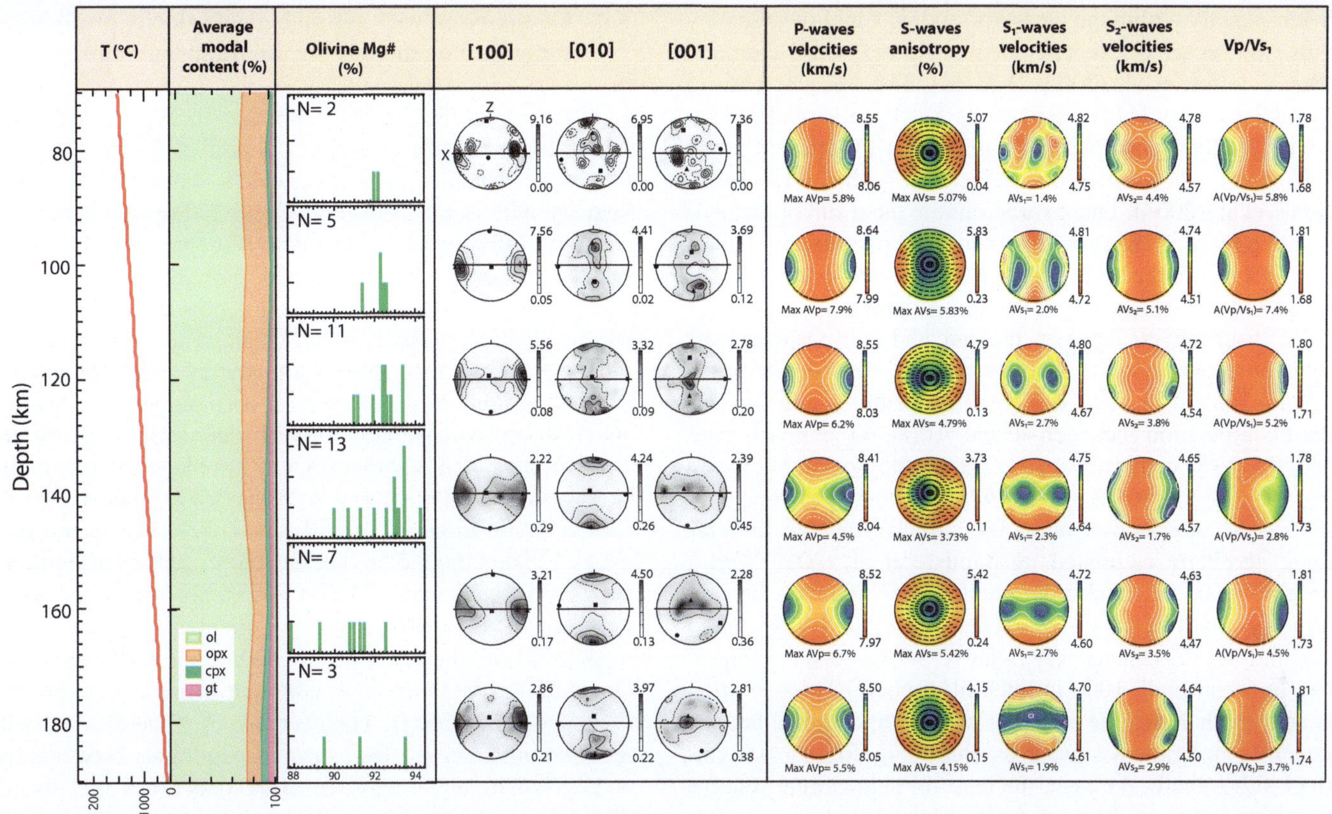

Fig. 6. Mean compositions, CPOs and seismic properties of five 20-km sections between 70 and 190 km depth calculated by averaging the olivine composition and modal contents of the samples originated from these depths. From left to right, are displayed: the geotherm of Baptiste et al. (2012), the average modal content of the samples, the number and the olivine Mg# of the samples in each depth section, the average olivine crystal-preferred orientations, and the 3-D distributions of P wave velocities (V_P), S wave polarization anisotropy (V_s), the orientation of the fast shear wave polarization plane, S_1 wave velocities (V_{S_1}), S_2 wave velocities (V_{S_2}), and V_p/V_{S_1} ratio.

($\sim 4.7\,\text{km}\,\text{s}^{-1}$) and low V_p/V_s ratios (1.65–1.70) observed locally above the flat slab in central Chile–Argentina were indeed interpreted by Wagner et al. (2008) as due to orthopyroxene enrichment. Yet, although extremely high orthopyroxene contents (> 30 %) are observed in the Kaapvaal mantle xenoliths, such an orthopyroxene enrichment can at most produce V_p/V_s ratios of 1.7. These results are consistent with V_p/V_s ratios in mantle xenoliths from subduction environments, in which reactive percolation of Si-rich fluids produced enrichment of orthopyroxene at the expenses of olivine in the supra-slab mantle (Soustelle and Tommasi, 2010). This led Soustelle and Tommasi (2010) to propose that the very low V_p/V_s ratios (1.65–1.70) observed locally in the mantle wedge do not result from Si enrichment in the mantle, but from the intrinsic anisotropy of the peridotites, representing regions in the mantle dominantly sampled by seismic waves propagating at a high angle to lineation. Analysis of the anisotropy of the V_p/V_{s1} ratio in the Kaapvaal peridotites (Figs. 2, 6) corroborates this conclusion.

Among the studied Kaapvaal peridotites, compositional changes result in density variations by up to 3 %, while V_p, V_s and the V_p/V_s ratio vary at most by 2.4 % (Fig. 4). These variation ranges are slightly higher than the ~ 2 % density and the 1.0–1.4 % V_s changes that Griffin et al. (2009) estimated to result from an increase of fertility in the mantle lithosphere by comparing data from domains where the crust has been stable since ≥2.5 Ga (Archons) and domains formed or modified at < 1 Ga (Tectons). They are, however, in agreement with the 2–3 % seismic velocity change due compositional variations estimated by Schutt and Lesher (2010) for the Kaapvaal mantle.

The P wave velocities calculated for the equilibration pressure and temperature of Kaapvaal xenoliths vary strongly at all depths (Fig. 5a). Between 90 and 150 km, the total variation of P wave velocity is of ~ 1.2 %, but reaches up to 2.3 % when the sheared peridotites, which were equilibrated at greater depths, are considered. Such lateral heterogeneity in P wave velocity may be explained by the compositional heterogeneity of the cratonic root, as discussed above. This interpretation is corroborated by the strong variability of density at all depths (Fig. 5d). A lower proportion of deep sheared peridotites and a stronger compositional heterogeneity within the presently studied xenolith suite may thus explain the discrepancy between our P wave data and that of James et al. (2004) (Fig. 5a). The 3 % decrease in S wave velocity, 3 % increase in the V_p/V_s ratio, and 2 % increase in density with depth observed in the present study (Fig. 5b, c, d), are in good agreement with the results of James et al. (2004). These variations are the result of the conjugate effects of changes in pressure, temperature, and composition within the cratonic mantle, and corroborate a higher sensitivity of S wave velocity to temperature.

To discriminate between thermal and compositional effects on P and S wave velocity profiles, we calculated the seismic properties and density of a sample with a standard composition (between 40 and 70 km: 65 % olivine with Mg#90, 23 % orthopyroxene, 12 % clinopyroxene; between 70 and 200 km: 65 % olivine with Mg#90, 20 % orthopyroxene, 10 % clinopyroxene, 5 % garnet), for both the cratonic geotherm estimated by Baptiste et al. (2012) and a "normal" 100 km-thick lithosphere (Fig. 5). The cooler cratonic geotherm results in an increase in V_p and V_s by up to 2.8 and 3.1 %, respectively, relative to the seismic velocities in a "normal" 100 km-thick lithosphere. This variation is smaller than the one resulting from compositional heterogeneity among the Kaapvaal xenoliths (Fig. 5). In a 100 km-thick lithosphere, the faster increase in temperature relative to pressure results in decrease of both P and S wave velocities with depth; this trend is changed at sublithospheric depths, where the pressure effect becomes dominant, leading to an increase in seismic velocities. Within the cratonic mantle lithosphere, the slower temperature increase with depth is largely compensated by the increase in pressure, leading to almost constant P wave velocities and a slower decrease of S wave velocity with depth. The introduction of garnet at 70 km depth results in a sharp increase in both P wave and S wave velocities for both geotherms. The stronger increase in P wave velocities indicates a stronger effect of this change in mineralogical composition (crystallization of garnet at the expenses of the pyroxenes and spinel) on the P wave velocities relative to S waves. This effect is not sampled in the present data set due to the small fraction of spinel peridotites analyzed and to the refractory composition of most shallow peridotites, which did not allow crystallization of garnet (Fig. 1, Table 1).

The marked variations in density and S wave velocity among the deepest xenoliths (Fig. 5), which are mainly sheared peridotites, further highlight the strong compositional control on the isotropic seismic velocities. They may be attributed to the strong variability in the garnet content and in the olivine Mg# that characterize the sheared peridotites (Fig. 4), and that has been attributed to heterogeneous refertilization shortly preceding mylonitization and extraction by the kimberlite (Smith and Boyd, 1987; Baptiste et al., 2012). At shallower depths, the variations in density can be attributed to the variability in orthopyroxene content and olivine Mg#.

5.2 Comparison with seismological velocity profiles and tomographic models for the Kaapvaal mantle

Comparison of the velocity profiles in Fig. 5 with one-dimensional P wave velocity (V_p) profiles for the Kaapvaal highlights that most P wave models show an increase of velocity with depth between 50 and 200 km depth, consistent with James et al. (2004) data (Fig. 5a). P wave velocities for our individual samples do not show a clear trend, due to the strong variability at all depths. However, except for the shallowest layer, which displays an average V_p significantly higher than those estimated in most tomographic models, but that is only based on two samples, the average V_p tends to increase slightly with depth. Between 90 and 180 km, the range of variation of V_p in the studied xenoliths is consistent with most P wave tomography models, except for the SATZ model of Zhao et al. (1999), which overestimates V_p, and the global model IASP91, which as expected strongly underestimates V_p in a cratonic root such as the Kaapvaal. The DataX model by Larson et al. (2006) is the model that fits the best our data, while the QPM model by Qiu et al. (1996) is the most coherent with the xenolith data of James et al. (2004). The decrease of V_p associated with the predominance of deep sheared peridotites between 180 and 200 km in the James et al. (2004) data is not fitted by any P wave tomographic model. This observation reinforces the idea that these sheared peridotites correspond to local modifications of the mantle lithosphere caused and oversampled by kimberlites.

The discrepancy between 1-D seismic profiles and xenoliths' velocities is still more marked for S waves (Fig. 5b). Most models do not fit neither our, nor the xenolith-based isotropic S wave velocity profiles of James et al. (2004), but there is a fit with the DataX model of Larson et al. (2006) and, in to lesser extent, with the model L&B06 of Li and Burke (2006). The QPM model of Qiu et al. (1996) and the model of Hansen et al. (2009) (H09) display a marked decrease of V_s at 110–120 km that is not observed in the xenolith data. The latter implies that the cratonic root is at least 180 km thick. The gradual increase in the isotropic S wave velocities calculated based on Kaapvaal xenolith data (Fig. 5b) is also at odds with the seismic discontinuity imaged by S wave receiver functions at $\sim$150 km depth (Wittlinger and Farra, 2007; Savage and Silver, 2008; Hansen et al., 2009). This suggests that this discontinuity has not a compositional origin.

P and S wave tomography studies reported lateral velocity perturbations of 0.5 % (James et al., 2001; Fouch et al., 2004a), while surface wave studies imaged velocity perturbations as high as 5–7 % within the Kaapvaal cratonic root (Ritsema and van Heijst, 2000; Priestley et al., 2006; Chevrot and Zhao, 2007; Fishwick, 2010). Priestley and Tilmann (2009) attributed the velocity variations within the Kaapvaal lithospheric mantle to thermal effects. On the contrary, Begg et al. (2009) highlighted that xenolith thermometry implies a very narrow geotherm range and suggested that lateral S wave velocity variations within the cratonic root are mainly due to compositional changes. A characteristic example of this intracratonic seismic velocity heterogeneity are the lower velocities ($\sim 0.5\,\%$ for V_p and $\sim 0.8\,\%$ for V_s) imaged below the Bushveld complex by body wave tomographies (James et al., 2001; Fouch et al., 2004a) (Fig. 1), but not by surface wave tomographies (e.g., Priestley and Tilmann, 2009), probably due to their poorer lateral resolution. The degree of variation of isotropic P and S wave velocities observed in the present study is higher, in particular for P waves, than any intracratonic velocity perturbation imaged by body wave tomographies. The latter may therefore be explained by compositional variations within the cratonic mantle lithosphere, such as, enrichment in Fe due to reactive melt percolation beneath the Bushveld.

The present data cannot, however, explain the much higher velocity anomalies inferred from surface wave data (7–8 %). Moreover, the variations in the xenolith P wave velocity at all depths are stronger than the S wave ones, what renders the explanation of the strong velocity gradients inferred from surface waves still more difficult. Partial melting within the lithosphere is a highly improbable cause for these velocity perturbations because of the cold geotherm beneath the Kaapvaal craton. The possible role of water content changes on seismic wave velocity is more difficult to access. Jacobsen et al. (2008) demonstrated that the combined effects of 3 mol % Fe and 0.8 wt. % H_2O reduce the olivine shear and bulk moduli. However, the effects of low water contents, similar to those measured in the Kaapvaal peridotites (≤ 150 wt. ppm H_2O, Baptiste et al., 2012; Peslier et al., 2010) on olivine elastic properties are poorly constrained.

At a given depth, the calculated variation in seismic velocity is almost as strong within xenoliths derived from a single pipe or from closely spaced localities as within the entire data set. Indeed, V_p and V_s varies by up to 1.1 % among xenoliths from Kimberley equilibrated between a 110 and 130 km depth (Fig. 5). V_p and V_s variations by up to 2.1 % are observed among Jagersfontein xenoliths equilibrated between a 150 and 170 km depth. V_p and V_s variations among xenoliths from closely spaced pipes (Jagersfontein, Kimberley and DeBeers, which are all within 100 km, Fig. 1) may reach 1.2 and 1.3 % between a 110 and 130 km depth. Such small-scale variations cannot be imaged by seismic tomography, which probably underestimates the heterogeneity of the subcratonic lithospheric mantle.

5.3 Seismic anisotropy

The changes in olivine CPO symmetry of the Kaapvaal peridotite xenoliths result in slight variations of the seismic anisotropy patterns (Fig. 2). Analysis of the individual sample's seismic anisotropy shows strong variations at all depths (Table 1), but average values for different depth ranges are rather homogeneous (Fig. 6). Maximum P and S wave average anisotropies (Fig. 6) are coherent with the values for the Kaapvaal mantle xenoliths reported by Long and Christensen (2000), but higher than those calculated by Ben Ismaïl et al. (2001). They are also coherent with the anisotropies calculated from Siberian xenoliths reported by Bascou et al. (2011).

The polarization and propagation anisotropies calculated for the Kaapvaal xenoliths may be compared to seismic anisotropy data obtained using both body and surface waves, which sample the Kaapvaal mantle root in different directions and with variable vertical and lateral resolutions. SKS splitting measurements sample the mantle along a vertical direction, and have no vertical but good lateral resolution ($\sim$50 km at lithospheric depths), whereas surface waves sample the mantle along horizontal directions. Moreover, the analysis of surface wave data with different periods allows discriminating between the lithospheric and sublithospheric contributions to seismic anisotropy.

The fast SKS polarization, which samples vertically the Kaapvaal's lithospheric and sublithospheric mantle, and the fast Rayleigh propagation at 70–120 s periods, which sample the Kaapvaal lower lithosphere, are characterized by a NNE–SSW orientation within the western and central part of the craton, and an E–W orientation in its northern part (Silver et al., 2001; Adam and Lebedev, 2012). This suggests that the deep mantle lithosphere beneath the Kaapvaal displays olivine CPO with consistent orientations at the length scales of tens to hundreds of kilometers.

We can therefore use the availability of both SKS splitting and surface wave anisotropy data in the Kaapvaal craton to try to determine the orientation of the foliation and of the lineation in the lithospheric mantle beneath the Kaapvaal craton. The idea is that both SKS and surface waves sample the same mantle lithosphere, which has a unique 3-D seismic anisotropy pattern, but SKS only probes the S wave polarization anisotropy in the vertical direction, while Rayleigh and Love waves travel along horizontal paths, allowing therefore to probe the S wave polarization and azimuthal anisotropies in the horizontal plane. To do so, we have estimated the anisotropy that would be measured by SKS, Rayleigh (S_V) and Love (S_H) waves for five endmember orientations of the foliation and lineation (Fig. 7). A delicate point in this analysis is to determine whether the horizontally S_H or the vertically polarized S wave S_V is the fast propagating S_1 wave, but this distinction can be made by analyzing the S wave polarization anisotropy in the horizontal plane for each case. S_1 is always polarized in a plane containing the lineation. Hence, when the lineation is vertical, S_1 corresponds to S_V (case 3). When it is horizontal, S_1 corresponds to S_H (cases 1, 2 and 5). For a 45° dipping lineation (case 4), the fastest wave is polarized in the horizontal or in the vertical plane depending on the propagation azimuth, but the birefringence is always higher when the fastest wave is polarized in the horizontal plane, implying that when significant anisotropy can be measured S_1 is equal to S_H.

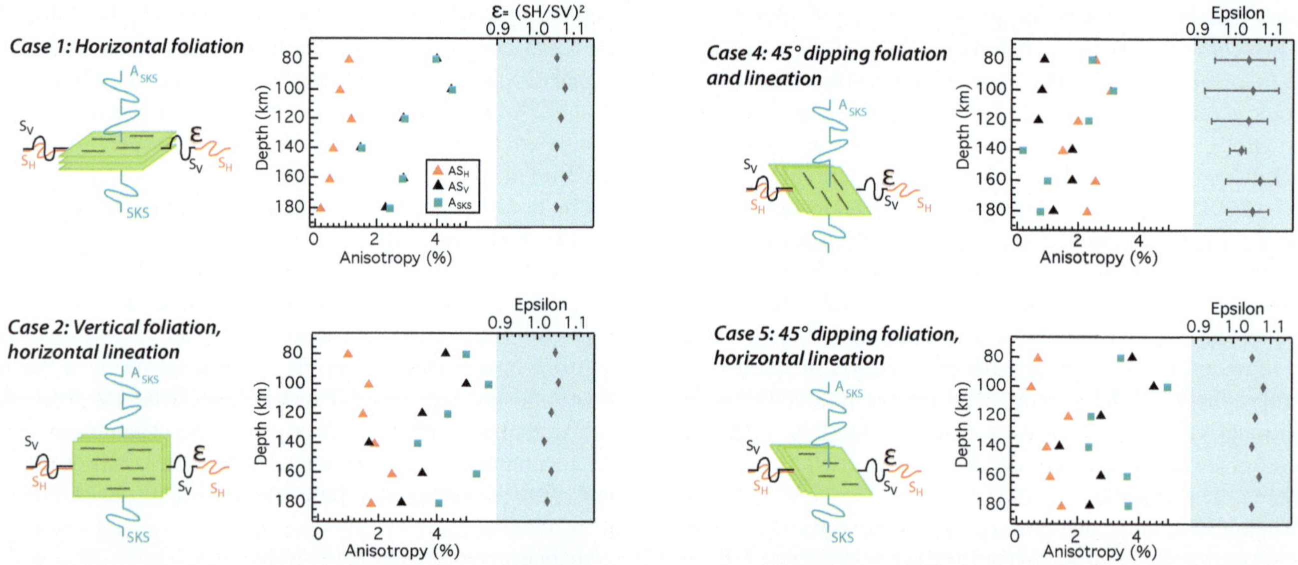

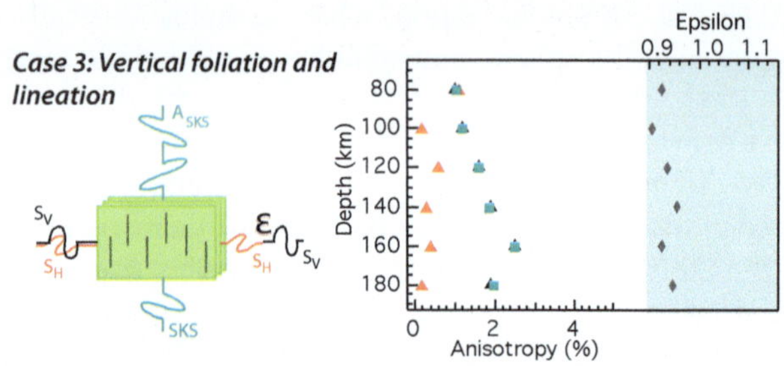

	Observed	Model 1	Model 2	Model 3	Model 4	Model 5
SKS splitting	~0.62s (SI01)	1.04s	1.51s	0.51s	0.58s	1.10s
Rayleigh Anis.	~ 0.1-2.3% (A&L12) ~ 0.6% (EK11)	1.5-4.5%	1.7-5.0%	1.0-2.5%	0.7-1.8%	1.4-4.5%
Love Anis.	~ 0.2% (EK11)	0.2-1.2%	1.0-2.5%	0.2-1.1%	1.5-3.1%	0.5-1.7%

Fig. 7. Calculated surface wave polarization anisotropy ($\varepsilon = (S_H/S_V)^2$), maximum azimuthal anisotropy of horizontally propagating S_H and S_V, and the maximum SKS anisotropy for the depth-averaged samples in five different "endmember" orientations of the foliation and the lineation: (1) horizontal foliation, (2) vertical foliation with a horizontal lineation, (3) vertical foliation and lineation, (4) 45° dipping foliation and lineation, and (5) 45° dipping foliation with a horizontal lineation. The table describes how well these different "endmember" models match SKS- and surface wave data (SI01: Silver et al., 2001; A&L12: Adam and Lebedev, 2012; EK11: Ekström, 2011).

If the foliation is horizontal (case 1), SKS waves and S_V waves will measure similar polarization and azimuthal anisotropies comprised between 1.5 and 4.5 % in the mantle lithosphere, while S_H waves will detect almost no azimuthal anisotropy (≤ 1 %) at all depths. Because S_H is faster than S_V, the surface wave polarization anisotropy (epsilon) is higher than 1 at all depths. If the foliation is vertical and the lineation horizontal (case 2), SKS waves sample a highly anisotropic direction (3.5 to 5.5 %) at all depths. The S_V wave propagation (2–4 %) anisotropy is, at all depths, twice as strong as that of the S_H waves (1–2 %). As in case 1, the surface wave polarization anisotropy is higher than 1 at all depths. If the foliation and the lineation are vertical (case 3), the surface wave polarization anisotropy is smaller than 1, indicating that S_V is always faster than S_H. SKS and S_V waves will measure low anisotropies, comprised between 1.0 and 2.5 %. The anisotropy measured by S_H waves should be even lower (<0.5 %). A 45° dipping foliation and lineation (case 4) should produce SKS anisotropies higher than 2 % and S_V wave azimuthal anisotropy lower than 1 % between 80 and 120 km. At greater depths, SKS wave anisotropy will be lower than 1 %, while S_V wave azimuthal anisotropy will vary between 1 and 2 %. The S_H wave azimuthal anisotropy will always be comprised between 1.5 and 2 %. The surface wave polarization anisotropy will be strongly dependent on the propagation direction. Finally, if we consider a 45° dipping foliation and a horizontal lineation (Case 5), the SKS and S_V waves should measure polarization and azimuthal anisotropies varying between 1.5 and 5 %, while S_H waves will measure azimuthal anisotropies lower than 2 %. As in cases 1 and 2, the surface wave polarization anisotropy is higher than 1, indicating that S_H is faster than S_V at all depths.

These predictions can be compared to the anisotropy detected using SKS and surface waves (Table in Fig. 7). Beneath the Kaapvaal craton, SKS delay times are low. Vinnik et al. (1995) first measured SKS delay times comprised between 0.4 and 1.4 s. Silver et al. (2001) obtained similar SKS delay times, between 0.3 and 1.1 s, with an average value of 0.62 s. Low delay times (0.15 to 0.75 s) were also obtained in a denser network near Kimberley by Fouch et al. (2004b). If we consider a 150 km-thick homogeneous anisotropic lithospheric mantle, these delay times yield anisotropies comprised between 0.9 and 3.4 %, with an average of 1.9 %. Such low SKS anisotropies are only obtained for either a vertical or a 45° dipping foliation and lineation (cases 3 and 4 in Fig. 7).

Surface wave data consistently points to S_H faster than S_V in the Kaapvaal lithospheric mantle (Freybourger et al., 2001; Sergei Lebedev, personal communication). This is consistent with receiver function data, which imaged a 160 km-thick anisotropic mantle keel with vertical slow axis ($V_{S_H} > V_{S_V}$) and decreasing anisotropy with depth due to a decrease in V_{S_H} while V_{S_V} remains constant (Wittlinger and Farra, 2007). These observations are not consistent with case 3 (Fig. 7), since vertical lineations always result in $V_{S_H} < V_{S_V}$. Moreover, coherent fast SKS polarization directions like those measured by Silver et al. (2001) are not expected if the lineation is vertical (case 3). They might however be produced by a 45° dipping foliation (case 4). Surface wave azimuthal anisotropies generated for case 4 range between 0.5 and 3 %, being consistent with the low azimuthal anisotropies inferred in both global and regional surface wave models (Ekström, 2011; Adam and Lebedev, 2012). However, case 4 produces a S_H azimuthal anisotropy that is stronger than the S_V one, which is not consistent with the stronger Rayleigh azimuthal anisotropy ($\sim 2\,\%$) relative to the Love azimuthal anisotropy ($\sim 1\,\%$) imaged beneath South Africa in the global surface wave model of Ekström (2011), and in the recent regional anisotropic tomography model for the Kaapvaal by Adam and Lebedev (2012). Such a relation is obtained in the cases where the lineation is horizontal (cases 1, 2, and 5 in Fig. 7), but too strong Rayleigh azimuthal and SKS polarization anisotropies accompany it.

Low SKS birefringence may result from vertical variations of the seismic anisotropy. Indeed, beneath the western Kaapvaal (Kimberley block), the Rayleigh wave azimuthal anisotropy shows a change of fast propagation directions from N–S in the crust to E–W in the mantle (Adam and Lebedev, 2012). A change with depth in the orientation of the fast direction from E–W to N–S was also detected at depths > 160 km beneath the Limpopo belt by a study associating P wave receiver functions and SKS waveforms inversion (Vinnik et al., 2012). The seismic discontinuity imaged by S wave receiver functions at ~150 km depth (Wittlinger and Farra, 2007; Savage and Silver, 2008; Hansen et al., 2009) also points to vertical variations in the deformation structure within the cratonic root, since drastic changes in composition or S wave velocities with depth are not observed in neither our data set nor in that of James et al. (2004). Most cases presented in Fig. 7 show a decrease in anisotropy at 140 km depth, which would be consistent with the observations of Wittlinger and Farra (2007), but the associated gradient in seismic velocities is too weak to produce a strong impedance contrast. Peslier et al. (2010) and Baptiste et al. (2012) did measure a marked decrease in OH concentrations in olivine at depths greater than 160 km. Yet the change in elastic properties produced by variations of OH contents from 150 ppm to a few ppm is probably too weak (Jacobsen et al., 2008) to explain the receiver function signal. A sharp change in the orientation of the foliation and lineation and the associated change in seismic anisotropy might produce the needed impedance contrast. It is also consistent with the anisotropic surface wave models and with the results of P receiver functions and SKS waveforms inversion, which propose change in the orientation of the fast direction at depths > 160 km (Adam and Lebedev, 2012; Vinnik et al., 2012). However, such a vertical variation in seismic anisotropy should produce a backazimuthal dependence of the shear wave splitting, which was not described in the regional studies (Silver et al., 2001; Fouch et al., 2004a; Silver et al., 2004; Vinnik et al., 1995).

In conclusion, no simple model may account for all seismic anisotropy data in the Kaapvaal. The two models that best explain the observations are (i) the presence of dipping foliations and lineations within the cratonic lithosphere, at ca. 160 km depth and (ii) a vertical variation in seismic anisotropy within the Kaapvaal mantle lithosphere, with dominant horizontal frozen flow directions in the shallower layer. In the second case, we may speculate that it might represent an interface marking a change in the geodynamical process responsible for the formation and thickening of the mantle root. A possible geodynamic scenario allowing for the development of 45° dipping foliations and lineations within the cratonic lithosphere would be the formation of the Kaapvaal craton keel by subduction stacking, as suggested by several geochemical and petrological studies (Shirey et al., 2001; Simon et al., 2007; Pearson and Wittig, 2008). However, in this case, fast SKS polarization directions normal to the main sutures should be expected, rather than the parallel to slightly oblique directions measured in the Kaapvaal craton (Silver et al., 2001). These predictions are also at odds with the parallelism between fast Rayleigh wave propagation directions and the Archean–Paleoproterozoic crustal structures in the Limpopo belt and in the northern Kaapvaal, but they may explain the fast Rayleigh wave propagation directions perpendicular to the crustal structures in the western part of the craton (Adam and Lebedev, 2012). On the other hand, if the cratonic root formed by episodic injection of small (100–200 km) diapiric upwellings (de Smet et al., 1998, 1999), a horizontal foliation and variable flow direction are to be expected across the craton. Such a variability of the structure within the root may explain the low SKS delay times measured in the craton, but would fail to explain the coherent fast polarization direction of SKS waves and fast propagation direction of Rayleigh waves over length scales of several hundreds of kilometers in the Kaapvaal mantle lithosphere (Silver et al., 2001; Adam and Lebedev, 2012).

6 Conclusions

The strong compositional heterogeneity of the Kaapvaal peridotite xenoliths results in up to 3 % variation in density, while V_p, V_s and the V_p/V_s ratio vary by up to 2.3 %. These variations are slightly higher than the velocity perturbations imaged by body wave tomography, but cannot explain the strong velocity anomalies reported by surface wave studies. Melt depletion tends to increase V_p and V_s and to decrease the V_p/V_s ratio and density. Orthopyroxene enrichment decreases the density and V_p, and strongly reduces the V_p/V_s ratio. Enrichment in garnet tends to increase the density strongly and slightly V_p and the V_p/V_s ratio, but has little to no effect on V_s. The V_p/V_s ratio is more sensitive to the orthopyroxene content than to olivine Mg# and garnet content. The present data also highlights that compositional heterogeneity in the Kaapvaal mantle root occurs at a scale much smaller than the one that may be sampled by seismic tomography, since at a given depth, the calculated variation in seismic velocity within xenoliths derived from a single pipe or from closely spaced localities is almost as strong as within the entire data set. Seismic tomography thus probably strongly underestimates the heterogeneity of the subcratonic lithospheric mantle.

Vertical density and seismic velocity profiles calculated using the xenoliths' compositions and equilibration conditions show that V_s decreases, whereas the V_p/V_s ratio and density increase with increasing depth. V_p does not show a clear trend; it is highly variable at all depths. Comparison of these xenolith-based seismic profiles with 1-D velocity profiles derived from seismological data in the Kaapvaal highlights clear discrepancies between seismological and xenolith-based data. For P waves, the DataX model by Larson et al. (2006) fits our data the best, while the QPM model by Qiu et al. (1996) is the most coherent with data of James et al. (2004). For S waves, most models, except for the DataX model of Larson et al. (2006) and, in a lesser extent, the model L&B06 of Li and Burke (2006), do not fit neither our, nor data of James et al. (2004), which imply that the cratonic root is at least 180 km thick. No seismological velocity model fits the decrease of V_p indicated by deep sheared peridotites between 180 and 200 km, reinforcing the idea that they represent very local modifications of the lithosphere caused and oversampled by kimberlites.

Changes in olivine CPO symmetry of the Kaapvaal peridotite xenoliths result only in slight variation of the seismic anisotropy patterns. Seismic anisotropy increase with the olivine J index and olivine content, but there is no simple relation with the olivine CPO symmetry (BA index). Maximum anisotropy values range between 2.5 and 10.2 % for P wave azimuthal anisotropy (AV_p) and between 2.7 and 8 % for S wave polarization anisotropy (AV_s) in agreement with existing xenolith data. The maximum P wave anisotropy of each 20 km interval varies between 4.5 and 7.9 %, whereas S wave anisotropy is comprised between 3.7 and 5.8 %. Models considering endmember orientations of the foliation and lineation in the subcratonic mantle lithosphere show that the simplest model that might produce both coherent fast directions over large domains, but low delay times imaged by SKS studies, and the low azimuthal surface wave anisotropy with S_H faster than S_V in the subcratonic mantle lithosphere is the presence of 45° dipping foliations and lineations. Yet, this model fails to account for the relative importance of the Rayleigh and Love azimuthal anisotropies. This observation and the discontinuity observed at ca. 160 km depth by receiver functions are best explained by a vertical variation in seismic anisotropy, due to a sharp change in orientation of the frozen deformation structures in the Kaapvaal mantle lithosphere. However, this scenario is at odds with the observation of coherent fast polarization direction of SKS waves and fast propagation direction of Rayleigh waves over scales of several hundreds of kilometers in the Kaapvaal mantle lithosphere (Silver et al., 2001; Adam and Lebedev, 2012).

Acknowledgements. The authors thank A. Vauchez and D. Mainprice for helpful discussions. D. Mainprice provided the programs for calculating seismic properties. C. Nevado and D. Delmas are thanked for providing high-quality polishing of sections for EBSD measurements. The two anonymous reviewers are thanked for their helpful comments. The EBSD-SEM national facility in Montpellier is supported by the Institut National de Sciences de l'Univers (INSU) du Centre National de la Recherche Scientifique (CNRS), France and by the Conseil Régional Languedoc-Roussillon, France.

Edited by: W. Geissler

References

Abramson, E. H., Brown, J. M., Slutsky, L. J., and Zaug, J.: The elastic constants of San Carlos olivine to 17 GPa, J. Geophys. Res., 102, 12253–12263, 1997.

Adam, J. M.-C.and Lebedev, S.: Azimuthal anisotropy beneath southern Africa from very broad-band surface wave dispersion measurements, Geophys. J. Int., 191, 155–174, 2012.

Adams, A. and Nyblade, A.: Shear wave velocity structure of the southern African upper mantle with implications for the uplift of southern Africa, Geophys. J. Int., 186, 808–824, 2011.

Allsopp, H. L., Bristow, J. W., and Skinner, E. M. W.: The Rb–Sr geochronology of the Colossus kimberlite pipe, Zimbabwe, Transactions of the Geological Society of South Africa, 88, 245–248, 1985.

Baptiste, V., Tommasi, A., and Demouchy, S.: Deformation and hydration of the lithospheric mantle beneath the Kaapvaal craton, South Africa, Lithos, 149, 31–50, 2012.

Bascou, J., Doucet, L. S., Saumet, S., Ionov, D. A., Ashchepkov, I. V., and Golovin, A. V.: Seismic velocities, anisotropy and deformation in Siberian cratonic mantle: EBSD data on xenoliths from the Udachnaya kimberlite, Earth Planet. Sc. Lett., 304, 71–84, 2011.

Begg, G. C., Griffin, W. L., Natapov, L. M., O'Reilly, S. Y., Grand, S. P., O'Neill, C. J., Hronsky, J. M. A., Poudjom Djomani, Y., Swain, C. J., Deen, T., and Bowden, P.: The lithospheric architecture of Africa: seismic tomography, mantle petrology, and tectonic evolution, Geosphere, 5, 23–50, 2009.

Bell, D. R., Rossman, G. R., Maldener, J., Endisch, D., and Rauch, F.: Hydroxide in olivine: a quantitative determination of the absolute amount and calibration of the IR spectrum, J. Geophys. Res., 108, B2, ECV8.1–ECV8.9, 2003.

Bell, D., Grégoire, M., Grove, T., Chatterjee, N., Carlson, R., and Buseck, P.: Silica and volatile-element metasomatism of Archean mantle: a xenolith-scale example from the Kaapvaal Craton, Contrib. Mineral. Petr., 150,251–267, 2005.

Ben Ismaïl, W., and Mainprice, D.: An olivine fabric database: an overview of upper mantle fabrics and seismic anisotropy, Tectonophysics, 296, 145–157, 1998.

Ben Ismaïl, W., Barruol, G., and Mainprice, D.: The Kaapvaal craton seismic anisotropy: petrophysical analyses of upper mantle kimberlite nodules, Geophys. Res. Lett., 28, 2497–2500, 2001.

Chai, M., Brown, J. M, and Slutsky, L. J.: The elastic constants of an aluminous orthopyroxene to 12.5 GPa, J. Geophys. Res., 102, 14770–14785, 1997a.

Chai, M., Brown, J. M., and Slutsky, L. J.: The elastic constants of pyrope-grossular-almandine garnets to 20 GPa, Geophys. Res. Lett., 24, 523–526, 1997b.

Chevrot, S. and Zhao, L.: Multiscale finite-frequency Rayleigh wave tomography of the Kaapvaal craton, Geophysical Journal International, 169, 1, 201–215, 2007.

Collins, M. D. and Brown, J. M.: Elasticity of an upper mantle clinopyroxene, Phys. Chem. Miner., 26, 7–13, 1998.

de Smet, J. H., van den Berg, A. P., and Vlaar, N. J.: Stability and growth of the continental shields in mantle convection models including melt production, Tectonophysics, 296, 15–29, 1998.

de Smet, J. H., van den Berg, A. P., and Vlaar, N. J.: The evolution of continental roots in numerical thermo-chemical mantle convection models including differentiation by partial melting, Lithos, 48, 153–170, 1999.

Ekström, G.: A global model of Love and Rayleigh surface wave dispersion and anisotropy, 25–250 s, Geophys. J. Int., 187, 1668–1686, 2011.

Fishwick, S.: Surface wave tomography: imaging of the lithosphere-asthenosphere boundary beneath central and southern Africa?, Lithos, 120, 63–73, 2010.

Fouch, M. J., James, D. E., VanDecar, J. C., and van der Lee, S.: Mantle seismic structure beneath the Kaapvaal and Zimbabwe Cratons, S. Afr. J. Geol., 107, 33–44, 2004a.

Fouch, M. J., Silver, P. G., Bell, D. R., and Lee, J. N.: Small-scale variations in seismic anisotropy near Kimberley, South Africa, Geophys. J. Int., 157, 764–774, 2004b.

Freybourger, M., Gaherty, J. B., Jordan, T. H., and the Kaapvaal Seismic Group: Structure of the Kaapvaal craton from surface waves, Geophys. Res. Lett., 28, 2489–2492, 2001.

Green, H. W. and Gueguen, Y.: Origin of kimberlite pipes by diapiric upwelling in upper mantle, Nature, 249, 617–620, 1974.

Grégoire, M., Bell, D. R., and Le Roex, A. P.: Garnet lherzolites from the Kaapvaal craton (South Africa): trace element evidence for a metasomatic history, J. Petrol., 44, 629–657, 2003.

Griffin, W. L., O'Reilly, S. Y., Afonso, J. C., and Begg, G. C.: The composition and evolution of lithospheric mantle: a re-evaluation and its tectonic implications, J. Petrol., 50, 7, 1185–1204, 2009.

Griffin, W. L., O'Reilly, S. Y., Natapov, L. M.,and Ryan, C. G.: The evolution of lithospheric mantle beneath the Kalahari craton and its margins, Lithos, 71, 215–241, 2003.

Hansen, S. E., Nyblade, A. A., Julià, J., Dirks, P. H. G. M., and Durrheim, R. J.: Upper-mantle low-velocity zone structure beneath the Kaapvaal craton from *S* wave receiver functions, Geophys. J. Int., 178, 1021–1027, 2009.

Hielscher, R. and Schaeben, H.: A novel pole figure inversion method: specification of the MTEX algorithm, J. Appl. Crystallogr., 41, 1024–1037, 2008.

Isaak, D., Graham, E. K., Bass, J. D., and Wang, H.: The elastic properties of single-crystal fayalite as determined by dynamical measurement techniques, Pure Appl. Geophys., 141, 393–414, 1993.

Jacobsen, S. D., Jiang, F., Mao, Z., Duffy, T. S., Smyth, J. R., Holl, C. M., and Frost, D. J.: Effects of hydration on the elastic properties of olivine, Geophys. Res. Lett., 35, L14303, doi:10.1029/2008GL034398, 2008.

James, D. E., Fouch, M. J., Vandecar, J. C., and Van der Lee, S.: Tectospheric structure beneath southern Africa: The Kaapvaal Project: formation and evolutions of cratons, Geophys. Res. Lett., 28, 2485–2488, 2001.

James, D. E., Boyd, F. R., Schutt, D., Bell, D. R., and Carlson, R. W.: Xenolith constraints on seismic velocities in the upper mantle beneath southern Africa, Geochem. Geophy. Geosy., 5, Q01002, doi:10.1029/2003GC000551, 2004.

Jordan, T. H.: Composition and development of continental tectosphere, Nature, 274, 5671, 544–548, 1978.

Kramers, J. D.and Smith, C. B.: A feasibility study of U–Pb and Pb–Pb dating of kimberlites using groundmass mineral fractions and whole-rock samples, Chem. Geol., 41, 23–38, 1983.

Kumazawa, M. and Anderson, O. L.: Elastic moduli, pressure derivatives, and temperature derivatives of single-crystal olivine and single-crystal forsterite, J. Geophys. Res., 74, 5961–5972, 1969.

Larson, M. A., Snoke, J. A., and James, D. E.: *S*-wave velocity structure, mantle xenoliths and the upper mantle beneath the Kaapvaal craton, Geophys. J. Int., 167, 171–186, 2006.

Li, A. and Burke, K.: Upper mantle structure of southern Africa from Rayleigh wave tomography, J. Geophys. Res., 111, B10303, doi:10.1029/2006JB004321, 2006.

Long, C. and Christensen, N. I.: Seismic anisotropy of South African upper mantle xenoliths, Earth Planet. Sc. Lett., 179, 551–565, 2000.

Mainprice, D.: A FORTRAN program to calculate seismic anisotropy from the lattice preferred orientation of minerals, Computat. Geosci., 16, 385–393, 1990.

Mainprice, D. and Humbert, M.: Methods of calculating petrophysical properties from lattice preferred orientation data, Surv. Geophys., 15, 575–592, 1994.

Mainprice, D. and Silver, P. G.: Interpretation of SKS-waves using samples from the subcontinental lithosphere, Phys. Earth Planet. In., 78, 257–280, 1993.

Mainprice, D., Barruol, G., and Ben Ismaïl, W.: The seismic anisotropy of the Earth's mantle: from single crystal to polycrystal, in: AGU Geophysical Monograph, edited by: Karato, A. M. F. S., Libermann, R. C., Masters, G., and Stixrude, L., AGU, Washington, DC, 117, 237–264, 2000.

Mainprice, D., Hielscher, R., and Schaeben, H.: Calculating anisotropic physical properties from texture data using the MTEX open source package, in: Deformation Mechanisms, Rheology and Tectonics: Microstructures, Mechanics and Anisotropy, edited by: Prior, D. J., Rutter, E. H., and Tatham, D. J., Geological Society, Special Publications, London, 175–192, 2011.

Pearson, D. G. and Wittig, N.: Formation of Archaean continental lithosphere and its diamonds: the root of the problem, J. Geol. Soc., 165, 895–914, 2008.

Pearson, D. G., Carlson, R. W., Shirey, S. B., Boyd, F. R., and Nixon, P. H.: Stabilisation of Archaean lithospheric mantle: A Re-Os isotope study of peridotite xenoliths from the Kaapvaal craton, Earth Planet. Sci. Lett., 134, 341–357, 1995.

Peslier, A. H., Woodland, A. B., Bell, D. R., and Lazarov, M.: Olivine water contents in the continental lithosphere and the longevity of cratons, Nature, 467, 78–81, 2010.

Phillips, D., Machin, K. J., Kiviets, G. B., Fourie, L. F., Roberts, M. A., Skinner, E. M. W.: A petrographic and $^{40}Ar/^{39}Ar$ geochronological study of the Voorspoed kimberlite, South Africa: implications for the origin of group II kimberlite magmatism, special issue, S. Afr. J. Geol., 101, 299–306, 1998.

Priestley, K.: Velocity structure of the continental upper mantle: evidence from southern Africa, Lithos, 48, 45–56, 1999.

Priestley, K. and Tilmann, F.: Relationship between the upper mantle high velocity seismic lid and the continental lithosphere, Lithos, 109, 112–124, 2009.

Priestley, K., McKenzie, D., and Debayle, E.: The state of the upper mantle beneath southern Africa, Tectonophysics, 416, 101–112, 2006.

Qiu, X., Priestley, K., and McKenzie, D.: Average lithospheric structure of southern Africa, Geophys. J. Int., 127, 563–587, 1996.

Ritsema, J. and van Heijst, H.: New seismic model of the upper mantle beneath Africa, Geology, 28, 63–66, 2000.

Savage, B. and Silver, P. G.: Evidence for a compositional boundary within the lithospheric mantle beneath the Kalahari craton from *S* receiver functions, Earth Planet. Sc. Lett., 272, 600–609, 2008.

Schutt, D. L. and Lesher, C. E.: Compositional trends among Kaapvaal Craton garnet peridotite xenoliths and their effects on seismic velocity and density, Earth Planet. Sc. Lett., 300, 367–373, 2010.

Sebai, A., Stutzmann, E., Montagner, J.-P., Sicilia, D., and Beucler, E.: Anisotropic structure of the African upper mantle from Rayleigh and Love wave tomography, Phys. Earth Planet. In., 155, 48–62, 2006.

Shirey, S. B., Carlson, R. W., Richardson, S. H., Menzies, A. H., Gurney, J. J., Pearson, D. G., Harris, J. W., and Wiechert, U.: Archean emplacement of eclogitic components into the lithospheric mantle during formation of the Kaapvaal Craton, Geophys. Res. Lett., 28, 2509–2512, 2001.

Silver, P. G., Gao, S. S., Liu, K. H., and the Kaapvaal Seismic Group: Mantle deformation beneath southern Africa, Geophys. Res. Lett., 28, 2493–2496, 2001.

Silver, P. G., Fouch, M. J., Gao, S. S., Schmitz, M., and the Kaapvaal Seismic Group: Seismic anisotropy, mantle fabric, and the magmatic evolution of Precambrian southern Africa, S. Afr. J. Geol., 107, 45–58, 2004.

Simon, R. E., Wright, C., Kgaswane, E. M., and Kwadiba, M. T. O.: The *P* wavespeed structure below and around the Kaapvaal craton to depths of 800 km, from traveltimes and waveforms of local and regional earthquakes and mining-induced tremors, Geophys. J. Int., 151, 132–145, 2002.

Simon, N. S. C., Carlson, R. W., Pearson, D. G., and Davies, G. R.: The origin and evolution of the Kaapvaal cratonic lithospheric mantle, J. Petrol., 48, 3, 589–625, 2007.

Smith, D. and Boyd, F. R.: Compositional heterogeneities in a high temperature lherzolite nodule and implications for mantle processes, in: Mantle Xenoliths, edited by: Nixon, P. H., John Wiley, Hoboken, NJ, 551–561, 1987.

Soustelle, V. and Tommasi, A.: Seismic properties of the suprasubduction mantle: constraints from peridotite xenoliths from the Avacha volcano, southern Kamchatka, Geophys. Res. Lett., 37, L13307, 1–5, 2010.

Suzuki, I., Anderson, O. L., and Sumino, Y.: Elastic properties of a single-crystal forsterite Mg_2SiO_4, up to 1,200 K, Phys. Chem. Miner., 10, 38–46, 1983.

Tommasi, A., Mainprice, D., Canova, G., and Chastel, Y.: Viscoplastic self-consistent and equilibrium-based modeling of olivine lattice preferred orientations: Implications for the upper mantle seismic anisotropy, J. Geophys. Res., 105, 7893–7908, 2000.

Tommasi, A., Godard, M., Coromina, G., Dautria, J.-M., and Barsczus, H.: Seismic anisotropy and compositionally induced velocity anomalies in the lithosphere above mantle plumes: a petrological and microstructural study of mantle xenoliths from French Polynesia, Earth Planet. Sc. Lett., 227, 539–556, 2004.

Tommasi, A., Vauchez, A., and Ionov, D. A.: Deformation, static recrystallisation, and reactive melt transport in shallow subcontinental mantle xenoliths (Tol Cenozoic volcanic field, SE Siberia), Earth Planet. Sc. Lett., 272,65–77, 2008.

Vinnik, L., Green, R. W. E., and Nicolaysen, L. O.: Recent deformation of the deep continental root beneath southern Africa, Nature, 375, 50–52, 1995.

Vinnik, L., Kiselev, S., Weber, M., Oreshin, S., and Makeyeva, L.: Frozen and active seismic anisotropy beneath southern Africa, Geophys. Res. Lett., 39,1–6, 2012.

Wagner, L. S., Anderson, M. L., Jackson, J. M., Beck, S. L., and Zandt, G.: Seismic evidence for orthopyroxene enrichment in the continental lithosphere, Geology, 36, 12, 935–938, 2008.

Wang, Y., Wen, L., and Weidner, D.: Upper mantle S_H- and P-velocity structures and compositional models beneath southern Africa, Earth Planet. Sc. Lett., 297, 596–608, 2008.

Wasch, L. J., van der Zwan, F. M., Nebel, O., Morel, M. L. A., Hellebrand, E. W. G., Pearson, D. G., and Davies, G. R.: An alternative model for silica enrichment in the Kaapvaal subcontinental lithospheric mantle, Geochim. Cosmochim. Ac., 73, 6894–6917, 2009.

Webb, S. L.: The elasticity of the upper mantle orthosilicates olivine and garnet to 3 GPa, Phys. Chem. Miner., 16, 684–692, 1989.

Wittlinger, G. and Farra, V.: Converted waves reveal a thick and layered tectosphere beneath the Kalahari super-craton, Earth Planet. Sc. Lett., 254, 404–415, 2007.

Zhao, M., Langston, C. A., Nyblade, A., and Owens, T. J.: Upper mantle velocity structure beneath southern Africa from modeling regional seismic data, J. Geophys. Res., 104, 4783–4794, 1999.

8

Review of some significant claimed irregularities in Scandinavian postglacial uplift on timescales of tens to thousands of years – earthquakes in Denmark?

S. Gregersen and P. H. Voss

Geological Survey of Denmark and Greenland, Ostervoldgade 10, 1350 Copenhagen K, Denmark

Correspondence to: S. Gregersen (sg@geus.dk)

Abstract. The postglacial uplift and surrounding subsidence in Scandinavia is well described as close to regular, and the phenomenon is similar on timescales of tens, hundreds and thousands of years studied via geodesy, seismology and geology. Searches for irregularities in the form of earthquakes claimed in the scientific literature have disclosed many earthquakes right after the Ice Age, 9000 yr ago, and some later cases for further evaluation. In a previous report, the present authors have mentioned doubts about the validity of some of the most significant claimed irregularities. In the present paper, a review is made of these significant claimed irregularities in the south-western flank of the Scandinavian postglacial uplift and subsidence via literature studies of geodetic and geological claims of earthquakes as well as discussions in the field. Geodetic observations exist for all of Scandinavia that describe the phenomenon on a scale of 10s–100s of years. Earthquake observations in seismology are of relevance in the same timescales. Geological studies of dated shore lines describe the postglacial vertical earth-surface motion in a quite different timescale of 100s–1000s of years. There is a need for integration of these observations geographically. This is happening in the various timescales in the DynaQlim project. The review finds the claims unlikely to be earthquakes concerning the following: (1) geodynamical motion in the Copenhagen area, (2) a palaeo-earthquake in Læsø and (3) the recently proposed water level discrepancy in the southern part of Denmark. The assessment is less certain, but falls to improbable when concerning (4) proposed palaeo-earthquakes by Hallandsåsen in south-western Sweden.

1 Introduction

Our geodesist colleagues are making more and more definitive observations of the uplift pattern of Scandinavia and the surrounding subsidence. Methods are improving from levelling and sea level observations to GPS measurements nowadays. Many DynaQlim papers have over the years substantiated the pattern of vertical motion on the timescales 10s to 100s of years. The central parts of such a good, regular reference pattern NKG2005LU (Vestøl, 2006; Ågren and Svensson, 2007; Engsager et al. 2006, personal communication) are in the present paper displayed in Fig. 1, overlayed on the earthquake activity. This geodetic pattern in essence represents also the geological pattern of a timescale of 1000s of years, which is established via dated beach lines in old as well as more recent data collections. De Geer gave in 1888 a summary based on geological data available at his time, shown by Mörner (1979) in a paper, where he presented his newer map with more data, but introducing some questionable assumptions on linearity. The overlapping geodetic and geological patterns more than indicate that it is the same phenomenon which uplifts Scandinavia in the timescales tens, hundreds and thousands of years. The pattern which can be well represented by a broad-scale upwelling bulge is used by other authors as data points for fitting in their theoretical modelling (e.g. Lambeck et al., 2010).

The first irregularity to be noticed in observations as well as modelling is that the uplift rate has changed drastically from very fast right after the Ice Age, 9000 yr ago, to much slower nowadays. The second observation is that the geographical patterns satisfying the observations are very regular, which means that irregularities are noticeable and worth a review. Irregularities have been found and evaluated in a number of papers for Norway (e.g. Olesen et al., 2004; Bungum et al., 2010). Here it is done for Denmark.

2 Stresses in the lithosphere of various timescales

In previous papers the stress development after the Ice Age has been treated (e.g. Gregersen and Voss, 2010). The stress field was right after off-loading of the ice, dominated by the uplift and pressure outwards from the central parts. Later, the present situation is that the uplift is less important while stresses caused by the present lithospheric plate movements are dominating with compression in a north-west–south-east direction (Gregersen and Voss, 2009; Olesen et al., 2004; Bungum et al., 1991; Lindholm et al., 2000; Gregersen, 1992). In several papers it is argued that there is no correlation between the uplift pattern and the pattern of earthquake locations (e.g. Gregersen, 2002; Gregersen and Voss, 2009, Zoback et al., 1989; Zoback, 1992; Poutanen et al., 2009), which for all of Scandinavia is displayed in Fig. 1. Like displays of earthquake activity in other parts of the world, it does not matter much which is the time period displayed. The general pattern is the same in any time period, while details can be different. The figure shows that the earthquake activity is most concentrated along the Norwegian coast and continental margin, along the Swedish east coast, and in and around the Oslo Fjord. It is well established that the earthquake activity in Denmark is the southern limitation of the Scandinavian seismicity (Gregersen et al., 1998). And there is less earthquake activity as we go south-west in Jylland in the middle of Denmark. It is also established with good credibility that the earthquake activity is smaller in northern Germany, in Poland and in the Baltic states (e.g. Grünthal et al., 2008). In the latter areas the seismograph coverage has until recently been significantly poorer, so the information in the map is influenced by less sensitivity to small earthquakes. But there is not much doubt that the Kaliningrad events in 2004 of magnitudes just over 5 are very unusual for the coast of Poland, Russia and Lithuania (Gregersen et al., 2007). The map should be complete down to magnitude 4.

3 Earthquake investigations of various timescales

Several investigations are available concerning the regional, recent earthquake activity around the south-western edge of the Fennoscandian Shield. First of all the prominent geological zone Sorgenfrei–Tornquist Zone is considered inactive, except for a small part in eastern Kattegat (e.g. Arvidsson et al. 1991; Gregersen et al. 1998). An illustration of this inactivity concerning earthquakes is shown in Fig. 2, supplementing the regional activity picture of Fig. 1. The earthquake dots in the two maps of Fig. 2 are the same, and they cover the same area. The purpose of the map without coasts and geology is to show the earthquake activity without prejudice from coast lines and geological zones. The Sorgenfrei–Tornquist Zone passes right through the middle of the illustration, but its trend can not be distinguished through earthquake activity. The Ringkobing–Fyn Ridge can not be distinguished either. This just emphasizes what has been pointed out before (e.g. Gregersen et al., 2011).

The summary of the small earthquake activity in Denmark (Figs. 1 and 2) is that it occurs not in one specific linear zone, but rather scattered in several zones in north-western Jylland and in Kattegat as well as completely unsystematic in eastern Denmark and southern Sweden (Gregersen et al. 1998). In a general sense, the seismicity pattern in Denmark shows us the irregular rheological edge zone of the stiff old Fennoscandian Shield, not as an active zone but rather as a broad area of transition from some earthquake activity in the shield north-east of the zone to quiet area south-west of the zone. In the European scale the Tornquist Zone is considered as present-day inactive, as presented in conference proceedings by Gregersen et al. (1995) and seen as lack of earthquakes in the lower right hand corner of Fig. 2, in Poland.

At least a part of Scandinavia has experienced postglacial earthquakes (e.g. Lagerbäck, 1990, 1991). These earthquakes (Fig. 1) in northern Scandinavia were caused by the Ice Age ice cap melting away within a short time. That the present-day stress regime is completely different is agreed by Mörner (2003), but additionally Mörner (2003, 2009) presents palaeo-seismological arguments on some large Swedish earthquakes further south in Sweden than those of Fig. 1. And in some cases, both in northern and southern Sweden, those earthquakes appear to be younger than 9000 yr, the time when the last ice cap melted away. Those arguments of Mörners (2003, 2009) are less convincing than those giving basis for the earthquake fault signatures of Fig. 1, where Mörner agrees with Lagerbäck (1990, 1991). An evaluation for Sweden is needed, just like that for Norway (Olesen et al., 2004). A first step is taken by the discussion concerning claim 4 of this paper.

4 Assessment of claims in the geologic and geodetic literature of active faulting near Denmark

The present assessment review includes both the long-term geological evidence and the short-term geodetic evidence. Additionally the assessment acknowledges the view, that the small present-day earthquake activity in Scandinavia is caused by a dominating stress field from lithospheric plate motion, supplemented by stresses from the postglacial uplift (Gregersen and Basham, 1989; Zoback et al., 1989; Gregersen, 1990 and 1992) as well as by other regional effects as lithospheric inhomogeneities (e.g. Hicks et al., 2000; Pascal and Cloetingh, 2009; Bungum et al., 2010; Olesen et al., 2012). Generalized maps of the very regular postglacial uplift of the investigated area near Denmark have recently been presented from geodetic sources (Knudsen et al.

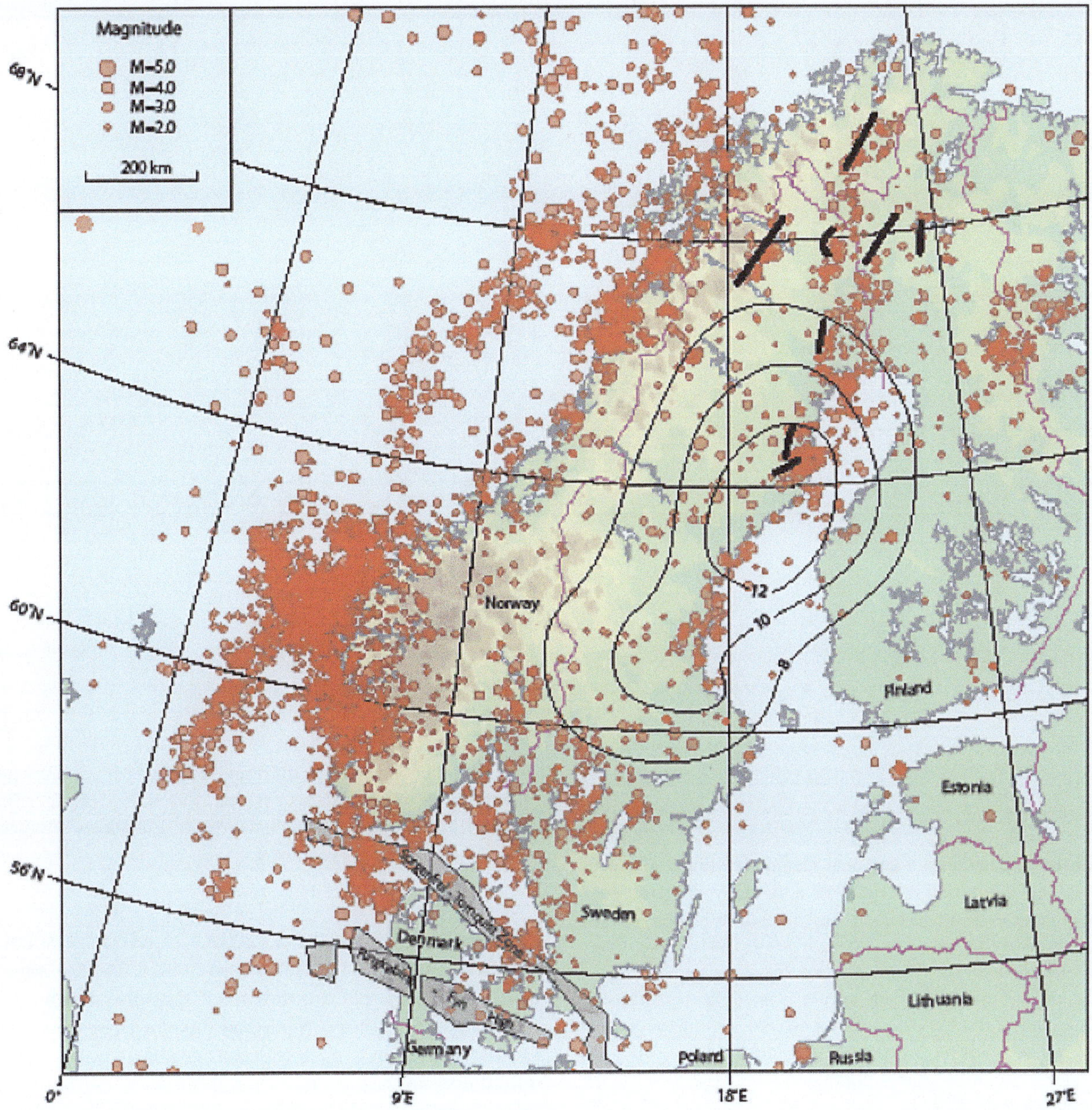

Fig. 1. Map of the earthquake geography in Scandinavia covering the years from January 1970 to December 2004. Earthquakes in the Danish area are extracted from the GEUS earthquake catalogue and earthquakes located outside this area are extracted from the Scandinavian catalogue from Helsinki University. The earthquake magnitude scale is given in the upper left corner. Contour lines show uplift in millimetres per year. Very thick black lines show the large postglacial faults of age close to 9000 yr (Lagerbäck, 1991). Updated earthquake files for Denmark are available in home page http://www.geus.dk/ under seismology, and for the rest of Scandinavia in home page http://www.seismo.helsinki.fi/. From Gregersen and Voss (2009).

2012; Hansen et al., 2011) as well as from geological sources (Mertz, 1924; Paasse, 1996; Christensen, 2001). These maps all show a very regular pattern in the area, with the exception of a line across southern Denmark introduced by Hansen et al. (2011), which will be treated later in this paper as claim 3. The most complete geodetic map, that by Knudsen et al. (2012), has been chosen as reference for the locations of the treated irregularities (Fig. 3). It takes into account permanent GPS, three campaigns of precise levelling over the last hundred years as well as ten sea level stations in the area.

4.1 Claim 1 – Carlsberg fault in the Copenhagen area

This fault is one of the most significant in the Copenhagen area (e.g. Rosenkrantz, 1937; Nielsen and Thybo, 2004;

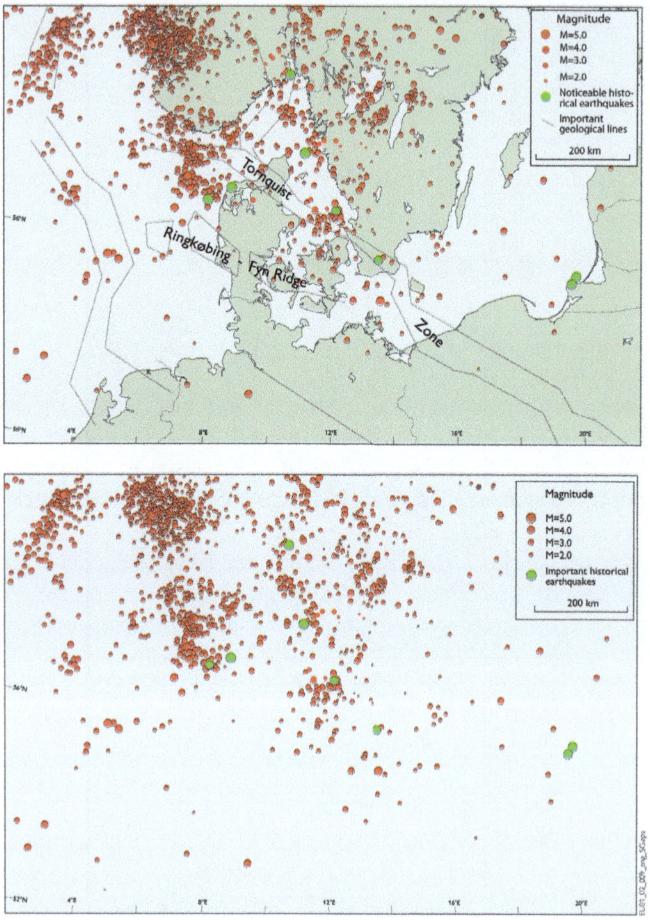

Fig. 2. Earthquake activity in and near Denmark without and with coast lines and geological zones. Red dots show modern well-located earthquakes of the last 75 yr, and green dots show the most important relatively larger earthquakes in 1759, 1841, 1904, 1985, 2004, 2008, and 2010. The two maps cover the same area. Denmark is in the middle. Upper right hand corner is inside the Fennoscandian Shield, and in the lower left corner we are outside the shield. A gradual and irregular transition from the shield is noted. This can be termed the edge of the shield in relation to stresses and strains. From Gregersen et al. (2011).

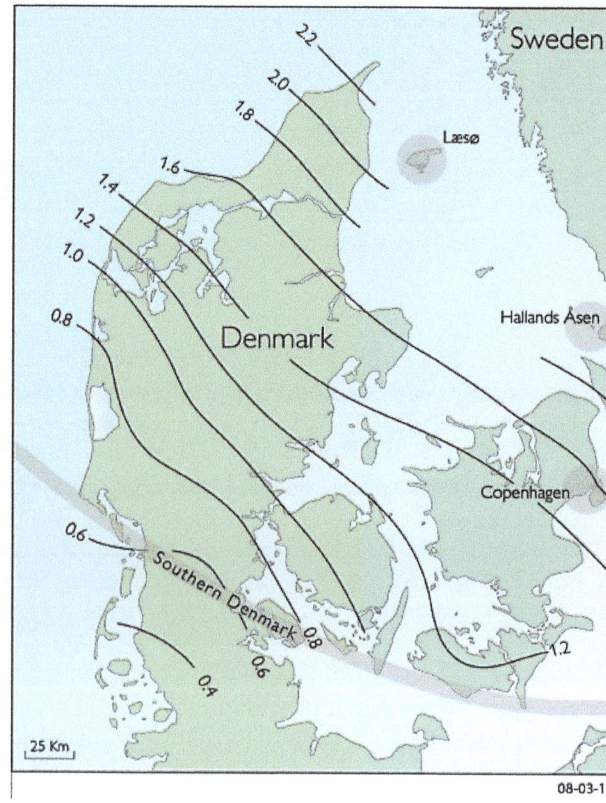

Fig. 3. Modern geodetic map (from Knudsen et al., 2012) with post-glacial uplift in centimetres per year. Based on GPS, levelling and sea level stations. Four treated claims of earthquakes or other irregularities shown by grey shading.

Nielsen et al. 2005) (marked as Copenhagen on Fig. 3). It is part of the stepwise edge zone between the Fennoscandian shield and the deep sedimentary Norwegian–Danish Basin with depth to bedrock under Copenhagen 3–4 km. The Carlsberg fault runs NNW–SSE. It is only known well near the Carlsberg Brewery, to which the excellent water flow in a 500 m wide zone is very important. The vertical displacement of the near-surface Cretaceous limestone is about 90 m. It is only known (Nielsen et al., 2005) to depths around 500 m. Its length is only well confirmed to 10–12 km (Blem, 2002), or tentatively about 20 km (Jakobsen et al., 2002). The fault trace on the surface has been claimed observed through extensive cracking of house walls (B. Larsen, personal communication, 2009; O. Winther Christensen, personal communication, 1990s; Ovesen et al., 2002; L. Nielsen, personal communication, 2009). A very basic remark in assessing these observations is that a comparison with the surroundings is necessary. And this comparison has not been done. Motion on the fault stands as a suggestion, not as a result, and the alternative suggestion that irregularities are caused by differential settling of the loose upper sediments is more probable. A project with satellite radar scanning will as time goes have a chance to localize patterns of surface motion (Jakobsen et al., 2013). Tentatively the Carlsberg fault is pointed out over a length of 25 km. Discussions on causes besides water pumping can be expected.

As seen in Fig. 2, the fault is not active with small earthquakes. It has been suggested by Rosenkrantz (1934, 1937) that the larger-than-usual earthquake which shook Copenhagen in 1930 can possibly be ascribed to this fault. Assessing this today, knowing the large uncertainties of earthquake locations in this area at that time, one must conclude that it could just as well be on another fault, easily ten kilometres away, and no indication has been observed for the choice.

The Carlsberg fault has also been held responsible for misfits in the Danish geodetic systems. The basis of the distance measuring system on the island Amager in Copenhagen has been described as deformed (Sand and Madsen, 1916; Ovesen et al., 2002). A large difference of 51 mm was reported between 1838 and 1911 of the basis line length of 2.7 km. The more probable explanation of this (M. Aarestrup, personal communication, 1991; K. Engsager, personal communication, 2009) is near-surface sediment motion of one of the end points by temperature effects, especially freezing and thawing, toward which modern geodesists protect their measurements by relying on a group of separated points at each end rather than just one vulnerable point. Comparisons to basis length measurements north-west of Copenhagen show standard errors in the old data, so large that the length change turns out to be insignificant (Karsten Engsager, personal communication 2010). Support for a conclusion on stable length of the basis is that the 1838 measurement stands alone, while the 1911 basis length has been confirmed with standard errors 3 mm in 1933, 1934 as well as 1980 and 1993 (O. B. Andersen, personal communication, 1993), i.e. before and after the 1930 earthquake.

Also another basic distance in Amager (in Danish called *prøvebane*, trial distance) has been reported deformed (e.g. Ovesen et al. 2002). For this the geodetic problem has been observationally identified as the above-mentioned end point trouble. One of the ends is sloping strangely, so the end must have been disturbed by winter freezing and thawing (Karsten Engsager, personal communication 2009). Recently doubt has been presented whether the extension southwards of the Carlsberg fault actually crosses the geodetic lines (Ovesen et al. 2002).

In a paper on the levelling in the area of Copenhagen, Mark and Jensen (1982) have the following conclusion for the period between the 1940s and the 1970s: "An adjustment of the vertical displacements shows a tilting about twice as big as the one you know for Denmark as a whole, but the direction of the isolines is about the same. [...]Possible displacements in connection with the so-called Carlsberg fault can't be recognized in the results, but maybe some systematical displacement in Amager can be explained by the graben-structure in the area between the Carlsberg fault and the Svanemølle fault."

The assessment of the above discussion will have to conclude that the Carlsberg fault is probably responsible neither for displacements in earthquakes nor for geodetic reference line problems. There is no reason to believe that the Carlsberg fault is active based on the presented observations as these can easily be explained by alternative phenomena in a non-seismic area. In the similar Norwegian evaluation by Olesen et al. (2004) this would be classified as E, very unlikely to be neotectonics.

4.2 Claim 2 – The island Læsø in the Kattegat Sea

Hansen (1977, 1980, 1986a, b, 1994, 1995) and Bahnson (1986) have made geological investigations of sand deposits of ages less than 7000 yr in Læsø and found indications of a large fault of vertical displacement of 4 m, from the surface to a depth of a few kilometres, and for an unknown length, which can at most be about 10 km in the terrain as it is. An earthquake has been proposed although no other indicators of earthquakes have been identified. The present assessment evaluates how this fault together with other indicators of postglacial movement fits into the tectonic setting close to the Sorgenfrei–Tornquist Zone.

In the 1980s the seismicity of the area was only marginally known (e.g. Lehmann, 1956; Gregersen, 1979), but tectonic activity was assumed by Bahnson (1986) and Hansen (1986a, b) with reference to structural studies showing many steep old faults (e.g. Baartman 1975). It was argued (Hansen, 1977; Bahnson, 1986) that Læsø does not fit in the postglacial uplift pattern of the neighboring Kattegat coasts. In a later paper this argument (Hansen 1994) was modified to: "Strangely enough the oldest beach line on Læsø [...] is at the same height as by Frederikshavn." How well the uplift curves of the highest beach lines in Læsø fit into the surroundings has later been illustrated by Christensen (2001) and quoted by Gregersen and Voss (2012).

Now assessing the Læsø observations 25 years later, we have more data on the region, both on the scale of Kattegat and for Scandinavia, and for other intraplate regions. This has changed the interpretation and understanding of the Læsø case. The question earlier was whether Læsø would fit into its assumed active surroundings. The question is now whether Læsø fits into its quiet, regular uplift surroundings. A review of the most significant arguments gives the following results:

1. As mentioned the maximum uplift of Læsø since the regional flooding after the Ice Age fits nicely into the regional picture. That the island of Læsø should have subsided 4 m and then uplifted the same number of metres within several thousands of years is improbable, since the effects of such an event would be recognisable in neighboring coasts, where nothing is observed. And even more improbable, subsidence and uplift several times as suggested in the book by Hansen (1994) is not a possibility.

2. The alternation between transgressions and regressions in a time interval of several thousand years right after the Litorina Sea transgression is found similar in other locations of the Kattegat area as referred to in the overview paper by Noe-Nygård (2006) as well as that by Christensen (2001) and by Bjørnsen et al. (2008). The phenomenon is called the Litorina transgressions (Christensen, 1995). The last one ended 4000–5000 yr ago.

3. The paper by Baartman (1975) spoke about structural differences. The reference does not give any argument on recent activity. The structural steps and faults are confirmed by the later projects EUGENO-S (EUGENO-S Working Group, 1989) and Tor

(e.g. Gregersen et al. 2009). But as an argument concerning recently active faulting, the reference is contradicted by recent evaluations of the earthquake activity (Figs. 1 and 2) (e.g. Gregersen et al. 1995 and 1998).

4. A curve showing changes of tilting of successive beach terraces was presented by Hansen (1980). When the slopes were calculated no evaluation was made of uncertainties (Hansen, 1980). The slopes were based on heights, averaged over 10–15 individual height measurements within a spread of approximately one metre. This one metre can be taken as a rough estimate of four times the standard error. The present assessment extends the data handling to take into account this estimated standard error of a quarter of a metre. This means intoducing standard errors of the slopes, over distances smaller than 5 km, at least as large as shown in Fig. 4, the basis of which is identical to Fig. 5 of Hansen (1980). It is here obvious that the tilting argument has become non-significant. The observed differences and the standard errors are of the same size, as shown in Fig. 4 of the present paper. The zigzag line of Hansen (1980) goes through the average values of the slopes. But the data may as well be explained by the thick straight line, that is within the belt of intervals determined by averages plus/minus one standard error. So the present assessment concludes that the differences in average slopes can not be used as an argument for discontinuous and occasionally reverse tilting. The data agree as well with steady uplift and tilting.

   The errors mentioned above are directly concerned with the old data of Hansen (1980), which were used to show tilting of the beach terraces with respect to horizontal water level. But it will also apply to modern data, which is now available for Læsø. The same size of standard errors in the use of geological markers for slope measurements and for height measurements with respect to water level, are found in recent work by Nielsen and Clemmensen (2009). They observe similar large measurement scatter in modern data on Anholt, in the Kattegat Sea. In addition one must take into account that the physical circumstances of ocean currents and stormy weather influence the slopes, so these observations in Fig. 4 can not give an argument for oscillating tectonic tilting by themselves.

5. An almost vertical fault displacement in the sediments of approx. 4 m, as said by Hansen (1986a, b), can be caused by displacement of various blocks in Læsø. The location of the fault is presented as a coastal cliff at the time of development, and coastal cliffs experience sliding by irregular erosion and gravity. The southern part of Læsø must have experienced a sediment slide along the observed fault, either fast or slow.

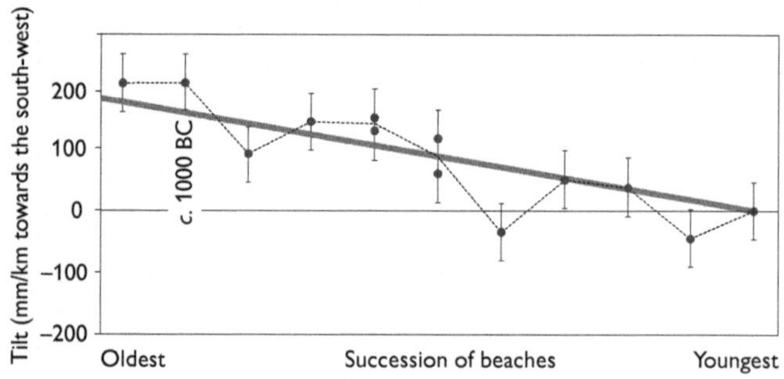

Fig. 4. Slopes of successive beach terraces of diminishing age to the right, with estimated standard errors. When standard errors are introduced in this Fig. 5 of Hansen (1980) it is obvious that abrupt changes of slopes are not significant. The slope readings are just as well explained by a steady change of slopes, approximated by the thick sloping line generally inside one standard error and certainly within two standard errors.

   Sediment cliff slides are more probable in Læsø than an earthquake, for which there is no other evidence in this aseismic area.

A whale in a height of 4.5 m, of age approx. 3000 yr as well as other dates have to fit into the regional frame. It may well show that the part of the island on which the whale rests did not follow the steady uplift of the oldest part of the island. A slide in a coastal cliff is naturally of great significance to the detailed geological description of Læsø, but of no significance to the tectonic evolution of the greater Kattegat region. The present topography is convincingly a succession of coastal terraces, which display a recent geological evolution explained in great detail by Hansen (1977, 1980, 1986a, b, 1994 and 1995) and followed up in recent further investigations (Hansen et al., 2011).

The present assessment of all of the old and new arguments concludes that the bedrock subsurface below Læsø has been uplifted just as the rest of the Kattegat region, regularly or with minor irregularities. This means that there was no Læsø earthquake 4000–5000 yr ago. In the classification used by Olesen et al (2004) this means class E, very unlikely to be neotectonics. The lower parts of Læsø with many well-documented beach terraces have their own exciting development history (e.g. Hansen 1994; Hansen et al., 2011), best explained by regular postglacial uplift together with coastal development by water currents and weather, which is beyond the scope of this paper.

4.3 Claim 3 – Line across southern Denmark

A new map has introduced a hypothetical irregularity line called southern Denmark in Fig. 3. It is based on the geodetic/oceanographic data of water level stations of the last 100 yr (Hansen et al., 2011). It includes a few more water level stations in the southern part than in Fig. 3, but is without

Fig. 5. Numbered beach ridges on Hallandsaasen seen from above. A major earth slide in the ringed area in the lower right side of the picture covers the older beach ridges (6–9) but not the younger ones (1–5). The view as seen during the field trip. The photograph is taken by the Russian seismologist Olga Pavlenko. Numbering is from Mörner (2003).

Fig. 6. Seismite? Disturbed water-filled sediment. Liquefaction. Photo by Soren Gregersen.

GPS and levelling, which in Fig. 3 supplemented the water level data. Instead of accepting irregularities in a few stations in southern Denmark as caused by weather and ocean currents, Hansen et al. (2011) focus arbitrarily on bedrock differences that show no motion in the seismological data or in the geodetic levelling and GPS. The uplift picture of Fig. 3 is more representative of the full data set, which does not have a discontinuity in southern Denmark. The claim can be classified by E in the classification used by Olesen et al. (2004), very unlikely to be neotectonics

South of Læsø in the Kattegat Sea (i.e. south of Claim 3 and north-west of Claim 4), the earthquake activity is interpreted from seismology (Gregersen et al., 1998). The seismicity picture as presented in Figs. 1 and 2 shows a concentration of earthquakes in a short part of the Sorgenfrei–Tornquist Zone, where the zone changes character (e.g. Graversen 2009). This has been described by Arvidsson et al. (1991), Gregersen et al. (1996) and Jensen et al. (2002) as connected to confirmed recently active basement faults. Here there is potential for earthquakes of magnitudes as large as around 5. This is not far from Hallandsaasen (Fig. 3). Does it mean that the claim of postglacial earthquakes there is supported?

4.4 Claim 4 – West coast of Sweden

Several locations on the Swedish west coast have been mentioned as postglacial earthquake sites by Mörner (2003, 2009). Signs of geological movement in these locations are variations in postglacial uplift of beach lines, faulting and rock deformation in a few cases, and in many incidences rock slides and liquefaction in several stratigraphic layers. In several places liquefaction deformations have been observed and in some also signs of tsunami disturbances of the sediments. The evidence for earthquake interpretation is not as overwhelming as in northern Sweden in the layers from the end of the Ice Age (e.g. Lagerbäck, 1991), but it seemed worthwhile investigating the claims of postglacial earthquakes by Mörner (2003, 2009). A field trip to Hallandsaasen of 12 international seismologists was carried out on 27 July 2013. Nils-Axel Mörner added excellent educational details to his publications (Mörner, 2003, 2009). Figure 5 is exactly like one from Mörner (2003). The view, as seen during the field trip, convinces that a rock slide has disturbed the metre high steps of the numbered beach ridges 6 to 9. The following pictures (Figs. 6 and 7) were taken of disturbances of dated water filled sediments (called seismites) and shattered rocks.

The field trip has given the evaluation that the signs shown of palaeo-earthquakes each could be caused by earthquake shaking, or by its own individual other cause. Misfit of beach lines in neighboring locations 400 m from each other could have been caused by slow non-earthquake motion or different sea wave and current systems. Talus falls and slides could occur without a starting earthquake. Liquefaction disturbance of sediment layers could be caused by water escape or slumping or bending of layers. Shattering of exposed bedrock could be caused by weather and climate, and tsunami signs in sediments are indications of disturbances, which could happen by non-earthquake slides as well as by earthquakes. Together one is left with a choice in the evaluation between coincidences or common cause: earthquake. The field trip convinced the participating seismologists that the claimed signs of earthquakes must be seriously taken into account.

Into the interpretation must go the overall pattern of gradual uplift of the whole region, into which only small irregularities fit. And when this is seen in connection with the drastic fall-off of stresses since the Ice Age (e.g. Gregersen and Voss, 2009), the present assessment results in a category D of

Fig. 7. Bedrock cliffs strongly fractured and moved by shaking or weather changes. Nils-Axel Mörner explaining the phenomenon. Photo by Soren Gregersen.

the classification used by Olesen et al. (2004), which means probably not neotectonics, because coincidence is the most probable explanation.

In Skåne, southern Sweden repeated GPS measurements by Pan et al. (1999) have indicated differences between the two sides of the Tornquist Zone. These measurements are not considered statistically significant by our Swedish geodesist colleagues, who now use permanent GPS stations. A small, seldom earthquake happened in Skåne 16 December 2008 on the north-eastern flank of the Tornquist Zone (Fig. 2). This earthquake was as small as the others that have been located in the surrounding region (Voss et al., 2009).

5 Conclusions

The overall conclusion is that a number of reports, claiming geologic or geodetic recent irregularity in and near Denmark, have been found and discussed. It has been identified that a supplementary geological comparison of the uplift phenomenon with identical counting of beach lines and identical age determinations across the Kattegat between the claims in Denmark and that in south-western Sweden will be worthwhile. The assessment has uncovered no signs of geologic recent faulting or recent crustal deformations, which contradict the seismicity picture of minor earthquake activity, up to a magnitude just below 6. The earthquake zones are those found by historical and instrumental-seismological investigations in the seas around Denmark as illustrated in Figs. 1 and 2. The activity is geographically spread out. Not even the very significant Sorgenfrei–Tornquist Zone is an irregularity zone with earthquakes.

Acknowledgements. We appreciate many good discussions with our colleagues Jens Morten Hansen, Peter Johannesen, Lars Henrik Nielsen, Tine B. Larsen, Tanni J. Abramovitz, Abbas Khan, Karsten Engsager, Holger Lykke-Andersen, Torben Bidstrup, Birger Larsen, Jørgen Leth, Lars Nielsen, Henrik Olsen, Hans Thybo and Lars B. Clemmensen.

Special Issue: "Lithosphere–cryosphere interactions"
Edited by: M. Poutanen, B. Vermeesen, V. Klemann, and C. Pascal

References

Ågren, J. and Svensson, R.: Postglacial Land Uplift Model and System Definition for the new Swedish Height System RH 2000, Reports in Geodesy and Geographical Information System Rapportserie, LMV-Rapport 2007:4, Lantmäteriet, Gävle, Sweden, 2007.

Arvidsson, R., Gregersen, S., Kulhanek, O., and Wahlström, R.: Recent Kattegat earthquakes – evidence of active intraplate tectonics in southern Scandinavia, Phys. Earth Plan. Int., 67, 275–287, 1991..

Baartman, H.: Some results of the interpretation of seismic data in the Kattegat, S. Øresund, and Østersø, W., in: Contributions to the interpretation of the Fennoscandian Border Zone, edited by: Baartman, H. and Christensen, O. B., Danmarks Geologiske Undersøgelse, 102, 6–20, 1975.

Bahnson, H.: Lithologisk undersøgelse af materiale fra seks boringer på Læsø, Danmarks Geologiske Undersøgelse serie D, 6, 7–27, 1986.

Bjørnsen, M., Clemmensen, L. B., Murray, A., and Pedersen, K.: New evidence of the Littorina transgressions in the Kattegat – Optically stimulated luminescence dating of a beach ridge system on Anholt, Denmark, Boreas, 37, 157–168, 2008.

Blem, H.: Carlsberg Fault – History, location and importance (in Danish), B. Danish Geotechnic. Soc., 19, 61–82, 2002.

Bungum, H., Alsaker, A., Kvamme, L. B., and Hansen, R. A.: Seismicity and seismotectonics of Norway and nearby continental shelf areas, J. Geophys. Res., 96, 2249–2265, 1991.

Bungun, H., Pascal, C., Olesen, O., Lindholm, C., Vestøl, O., and Gibbons, S.: To what extent is the present seismicity of Norway driven by postglacial rebound?, J. Geol. Soc. London, 167, 373–384, 2010.

Christensen, C.: The Litorina transgressions in Denmark, in: Man and sea in the Mesolithic, edited by: Fischer, A., Oxbow Monograph, Oxford, 53, 15–22, 1995.

Christensen, C.: Coastal settlement and sea level change in the Stone Age, in: Jensen, O. L., Soerensen, S. A., and Hansen, K. M., Denmarks Hunting Stone Age – status and perspectives, Hoersholm Egns Museum 2001, 183–193, 2001.

EUGENOS Working Group: Crustal structure and tectonic evolution of the transition between the Baltic Shield and the North German Caledonides (the EUGENOS Project), Tectonophysics, 150, 253–348, 1988.

Graversen, O.: Structural analysis of superposed fault systems of the Bornholm horst block, Tornquist Zone, Denmark, Bull. Geol. Soc. Denmark, 57, 25–49, 2009.

Gregersen, S.: Earthquakes in the Skagerrak recorded at small distances. Bulletin of the Danish Geological Society, 28, 5–9, 1979.

Gregersen, S.: Earthquake activity in the North Sea area, Dansk Geologisk Forening, Årsskrift for 1981, 157–159, 1982 (in Danish).

Gregersen, S.: Crustal stress in northern Europe, in: Proceedings of the VI workshop on the European Geotraverse Project, edted by: Freeman, R. and Müller, S., ESF Strasbourg, 357–360, 1990.

Gregersen, S.: Crustal stress regime in Fennoscandia from focal mechanisms, J. Geophys. Res., 97, 11821–11827, 1992.

Gregersen, S.: Earthquakes and change of stress since the ice age in Scandinavia, Bull. Geol. Soc. Denmark, 49, 73–78, 2002.

Gregersen, S. and Basham, P. W. (Eds.): Earthquakes at North Atlantic Passive Margins: Neotectonics and Postglacial Rebound, Kluwer Academic Press, 716 pp., 1989.

Gregersen, S. and working group on recent crustal movements around the Tornquist Zone: Recent crustal movements and earthquakes in the area of the Tornquist Zone, Studia Geophysica et geodaetica, 39, 257–261, 1995.

Gregersen, S. and Voss, P.: Stress change over short geological time: case of Scandinavia over 9000 years since the Ice Age, in: Paleoseismology: Historical and prehistorical records of earthquake ground effects for seismic hazard assessment, edited by: Reicherter, K., Michetti, A. M., and Silva Barroso, P. G., Geological Society of London, Special Publications, 316, 173–178, 2009.

Gregersen, S. and Voss, P.: Irregularities in Scandinavian postglacial uplift/subsidence in time scales tens, hundreds, thousands of years, J. Geodyn., 50, 27–31, 2010.

Gregersen, S. and Voss, P. H.: Efforts to include geological and geodetic observations in the assessment of earthquake activity in Denmark, Geological Survey of Denmark and Greenland Bulletin, 26, 41–44, 2012.

Gregersen, S., Leth, J., Lind, G., and Lykke-Andersen, H.: Earthquake activity and its relationship with geologically recent motion in Denmark, Tectonophysics, 257, 265–273, 1996.

Gregersen, S., Hjelme, J., and Hjortenberg, E.: Earthquakes in Denmark, Bull. Geological Soc. Denmark, 44, 115–127, 1998.

Gregersen, S., Wiejacz, P., Dębski, W., Domanski, B., Assinovskaya, B., Guterch, B., Mäntyniemi, P., Nikulin, G., Pacesa, A., Puura, V., Aronov,A. G., Aronova, T. I., Grünthal, G., Husebye, E. S., and Sliaupa, S.: The exceptional earthquakes in Kaliningrad District, Russia on September 21, 2004, Phys. Earth Planet. Inter., 164, 63–74, 2007.

Gregersen, S., Voss, P., Nielsen, L. V., Achauer, U., Busche, H., Rabbbel, W., and Shomali, Z. H.: Uniqueness of modelling results from teleseismic P-wave tomography in Project Tor. Tectonophysics, 481, 99–107, doi:10.1016/j.tecto.2009.01.020, 2009.

Gregersen, S., Larsen, T. B., and Voss, P. H.: Earthquakes on the Sorgenfrei-Tornquist Zone, Geoviden 2011 no. 1. GEUS Copenhagen, 2011 (in Danish).

Grünthal, G., Stromeyer, D., Wylegalla, K., Kind, R., Wahlström, R., Yuan, X., and Bock, G.: The Mw 3.1–4.7 earthquakes in the southern Baltic Sea and adjacent areas in 2000, 2001 and 2004, J. Seismology, 12, 413–429, doi:10.1007/s10950-008-9096-0, 2008.

Hansen, J. M.: Sedimentary history of the island Læsø, Denmark, Bull. Geol. Soc. Denmark, 26, 217–236, 1977.

Hansen, J. M.: Læsøs postglaciale udvikling i relation til den Fennoskandiske Randzone, Dansk Geologisk Forening, Årsskrift for 1979, 23–30, 1980.

Hansen, J. M.: Læsø: A result of fault displacements, earthquakes and level changes, Danmarks Geologiske Undersøgelse serie D, 6, 47–72, 1986a (in Danish).

Hansen J. M.: A prehistoric earthquake, Museumsforeningen for Læsø, Årbog for 1985, 17–23, 1986b (in Danish).

Hansen, J. M.: Læsø's evolution and landscapes – about the island that rocks and jumps , Danish Geological Survey, Geografforlaget, Brenderup, Denmark, 55 pp., 1994 (in Danish).

Hansen, J. M.: The appearance of an island, its beach development and vegetation evolution: Natural and human-made landscapes on Læsø, Geologisk Tidsskrift, 2, 1–74, 1995 (in Danish).

Hansen, J. M., Aagaard, T., and Binderup, M.: Absolute sea levels and isostatic changes of the eastern North Sea to central Baltic region during the last 900 years, Boreas, 41, 180–208, doi:10.1111/j.1502-3885.2011.00229.x, 2011.

Hicks, E. C., Bungum, H., and Lindholm, C. D.: Stress inversion of earthquake focal mechanism solutions from onshore and offshore Norway, Norsk Geol. Tidsskr., 80, 235–250, 2000.

Jakobsen, P. R., Fallesen, J., and Knudsen, C. Structures in the Copenhagen subsurface – folds, faults and cracks (in Danish), B. Danish Geotechnic. Soc., 19, 19–29, 2002.

Jakobsen, P. R., Wegmuller, U., Capes, R., and Pedersen, S. A. S.: Terrain subsidence detected by satellite radar scanning of the Copenhagen area, Denmark, and its relation to the tectonic framework, Geol. Surv. Den. Greenl., 26, 25–28, 2013.

Jensen, J. B., Petersen, K. S., Konradi, P., Kuijpers, A, Bennike, O., Lemke, W., and Endler, R.: Neotectonics, sea-level changes and biological evolution in the Fennoscandian Border Zone of the southern Kattegat Sea, Boreas, 31, 133–150, 2002.

Knudsen, P., Engsager, K. E., and Khan A. K.: Documentation of calculation of new uplift model 2011, Report from DTU-Space to the Danish Geodata-authority. Home page of DTU-Space, 2012 (in Danish).

Lagerbäck, R.: Late Quaternary faulting and paleoseismicity in northern Scandinavia, with particular reference to the Lansjärv area, northern Sweden. Geologiska Föreningens i Stockholm Förhandlinger, 112, 333–354, 1990.

Lagerbäck, R.: Seismically deformed sediments in the Lansjärv area, northern Sweden. SKB Technical Report 91-17, Svensk Kärnbränslehandtering AB, Stockholm, Sweden, 58 pp., 1991.

Lambeck, K., Purcell, A., Zhao, J., and Svensson, N.-O.: The Scandinavian Ice Sheet: from MIS 4 to the end of the Last Glacial Maximum, Boreas, 39, 410–435, doi:10.1111/j.1502-3885.2010.00140.x, 2010.

Lehmann, I.: Danish earthquakes, Meddelelser for Dansk Geologisk Forening, 13, 88–103, 1956 (in Danish).

Lindholm, C. D., Bungum, H., Hicks, E., and Villagran, M.: Crustal stress and tectonics in Norwegian regions determined from earthquake focal mechanisms, Geol. Soc. London, 167, 429–439, 2000.

Mark, A. N. and Jensen, J. E. K.: Level changes in Copenhagen, Landinspektøren, 12 pp., 1982.

Mertz, E. L.: Late and post-glacial height changes in Denmark, Danish Geological Survey (DGU) 2. række, 41, 50 pp., 1924 (in Danish).

Mörner, N.-A.: The Fennoscandian uplift and late Cenozoic geodynamics: Geological evidence, GeoJournal, 3, 287–318, 1979.

Mörner, N.-A.: Paleoseismicity of Sweden. A novel paradigm, JOFO Grafiska AB, Stockholm, 2003.

Mörner, N.-A.: Late Holocene earthquake geology in Sweden, in: Paleoseismology: Historical and prehistorical records of earthquake ground effects for seismic hazard assessment, edited by: Reicherter, K., Michetti, A. M., and Silva Barroso, P. G., Geol. Soc. London, Special Publications, 316, 179–188, doi:10.1144/SP316.10, 2009.

Nielsen, L. and Clemmensen, L. B.: Sea-level markers identified in ground-penetrating radar data collected across a modern beach ridge system in a microtidal regime, Terra Nova, 21, 474–479, 2009.

Nielsen, L. and Thybo, H.: Location of the Carlsberg Fault zone from seismic controlled-source fan recordings, Geophys. Res. Lett., 31, L07621, doi:10.1029/2004GL019603, 2004.

Nielsen, L. Thybo, H., and Jørgensen, M. I.: Integrated seismic interpretation oft he Carlsberg Fault zone, Copenhagen, Denmark, Geophys. J. Int., 162, 461–478, 2005.

Noe-Nygård, N.: From Ice Age to Stone Age, in: The Nature of Denmark, Geology, Gyldendal, Copenhagen, 2006 (in Danish).

Olesen, O., Blikra, L. H., Braathen, A., Dehls, J. F., Olsen, L., Rise, L., Roberts, D., Riis, F., Faleide, J. I., and Anda, E.: Neotectonic deformation in Norway and its implications: a review, Norwegian J. Geol., 84, 3–34, 2004.

Olesen, O., Bungum, H., Dehls, J., Lindholm, C., Pascal, C., and Roberts, D.: Neotectonics, seismicity and contemporary stress field in Norway – mechanisms and implications, NGU Special Publication, 13, 143–172, 2012.

Ovesen, N. K., Blem, H., Gregersen, S., Møller, H. M. F., and Frederiksen, J. K.: Recent terrain-movements in Copenhagen, Bull. Danish Geotechn. Soc., 19, 183–192, 2002 (in Danish).

Pan, M., Sjöberg, L. E., Talbot, C.. and Asenjo, E.: GPS measurements of crustal deformation in Skåne, Sweden between 1989 and 1996, Geofys. Fören. Forhandl., 121, 67–72. 1999.

Pascal, C. and Cloetingh, S. A. P. L.: Gravitational potential stresses and stress field of passive continental margins: Insights from the south-Norway shelf, Earth Planet. Sc. Lett., 277, 464–473, 2009.

Poutanen, M., Dransch, D., Gregersen, S., Haubrock, S., Ivins, E. R., Klemann, V., Kozlovskaya, E., Kukkonen, I., Lund, B., Lunkka, J.-P., Milne, G., Müller, J., Pascal, C., Pettersen, B.-R., Scherneck, H.-G., Steffen, H., Vermeersen, B., and Wolf, D.: DynaQlim – Upper mantle dynamics and Quaternary climate in cratonic areas, in: New Frontiers in Integrated Solid Earth Sciences, edited by: Cloetingh, S. and Negendank, J., International Year of Planet Earth Book Series, Springer Verlag, 349–372, 2009.

Paasse, T.: A mathematical model of the shore level displacement in Fennoscandia, Swedish Nuclear Waste management Company (SKB) Technical report 96-24, Stockholm Sweden, 92 pp., 1996.

Rosenkrantz, A.: Remark considering a fault in the subsurface of southeastern Copenhagen, Bull. Geol. Soc. Denmark, 8, 407–410, 1934 (in Danish).

Rosenkrantz, A.: Remarks on the Danian stratigraphy and tectonics on eastern Sealand, Bull. Geol. Soc. Denmark, 9, 199–212, 1937 (in Danish).

Sand, M. J. and Madsen, V. H. O.: New Basis measurements in Denmark. Den Danske Gradmaaling, Ny række, hefte 15, Copenhagen, 1916.

Vestøl, O.: Determination of postglacial land uplift in Fennoscandia from levelling, tide-guages and continuous GPS stations using least squares collocation, J. Geodynamics, 80, 248-258, 2006.

Voss, P., Larsen, T. B. Ottemöller, L., and Gregersen, S.: Earthquake in southern Sweden wakes up Denmark on 16 December 2008, GEUS Review of Survey Activities (ROSA) 2008, 9–12, 2009.

Zoback, M. L.: First and second-order patterns of stress in the lithosphere: The World Stress Map Project, J. Geophys. Res., 97, 11703–11728, 1992.

Zoback, M. L., Zoback, M. D., Adams, J., Assumpcao, M., Bell, S., Bergmann, E. A., Bluemling, P., Brereton, N. R., Denham, D., Ding, J., Fuchs, K., Gay, N., Gregersen, S., Gupta, H. K., Gvishiani, A., Jacob, K., Klein, R., Knoll, P., Magee, M., Mercier, J. L., Müller, B. C., Paquin, C., Rajendran, K., Stephansson, O., Suarez, G., Suter, M., Udias, A., Xu, Z. H., and Zhizhin, M.: Global patterns of tectonic stress, Nature, 341, 291–298, 1989.

New insights on the occurrence of peperites and sedimentary deposits within the silicic volcanic sequences of the Paraná Magmatic Province, Brazil

A. C. F. Luchetti[1], **A. J. R. Nardy**[1], **F. B. Machado**[2], **J. E. O. Madeira**[3], **and J. M. Arnosio**[4]

[1]Instituto de Geociências e Ciências Exatas, Universidade Estadual Paulista, Rio Claro, SP, 13506-900, Brazil
[2]Departamento de Ciências Exatas e da Terra, Universidade Federal de São Paulo, São Paulo, SP, 09920-540, Brazil
[3]Departamento de Geologia da Faculdade de Ciências (GeoFCUL), Instituto Dom Luiz, Universidade de Lisboa, Lisboa, Portugal
[4]Instituto Geonorte, Facultad de Ciencias Naturales, Universidad Nacional de Salta, Salta, Argentina

Correspondence to: A. C. F. Luchetti (caroluch@rc.unesp.br)

Abstract. The PMP (Paraná Magmatic Province) is characterized by lava flows of the Early Cretaceous Serra Geral Formation which covers about 75 % of the Paraná Basin (southern and southeastern Brazil), composed of a thick (up to 1600 m) volcanic sequence formed by a succession of petrographically and geochemically distinct units of basic and silicic composition. The whole package must have been emplaced during approximately 3 million years of nearly uninterrupted activity. A few aeolian sandstone layers, indicating arid environmental conditions (the Botucatu Formation), are interlayered in the lower basalts. Above the basalts, the Palmas and Chapecó Members are composed of silicic volcanic rocks (quartz latites, dacites, rhyodacites and rhyolites) and basalts. This paper presents new evidence of sedimentation episodes separating silicic volcanic events, expressed by the occurrence of sedimentary deposits. Interaction between the volcanic bodies and the coeval unconsolidated sediments formed peperites. The sediments were observed between basaltic lava flows and silicic rocks or interlayered in the Palmas-type rocks, between the Chapecó-type rocks and overlying basaltic flows, between silicic bodies of the Palmas and Chapecó types, and interlayered within Palmas-type units. The observed structures indicate that the sediments were still wet and unconsolidated, or weakly consolidated, at the time of volcanism, which, coupled with the sediment features, reflect environmental conditions that are different from those characterizing the Botucatu arid conditions.

1 Introduction

The Early Cretaceous Serra Geral Formation is the result of a major volcanic phase that covered about 917 000 km^2, about 60 % of the surface of the Paraná Basin (Frank et al., 2009; Fig. 1). Three main petrographic types can be distinguished from macroscopic observation of these rocks. The most common type are basalts, presenting predominant intergranular texture and its variations, including subophitic, intersertal and hialophitic. The other two types of silicic rocks present massive and aphyric textures (Palmas type – ATP) and porphyritic textures (Chapecó type – ATC).

Macroscopic characteristics allowed easy separation in the field of these two members of the Serra Geral Formation and their geological mapping (Bellieni et al., 1983; Piccirillo et al., 1988). The Palmas and Chapecó rock types occur in association with basaltic flows that are more common near the top and bottom of these two lithostratigraphic units. Geological mapping also shows that the Palmas and Chapecó Members cover 63 000 km^2, in the states of Paraná, Santa Catarina and Rio Grande do Sul. The volume of the two members amounts to approximately 14 500 km^3, which corresponds to 2.5 % of the total volume of the Serra Geral Formation (Nardy et al., 2002, 2008). Geochronological dating by $^{40}Ar/^{39}Ar$ shows that the age of volcanic rocks of the Serra Geral Formation ranges from 133.6 to 131.5 Ma in its northern sector, and from 134.6 to 134.1 Ma in the southern (Renne et al., 1992, 1996a, b; Turner et al., 1994; Ernest et al., 1999, 2002;

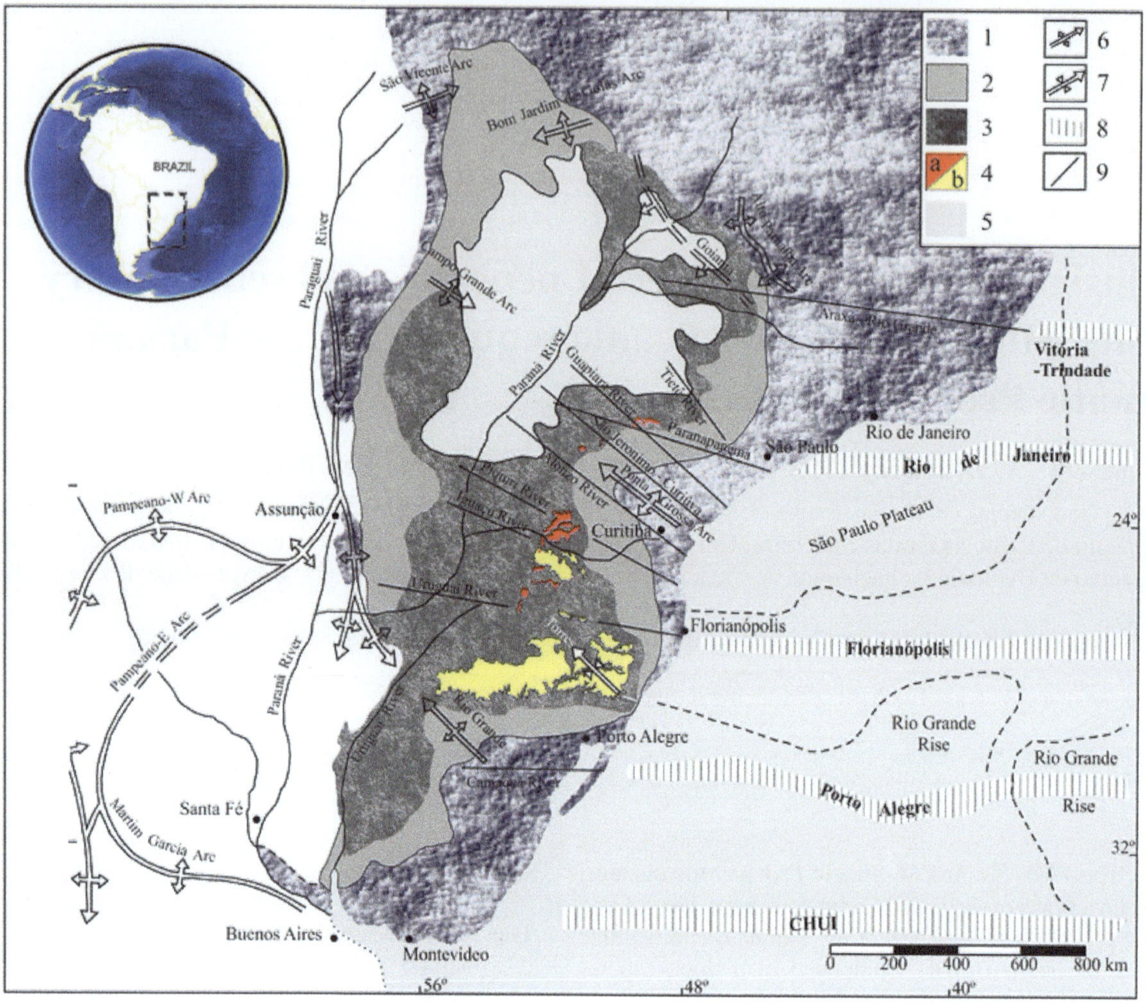

Fig. 1. Map of the Paraná Basin with the location of the acidic members of the Serra Geral Formation according to Nardy et al. (2008). Legend: 1 – pre-volcanic sedimentary rocks; 2 – basalts (the Serra Geral Formation); 3– silicic **(a)** Chapecó and **(b)** Palmas members (the Serra Geral Formation); 4 – sedimentary post-volcanic sequences (Bauru Basin); 5 – anticline structures; 6 – syncline structures; 7 – oceanic lineaments; 8 – continental lineaments.

Mincato et al., 2003; Thiede and Vasconcelos, 2010; Pinto et al., 2010). More recently, Janasi et al. (2011), using U / Pb ratios from baddeleyite/zircon crystals determined by isotope dilution thermal ionization mass spectrometry from rocks of the Chapecó Member, obtained an age of 134.3 ± 0.8 Ma, compatible with previous age determinations. However, ages obtained in the basaltic flows indicate the duration of the volcanism to be around 3 Ma, which is consistent with paleomagnetic data presented by Ernesto and Marques (2004).

Thus far, the presence of sediments (sandstones of the Botucatu Formation) intercalated in the volcanic sequence was only reported in the lower basaltic pile. These consist of sand bodies presenting aeolian structures such as bypass surfaces, single dunes, sand-filled cracks and multi-dune ergs (Jerram and Stollhofen, 2002; Petry et al., 2007; Waichel et al., 2008), representing a desert environment that persisted during the voluminous initial phase of basaltic volcanism. In this work, the occurrence of sedimentation and development of associated peperites in the final stage of the PMP (Paraná Magmatic Province) volcanic event is presented. The sediments, predominantly sandy–silty, and unrelated to the Botucatu Formation, indicate a change in the environmental conditions in the Paraná Basin, and attest to the occurrence of significant periods of quiescence during the final stages of the magmatic activity.

2 Petrographic and geochemical aspects of silicic volcanic rocks

The Palmas-type silicic volcanic rocks (ATP) are characterized by light-gray to brownish-red color, hypohyaline–holohyaline, aphyric to micro-porphyritic textures and a striking salt-and-pepper aspect, with some holohyaline units (pichstones) displaying black color, glassy luster and prominent conchoidal fractures. However, due to its amorphous nature, the glass is easily altered and thus in most outcrops the rock is completely weathered, presenting a brownish color and (often resembling sedimentary deposits) dotted with abundant vesicles and quartz-filled amygdales up to 10 mm in length. The mineralogy is composed of microphenocrysts (granularity smaller than 0.2 mm) of plagioclase (labradorite) – which comprise up to 16 % of the total volume of the rock – 11 % of augite, 3 % of pigeonite, 5 % of magnetite and less than 1 % of apatite. These crystals may exhibit rapid cooling structures (quenching), developing skeletal, lath and hollow shapes, or swallowtail terminations. The matrix reaches 63 % of the rock volume on average, and is composed of dark-brown, slightly birefringent glass, characterized by a granophyric texture of abundant microlites, and alkali feldspar and quartz intergrowth that surrounds the crystal phases.

The Chapecó-type silicic volcanic rocks (ATC) are porphyritic, with an average of 24 % of plagioclase phenocrysts up to 2 cm long, in a light-gray (when fresh) to brown (when weathered) aphanitic matrix. The mineralogy consists of euhedral andesine phenocrysts in a matrix composed of 4.5 % of augite, 2.2 % of pigeonite, 3.7 % of magnetite, and 1.7 % of apatite (average composition) surrounded by quartz and alkali feldspar fabric in felsitic, locally granophyric, arrangement (vitrophyric texture).

The chemical composition of the silicic volcanic rocks of the PMP, according to Nardy et al. (2008), show two main groups that may be observed in an R1 × R2 diagram (Fig. 2; De La Roche et al., 1980). The first one, the low-Ti suite, belongs to the tholeiitic field (tholeiitic basalts, andesi-basalts and andesites) associated with the Palmas-type silicic volcanic rocks, which belong to the rhyodacite and rhyolite fields. The second group, the high-Ti suite, is displaced towards the transitional field (transitional basalt, lati-basalt and latites). The Chapecó-type silicic volcanic rocks belong to this group in the rhyodacite and quartz latite fields (Fig. 2). The bulk-rock representative compositions for both Palmas and Chapecó types are listed in Table 1.

According to Bellieni et al. (1984a) the chemistry of the volcanic rocks and their spatial distribution allow the Paraná Basin to be schematically subdivided into three main regions: (1) southern, encompassing the tholeiitic suite in the southern Uruguay River alignment; (2) northern, where tholeiitic-transitional rocks occur in the northern Piquiri River alignment; and (3) central, located between the Piquiri and Uruguay River alignments, where both rock types are present (Fig. 1). The spatial distribution of the tholeiitic – ATP and transitional-tholeiitic-ATC suites indicates that acidic volcanic rocks may have been derived from the associated basalts, or ATC melts are derived from tholeiitic-transitional basalts and ATP from tholeiitic basalts as suggested by Bellieni et al. (1986) and Garland et al. (1995).

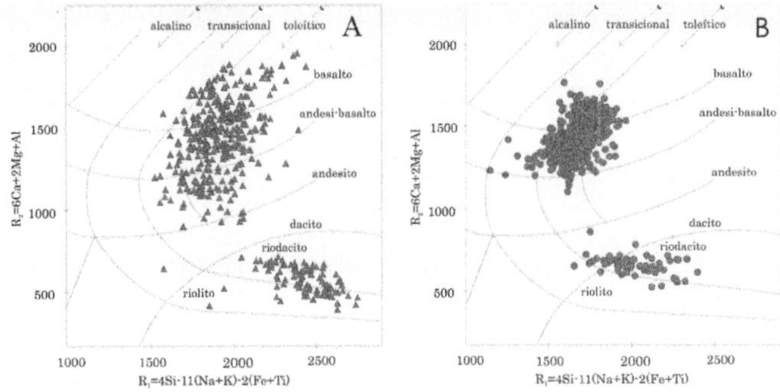

Fig. 2. R1 × R2 diagram (De La Roche et al., 1980). **(A)** Tholeiitic suite (Low-Ti basalts and ATP), and **(B)** tholeiitic-transitional suite (High-Ti basalts and ATC).

3 Stratigraphy

The 1600 m thick PMP volcanic sequence consists of up to 32 lava flows of predominantly basic to intermediate compositions (tholeiitic basalts, andesi-basalts and andesites), as well as felsic volcanic rocks (dacites, rhyodacites and rhyolites; Bellieni et al., 1984, 1986).

The base of the stratigraphic column is composed of a thick sequence of basic to intermediate flows that overlap the aeolian sandstones of the Botucatu Formation. The sandstones may also occur interlayered in the first hundred meters of the basaltic pile. The Palmas and Chapecó Members overlap the basalt flows. The Palmas Member is characterized by silicic volcanic bodies (ATP type) associated with a few basaltic lava flows and crops out from the central region of the basin southwards, where it may reach a thickness of 270 m. The Chapecó Member, composed of silicic volcanic rocks (ATC type), occurs in the northern and central regions of the Paraná Basin, the largest thickness, reaching 250 m, being present in the central region. This Member overlaps the basalts, but in the northern portion of the basin (Paranapanema River region – SP), it is found directly on the sandstones of the Botucatu Formation.

In the center of the basin the two silicic Members overlap, indicating that the Palmas Member is older than Chapecó, although ATP-type rocks may be found interlayered in the Chapecó Member.

The last pulses of PMP volcanism emplaced basalt flows that cover both the Palmas- and Chapecó-type rocks and become thicker towards the northern area of the basin.

3.1 Peperites and sedimentary deposit

The literature (Marques and Ernesto, 2004; Thiede and Vasconcellos, 2011) indicates that the PMP magmatism occurred quickly, during a time interval that did not exceed 3 Ma, and in a rather continuous way, based on a few occurrences of sedimentary intercalations or paleosoils within the volcanic sequence. However, recent field work has revealed the

Table 1. Representative analysis of Palmas and Chapecó silicic volcanics of the PMP (major elements, in oxide wt %; trace elements, in ppm).

	Palmas					Chapecó		
Sample	KC	KSE	KSE	KS	KSU	KC	PU	KNO
No.	505	406	419	319	237	482	1011	442
SiO_2	65.48	66.44	67.58	68.70	70.46	64.38	65.44	66.93
TiO_2	1.11	0.96	0.95	0.95	0.71	1.46	1.57	1.24
Al_2O_3	12.96	12.74	12.35	12.12	12.38	12.83	13.03	13.03
Fe_2O_3	6.78	6.15	6.15	5.69	5.22	7.01	7.57	6.54
MnO	0.08	0.11	0.20	0.10	0.09	0.16	0.12	0.10
MgO	1.08	1.70	1.22	1.28	0.61	1.36	1.28	0.75
CaO	2.78	2.93	3.09	2.68	2.20	2.91	2.94	2.07
Na_2O	3.64	2.87	2.78	2.30	2.92	3.32	3.61	3.36
K_2O	4.01	3.89	4.17	4.69	4.74	4.45	4.33	4.61
P_2O_5	0.33	0.27	0.26	0.26	0.20	0.48	0.46	0.33
LOI	1.93	1.91	1.01	0.99	0.57	1.05	0.61	1.63
SUM	100.17	99.98	99.75	99.76	100.09	99.41	100.96	100.60
Cu	128	75	78	63	18	7	9	14
Ni	7	8	7	8	3	4	5	5
Ba	610	706	694	588	613	1076	1003	1199
Rb	169	160	165	175	206	100	101	136
Sr	135	143	127	137	102	360	337	318
Zr	279	258	252	266	319	633	670	592
Y	63	41	57	42	55	65	66	60
Nb	22	20	20	21	23	48	51	44
U	3.51	4.07	4.04	4.17	3.25	1.74	1.91	2.68
Th	11.53	11.30	11.64	12.09	12.25	8.76	8.62	12.69
La	42.0	35.0	36.1	36.0	43.2	60.4	63.7	67.8
Ce	88.0	76.0	78.0	73.0	95.5	144.6	147.0	145.0
Nd	46.0	36.0	34.8	34.0	42.7	68.4	71.7	70.9
Sm	9.40	7.00	7.47	7.10	8.78	15.78	15.10	14.50
Eu	2.02	1.66	1.53	1.53	1.60	3.60	3.52	3.11
Gd	9.80	7.30	7.28	8.10	8.78	14.04	13.10	11.90
Dy	8.30	6.90	7.43	7.40	8.32	10.83	11.50	10.20
Ho	1.70	1.40		1.60				
Er	4.70	4.00	4.54	4.60	5.25	5.97	6.21	5.42
Yb	4.50	3.40	3.92	3.70	4.61	4.79	5.03	4.39
Lu	0.60	0.50	0.61	0.60	0.71	0.65	0.76	0.65

presence of frequent sediment lenses and peperites in various stratigraphic levels within the silicic volcanic sequence.

Peperite is a genetic term used for rocks formed in situ by the interaction between hot magma (intrusive bodies, lava or pyroclastic flows) and coeval wet sediments (Fisher, 1960; Williams and McBirney, 1979; White et al., 2000; Skilling et al., 2002). Nonetheless, some authors (e.g., Jerram et al., 1999; Jerram and Stollhofen) have also described an interaction with dry sediments. Peperite is classified into two basic types according to the shape of its elements (Busby-Spera and White, 1987): *blocky*, in which the volcanic clasts present angular shapes and show a jigsaw-fit texture reflecting in situ quench fragmentation in a brittle state, and *fluidal* in which volcanic clasts present irregular, fluid (amoeboid), globular to undefined shapes, reflecting a ductile state during fragmentation, with the sediment often filling vesicles and being injected into cracks in the volcanic clasts/rocks, although more complex shapes may also be found (McPhie et al., 1993; Skilling et al., 2002).

Sedimentary deposits and peperites found in the Chapecó and Palmas members present a wide distribution in the Paraná Basin, as shown in Fig. 3, and are described below.

In the São Jerônimo da Serra (Paraná State) region, in the northern sector of the Paraná Basin, ATC-type rocks overlap a sandstone forming blocky peperite (Fig. 4). The blocky peperite is comprised of clastic dikes, which can be from a few centimeters to just over 1 m thick (Fig. 4a), and breccias, composed of matrix-supported angular to rounded volcanic blocks of variable size (Fig. 4b). The sandstone is poorly sorted with angular to rounded quartz grains (Fig. 4c and d),

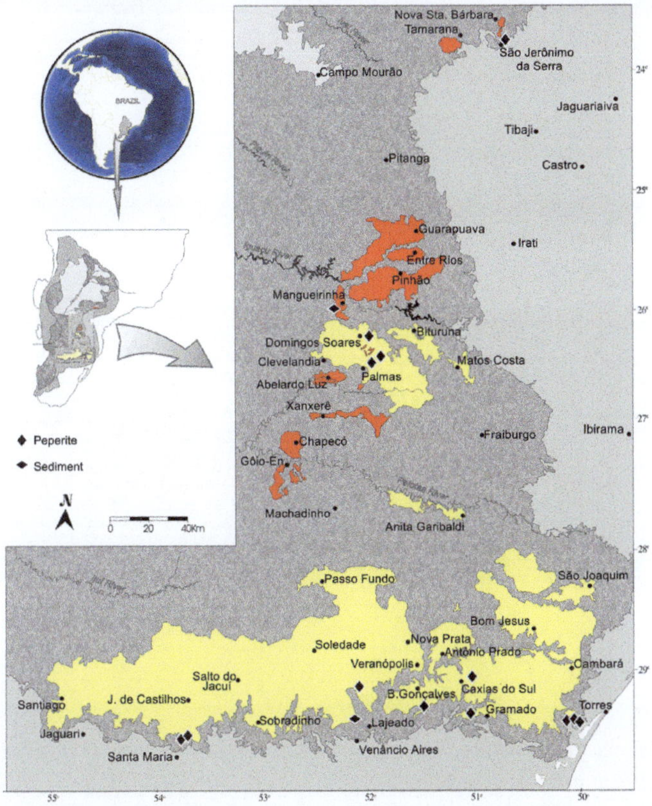

Fig. 3. Map with the location of sediment and peperite outcrops (lozenges) within the Palmas and Chapecó Members: yellow – ATP units; red – ATC units.

and it was silicified by thermal metamorphism by the overlying volcanic material.

In the region of Mangueirinha and Palmas (Paraná State), in the center of the basin, both ATC- and ATP-type rocks crop out as either overlying basalts or overlapping each other. A sandstone layer was observed intercalated between ATC-type silicic body and an overlying basaltic lava flow. The base of the sediment is a breccia formed by vesicular ATC clasts set in a sandy matrix, implying some erosion degrees of the top of the volcanic body during sedimentation. In another location, an ATC-type silicic unit overlies a vesicular ATP-type silicic body, with a red clayey–silty sediment intercalated between the two units. The sediment was injected upwards into fractures in the overlying ATC unit forming peperite with both fluidal and blocky features (Fig. 5a and b), while in the underlying ATP unit, the sediment filled cooling cracks without any peperitic interaction (figure 5c and d). In a third exposure, a reddish-brown silty sediment, intercalated between two ATP volcanic units, formed peperite with blocky jointing morphology (Skilling et al., 2002), characterized by the injection of sediment into centimeter- to millimeter-spaced joints in the base of the overlying volcanic unit (Fig. 6).

In the southern region of the Paraná Basin (Rio Grande do Sul State), sediments and peperites were observed between

Fig. 4. Peperite from São Jerônimo da Serra: **(a, b)** photomicrographs of the immature and poorly sorted sediment (**c** = // polarizers; **d** = *X* polarizers), **(c)** clastic dike more than 1 m thick, and **d** angular to sub-angular shaped clasts set in a sandstone matrix. S – sandstone; V – volcanic rock.

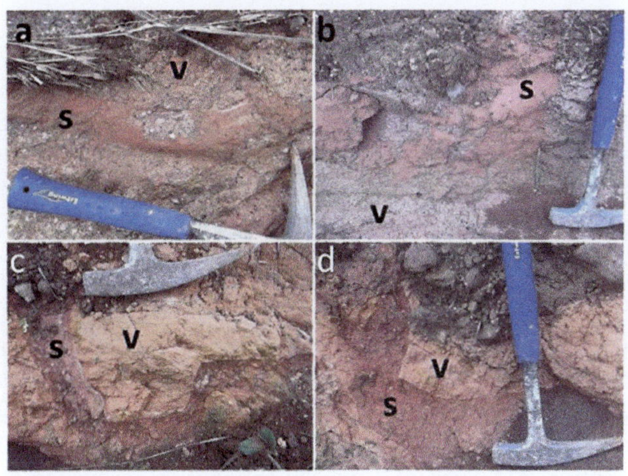

Fig. 5. Silicic ATC- and ATP-type rocks (V) in contact with red clayey siltstone (S): **(a, b)** peperites in the base of an ATC-type rock; (**c** and **d**) fractures in the top of an ATP-type rock filled with sediment, in the central region of the Paraná Basin.

some basaltic lower units and the overlying Palmas silicic sequence. In the Santa Maria region, blocky peperite was observed in the base of the lowermost silicic volcanic unit. The peperite displays a well-developed jigsaw-fit texture, and closely packed, blocky to cuneiform juvenile clast shapes of several sizes, separated by an orange-colored sedimentary material (Fig. 7). The sediment exhibits a high degree of baking (thermal metamorphism), and the juvenile clasts show intense devitrification, indicating very high temperature of the volcanic material interacting with the sediment.

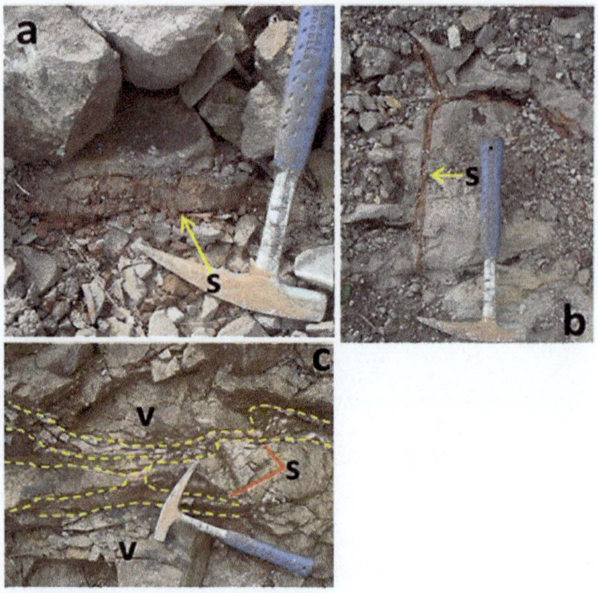

Fig. 6. Blocky jointed peperite in the base of an ATP-type unit (Domingos Soares city – Paraná State, central region of the Paraná Basin): V – ATP rock; S – sediment.

Fig. 7. Blocky peperite formed by interaction of an ATP-type volcanic body with sediment in the Santa Maria region (Rio Grande do Sul State): well-developed jigsaw-fit texture, closely packed, gray blocky to cuneiform clasts of variable sizes, separated by orange sedimentary material, displaying intense baking.

Near the city of Venâncio Aires, a 10 m thick sedimentary layer underlies an ATP silicic volcanic unit. The sediment is a reddish-colored sandstone composed of sub-angular to rounded quartz grains (Fig. 8). The volcanic unit presents a well-developed horizontal jointed base and the contact with the sandstone is sharp, apparently lacking peperitic interaction (Fig. 8).

A road cut on the Soledade to Lajeado highway exposes a peperite in the base of a 25 m thick basaltic lava flow interlayered within the ATP silicic volcanic sequence. Volcanic clasts in the peperite are vesicular and display a variety of morphologies, from blocky to fluidal. The sediment is a red, poorly sorted sandstone, which partially fills vesicles in the volcanic clasts.

Eastward, in the area between Bento Gonçalves and Cambará do Sul (Rio Grande do Sul), all the observed peperites are interlayered within the silicic volcanic sequence and the sedimentary material becomes finer and predominantly silty. A peperite was observed in a quarry floor in Nova Petrópolis, near the city of Gramado, resulting from the interaction of sediment with a thin (~ 1 m thick) amygdaloidal basalt interlayered in the silicic volcanic sequence. It displays blocky juvenile clasts with irregular shapes separated by orange to reddish, poorly sorted fine-grained sandstone. The sediment presents vesicles that probably resulted from volatilization of sediment water by heating (Fig. 9).

On the Rota do Sol highway, connecting Caxias do Sul to the coast, three peperite levels were observed interlayered in a sequence of black and glassy volcanic units of the Palmas Member. The stratigraphically lower two are similar and characterized by vesicular, pale to greenish and glassy volcanic clasts, presenting angular to rounded irregular shapes (Fig. 10). The sedimentary material is a brown, moderate to poorly sorted siltstone that also fills vesicles in volcanic clasts. The uppermost peperite presents juvenile volcanic clasts with a wide variety of morphologies, the fluidal shapes being more frequent (Fig. 11). The green and vesicular volcanic clasts are surrounded by a reddish-brown siltstone, which also fills vesicles and fractures in volcanic clasts (Fig. 11d). It is slightly coarser and more poorly sorted than the previous two sediments (Fig. 12), and many clasts of the stratigraphically lower peperite display perlitic fractures (Fig. 12c).

4 Discussion and concluding remarks

The PEMP (Paraná–Etendeka Magmatic Province) is considered one of the largest LIPs (large igneous provinces) of continental crust in the world, encompassing an area of nearly 1 million cubic kilometers (Bryan et al, 2010). Of total volume of the volcanic products, 95 % are preserved in the South American continent, in the PMP. All volcanic material

Fig. 8. Outcrop in the Venâncio Aires area (Rio Grande do Sul State), where a silicic volcanic unit of the ATP type (V) rests on a reddish sedimentary layer (S) lacking peperitic interaction. On the right lower corner a photomicrograph of the sandstone (*X* polarizers).

Fig. 10. Aspects of the two lowermost peperites in the Palmas volcanic sequence, along the Rota do Sol highway (Rio Grande do Sul State): images of the lowermost peperite (**a**, **b**, and **c**) and of the middle peperite (**d**). The volcanic clasts are pale and light or dark green, while the siltstone is dark brown.

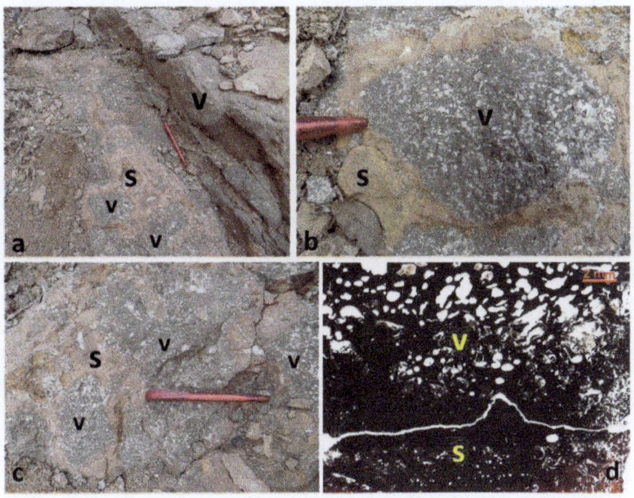

Fig. 9. Aspects of a peperite produced by interaction of a thin basalt flow (within the silicic Palmas Member) and fine-grained sediment near Nova Petrópolis (Rio Grande do Sul State), southern part of the basin: **(a–c)** irregular blocky volcanic clasts (V) in a reddish sandstone matrix (S), and **(d)** photomicrograph showing the amygdaloidal basalt (V) in the upper part and a poorly sorted fine-grained sandstone in the bottom (S).

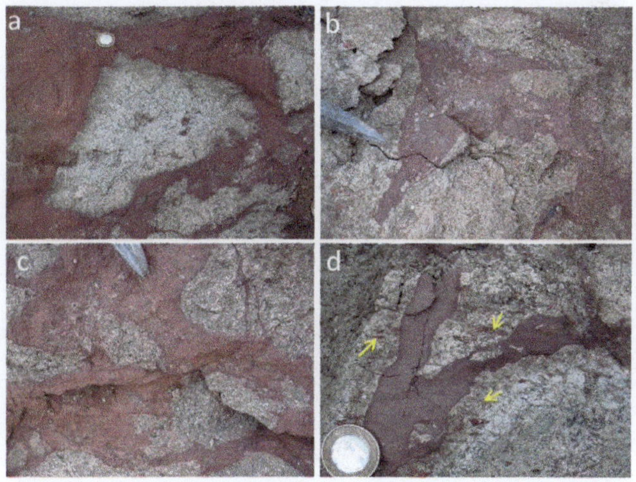

Fig. 11. Peperite from the stratigraphically higher level on the section exposed in the Rota do Sol highway: green to pale volcanic clasts, displaying sub-angular to irregular morphologies, set in a reddish-brown siltstone matrix, which also fills vesicles in volcanic clasts (yellow arrows in **d**).

was erupted in a short period of time (~3 million years) without significant interruption, as deduced from the scarcity of sediments interlayered within the volcanic sequence. In fact, and up to now, the only references to the presence of sediments interbedded within the volcanics corresponded to layers or lenses of sandstones (intertraps), a few centimeters to several meters thick, from the Botucatu Formation. These occur only in the base of the lava flow pile in both the African (Jerram et al., 1999, 2000; Jerram and Stollhofen, 2002) and South American continents (Petry et al., 2007; Waichel et al., 2008). Sandstone intertraps in rhyodacites in the Piraju–Ourinhos region (São Paulo State; Janasi et al., 2007; Luchetti, 2010), where the thick basaltic sequence is missing, were also known. However, until now, there were no reports of features indicating significant time breaks in the upper part of the PMP stratigraphic sequence.

New observations of the occurrence of sedimentary lenses and peperites, resulting from volcano-sedimentary interaction at the base and within the upper silicic sequences of the PMP are presented in this paper. Throughout the Paraná

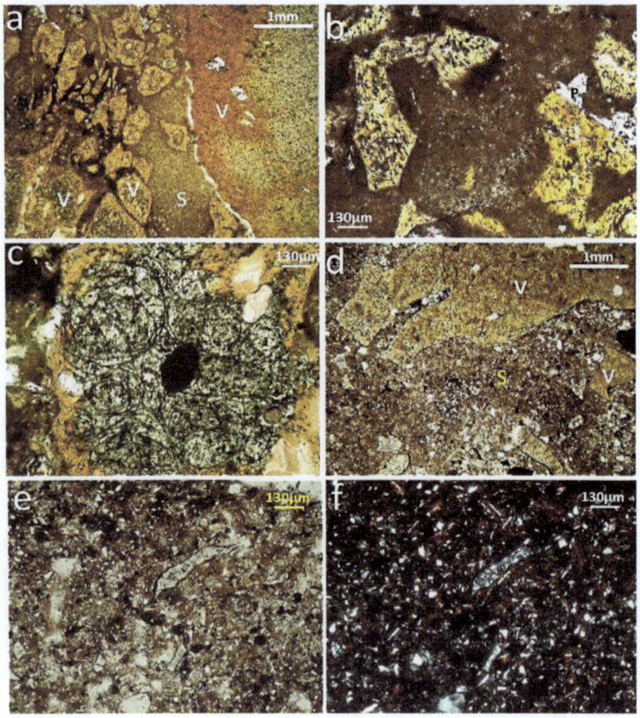

Fig. 12. Photomicrographs of peperites from the Palmas volcanic sequence, along Rota do Sol highway: (**a**, **b** and **d**) volcanic glassy clasts (V) set in a brown siltstone matrix (S); (**c**) volcanic clast displaying concentric fractures – perlitic texture - common in the two lowermost peperites; (**e** and **f**) detail of the poorly sorted, slightly coarser siltstone from **(d)** (**a**, **b**, **c**, **d** and **e** – // polarizers; **f** – *X* polarizers; P = plagioclase).

Basin these were observed associated with silicic and basaltic units of the Palmas and Chapecó Members.

Some works on the PMP report peperites derived from the basalt-flow–sediment interaction (Petry et al., 2007; Waichel et al., 2007, 2008), the most common types found in worldwide volcanic sequences. Nevertheless, structures observed in this study reveal that this interaction also occurred between silicic extrusive bodies and wet sediment. These "silicic peperites" are mostly blocky, highlighting a more viscous magma (Dadd and Wagoner, 2002). However, they may also display fluidal features, which indicate less viscous melts, likely due to higher temperatures (> 1000 °C; Piccirillo and Melfi, 1988).

The sediments display immature features, suggesting limited transport and, depending on the location, range from moderately to poorly sorted sandstones to siltstones. There was often lava–sediment interaction producing peperites or sediment deformation by the weight of overlying volcanic units. Evidence for the presence of wet and unconsolidated or poorly consolidated sediments is the variety of morphologies found, injections of sediment into fractures, vesicles in the juvenile clast filled by sediment and vesiculated sediment (Skilling et al., 2002).

Paleoenvironment climatic conditions at the time of the PMP volcanism were quite dry, since the Serra Geral Formation overlies the Botucatu Formation, described as a dry aeolian system (Sherer, 2000). Sandstone lenses up to 20 m thick interlayered in the first basalt lava flows (Jerram et al., 1999, 2000; Jerram and Stollhofen, 2002; Petry et al., 2007; Waichel et al., 2008) reveal that this system remained active during the magmatic event. However, new evidence of wet and locally silty sediment between basalt and silicic units and interlayered in the silicic units (end of the volcanic sequence) presented here reflect a paleoenvironmental change, from a dry climate in the beginning of the PMP volcanic activity to a more humid environment (fluvial–lacustrine) during the latest phases of the magmatism, as also evidenced by Waichel et al. (2007) in peperites found in some basalt units of the PMP in contact with lacustrine sediments. In the São Jerônimo da Serra region, the northern area of the basin, the change must have started before the volcanism. Furthermore, this seems to be common in this type of setting in which humid and dry aeolian systems may alternate over time due to climate changes (Assine et al., 2004).

The deposition of sediments must have taken place in depressed portions of the paleorelief (small valleys or depressions of the original volcanic morphology), and suggests a decrease in eruptive frequency towards the end of the volcanic activity, allowing time for the deposition of sedimentary material between individual volcanic events.

In conclusion, the PMP volcanism was not completely continuous, but presented significant pauses, mainly in the initial and terminal phases. On the other hand, the occurrence of sediments separating the top of the lower basaltic sequence and the beginning of the silicic extrusions may represent a pause in the volcanic activity that coincides with the compositional change in the magmatism of the Paraná Magmatic Province.

Acknowledgements. This study was supported by São Paulo Research Foundation (FAPESP) and National Council of Technological and Scientific Development (CNPq). We thank Francisco Negri for help in fieldwork in the central basin and Dr. Scarlato, Piergiorgio, Dr. Giacomoni, Pier Paolo and an anonymous referee for valuable suggestions and review.

Edited by: P. Scarlato

References

Assine, M. L., Piranha, J. M., Carneiro, C. D. R.: Os paleodesertos Pirambóia e Botucatu, in: Geologia do Continente Sul-Americano: evolução da obra de Fernando Flávio Marques de Almeida, edited by: Mantesso-Neto, V., Bartorelli, A., Carneiro, C. D. R., and Brito-Neves, B. B., SP, Brasil, Beca Prod Cult Ltda, 77–92, 2004.

Bellieni, G., Brotzu, P., Comin-Chiaramonti, P., Ernesto, M., Ernesto, M., Melfi, A. J., Pacca, I. G., and Piccirillo, E. M.:

Flood basalt to rhyolites suites in the southern Paraná plateau (Brazil): paleomagnetism, petrogenis and geodynamic implications, J. Petrol., 25, 579–618, 1984.

Bellieni, G., Comin-Chiaramonti, P., Marques, L. S., Melfi, A. J., Nardy, A. J. R., Papatrechas, C., Piccirillo, E. M., and Roisenberg, A.: Petrogenetic aspects of acid and basaltic lavas from the Paraná plateau (Brazil): geological, mineralogical and petrochemical relationships, J. Petrol., 27, 915–944, 1986.

Bryan, S. E., Ukstins-Peate, I., Peate, D. W., Self, S., Jerram, D. A., Mawby, M. R., Marsh, J. S., and Miller, J. A.: The largest volcanic eruptions on Earth. Earth-Sci. Rev., 102, 207–229, 2010.

Busby-Spera, C. J. and White, J. D. L.: Variation in peperite textures associated with differing host sediment properties, B. Volcanol., 49, 765–775, 1987.

Dadd, K. A. and Van Wagoner, N. A.: Magma composition and viscosity as controls on peperite texture: an example from Passamaquoddy Bay, southeastern Canada, in: Peperite: Processes and Products of Magma-Sediment Mingling, edited by: Skilling, I. P., White, J. D. L., and McPhie, J., J. Volcanol. Geoth. Res., 114, 63–80, 2002.

De La Roche H., Leterrier P., Grandclaude P., and Marshal M.: A classification of volcanic and plutonic rocks using R1 × R2 diagram and major element analysis. Its relationships with current nomenclature, Chem. Geol., 28, 183–210, 1980.

Ernesto, M., Raposo, M. I. B., Marques, L. S., Renne, P. R., Diogo, L. A., and De Min, A.: Paleomagnetism, geochemistry and 40ar/39ar dating of the north-eastern Paraná Magmatic Province, Tectonic implications, J. Geodyn., 28, 321–340, 1999.

Fisher, R. V.: Classification of volcanic breccias, Geol. Soc. Am. Bull., 71, 973–982, 1960.

Frank, H. T., Gomes, M. E. B., and Formoso, M. L. L.: Review of the areal extent and the volume of the Serra Geral Formation, Paraná Basin, South America, Pesquisas em Geociências, 36, 49–57, 2009.

Garland, F., Hawkesworth, C. J., and Mantovani, M. S. M.: Description and Petrogenesis of the Paraná Rhyolites, Southern Brazil, J. Petrol., 36, 1193–1227, 1995.

Janasi, V. A., Montanheiro, T. J., Freitas, V. A., Reis, P. M., Negri, F. A., and Dantas, F. A.: Geology, petrography and geochemistry of the acid volcanism of the Paraná Magmatic Province in the Piraju–Ourinhos region, SE Brazilian, J. Geosci., 37, 745–759, 2007.

Janasi, V. A., Freitas, V. A., and Heaman, L. H.: The onset of ?ood basalt volcanism, Northern Paraná Basin, Brazil: A precise U–Pb baddeleyite/zircon age for a Chapecó-type dacite, Earth. Planet. Sc. Lett., 302, 147–153, 2011.

Jerram, D. A., Mountney, N., and Stollhofen, H.: Facies architecture of the Etjo Sandstone Formation and its interaction with the Basal Etendeka food basalts of NW Namibia: Implications for offshore analogues, in: The Oil and Gas Habitats of the South Atlantic, edited by: Cameron, N., Bate, R., and Clure, V., Geological Society, London, Special Publications, 153, 367–380, 1999.

Jerram, D. A., Mountney, N., Howell, J., Long, D., and Stollhofen, H.: Death of a Sand Sea: an active erg systematically buried by the Etendeka flood basalts of NW Namibia, J. Geol. Soc. London, 157, 513–516, 2000.

Jerram, D. A. and Stollhofen, H.: Lava-sediment interaction in desert settings; are all peperite like textures the result of magma-water interaction?, J. Volcanol. Geoth. Res., 114, 231–49, 2002.

Luchetti, A. C. F.: Aspectos vulcanológicos dos traquidacitos da região de Piraju–Ourinhos (SP) [Master Dissertation], São Paulo, SP, Brazil, Universidade de São Paulo, 92 pp., 2010.

Malagutti, M. I. A., Bahia F, Moreno, M. M. T., and Nardy, A. J. R.: Determinação de elementos terras raras e ítrio em rochas silicáticas por ICP-OES com separação em resina trocadora de íons, Geochem. Bras., 12, 75–80, 1998.

Marques, L. S. and Ernesto, M.: O magmatismo toleítico da Bacia do Paraná, in: Geologia do Continente Sul-Americano, edited by: Mantesso-Neto, V., Bartorelli, A., Carneiro, C. D. R., and Brito-Neves, B. B., Beca Produções Culturais Ltda, 2004.

McPhie, J., Doyle, M., and Allen, R.: Volcanic textures: a guide to the interpretation of textures in volcanic rocks, University of Tasmania, 197 pp., 1993.

Milner, S. C., Duncan, A. R., and Ewart, A.: Quartz latite rheoignimbrite flows of the Etendeka Formation, north western Namibia, B. Volcanol., 54, 200–219, 1992.

Mincato, R. L., Enzweiler, J., and Schrank, A.: Novas idades $^{39}Ar/^{40}Ar$ e implicações na metalogênese dos depósitos de sulfetos magmáticos de Ni-Cu-EPG na Província Ígnea Continental do Paraná, in: 9th Brazilian Congress of Geochemistry, Belém (Pará), Brazil, 2–9 November 2003, 67–92, 2003.

Nardy, A. J. R., Enzweiler, J., Bahia F, O., Oliveira, M. A. F., and Peneiro, M. A. V.: Determinação de elementos maiores e menores em rochas silicáticas por espectometria de fluorescência de raios-x: resultados preliminares, in: 6th Brazilian Congress Of Geochemistry, Salvador (Bahia), Brazil, 346–348, 1997.

Nardy, A. J. R., Oliveira, M. A. F., Betancourt, R. H. S., Verdugo, D. R. H., and Machado, F. B.: Geologia e estrtigrafia da Formaçõ Serra Geral, Revista Geociências, 21, 15–32, 2002.

Nardy, A. J. R., Machado, F. B., and Oliveira, M. A. F.: As rochas vulcânicas mesozóicas ácidas da Bacia do Paraná: litoestratigrafia e considerações geoquímico-estratigráficas, Brazil. J. Geosci., 38, 178–195, 2008.

Peate, D., Hawkesworth, C. J., and Mantovani, M. S. M.: Chemical Stratigraphy of the Paraná Lavas (South America): Classification of Magma Types and their Spatial Distribution, B. Volcanol., 55, 119–139, 1992.

Petry, K., Jerram, D. A., De Almeida, D. P. M., Zerfass, H.: Volcanic-sedimentary features in the Serra Geral Fm., Paraná Basin, southern Brazil: Examples of dynamic lava-sediment interactions in an arid setting, J. Volcanol. Geoth. Res., 159, 313–325, 2007.

Piccirillo, E. M. and Melfi, A. J. (Eds.): The Mesozoic Flood Volcanism of the Paraná Basin: Petrogenetic and Geophysical Aspects, Instituto Geofísico, Astronômico e Ciências Atmosféricas, Universidade de São Paulo, São Paulo, Brazil, 1988.

Renne, P. R., Ernesto, M., Pacca, I. G., Coe, R. S., Glen, J. M., Prévot, M., and Perrin, M.: The age of Parana flood volcanism, rifting of Gondwanaland, and the Jurassic–Cretaceous boundary, Science 258, 975–979, 1992.

Renne, P. R., Deckart, K., Ernesto, M., Féraud, G., and Piccirillo, E. M.: Age of the Ponta Grossa dyke swarm (Brazil), and implications to Paraná flood volcanism, Earth. Planet. Sc. Lett. 144, 199–211, 1996a.

Renne, P. R., Glen, J. M., Milner, S. C., Duncan, A. R.: Age of Etendeka flood volcanism and associated intrusions in south-western Africa, Geology 24, 659–662, 1996b.

Santos, R. N., Marques, L. S., Brenha-Ribeiro, F., and Nicola, S. H. A.: Determination of uranium concentrations and activity ratios in silicate rocks by alpha spectrometry: application to volcanic rocks from Trindade and Martin Vaz Islands, Appl. Radiat. Isot., 56, 741–750, 2002.

Santos, R. N., Marques, L. S., Brenha-Ribeiro, F., and Nicola, S. H. A.: Determination of thorium concentrations and activity ratios in silicate rocks by alpha spectrometry, Appl. Radiat. Isot., 56, 741–750, 2004.

Scherer, C.: Aeolian dunes of the Botucatu Formation (Cretaceous) in southernmost Brazil: morphology and origin, Sed. Geol., 137, 63–84, 2000.

Skilling, I. P., White, J. D. L., and Mcphie, J.: Peperite: a review of magma-sediment mingling, J. Volcanol. Geoth. Res., 114, 1–17, 2002.

Sun, S. and Mcdonough, W. F.: Chemical and isotopic systematics of oceanic basalts: implications for mantle composition and processes, in: Magmatism in the Ocean Basins, edited by: Saunders A. D. and Norry M. J., Geol. Soc. London, 1, London, 313–345, 1989.

Thiede, D. S. and Vasconcelos, P. M.: Paraná flood basalts: rapid extrusion hypothesis confirmed by new $^{40}Ar/^{39}Ar$ results, Geology, 38, 747–750, 2010.

Turner, S., Regelous, M., Kelley, S., Hawksworth, C., and Mantovani, M. M. S.: Magmatism and continental break-up in the South Atlantic: high precision $^{40}Ar/^{39}Ar$ geochronology, Earth Plan. Sci. Lett., 121, 333–348, 1994.

Waichel, B. L., Lima, E. F., Sommer, C. A., and Lubachesky, R.: Peperite formed by lava flows over sediments: An example from the central Paraná Continental Flood Basalts, Brazil, J. Volcanol. Geoth. Res., 159, 343–354, 2007.

Waichel, B. L., Scherer, C. M. S., and Frank, H. T.: Basaltic lava flows covering active aeolian dunes in the Paraná Basin in southern Brazil: Features and emplacement aspects, J. Volcanol. Geoth. Res., 171, 59–72, 2008.

Waichel, B. L., Lima, E. F. de, Viana, A. R., Scherer, C. M., Bueno, G. V., and Dutra, G.: Stratigraphy and volcanic facies architecture of the Torres Syncline, Southern Brazil, and its role in understanding the Paraná–Etendeka Continental Flood Basalt Province, J. Volcanol. Geoth. Res., 215–216, 74–82, 2012.

Williams, H. and McBirney, A. R.: Volcanoes, San Francisco, Freeman, Cooper and Company, 397 pp., 1979.

White, J. D. L., Mcphie, J., and Skilling, I. P.: Peperite: a useful genetic term, B. Volcanol., 62, 65–66, 2000.

10

Seismic visibility of a deep subduction channel – insights from numerical simulation of high-frequency seismic waves emitted from intermediate depth earthquakes

W. Friederich[1], **L. Lambrecht**[1], **B. Stöckhert**[1], **S. Wassmann**[1], **and C. Moos**[2]

[1]Ruhr-University Bochum, Institute of Geology, Mineralogy and Geophysics, Bochum, Germany
[2]Ruhr-University Bochum, Institute for Computational Engineering, Bochum, Germany

Correspondence to: W. Friederich (wolfgang.friederich@rub.de)

Abstract. Return flow in a deep subduction channel (DSC) has been proposed to explain rapid exhumation of high pressure–low temperature metamorphic rocks, entirely based on the fossil rock record. Supported by thermo-mechanical models, the DSC is envisioned as a thin layer on top of the subducted plate reaching down to minimum depths of about 150 km. We perform numerical simulations of high-frequency seismic wave propagation (1–5 Hz) to explore potential seismological evidence for the in situ existence of a DSC. Motivated by field observations, for modeling purposes we assume a simple block-in-matrix (BIM) structure with eclogitic blocks floating in a serpentinite matrix. Homogenization calculations for BIM structures demonstrate that effective seismic velocities in such composites are lower than in the surrounding oceanic crust and mantle, with nearly constant values along the entire length of the DSC. Synthetic seismograms for receivers at the surface computed for intermediate depth earthquakes in the subducted oceanic crust for models with and without DSC turn out to be markedly influenced by its presence or absence.

While for both models P and S waveforms are dominated by delayed high-amplitude guided waves, models with DSC exhibit a very different pattern of seismic arrivals compared to models without DSC. The main reason for the difference is the greater length and width of the low-velocity channel when a DSC is present. Seismic velocity heterogeneity within the DSC or oceanic crust is of minor importance. The characteristic patterns allow for definition of typical signatures by which models with and without DSC may be discriminated.

The signatures stably recur in slightly modified form for earthquakes at different depths inside subducted oceanic crust. Available seismological data from intermediate depth earthquakes recorded in the forearc of the Hellenic subduction zone exhibit similar multi-arrival waveforms as observed in the synthetic seismograms for models with DSC. According to our results, observation of intermediate depth earthquakes along a profile across the forearc may allow to test the hypothesis of a DSC and to identify situations where such processes could be active today.

1 Introduction

Since the establishment of plate tectonics, most ideas on structure and properties of the plate interface along subduction zones were inspired by textbook cartoons, showing a giant thrust fault with its position being marked by Benioff zone seismicity. The narrow fault or shear zone concept also underlies thermal models of subduction zones (e.g. Turcotte and Schubert, 2002; Van Den Beukel and Wortel, 1988; Peacock, 1996; Syracuse et al., 2010), with a downward single pass flow beneath a rigid mantle wedge in the forearc. This starting model became challenged by the finding that high pressure over temperature (P/T) metamorphic rocks are very widespread within continental crust (e.g. Ernst and Liou, 1999; Tsujimori et al., 2006; Agard et al., 2009; Guillot et al., 2009; Ota and Kaneko, 2010), obviously having been exhumed from a subduction zone. While in many cases some cooling associated with decompression requires

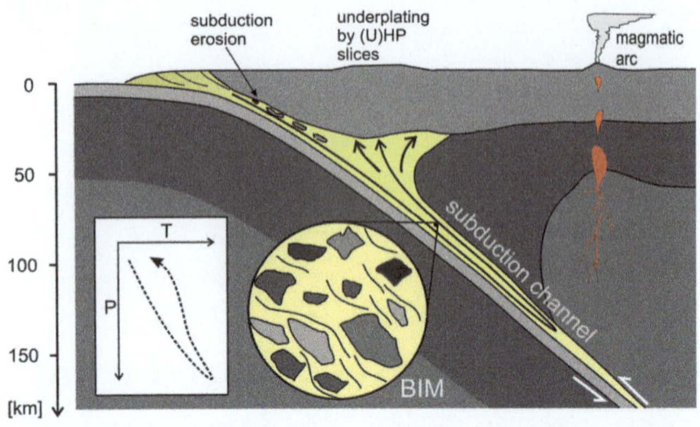

Fig. 1: Envisaged geometry of deep subduction channel (DSC), including subduction erosion and underplating (after Stöckhert, 2002). Insets show scale-invariant block in matrix (BIM) structure and typical pressure–temperature (*P–T*) path derived for rapidly exhumed high P/T metamorphic slices.

exhumation as relatively small bodies with heat transfer into a cooler environment (e.g. Schertl et al., 1991; Stöckhert et al., 1997; Reinecke, 1998; Groppo et al., 2007; Groppo and Castelli, 2010; Krebs et al., 2011; Angiboust et al., 2012), geochronological results pose narrow time windows for typical timescales of exhumation, asking for displacement rates comparable to plate velocity (e.g. Gebauer et al., 1997; Rubatto and Hermann, 2001; Baldwin et al., 2004; Little et al., 2011). Rapid exhumation with concomitant cooling reconciles neither with uplift and erosion nor with crustal extension (Platt, 1993; Ring and Brandon, 1999). Moreover, geochronological results suggest that in some cases a large part of exhumation has taken place during ongoing subduction, prior to collision (e.g. Lardeaux et al., 2001; Rubatto et al., 2011). These observations have led to postulation of a deep subduction channel (DSC) with forced return flow as a possible mechanism for exhumation (Fig. 1).

The concept of a subduction channel was first applied to explain structure and $P–T$ paths of the high P/T metamorphic Franciscan mélange in California by Cloos (1982), and elaborated by Shreve and Cloos (1986) and Cloos and Shreve (1988a, b). It was inferred to play a role in the frontal forearc beneath an accretionary wedge, competing with models based on critical wedge geometry (e.g. Platt, 1993). Identification of ultra-high-pressure (UHP) metamorphic rocks starting in the 1980s (e.g. Chopin, 1984; Schertl et al., 1991; Schreyer, 1995; Harley and Carswell, 1995), and in particular the finding that these rocks undergo very rapid exhumation (e.g. Gebauer et al., 1997; Rubatto and Hermann, 2001; Baldwin et al., 2004; Parrish et al., 2006), inspired tectonic models invoking a DSC reaching to greater depth. Forced return flow of material from depths exceeding 100 km, as schematically shown in Fig. 1, seemed to provide a feasible alternative for generation and exhumation of smaller UHP metamorphic slices, distinct from coherent crustal or lithospheric sections of the downgoing plate during collision (e.g. Chemenda et al., 1995; Hacker et al., 2000; Kylander-Clark et al., 2012). Numerical simulations (e.g. Burov et al., 2001; Gerya et al., 2002; Warren et al., 2008) generating bulk flow patterns and $P–T$ paths for individual material points, grossly consistent with natural record and structures of orogenic belts (e.g. Engi et al., 2001; Stöckhert and Gerya, 2005), support the principal feasibility of the DSC concept.

Geometry and width of a DSC remain poorly constrained by natural observations (e.g. Warren, 2013). A lower bound to thickness is given by the size of high P/T metamorphic slices supposed to be carried by forced return flow in a low-viscosity matrix, and an upper bound by the requirement of high rate of return flow (e.g. Gerya and Stöckhert, 2006). These bounds are assumed to be on the order of 1 and 10 km, respectively. The fact that high P/T metamorphic rocks can undergo cooling during extremely rapid exhumation requires slices or blocks of limited size, allowing effective heat transfer into a cooler environment. Embedding into a cooler environment, by whatever mechanism and kinematics, requires large amounts and rates of displacement, which in turn imply high strain rates and a complex structural evolution. A DSC is envisaged to obey these conditions, invoking chaotic flow in a heterogeneous rock assemblage, and effective mixing of material of different provenance on different length scales (e.g. Gerya and Stöckhert, 2002, 2006). High P/T metamorphic tectonic mélanges are therefore likely candidates to represent material properties in a fossil subduction channel.

When viewed on mesoscopic scale, mélanges frequently exhibit block-in-matrix (BIM) structures. At best slightly deformed blocks of variable provenance are embedded in a highly strained matrix, which has undergone prolonged deformation (Bebout, 2007; Festa et al., 2010; Grigull et al., 2012). The evolution of this spectacular structure is still poorly understood. Importance of sedimentary processes at the surface versus tectonic processes at depth, or both in combination, remains a matter of debate (Dilek et al., 2012). In any case, BIM structures imply contrasts in mechanical properties – the blocks undergoing no notable permanent deformation while the matrix must deform at very high strain rate, despite identical conditions in terms of temperature and stress. The mechanism allowing for deformation of fine-grained polyphase rocks at very low differential stress and high strain rates is dissolution precipitation creep (Wassmann and Stöckhert, 2012; Grigull et al., 2012). Consequently, mélanges with BIM structures reflect low viscosity, which may be an important factor to enable forced return flow in a DSC.

Typical bulk material properties in a subduction channel at an arbitrary depth can hardly be predicted. Exhumed high P/T metamorphic mélanges, inferred to represent a fossil DSC, demonstrate that a variety of rock types, both in terms of bulk composition and metamorphic mineral assemblage, occur intermingled on various length scales (Krebs et al.,

2011; Grigull et al., 2012), ranging from map scale (e.g Angiboust et al., 2012) down to the scale of the individual outcrop (e.g. Federico et al., 2007). As basic eclogite and serpentinite probably represent extremes, the amplitude of heterogeneity is constrained by the contrast in mechanical properties between these rock types. Therefore, we assume a composite material with eclogite blocks embedded in a serpentinite matrix to account for heterogeneity in the DSC, with a block size roughly 1 order of magnitude smaller than the wavelength of P waves. Taking these extremes, other heterogeneities are assumed to produce a much weaker effect in terms of seismic velocity.

Even for a uniform and specified bulk composition, it would not be possible to predict physical rock properties, as adaption of phase assemblages to changing temperatures and pressures is controlled by kinetics and generally delayed both during burial and exhumation. Blueschists, eclogites, or in particular ultrahigh pressure metamorphic rocks at the surface demonstrate substantial disequilibrium. As such, material in a DSC must be highly heterogeneous, both in bulk composition and in mineral phase assemblage.

The DSC concept was developed to explain the rock record (i.e. structural aspects of high P/T metamorphic terranes, petrologic observations, and geochronological constraints). Whether forced return flow is commonplace in subduction zones, or whether it is a special case restricted to certain stages or geometrical properties, so far remains an unresolved question. While structural geology and metamorphic petrology deal with the product of long-term processes taking millions of years, seismological observations may provide evidence for the presence of a currently active DSC in situ. Here, we investigate by numerical simulations whether a DSC can be detected using high-frequency seismic waves (up to 5 Hz) recorded at the surface.

Guided waves in a subduction zone have previously been simulated by Martin et al. (2003), Martin et al. (2005) and Martin and Rietbrock (2006) using finite-difference modeling. They studied the influence of slab geometry and earthquake locations on the properties of guided waves but their model did not consider the presence of a DSC. Our simulations extend previous work by Essen et al. (2009) who investigated the influence of a DSC on comparatively low-frequency (< 1 Hz) guided waves and assumed constant seismic velocities in predefined subregions of the model. As effect of a DSC, the authors identified a substantial shift of the maximum amplitude of guided waves from the area where the oceanic crust enters the mantle towards an area in the forearc located above the mantle wedge. Owing to the low-frequency signals, differences in the internal structure of the guided wave signals could not be analyzed.

The geometry of our simple seismic model is based on that of the Hellenic subduction zone, for which geological and seismological phenomena (Meier et al., 2007) are tentatively ascribed to anomalous material extruded from a DSC beneath Crete. Seismic velocities are calculated based on predictions on temperature and pressure field and assumptions on rock composition and mineral assemblage. For the subduction channel, a BIM structure is assumed composed of a serpentinite matrix and eclogite blocks, for simplicity. In the simulations, we consider seismic waves recorded in the forearc and emitted from intermediate depth earthquakes in the subducted oceanic crust or the lithospheric mantle below. These waves, spending the major part of their travel path in subducted lithosphere and overlying mantle wedge, are expected to be affected by presence of an intervening DSC with distinct properties. For comparison, we also show examples of observed waveforms from intermediate depth earthquakes recorded in the forearc of the Hellenic subduction zone.

2 A seismic velocity model of a subduction zone

Seismic velocities of mantle rocks are controlled by the physical properties of their mineralogical constituents which in turn depend on temperature and stress state. Due to phase transitions during subduction which may be delayed by kinetic constraints on the mineral reactions, physical rock properties may also depend on the dynamics of the subduction process.

While seismic velocities in subducted oceanic crust or within an inferred DSC are subject to great uncertainty, typical seismic velocities in both subducted mantle lithosphere and mantle wedge are fairly well known from seismic tomography. In order to obtain consistent, pressure and temperature dependent values of P velocities, S velocities and density in our model, we chose to calculate these values for presumed standard rock compositions.

2.1 Temperature model

As a first step towards seismic velocities, a temperature model of the Hellenic subduction zone is set up. We consider a vertical cross section running in north–south direction from the Cyclades to south of Crete. The structure along this transect is quite well known from previous seismological studies (Meier et al., 2007; Endrun et al., 2004; Snopek et al., 2007). Basic building blocks are from south to north (Fig 2, top): a 90 km thick continental African lithosphere in the south with 32 km thick crust on top, an accretionary wedge underlain by subducting oceanic crust and oceanic mantle lithosphere dipping at an angle of 15 degrees down to a depth of 60 km and then bending to an angle of 30 degrees at greater depths, and 60 km thick Aegean lithosphere with 20 km thick crust in the north. The subducted oceanic crust is assumed to be 7 km thick.

Based on this structural model a temperature model was developed using analytic expressions for temperature in different regions of the subduction zone. Although it is not as accurate as a numerical fluid dynamical model it approaches such models very well as shown by Davies (1999).

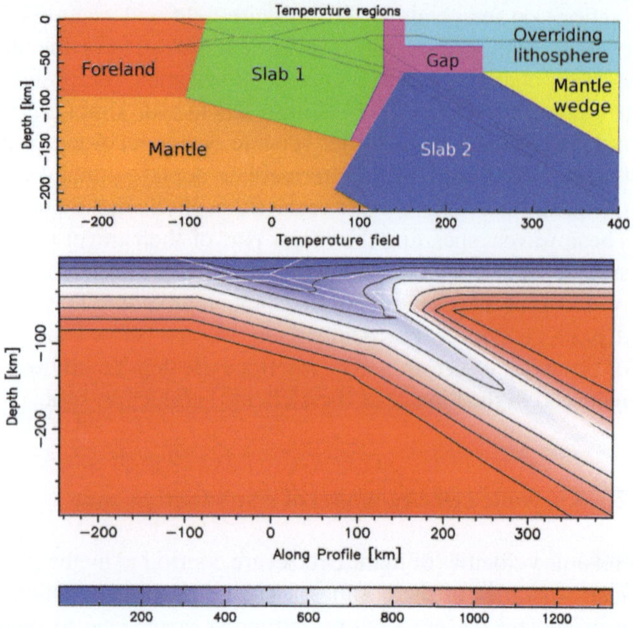

Fig. 2: (Top) Regions within which different laws for calculating temperature were applied, overlain by black lines indicating structural regions of subduction zone model derived from seismological studies, (bottom) temperature distribution on a vertical section through subduction zone model. Structural regions now indicated by white lines. Temperature at the base of the lithosphere has been assumed as 1250° C. Units: degrees Celsius.

To calculate the complete temperature field using analytical expressions, the subduction zone is divided into several regions (Fig 2, top)

- regular crust and lithosphere of overriding and subducted plate (foreland and overriding lithosphere): use of a parabolic temperature–depth relation based on static heat conduction with internal heating down to a depth of 24 km and a temperature at the base of the lithosphere of 1250 ° C;
- the mantle below the subducting plate as well as the mantle wedge below the overriding Aegean lithosphere and above the slab: temperature increase following an adiabatic gradient of 0.4 K km^{-1};
- the forearc zone including the accretionary wedge and the upper part of the slab dipping at an angle of 15° (slab 1): here we use analytic expressions derived by Royden (1993) for the calculation of temperature in eroding orogenic belts and accretionary prisms with prescribed erosion and accretion rates and a velocity of the ingoing lithosphere of 40 mm yr^{-1};
- the region inside the more steeply dipping slab including a boundary layer surrounding the slab to allow for diffusion of heat (slab 2): here we assume time-dependent heat conduction perpendicular to the slab dip and advection of temperature parallel to slab dip. An analytical solution to this problem has been given by Davies (1999);
- a region between the forearc zone and the steeper part of the slab where none of the above expressions are valid (gap): here we interpolate temperature using neighboring values.

With typical values for thermal diffusivity of $10^{-6}\,m^2\,s^{-1}$ and thermal conductivity of 2.5 W/(K m), we obtain the temperature field shown in Fig. 2, bottom.

2.2 Predicting seismic velocities from mineral properties

We follow an approach by Stixrude and Lithgow-Bertelloni (2005) (SLB2005). They use an expression of the mineral's free energy consisting of two parts: a "cold" part which contains an expansion of free energy to fourth order in the finite strain tensor and does not depend on temperature, and a "thermal" part which specifies the free energy of the vibrations of the crystal lattice in the quasi-harmonic approximation. Besides its explicit temperature dependency the thermal part also varies with strain via the lattice vibrational frequencies. The formulation is completely anisotropic and allows for the calculation of the fully anisotropic tensor of elastic constants if corresponding mineral data are available. The first derivative of free energy with respect to a small change of finite strain due to the passage of seismic waves yields the stress tensor and a further derivative gives the tensor of elastic constants. By evaluating the expressions for the elastic constants at conditions where measurements are available (zero finite strain), the expansion coefficients of the free energy can be determined. For the current application, we assume isotropic finite strains and calculate the equivalent isotropic elastic constants of the medium expressed as the bulk (K) and shear modulus (G). To quantify the thermal part of free energy, SLB2005 assume a parabolic Debye density of vibrational states with a maximum frequency that determines the Debye temperature Θ_D. This permits an explicit representation of the internal energy, entropy and specific heat of a mineral by integrals which can be numerically evaluated. Strain dependency of the thermal part comes in via the Grüneisen tensor which is the derivative of the lattice vibrational frequencies with respect to strain. For isotropic strains we get the well-known Grüneisen constant. A complete description also requires strain derivatives of the Grüneisen tensor which possesses the same symmetries as the elasticity tensor. In the isotropic case it can be represented by two constants (similar to K and G) which we denote by q and η_S. In summary, the SLB2005 theory requires the specification of the following nine parameters at ambient conditions: molar volume V, bulk modulus K and its pressure derivative

dK/dP, shear modulus G and pressure derivative dG/dP, Debye temperature Θ_D, Grüneisen constant γ, Grüneisen strain derivatives q and η_S.

Calculating anisotropic elastic properties from mineral properties as described above using SLB2005 theory is possible but requires much more detailed data on mineral properties. For the cold part, the full tensor of elastic constants and their pressure derivatives at ambient conditions are needed. For the thermal part, data for the Grüneisen tensor and its strain derivatives at ambient conditions are required. Since anisotropy is typically small, a more practical procedure would be to derive isotropic elastic constants from mineral properties and then add anisotropic perturbations with the desired symmetry properties.

A mineral physics approach to calculate seismic velocities for a subduction zone model may be regarded as of limited value, given the uncertainties in terms of mineral assemblage. It should be noted, however, that these uncertainties mainly affect the subducted oceanic crust, a phase diagram for MORB being depicted in Fig. 3. At any rate, effects related to sluggish kinetics in metamorphic reactions or temperature gradients are expected to be much more prominent compared to those related to uncertainties in the mineral database. Low velocity of oceanic crust compared to surrounding mantle as well as the steady increase of velocities with metamorphic grade leading to reduction of the velocity contrast with depth, are essential for the present modeling of seismic wave propagation.

2.3 Mineral database

Mineral physics data are taken from the very comprehensive database collected by Holland and Powell (1998) (HP98), which, however, does not directly list the nine parameters needed by the SLB2005 theory. Data associated with shear deformations are completely missing and instead of Debye temperature, Grüneisen constant and q, it provides the entropy at ambient conditions, the thermal expansivity α and the specific heat c_P as a function of temperature at atmospheric pressure. Since there is a strong dependency of $\alpha(T)$ and $c_P(T)$ on the Grüneisen constant γ and its bulk strain derivative q, we applied a grid search to find those values of γ and q that best reproduced thermal expansivity and specific heat given in the HP98-database. Similarly, Debye temperature was determined by a fit to the given entropy value in HP98. This procedure worked well for most minerals (with few exceptions) with errors of less than 5 percent. Finally, to obtain values for the parameter η_S associated with the shear part of the Grüneisen strain derivative tensor, we again performed a grid search for that value of η_S which best reproduces the quantity $1/(-G\alpha)(\partial G/\partial T)$ given in tables by Hacker et al. (2003b). In this way, a database of mineral parameters was compiled from which physical properties of subduction zone minerals were calculated using the SLB2005 theory.

2.4 Phase diagrams

A major difficulty of estimating physical properties of mantle rocks is the uncertainty about which mineral assemblages (phases) are stable at given temperature and pressure. Usually, it is assumed that only thermodynamically stable phases are present. In that case, it is sufficient to determine phase diagrams for the relevant rock facies. However, even this step is plagued by uncertainties because the rock compositions are not known very well. Even worse, observations of ultra-high-pressure rocks at the surface clearly testify that metastable phases may persist in earth's mantle for a long time and that the hypothesis of thermodynamic stability is violated. This is particularly true for subduction zones where exceptionally low temperatures inside the slab may strongly hamper phase transformations as the material moves to greater depths. For simplicity, we chose to determine the stable phase using phase diagrams calculated on the assumption of thermodynamic equilibrium. Hacker et al. (2003b) presents phase diagrams including volume fractions of the minerals contained in the phase for a typical mid-ocean ridge basalt composition, which we apply to rocks in the subducted oceanic crust, and for a harzburgitic composition which we use for mantle rocks (Fig. 3). We do not perform calculations for rocks in the shallow crust since usually density and seismic velocities are fairly well known there from seismic experiments.

It should be noted that the phase diagrams by Hacker et al. (2003b) assume MORB and harzburgite to be 100 percent water saturated. This leads to the appearance of many hydrated phases in the lower left part of the phase diagrams. To avoid these phases appearing in the subducting mantle part, we instead use dry harzburgite in this region of the model. Dry harzburgite only exhibits the phases F, G and K of the Harzburgite phase diagram shown in Fig. 3.

The geological record of exhumed high P/T metamorphic rocks (Stöckhert, 2002) indicates that a DSC could reach down to a depth of at least 130 km. However, since the existence of a DSC is linked to the release of water from the underlying oceanic crust, a consistent petrological model should allow the presence of some water-bearing minerals at these depths. The phase diagram by Hacker et al. (2003b) predicts the transition from the water-bearing zoisite eclogite to dry diamond eclogite at a pressure of about 3.3 GPa for temperatures between 500 and 700 °C. This pressure corresponds to a depth of about 105 km which is much less than required for a deep-reaching subduction channel. Assuming a kinetic delay of the transition owing to the low temperatures in the oceanic crust, we shifted the zoisite–diamond eclogite transition in the MORB phase diagram to a pressure of 4.7 GPa or equivalent depth of 150 km.

2.5 Evaluation of seismic wave velocities and density

Seismic velocities are determined on a dense grid that covers the cross section of the Hellenic subduction zone shown in

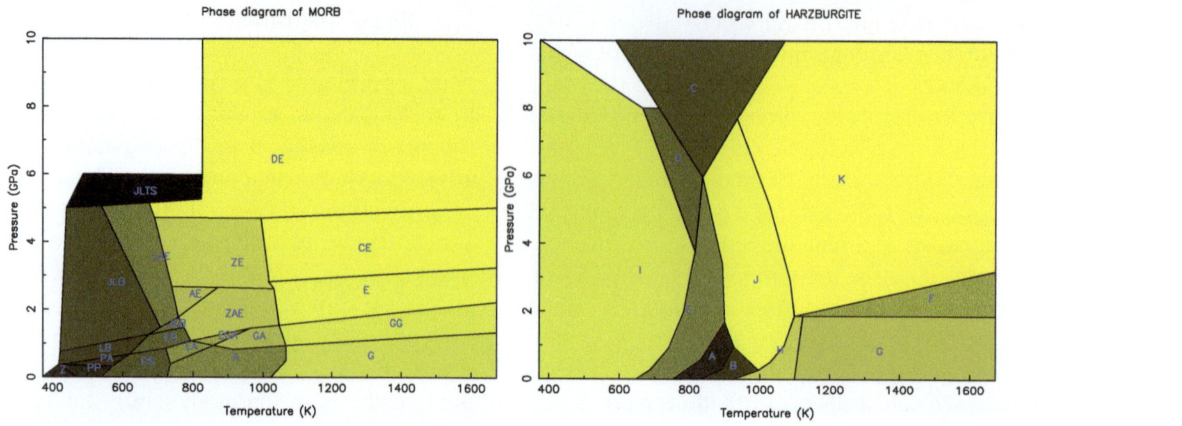

Fig. 3: Phase diagrams for a typical MORB composition (left) and a harzburgite composition (right) according to Hacker (2003). Phase regions have been extrapolated to higher pressures and temperatures. Abbreviations (left): Z: zeolite; PP: prehnite pumpellyite; PA: prehnite actinolite; GS: greenschist; EA: epidote amphibolite; GA: garnet amphibolite; EGA: eclogite garnet amphibolite; A: amphibolite; G: granulite; GG: garnet granulite; LB: lawsonite blueschist; JLB: jadeite lawsonite blueschist; CJLB: coesite jadeite lawsonite blueschist; JLTS: jadeite lawsonite talc blueschist; EB: epidote blueschist; JEB: jadeite epidote blueschist; ZAE: zoisite-amphibole eclogite; AE: amphibole eclogite; ZE: zoisite eclogite; LAE: lawsonite amphibole eclogite; E: eclogite; CE: coesite eclogite; DE: diamond eclogite. (Right) A: talc-chlorite-dunite; B: anthophyllite-chlorite-dunite; C: phaseA-chlorite-peridotite; D: serpentine-chlorite-phaseA; E: serpentine chlorite dunite; F: garnet harzburgite-1; G: spinel harzburgite-1; H: spinel-harzburgite-2; I: serpentine chlorite brucite; J: chlorite harzburgite; K: garnet harzburgite-2.

Fig. 2. At the grid nodes, lithostatic pressure is calculated using the vertical density distribution of a standard earth model. Temperature at the grid nodes is calculated as described above (see Fig. 2). Depending on the bulk compositions at a given grid point (continental crust, MORB, harzburgite, dry harzburgite), we look up the stable phase and its modal composition in the corresponding phase diagram (Fig. 3) and calculate elastic moduli and density for each constituting mineral using the theory of SLB2005 and the corresponding parameters in the mineral database. Effective elastic moduli of the phase are obtained by a volume fraction weighted average of the individual reciprocal elastic moduli (Reuss average). Density is calculated by averaging the individual density values according to the volume fractions. This procedure is certainly not satisfactory as it is known that the effective elastic moduli may have varying values depending on the spatial distribution of the modal constituents inside the phase. We have chosen to display here the Reuss averaged values to give an impression of the variations of seismic wave speeds inside a subduction zone. Hashin–Shtrikman bounds (Hashin and Shtrikman, 1963) can be readily calculated and provide an estimate of the variability of these values. Since, however, seismic wave speeds derived from tomographic imaging represent averages over rock volumes of kilometer size, a homogenization scheme is needed to obtain accurate estimates for the effective elastic moduli of such a rock volume. This is particularly true in subduction zones where large changes of temperature and also stress may occur within small distances. Clues of the spatial distribution of rocks inside a seismically "visible" rock volume may be taken from numerical models or geological outcrops.

A weak but ubiquitous anisotropy of the oceanic mantle lithosphere is observed in seismological studies represented by a 3 % radial anisotropy in standard earth model PREM (Dziewonski and Anderson, 1981), implying anisotropy of the mantle part of the subducted slab. Also, deformation in the mantle wedge may create anisotropy. In this model, anisotropy is neglected to avoid complications in the seismic wave field that could be erroneously attributed to the presence of a DSC. Moreover, anisotropy of slab and mantle wedge will not notably affect guided waves, in contrast to anisotropy in a DSC or subducted oceanic crust.

In Fig. 4 we show the resulting distribution of compressional and shear wave speed as well as density and quality factor down to a depth of 150 km. Velocities vary from values below 6.0/3.0 $km\,s^{-1}$ mostly in the crust to values of 8.7/4.8 $km\,s^{-1}$ in the fastest parts of the slab, respectively. We can clearly recognize the gradual velocity increase in the oceanic crust due to release of water. In the wedge above the oceanic crust we obtain very low wave speeds caused by the presence of hydrated phases. They are partially slower than in the oceanic crust leading to a positive jump of velocity at the plate contact down to a depth of 40 km. For density, we obtain a picture that correlates very well with the seismic wave speeds. Hydrous phases now show up as regions of lower density. At about 80 km depth the oceanic crust becomes increasingly dense due to the appearance of various eclogite phases.

Table 1: List of constants used to compute temperature and pressure dependent values for the inverse quality factor according to Eq. 1

A_0	Pre-exponential factor	12.5
f_0	Reference frequency Hz	0.1
α	Frequency exponent	0.25
A_D	Dry constant	1.25×10^6
A_W	Wet constant	3.65
E_D	Dry activation energy kJ Mol^{-1}	510
V_D	Dry activation volume ccm Mol^{-1}	14
E_W	Wet activation energy kJ Mol^{-1}	410
V_W	Wet activation volume ccm Mol^{-1}	11
C_{OH}	Water content in ppm H/Si	100
r	Water content exponent	1.2

For some numerical calculations of wave propagation shown later, we also accounted for attenuation. Attenuation in the mantle primarily depends on temperature, pressure and water content and the activation energies and activation volumes of the dissipating mechanisms. We use the following formula to calculate the inverse quality factor for our model (Karato and Jung, 1998):

$$Q^{-1} = A_0\left(\frac{f_0}{f}\right)^{\alpha}\left(A_D e^{-(E_D+pV_D)/(RT)} + A_W C_{OH}^r e^{-(E_W+pV_W)/(RT)}\right)^{\alpha}, \tag{1}$$

where E_D and V_D are the activation energy and volume for dissipation of dry rock, respectively, and E_W and V_W are their counterparts for water-induced dissipation mechanisms. C_{OH} is the water content in ppm H/Si, r the water content exponent and A_D and A_W, respectively, are factors determining the contributions of the two terms to attenuation. α is the frequency exponent and A_0 is a pre-exponential factor chosen in a way to produce seismologically reasonable values of the quality factor. Values chosen for the various constants are listed in Table 1. The quality factor is generally high in the area of the subducted slab due to the low temperatures prevailing there (Fig. 4). For this reason, we do not expect strong effects of attenuation on the wave propagation characteristics in the slab.

2.6 Effective seismic velocities in a deep subduction channel

Seismic velocities in the DSC require separate considerations. As mentioned in the introduction, we make the simplifying but reasonable assumption that the material inside the DSC is composed of eclogite blocks of various sizes floating in a low-viscosity serpentinite matrix (block in matrix structure). Since seismic waves recorded at the surface have wavelengths of a few kilometers and the typical size of the blocks is about one order of magnitude smaller, it is sufficient to consider effective seismic velocities of the composite material. They are computed using a theory of Sabina and Willis (1988) which is applicable up to frequencies where the effective wavelength approaches the size of the blocks. The theory delivers complex values for the isotropic elastic constants of an equivalent effective medium from which frequency-dependent values for wave velocities and scattering attenuation can be calculated. Controlling input parameters for the theory are the elastic constants and density of the constituting materials eclogite and serpentinite and the frequency–size distribution of the blocks. For eclogite, we used a P wave velocity of 8.36 km s^{-1}, an S wave velocity of 4.67 km s^{-1} and a density of 3640 kg m^{-3}. P and S velocities as well as density of serpentinite were taken from Hacker et al. (2003b) as 6.0 km s^{-1}, 2.73 km s^{-1} and 2680 kg/m^3, respectively. Recently, Bezacier et al. (2010) obtained isotropic Reuss average aggregate P and S velocities of antigorite using Brillouin-scattering of 6.17 km s^{-1} and 3.32 km s^{-1}. While the value for P velocity is consistent with Hacker's value, S velocity of Bezacier et al. (2010) is significantly higher. Moreover, antigorite and deformed serpentinite exhibit strong anisotropy (Bezacier et al., 2010) which is not accounted for in our simulations. Since samples taken from outcrops of exhumed fossil DSCs indicate a complex flow of the serpentinite matrix around eclogite blocks, we argue that the large-scale anisotropy of the BIM structure is much smaller than the one measured on pure samples in the laboratory. Nevertheless, remaining anisotropy may be a further important indicator of the existence of a DSC.

We have done several calculations using different frequency–size distributions of the blocks that were determined by Grigull et al. (2012) from outcrops of exhumed fossil DSCs. The results show that the seismic velocities of a BIM compound lie between the end-member values for blocks and matrix and mainly depend on the relative amount of blocks and their size distribution. Since the velocity of serpentinite is relatively low and the fraction of eclogite blocks small, the effective velocity of the DSC is found to be less than the velocities of the surrounding mantle and oceanic crust. In addition, petrophysical calculations suggest that seismic velocities of serpentinite matrix are nearly constant over the entire length of the DSC because the effects of temperature and pressure approximately cancel each other. All these considerations lead to a DSC model with P wave speed of around 6.6 km s^{-1} and S wave speed of 3.3 km s^{-1} (Fig. (5).

The upper interface of the DSC is assumed to be planar although numerical simulations (Gerya et al., 2006) suggest that it may be corrugated by plumes developing in the DSC. A markedly non-planar interface would certainly produce scattering depending on length scale and shape. Our model of the DSC should be considered as an end-member model that avoids such complexity by starting from the simplest assumptions.

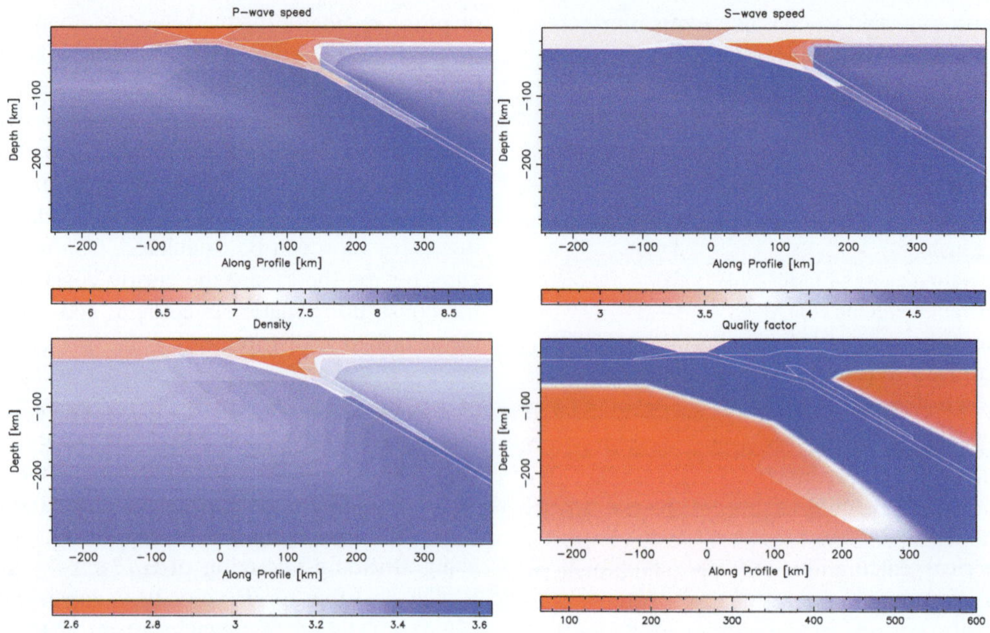

Fig. 4: Distribution of P wave velocities (top left) and S velocities (top right) in km s^{-1}, density in g cm^3 (bottom left) and quality factor Q (bottom right), calculated from the temperature model using SLB2005 theory, the phase diagrams of Hacker (2003) and our mineral database which was derived from those of Holland and Powell (1998) and Hacker (2003).

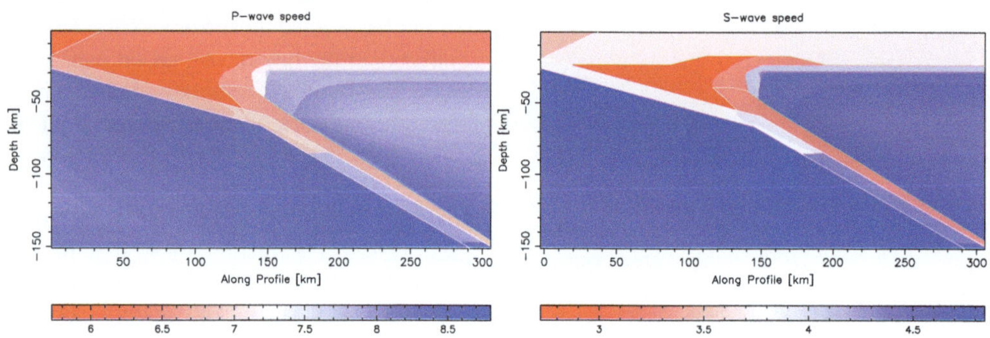

Fig. 5: Distribution of P wave velocity and S wave velocity for the subduction zone model including a DSC.

3 Effects of a deep subduction channel on wave propagation

Can a subduction channel presumably less than 10 km thick be detected by seismic methods? Tomographic studies appear not to be suitable because their resolution is generally insufficient to image such a thin channel. Methods using scattered or converted teleseismic waves (Bostock, 2002; Suckale et al., 2009) have potential to reach a resolution on the order of 10 km, but it is questionable whether they could distinguish the subduction channel from the oceanic crust. An alternative approach could be to search for specific phases in the seismic records which may serve as indicators for the existence of a channel. Since the channel is a small scale feature, visible effects on the seismic wave field can only be expected if the waves spend a considerable fraction of their propagation time inside the channel. For this reason, we consider here intermediate-depth earthquakes located in the subducted oceanic crust which directly radiate seismic energy into the subduction channel. Since both oceanic crust and channel are low-velocity regions for the major part of the propagation distance, we expect the formation of delayed high-amplitude guided waves. In the following, we will show by numerical simulations that a thin low-velocity channel on top of the subducted oceanic crust can significantly alter the seismic wave field observed at the surface.

3.1 Model setup

We use the 2-D Spectral Element Code by Komatitsch and Vilotte (1998) to simulate wave propagation in a 650 km wide and 300 km deep rectangular domain (Fig. 6). The domain is discretized into 835×96 quadrilateral elements with four internal Gauss–Legendre–Lobatto points per element. Source time function is a Ricker wavelet with a dominant source frequency of 2 Hz implying a significant spectral bandwidth from 1 to 5 Hz. We computed 120 000 time steps with a sampling interval of 0.0011 s. Synthetic seismograms are calculated for a line of 130 receivers at the surface spanning the entire width of the domain with a spacing of 5 km. Receivers are numbered consecutively beginning with 1 from left (south) to right (north). Paraxial absorbing boundary conditions are applied at all boundaries except for the surface which is free. The computational domain has been chosen much deeper and wider than required to avoid disturbing reflections from the boundaries.

The model is only 2-D implying excitation by line sources instead of point sources leading to reduced geometrical spreading and phase shifts. For comparison with observations, 3-D computations would be an option for the future.

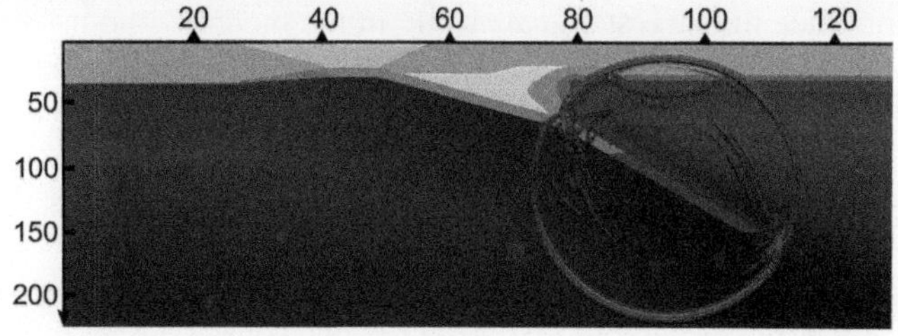

Fig. 6: Snapshot of vertical displacement illustrating general setup of the numerical model. Propagation time is 13.2 s. An explosive source radiates *P* waves travelling up the slab. Inside the oceanic crust, the *P* wave is already slightly delayed but the wavefront is still attached to the waves propagating in the surrounding mantle. At the upper and lower interface, PS conversions have developed. They can be identified by their smaller wavelength. Vertical axis annotation gives depth in km. Horizontal axis specifies index of receivers. Color shading for displacement displays negative values as blue and positive values as red shades. Color scale is nonlinear. RGB values are proportional to normalized absolute value of displacement to the power of 0.3 enhancing low-amplitude displacements. Displacements less than 1 percent of the maximum value are suppressed.

3.2 Guided *P* waves

At first, we concentrate on the essential features of the compressional (*P*) wave field. An explosive point source is put inside the oceanic crust at a depth of 108 km. Although an explosive source can not represent a real earthquake, it proves useful here because it does not radiate shear waves which, owing to their large amplitudes, hamper the proper visualization of details in the *P* wave field by means of snapshots. Moreover, an explosive source well approximates *P* waves from a double couple source with maximum radiation into the low-velocity waveguide formed by oceanic crust and DSC. This is confirmed by computations with a double-couple source shown in context with *S* wave propagation below (Fig. 14).

The general propagation situation is displayed in Fig. 6. A *P* wave with nearly spherical wavefront has expanded from the source. In the oceanic crust, a guided wave has formed which is already delayed compared to the wavefronts in the surrounding mantle. In addition, some *PS* conversions have developed at the upper and lower boundary of the oceanic crust.

The wave field develops more structure while it passes through regions with significantly lower velocities than in the surrounding mantle. When the guided waves reach the bend of the slab and propagate into the serpentinized mantle wedge (Fig. 7) two distinct wave packages can be identified: a delayed high-amplitude guided wave train made up of the delayed direct *P* wave followed by PSP-conversions created at the crust–mantle interface, and an advancing refracted *P* wave generated by the fast wave travelling in the high-velocity mantle below the interface.

Later, these wave trains appear as prominent arrivals at the surface. The models with and without DSC mostly differ with respect to the guided waves while the direct *P* wave (2) and the slab-refracted *P* wave (1) appear in both models. In the model without DSC, there is only one guided wave radiating into the wedge (3) and one attached to the oceanic crust (4), whereas in the model with DSC we find two guided waves emanating from the upper boundary of the DSC (3 and 3′) and two guided waves attached to the lower boundary of the oceanic crust (4 and 4′).

The snapshots shown in Fig. 8 allow a better understanding of the resulting signals at the surface receivers. In the first one, at 24.2 s after origin time (Fig. 8a), the guided waves reach the surface for the first time. The direct *P* wave is still the first arrival but about to be overtaken by the slab refracted *P* wave. In the model with DSC (left panel), the two guided wave trains 3 and 3′ follow the direct wave with high amplitudes. Phases 4 and 4′ are visible as weak signals in the oceanic crust. In the model without DSC (right panel), phase 3 generates a surface arrival while guided phase 4, attached to the oceanic crust, has developed a strong straight wavefront inside the wedge. Moreover, a secondary phase (5) has formed during travel through the serpentinized mantle wedge.

At 27.5 s after origin time, (Fig. 8b), the direct *P* wave (2) has just been overtaken by the slab refracted wave (1). In the model with DSC (left panel), the guided phases 3 and 3′ still produce the largest signals while in the model without DSC (right panel) phase 4 becomes dominant and the secondary phase 5 starts to split off the main guided phase 4.

At 36.3 s after origin time (Fig. 8c), the slab-refracted wave (1) has left behind the slower waves and forms the first arrival. The direct *P* wave (2) has been overtaken by the guided waves and is submerged by the strong phase 4. In the model without DSC (right panel), phase 4 now forms the major onset closely followed by the slower phase 5. A PS-conversion of phase 5 at the boundary between crust and serpentinized mantle wedge has also reached the surface as phase 7. In the model with DSC (left panel), the guided phases 3 and 3′ are still dominant but partly overtaken by the guided phases 4 and 4′ which were previously attached to the oceanic crust. The latter only form weak arrivals in the seismograms compared to the strong phase 4 in the model without DSC. The PS-conversion of phase 3 at the crust–wedge boundary (phase 7) is still very weak. In addition, a converted PS-refraction from the slab starts to move towards the surface (phase 6) and a reflection at the slab interface of the surface-reflected direct *P* wave develops (phase 8).

With help of the snapshots, synthetic seismogram sections computed for the receivers at the surface (Fig. 9 and 10) can be readily interpreted. Remarkably, though the velocity models just differ by the presence or absence of the thin DSC, these record sections differ in the major arrivals. The pattern of onsets in the sections indeed define tell-tale signatures for the presence or absence of a DSC. The "no-DSC" signature (Fig. 9) is characterized by the dominating phase 4 originating as guided wave from the oceanic crust and its follow-up phase 5 that emerged during propagation through the slow mantle wedge. An arrival between phase 2 and 3 is missing. On the other hand, the "DSC"-signature is characterized by the strong phase pair 3 and 3′ originating as guided waves from the DSC. Phase 4 is essentially missing and replaced by the weak phase pair 4 and 4′. It should also be noted that in the model with DSC the combination of direct wave (1), slab refracted (2) and guided waves (3, 3′) forms uninterrupted wave trains of about 4 s length between receivers 60 and 70, a location corresponding to Crete in our subduction zone model.

To investigate whether the tell-tale signatures with respect to the presence or absence of a DSC are preserved when the source is moved up- or downwards inside the oceanic crust, we performed numerical simulations for varying source positions inside the oceanic crust (Fig. 11) for the model with DSC. If the source is too shallow, the phases 2, 3′ and 3 merge and are no longer distinguishable. In our model, this occurs at source depths less than 70 km. For greater source depths, the "DSC"-signature can be clearly recognized. The additional phase 3′ is present in all cases. The deeper the position of the source, the longer are the guided wave trains. This behavior is a consequence of the longer travel distance inside the waveguide formed by oceanic crust and DSC. For a source depth of 155 km the arrivals created by the guided waves become extremely long. Wave trains of up to 10 s length can be generated without any need for small-scale scatterers inside the medium. Due to the low velocity in the DSC, multiple *P* reflections and PS-conversions can build up in the waveguide and are able to conserve their amplitudes because of the velocity contrasts on both sides. Without DSC, there is no real difference between a deep or a shallow source with respect to the length of the guided wave. This behavior is caused by the steady velocity increase inside the oceanic crust with depth due to progressive transformation of basalts into less water-bearing, denser and faster mineral phases. The oceanic crust alone starts to become an efficient waveguide only at rather shallow depths.

What happens if the source is not inside the oceanic crust but either in the mantle below or in the DSC above? We performed two additional simulations where we moved the source from its original location within the oceanic crust at 108 km depth to a depth of 118 km in the mantle part of the slab and to a depth of 98 km in the DSC, respectively (Fig. 12). If the source is in the slab mantle, we observe a clear "no-DSC"-signature despite the presence of the DSC in the model. If the source is moved into the DSC, the record section is still of "DSC"-type. This finding may prove unfavourable for a detection of a DSC from seismic records. Earthquakes may well move into the mantle part of the slab because the temperature minimum also moves there during slab descent. On the other hand, if intermediate depth earthquakes are caused by dehydration reactions, it is highly probable that they occur in the oceanic crust which contains most of the subducted water and from where fluids are discharged into the overlying mantle wedge (Hacker et al., 2003a).

One could argue that errors in earthquake location may inhibit the identification of a DSC, for example, when a "no-DSC"-signature is observed for an earthquake erroneously located in the oceanic crust. Hence, seismic evidence for a DSC should always relate to observation of both signatures. If both are present, in the ideal case this could even be used to discriminate between earthquakes located within oceanic crust and DSC and those located deeper in the mantle part of the slab.

For efficiency reasons, the numerical simulations were carried out neglecting attenuation of seismic waves. To ensure the validity of the conclusions derived from the purely elastic simulations, we performed one simulation for a model with DSC assuming a spatial distribution of the quality factor as depicted in Fig. 4. The resulting record section for a source at 108 km depth does not significantly differ from the one shown in Fig. 10. In particular, the "DSC"-signature is fully preserved with attenuation.

3.3 Guided shear waves

Can shear waves help distinguish a subduction zone with DSC from one without? We performed further simulations for a double-couple source with a rupture plane striking

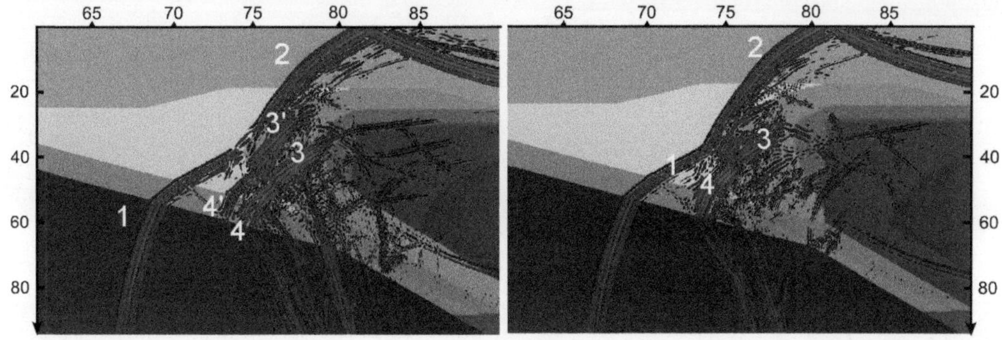

Fig. 7: Snapshot of P wave propagation for a model with DSC (left) and without DSC (right). Propagation time is 18.15 s. Phase 1 is the slab refracted wave, phase 2 the direct P phase and 3 and 4 are guided waves. Note that in the model with DSC there are two guided waves (3, 3′ and 4, 4’). Phases 4 and 4′ in the model with DSC are much weaker and phases 3 and 3′ are much stronger than their counterparts in the model without DSC. Numbering of phases is the same as in Fig. 9. See Fig. 6 for explanation of axis labels and color shading.

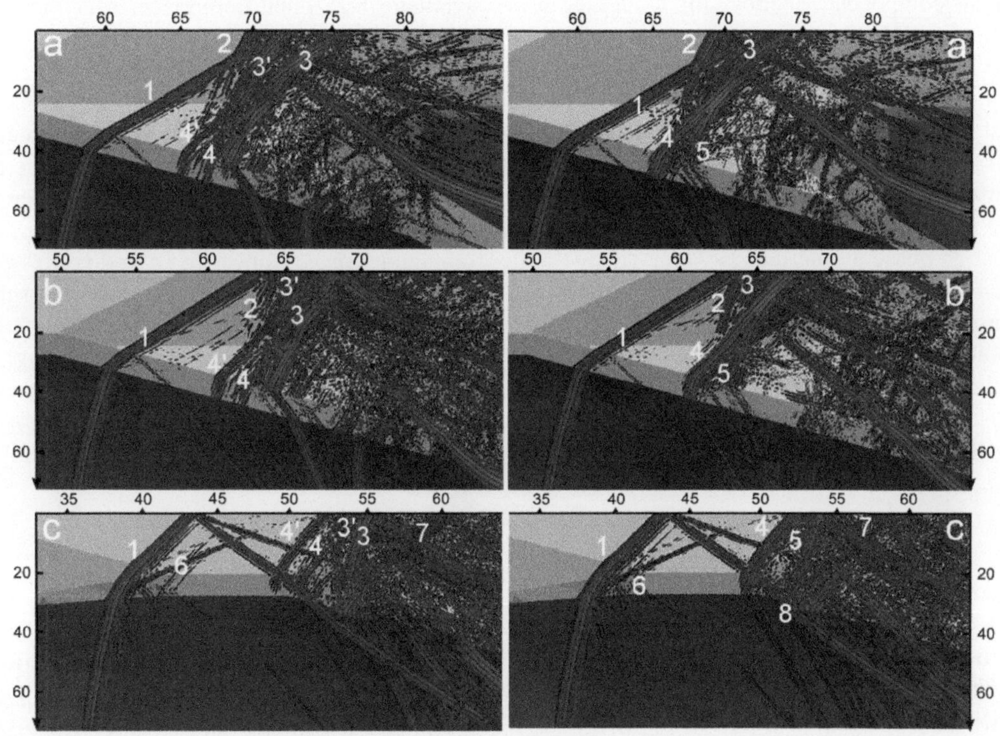

Fig. 8: Sequence of snapshots illustrating the evolution of the wave field. Left column: model with DSC; Right column: model without DSC. Vertical axis annotation gives depth in km. Horizontal axis specifies index of receivers. **(a)** $t = 24.2$ s, guided phases 3 have reached the surface. The direct phase still forms the first onset. **(b)** $t = 27.5$ s, slab-refracted wave 1 overtakes direct wave 2; phase 4 forms dominant signal in model without DSC and a secondary guided phase 5 develops in the slow serpentinized wedge; on the contrary, in the model with DSC, phases 3 form distributed signals at the surface and phases 4 stay weak. **(c)** $t = 36.3$ s. In the model with DSC, phases 4 overtake phases 3 and form weak precursory signals; an S conversion of phase 5 reaches the surface (phase 7). The slab refracted S conversion (phase 6) becomes visible and a reflection from the slab interface of the P surface reflection emerges. Strongest signal in model with DSC is phase 3. Phase 4 is strongest signal in model without DSC. Numbering of phases is the same as in Fig. 9. See Fig. 6 for explanation of axis labels and color shading.

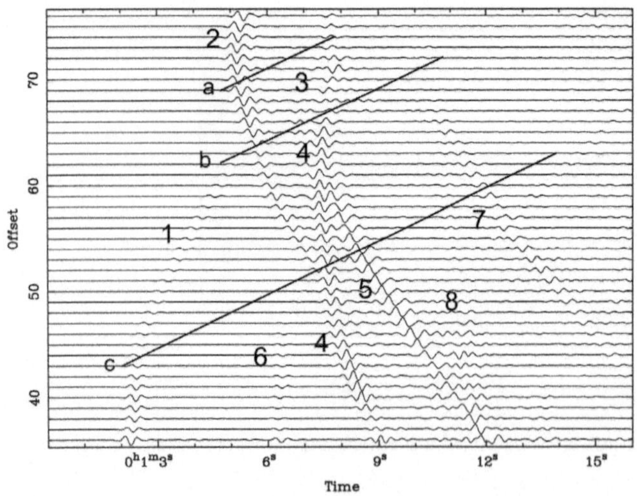

Fig. 9: Model without DSC: time-reduced record section of synthetic horizontal component seismograms computed at receivers at the surface. Numbers from 1 to 8 mark major seismic phases. (1) *P* wave refracted at the interface between subducted oceanic crust and lithospheric mantle, (2) direct *P* phase, (3) first guided *P* phase emerging from the deep oceanic crust, (4) second guided *P* phase emerging from the deep oceanic crust, (5) third guided *P* phase developing in the serpentinized mantle wedge, (6) refracted *S* phase converted from direct *P* phase at interface between subducted oceanic crust and lithospheric mantle, (7) *S* conversion of upgoing guided phase 5 at serpentinized wedge–crust interface, (8) surface reflection of direct *P* phase reflected back at slab interface. Solid lines crossing seismograms indicate lines of constant time for comparison with snapshots a–c of Fig. 8 on the right. Associated times from top to bottom are $t = 24.2$ s, 27.5 s and 36.3 s. Source is explosive and at 108 km depth.

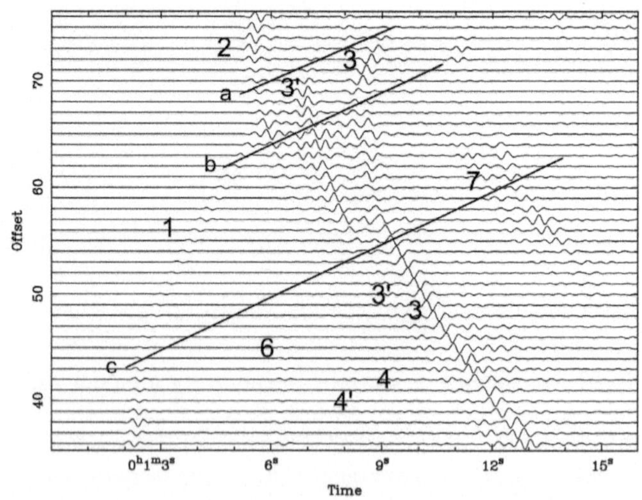

Fig. 10: Model with DSC: time-reduced record section of synthetic horizontal component seismograms computed at receivers at the surface. Numbers from 1 to 7 mark major seismic phases. (1) *P* wave refracted at the interface between subducted oceanic crust and lithospheric mantle, (2) direct *P* phase, (3) and (3') two guided *P* phases emerging from the deep oceanic crust and radiated into the serpentinized mantle wedge, (4) and (4') guided *P* phases emerging from the deep oceanic crust staying attached to the oceanic crust, (6) refracted *S* phase converted from direct *P* phase at interface between subducted oceanic crust and lithospheric mantle, (7) *S* conversion of upgoing guided phases 3 at serpentinized wedge–crust interface. Solid lines crossing seismograms indicate lines of constant time for comparison with snapshots a–c of Fig. 8 on the left. Associated times from top to bottom are $t = 24.2$ s, 27.5 s, 36.3 s. Source is explosive at 108 km depth.

perpendicular to the subduction direction (profile direction) and dipping 15 degrees downwards. Since the deeper part of the slab dips at 30 degrees, the angle of 15 degrees was chosen to achieve both sufficient *P* and *S* wave radiation into the oceanic crust. The source depth is again 108 km as in our reference case for *P* waves. The synthetic record sections are slightly more complicated because *SP* conversions arrive now before the guided shear waves. As for the *P* waves, again tell-tale signatures can be observed in the record sections which allow a discrimination between models with and without DSC by one quick glance (Fig. 13). The guided shear wave signals appear within the rectangular boxes inserted into the record sections. Beyond receiver number 65, three distinct phases (3a, 3b, 3c) can be recognized for the model with DSC whereas there is only one for the model without. At receivers 40 to about 65, the situation reverses: the model without DSC now exhibits a package of three phases while there is only one in the model with DSC. The phases preceding those in the rectangular box appear for both models and are not suitable for discrimination purposes. They are the direct *S* wave (1), a slab-refracted *S* wave (2), a slab-refracted *SP* surface conversion (4) and a slab-reflected *SP* conversion (5). The snapshots in Fig. 14 provide some additional understanding of guided shear wave propagation: in the model with DSC, an extended M-shaped guided wave package develops in the steeper part of the slab from which phases 3a–c develop later. As for *P* waves, the guided wave train in the model without DSC is rather compact (Fig. 14a). Once these waves move into the serpentinized mantle wedge, they develop strong wavefronts (Fig. 14b) which keep their identity while passing through the wedge and arrive as strong individual phases at the surface (Fig. 14c).

3.4 Observations of guided *P* waves in the Aegean

P phases of typical seismograms from intermediate-depth earthquakes recorded in the forearc of the Hellenic subduction zone are shown in Fig. 15. The data were acquired between October 2005 and March 2007 during the

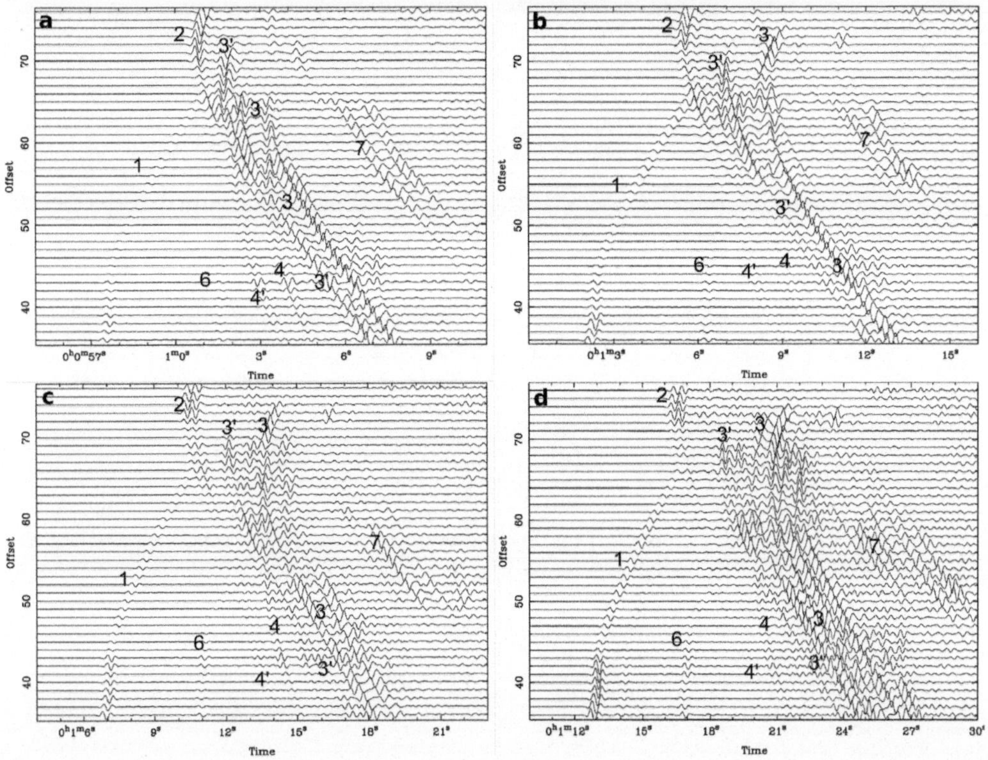

Fig. 11: Time-reduced synthetic record section for the model with DSC for different source depths inside the oceanic crust. Depths: **(a)** 88 km, **(b)** 108 km, **(c)** 128 km, **(d)** 155 km. The additional phase 3′ is present for all depths. Note the increasing length of the guided wave train with increasing source depth.

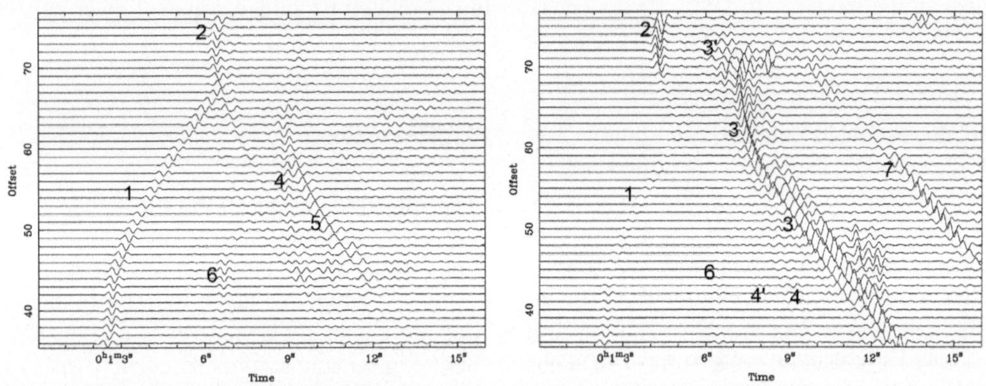

Fig. 12: Time-reduced synthetic horizontal component record sections for the model with DSC and explosive sources located below the oceanic crust (left panel) and inside the DSC (right panel). In the former case, the record section exhibits a "no DSC"-signature (compare to Fig. 9), in the latter case, it exhibits a clear "DSC" signature (compare to Fig. 10). Phases are numbered consistent with Figs. 9 and 10.

EGELADOS experiment (Friederich and Meier, 2008) in the southern Aegean. The seismograms are band-passed filtered in the frequency range from 2 to 5 Hz in order to mimic the effect of the Ricker wavelet used in the numerical modeling. We selected seismograms from stations located on Crete for the most part, complemented by some low-noise ocean-bottom hydrophone records. The data records are sorted according to epicentral distance. The *P* signals are characterized by multiple high-amplitude wave trains of several seconds length that are also found in the synthetic seismograms where they are produced by guided waves in oceanic crust and DSC.

As the installed stations are not aligned along a section plane containing the hypocenter, and their spacing is too

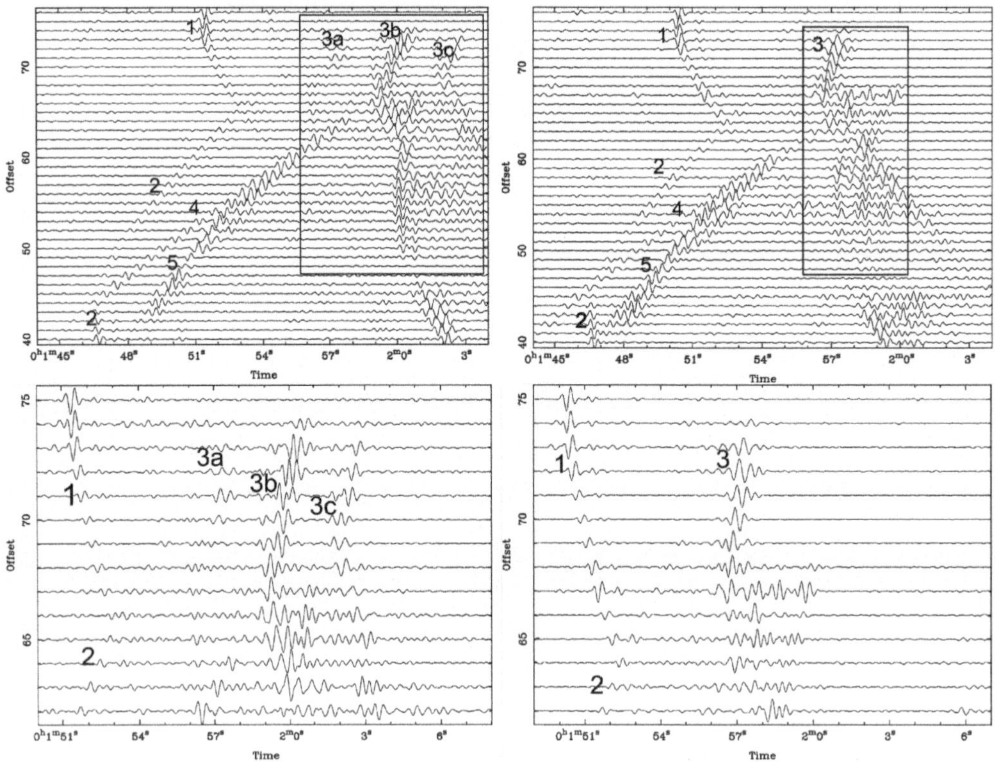

Fig. 13: Time-reduced synthetic record sections for a double-couple source at 108 km depth showing S waves. Left panel: model with DSC. Right panel: model without DSC. Marked phases: (1) direct S wave, (2) S wave refracted at slab interface, (3) guided wave, (3a–c) multiple guided waves in model with DSC, (4) SP surface conversion refracted at slab interface, (5) SP surface conversion reflected at slab interface. Phases 1, 2, 4 and 5 appear in models with and without DSC. Note the three distinct guided waves in the model with DSC. Boxes enclose signals related to guided waves. Lower panels: zoom into the receiver range 62 to 75 where the three guided arrivals are best visible.

large and irregular, the available data are not suitable for recognizing the DSC or no-DSC signature identified within our simulations. New dedicated seismic field experiments are required to clarify this issue. The MEDUSA experiment (Suckale et al., 2009) carried out in the Ionian region may provide the required station density and alignment to reveal DSC signatures.

Long wave trains could alternatively be generated by scatterers of a size on the order of a wavelength. However, numerical simulations done by the authors (unpublished) indicate that the length and amplitude of the observed P wave trains can only be reproduced by placing strong, kilometer-sized velocity anomalies of 20 to 40 percent contrast into the oceanic crust. The beauty of our model is that the long, high amplitude wave trains can be effectively generated by a simple, thin, homogeneous low-velocity channel on top of the oceanic crust.

4 Discussion

The fundamental question is to which extent the results of our numerical simulations can be applied to natural situations. More specifically, can seismic records that display the "DSC signature" be taken to indicate the presence of a subduction channel, or, vice versa, can the absence of this signature in seismic records be taken to discard the hypothesis? At this point, it must be emphasized that the structural model of a DSC underlying our simulations is extremely simplistic, although it incorporates presumably realistic structural elements, phase assemblages and elastic properties based on mineral physics. These properties are poorly constrained, mainly due to uncertainties with respect to bulk composition and mineral assemblage, widespread disequilibrium and metastability, heterogeneity on all length scales, probably steep temperature gradients and possibly complex internal structures which can only tentatively be inferred from geological observations in exhumed high P/T metamorphic terranes, leaving ample space for interpretation.

Moreover, at present it seems impossible to identify classes or styles of DSCs, as a "quenched" DSC structure

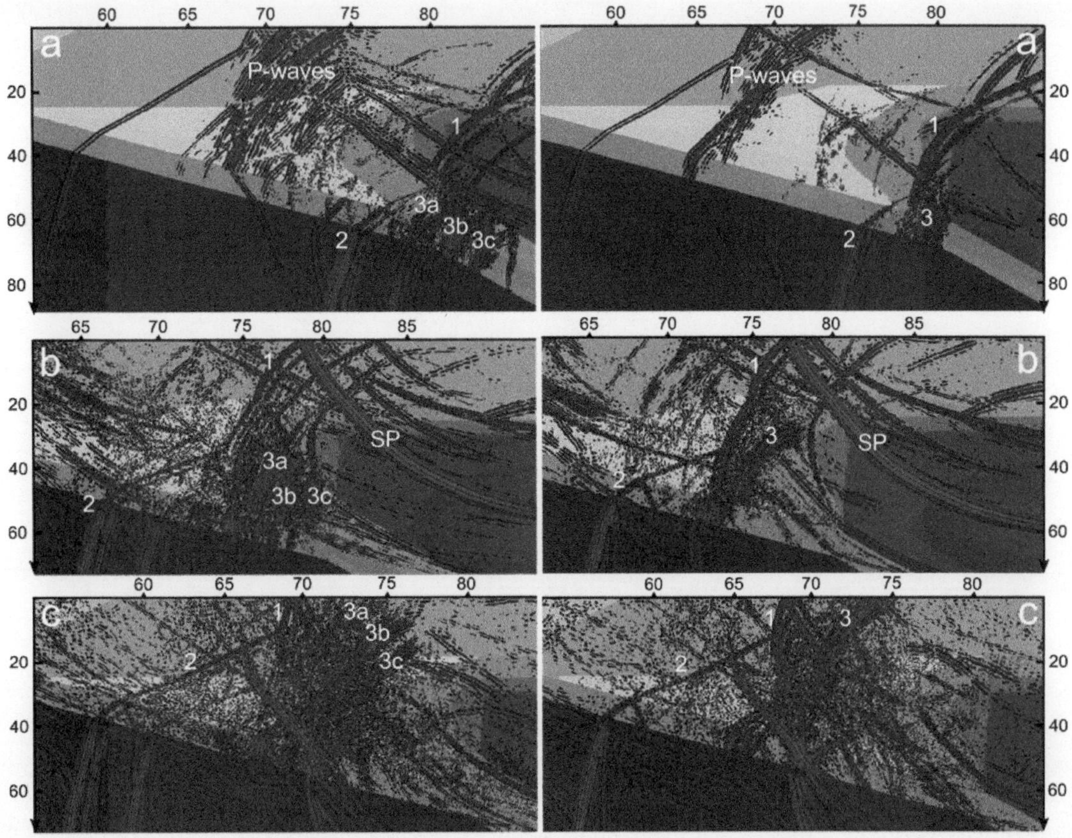

Fig. 14: Sequence of snapshots of vertical displacement for double couple source. Left panels: model with DSC. Right panels: model without DSC. Vertical axis annotation gives depth in km. Horizontal axis specifies index of receivers. Phase identification is given in Fig. 13. **(a)** $t = 24.75$ s. Guided S waves about to enter the serpentinized mantle wedge; several wave packets have developed in model with DSC (phases 3a, 3b, 3c). **(b)** $t = 34.1$ s. Guided wave already in serpentinized area, direct S wave already at surface, note strong SP surface conversion. **(c)** $t = 44$ s. Guided waves about to reach the surface. See Fig. 6 for explanation of axis labels and color shading.

is not available for analysis at the surface. Numerical simulations are probably the only way to depict possible histories of DSCs, choosing starting and boundary conditions which at least appear realistic, corresponding to common sense. This situation could improve when presently active subduction channels at depth can be identified by seismological means. Only in this case, true snapshots may be correlated with current state and history of the respective subduction system, and eventually open the chance to identify different types of DSC. Our present contribution is thought to provide a small step into this direction.

Limited by the wavelength of recordable seismic signals and the resolution of seismic experiments, our model of a subduction channel accordingly integrates over a large volume. Most importantly, it substitutes highly heterogeneous material in a real subduction channel by a homogenized, effective material which neglects heterogeneities below the kilometer-scale and temperature gradients within the subduction channel. The temperature field is expected to be controlled by the overall flow pattern which may change in space and time with a potentially strong feedback. In addition, our model adopts a simple crustal structure and is only two-dimensional. In contrast, the Hellenic subduction zone (HSZ) to which it is applied exhibits considerable curvature and a possibly corrugated slab shape (Knapmeyer, 1999; Papazachos et al., 2000). Finally, anisotropy inside the subduction channel is neglected, as many exposed high P/T metamorphic rock assemblages show structural patterns which cannot necessarily be expected to produce anisotropy on the length scale resolved by seismic experiments. The same holds for BIM structures taken to represent a subduction channel here.

Available data from the HSZ are not suitable to search for "DSC signatures" by tracking seismic phases over epicentral distance, because the seismic stations in the forearc of the HSZ are not located on a straight profile but exhibit varying azimuths with respect to the intermediate-depth hypocenters (map in Fig. 15d). A passive seismic experiment with stations aligned parallel to the dip direction of the subducted

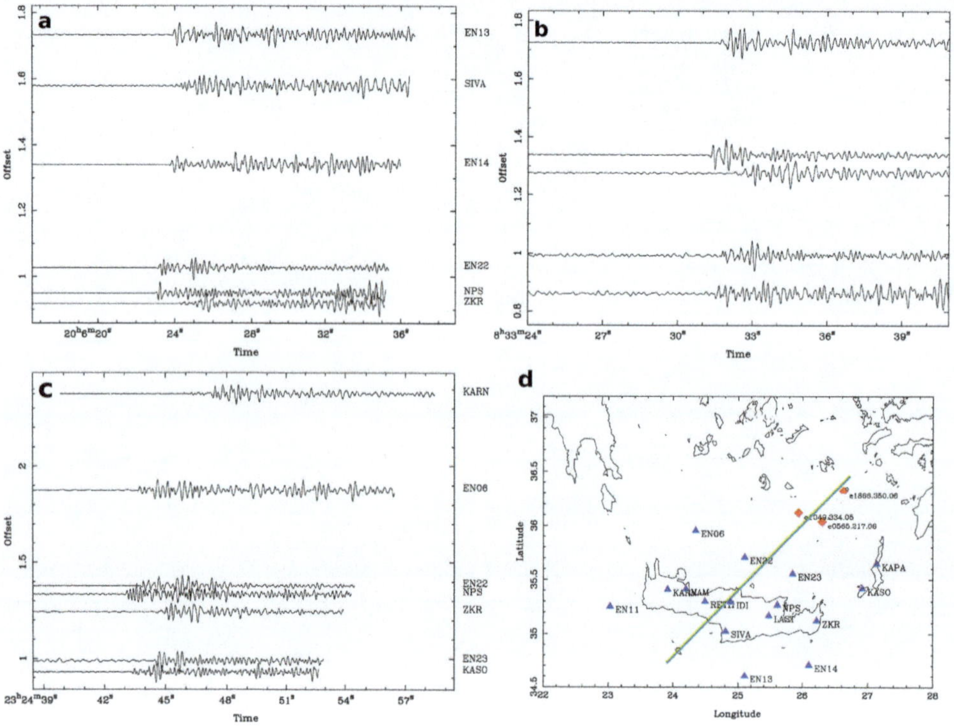

Fig. 15: Examples of time-reduced seismograms for intermediate depth events recorded at stations located in the forearc of the Hellenic subduction zone as shown in the event and station map in **(d)**. **(a)** Event e0565.217.06, depth 87 km; **(b)** event e1049.334.05, depth 99 km; **(c)** event e1866.350-06, depth 110 km. Note the multi-onset character of the *P* phases and the generally long complex wave trains. Red diamonds in **(d)** show event epicenters and green line indicates possible profile for a seismic experiment.

slab, reaching from the outer rise to the backarc, could remedy the situation. In such an experiment, stations should be linearly aligned with a spacing of less than 5 km. Even though intermediate-depth earthquake data generally exhibit more complicated waveforms than those obtained from corresponding simulations, the characteristic guided *P* wave trains of several seconds length are well discernible. These strong signals, tracked and correlated over epicentral distance, would allow identification of tell-tale signatures for the presence or absence of a subduction channel. In the Aegean, as probably in most subduction zones worldwide, ocean bottom seismometers would be required besides land stations to provide the required data.

The HSZ is considered an attractive target region for this type of experiment, as an active DSC has been proposed based on independent evidence (Meier et al., 2007). These authors found that the outline of a region lacking microseismicity at a depth of about 20 to 40 km coincides with the coast line of Crete. Significant uplift of Crete since 4 Ma in an overall extensional environment is therefore suspected to be correlated with anomalous material at depth. Based on position within the HSZ, the properties are tentatively proposed to reflect extrusion of material from a DSC, which became fully established at about 4 Ma after Miocene microcontinent collision (Meier et al., 2007). The fact that only discrete parts of the forearc, forming the islands, undergo uplift may be explained by disintegration of the DSC into distinct fingers, as the radius of the curved plate interface increases towards the surface. If signatures of a DSC could be observed in a seismic experiment at the HSZ, they could be taken to add another piece of evidence in favour of the above hypothesis..

A limiting factor for any such passive seismic experiment is the occurence of suitable intermediate depth earthquakes. For example, observations of seismicity in the Aegean (Hatzfeld and Martin, 1992) indicate that intermediate depth earthquakes cluster in a small region close to the islands Nisyros and Astypalea in the southeastern corner of the Aegean. Using these earthquakes would imply a profile across Karpathos, for which the layout requires a large share of ocean bottom seismometers. For a more convenient seismic profile across Crete, intermediate depth earthquakes further to the west along the strike of the subduction zone would be prerequisite, which are rarely observed. As a compromise, a profile oriented towards the Nisyros earthquake cluster could be feasible (green line in Fig. 15d), albeit being oblique to the dip of the slab. In that case, one would have to cope with three-dimensional propagation effects owing to slab curvature.

5 Conclusions

We performed numerical simulations of seismic wave propagation in a generic subduction zone model to explore potential seismological evidence for an active deep subduction channel. The distribution of seismic velocities in our model is based on geometric and structural information, predicted temperature field, phase diagrams for MORB and harzburgite, and properties of minerals at high pressures and temperatures predicted using equations of state.

Although suffering from various uncertainties, the model produces self-consistent values for P, S velocity and density and their variation with pressure and temperature. It produces velocity gradients in the slab and the mantle wedge and a steady velocity increase in the oceanic crust, an essential feature for the development of guided waves. It thus represents a more realistic model of a subduction zone than models assuming subregions with constant velocities gleaned from tomographic images.

The deep subduction channel itself is modeled as a thin layer located on top of the subducted oceanic crust. This layer is composed of eclogitic blocks floating in a serpentinite matrix. Homogenization calculations show that such a BIM structure exhibits frequency-dependent seismic velocities with values intermediate between those of block material and pure matrix. Since seismic velocities of serpentinite are low compared to those of typical mantle rock, and the volume fraction of eclogitic blocks is small (between 5 and 50 %), the subduction channel is predicted to represent a low-velocity zone acting as wave guide. Petrophysical calculations suggest that seismic velocities of serpentinite matrix are nearly constant over the entire length of the subduction channel, the effects of temperature and pressure approximately cancelling each other.

Our numerical simulations focus on high-frequency seismic waves emitted from intermediate depth earthquakes. Analysis of synthetic seismograms for receivers located in the forearc indicate significant differences of P and S waveforms for models with and without deep subduction channel. Although only a very small volume of the model is covered by the subduction channel, the presence or absence of a channel is found to affect major seismic arrivals. Record sections along profiles crossing the forearc parallel to slab dip exhibit tell-tale patterns which allow to discriminate between models with and without deep subduction channel. The "DSC" and "no-DSC" signatures are found to represent stable features which remain preserved when the earthquake is shifted up or down within the oceanic crust. The signature also remains visible when the source mechanism is changed. Only in case the earthquake occurs below the oceanic crust, the "DSC" signature is lost and a "no-DSC" signature appears. Our results suggest that the hypothesis of deep subduction channel, explaining rapid exhumation of high P/T metamorphic rocks, might be tested by an appropriately designed seismic experiment.

Ideally, such an experiment would provide observations of both DSC and no-DSC signatures from different earthquakes. This way, it would become unnecessary to discriminate between earthquakes located within DSC or oceanic crust and those located in the slab mantle, but the position could be inferred from the signature.

Acknowledgements. This study was supported by the German Research Foundation through the Collaborative Research Center 526 "Rheology of the Earth – from the upper crust to the subduction zone". Thanks to the SPECFEM community for providing great software.

References

Agard, P., Yamato, P., Jolivet, L., and Burov, E.: Exhumation of oceanic blueschists and eclogites in subduction zones: timing and mechanisms, Earth-Sci. Rev., 92, 53–79, 2009.

Angiboust, S., Agard, P., Yamato, P., and Raimbourg, H.: Eclogite breccias in a subducted ophiolite: A record of intermediate-depth earthquakes?, Geology, 40, 707–710, 2012.

Baldwin, S. L., Monteleone, B. D., Webb, L. E., Fitzgerald, P. G., Grove, M., and Hill, E. J.: Pliocene eclogite exhumation at plate tectonic rates in eastern Papua New Guinea, Nature, 431, 263–267, 2004.

Bebout, G. E.: Metamorphic chemical geodynamics of subduction zones, Earth Planet. Sci. Lett., 260, 373–393, 2007.

Bezacier, L., Reynard, B., Bass, J. D., Sanchez-Valle, C., and de Moortèle, B. V.: Elasticity of antigorite, seismic detection of serpentinites, and anisotropy in subduction zones, Earth Planet. Sci. Lett., 289, 198–208, 2010.

Bostock, M.: Kirchhoff-approximate inversion of teleseismic wavefields, Geophys. J. Int., 149, 787–795, 2002.

Burov, E., Jolivet, L., Le Pourhiet, L., and Poliakov, A.: A thermomechanical model of exhumation of high pressure (HP) and ultra-high pressure (UHP) metamorphic rocks in Alpine-type collision belts, Tectonophysics, 342, 113–136, 2001.

Chemenda, A. I., Mattauer, M., Malavieille, J., and Bokun, A. N.: A mechanism for syn-collisional rock exhumation and associated normal faulting: Results from physical modelling, Earth Planet. Sci. Lett., 132, 225–232, 1995.

Chopin, C.: Coesite and pure pyrope in high-grade blueschists of the Western Alps: a first record and some consequences, Contribut. Mineral. Petrol., 86, 107–118, 1984.

Cloos, M.: Flow melanges: Numerical modeling and geologic constraints on their origin in the Franciscan subduction complex, California, Geol. Soc. Am. Bull., 93, 330–345, 1982.

Cloos, M. and Shreve, R. L.: Subduction-channel model of prism accretion, melange formation, sediment subduction, and subduction erosion at convergent plate margins: 1. Background and description, Pure Appl. Geophys., 128, 455–500, 1988a.

Cloos, M. and Shreve, R. L.: Subduction-channel model of prism accretion, melange formation, sediment subduction, and subduction erosion at convergent plate margins: 2. Implications and discussion, Pure Appl. Geophys., 128, 501–545, 1988b.

Davies, J. H.: Simple analytic model for subduction zone thermal structure, Geophys. J. Int., 139, 823–828, 1999.

Dilek, Y., Festa, A., Ogawa, Y., and Pini, G. A.: Chaos and geodynamics: Mélanges, mélange-forming processes and their significance in the geological record, Tectonophysics, 568, 1–6, 2012.

Dziewonski, A. M. and Anderson, D. L.: Preliminary reference Earth model, Phys. Earth Planet. Int., 25, 297–356, 1981.

Endrun, B., Meier, T., Bischoff, M., and Harjes, H.-P.: Lithospheric structure in the area of Crete constrained by receiver functions and dispersion analysis of Rayleigh phase velocities, Geophys. J. Int., 158, 592–608, 2004.

Engi, M., Berger, A., and Roselle, G. T.: Role of the tectonic accretion channel in collisional orogeny, Geology, 29, 1143–1146, 2001.

Ernst, W. and Liou, J.: Overview of UHP metamorphism and tectonics in well-studied collisional orogens, Int. Geol. Rev., 41, 477–493, 1999.

Essen, K., Braatz, M., Ceranna, L., Friederich, W., and Meier, T.: Numerical modeling of seismic wave propagation along the plate contact of the Hellenic Subduction Zone - the influence of a deep subduction channel, Geophys. J. Int., 179, 1737–1756, 2009.

Federico, L., Crispini, L., Scambelluri, M., and Capponi, G.: Ophiolite mélange zone records exhumation in a fossil subduction channel, Geology, 35, 499–502, 2007.

Festa, A., Pini, G. A., Dilek, Y., and Codegone, G.: Mélanges and mélange-forming processes: a historical overview and new concepts, Int. Geol. Rev., 52, 1040–1105, 2010.

Friederich, W. and Meier, T.: Temporary Seismic Broadband Network Acquired Data on Helenic subduction Zone, EOS Transactions AGU, 89, 378, 2008.

Gebauer, D., Schertl, H.-P., Brix, M., and Schreyer, W.: 35 Ma old ultrahigh-pressure metamorphism and evidence for very rapid exhumation in the Dora Maira Massif, Western Alps, Lithos, 41, 5–24, 1997.

Gerya, T. and Stöckhert, B.: Two-dimensional numerical modeling of tectonic and metamorphic histories at active continental margins, Int. J. Earth Sci., 95, 250–274, 2006.

Gerya, T. V. and Stöckhert, B.: Exhumation rates of high pressure metamorphic rocks in subduction channels: The effect of Rheology, Geophys. Res. Lett., 29, 8, 102–1–102–4, 2002.

Gerya, T. V., Stöckhert, B., and Perchuk, A. L.: Exhumation of high-pressure metamorphic rocks in a subduction channel: A numerical simulation, Tectonics, 21, 6, 6–1–6–19, 2002.

Gerya, T. V., Connolly, J. A., Yuen, D. A., Gorczyk, W., and Capel, A. M.: Seismic implications of mantle wedge plumes, Phys. Earth Planet. Int., 156, 59–74, 2006.

Grigull, S., Krohe, A., Moos, C., Wassmann, S., and Stöckhert, B.: "Order from chaos": A field-based estimate on bulk rheology of tectonic mélanges formed in subduction zones, Tectonophysics, 568–569, 86–101, 2012.

Groppo, C. and Castelli, D.: Prograde *P-T* evolution of a lawsonite eclogite from the Monviso meta-ophiolite (Western Alps): dehydration and redox reactions during subduction of oceanic FeTi-oxide gabbro, J. Petrol., 51, 2489–2514, 2010.

Groppo, C., Lombardo, B., Rolfo, F., and Pertusati, P.: Clockwise exhumation path of granulitized eclogites from the Ama Drime range (Eastern Himalayas), J. Metamor. Geol., 25, 51–75, 2007.

Guillot, S., Hattori, K., Agard, P., Schwartz, S., and Vidal, O.: Exhumation processes in oceanic and continental subduction contexts: a review, Subduction Zone Geodynamics, Front. Earth Sci., Springer, 175–205, 2009.

Hacker, B., Peacock, S., Abers, G., and Holloway, S.: Subduction factory: 2. Are intermediate-depth earthquakes in subducting slabs linked to metamorphic dehydration reactions?, J. Geophys. Res., 108, doi:10.1029/2001JB001129, 2003a.

Hacker, B. R., Ratschbacher, L., Webb, L., McWilliams, M. O., Ireland, T., Calvert, A., Dong, S., Wenk, H.-R., and Chateigner, D.: Exhumation of ultrahigh-pressure continental crust in east central China: Late Triassic-Early Jurassic tectonic unroofing, J. Geophys. Res., 105, 13339–13364, 2000.

Hacker, B. R., Abers, G. A., and Peacock, S. M.: Subduction factory 1. Theoretical mineralogy, densities, seismic wave speeds, and H_2O contents, J. geophys. Res, 108, doi:10.1029/2001JB001127, 2003b.

Harley, S. and Carswell, D.: Ultradeep crustal metamorphism: A prospective view, Journal of Geophysical Research, 100, 8367–8380, 1995.

Hashin, Z. and Shtrikman, S.: A variational approach to the theory of the elastic behaviour of multiphase materials, J. Mech. Phys. Sol., 11, 127–140, 1963.

Hatzfeld, D. and Martin, C.: Intermediate depth seismicity in the Aegean defined by teleseismic data, Earth Planet. Sci. Lett., 113, 267–275, 1992.

Holland, T. and Powell, R.: An internally consistent thermodynamic data set for phases of petrological interest, J. Metamor. Geol., 16, 309–343, 1998.

Karato, S.-I. and Jung, H.: Water, partial melting and the origin of the seismic low velocity and high attenuation zone in the upper mantle, Earth Planet. Sci. Lett., 157, 193–207, 1998.

Knapmeyer, M.: Geometry of the Aegean Benioff zones, Annali di Geofisica, 42, 27–38, 1999.

Komatitsch, D. and Vilotte, J.-P.: The spectral element method: An efficient tool to simulate the seismic response of 2-D and 3-D geological structures, Bull. Seismol. Soc. Am., 88, 368–392, 1998.

Krebs, M., Schertl, H.-P., Maresch, W., and Draper, G.: Mass flow in serpentinite-hosted subduction channels: P–T–t path patterns of metamorphic blocks in the Rio San Juan mélange (Dominican Republic), J. Asian Earth Sci., 42, 569–595, 2011.

Kylander-Clark, A. R., Hacker, B. R., and Mattinson, C. G.: Size and exhumation rate of ultrahigh-pressure terranes linked to orogenic stage, Earth Planet. Sci. Lett., 321, 115–120, 2012.

Lardeaux, J., Ledru, P., Daniel, I., and Duchene, S.: The Variscan French Massif Central – a new addition to the ultra-high pressure metamorphic "club": exhumation processes and geodynamic consequences, Tectonophysics, 332, 143–167, 2001.

Little, T., Hacker, B., Gordon, S., Baldwin, S., Fitzgerald, P., Ellis, S., and Korchinski, M.: Diapiric exhumation of Earth's youngest (UHP) eclogites in the gneiss domes of the D'Entrecasteaux Islands, Papua New Guinea, Tectonophysics, 510, 39–68, 2011.

Martin, S. and Rietbrock, A.: Guided waves at subduction zones: dependencies on slab geometry, receiver locations and earthquake sources, Geophys. J. Int., 167, 693–704, 2006.

Martin, S., Rietbrock, A., Haberland, C., and Asch, G.: Guided waves propagating in subducted oceanic crust, J. geophys. Res., 108, 2536, doi:10.1029/2003JB002450, 2003.

Martin, S., Haberland, C., and Rietbrock, A.: Forearc decoupling of guided waves in the Chile–Peru subduction zone, Geophys. Res. Lett., 32, L23309, doi:10.1029/2005GL024183, 2005.

Meier, T., Becker, D., Endrun, B., Rische, M., Bohnhoff, M., Stöckert, B., and Harjes, H.-P.: A model for the Hellenic subduction zone in the area of Crete based on seismological investigations, Geol. Soc. Lond. Spec. Publ., 291, 183–199, 2007.

Ota, T. and Kaneko, Y.: Blueschists, eclogites, and subduction zone tectonics: Insights from a review of Late Miocene blueschists and eclogites, and related young high-pressure metamorphic rocks, Gondwana Res., 18, 167–188, 2010.

Papazachos, B., Karakostas, V., Papazachos, C., and Scoridilis, E.: The geometry of the Wadati-Benioff zone and lithospheric kinematics in the Hellenic arc, Tectonophysics, 319, 275–300, 2000.

Parrish, R. R., Gough, S. J., Searle, M. P., and Waters, D. J.: Plate velocity exhumation of ultrahigh-pressure eclogites in the Pakistan Himalaya, Geology, 34, 989–992, 2006.

Peacock, S. M.: Thermal and petrologic structure of subduction zones, Geophys. Monogr. Ser., 96, 119–133, 1996.

Platt, J.: Exhumation of high-pressure rocks: A review of concepts and processes, Terra nova, 5, 119–133, 1993.

Reinecke, T.: Prograde high-to ultrahigh-pressure metamorphism and exhumation of oceanic sediments at Lago di Cignana, Zermatt-Saas Zone, western Alps, Lithos, 42, 147–189, 1998.

Ring, U. and Brandon, M. T.: Ductile deformation and mass loss in the Franciscan subduction complex: implications for exhumation processes in accretionary wedges, Geol. Soc. Lond. Spec. Publ., 154, 55–86, 1999.

Royden, L. H.: The steady state thermal structure of eroding orogenic belts and accretionary prisms, J. Geophys. Res., 98, 4487–4507, 1993.

Rubatto, D. and Hermann, J.: Exhumation as fast as subduction?, Geology, 29, 3–6, 2001.

Rubatto, D., Regis, D., Hermann, J., Boston, K., Engi, M., Beltrando, M., and McAlpine, S. R.: Yo-yo subduction recorded by accessory minerals in the Italian Western Alps, Nat. Geosci., 4, 338–342, 2011.

Sabina, F. J. and Willis, J.: A simple self-consistent analysis of wave propagation in particulate composites, Wave Motion, 10, 127–142, 1988.

Schertl, H.-P., Schreyer, W., and Chopin, C.: The pyrope-coesite rocks and their country rocks at Parigi, Dora Maira Massif, Western Alps: detailed petrography, mineral chemistry and PT-path, Contributions to Mineralogy and Petrology, 108, 1–21, 1991.

Schreyer, W.: Ultradeep metamorphic rocks: The retrospective viewpoint, J. Geophys. Res.-Sol. Ea. (1978–2012), 100, 8353–8366, 1995.

Shreve, R. L. and Cloos, M.: Dynamics of sediment subduction, melange formation, and prism accretion, J. Geophys. Res., 91, 10229–10245, 1986.

Snopek, K., Meier, T., Endrun, B., Bohnhoff, M., and Casten, U.: Comparison of gravimetric and seismic constraints on the structure of the Aegean lithosphere in the forearc of the Hellenic subduction zone around Crete, J. Geodynamics, 44, 173–185, 2007.

Stixrude, L. and Lithgow-Bertelloni, C.: Thermodynamics of mantle minerals–I. Physical properties, Geophys. J. Int., 162, 610–632, 2005.

Stöckhert, B.: Stress and deformation in subduction zones: insight from the record of exhumed metamorphic rocks, Geol. Soc. Lond. Spec. Publ., 200, 255–274, 2002.

Stöckhert, B. and Gerya, T. V.: Pre-collisional high pressure metamorphism and nappe tectonics at active continental margins: A numerical simulation, Terra Nova, 17, 102–110, 2005.

Stöckhert, B., Massonne, H.-J., and Ursula Nowlan, E.: Low differential stress during high-pressure metamorphism: The microstructural record of a metapelite from the Eclogite Zone, Tauern Window, Eastern Alps, Lithos, 41, 103–118, 1997.

Suckale, J., Rondenay, S., Sachpazi, M., Charalampakis, M., Hosa, A., and Royden, L. H.: High-resolution seismic imaging of the western Hellenic subduction zone using teleseismic scattered waves, Geophysical Journal International, 178, 775–791, 2009.

Syracuse, E. M., van Keken, P. E., and Abers, G. A.: The global range of subduction zone thermal models, Phys. Earth Planet. Int., 183, 73–90, 2010.

Tsujimori, T., Sisson, V. B., Liou, J. G., Harlow, G. E., and Sorensen, S. S.: Very-low-temperature record of the subduction process: A review of worldwide lawsonite eclogites, Lithos, 92, 609–624, 2006.

Turcotte, D. L. and Schubert, G.: Geodynamics, Cambridge University Press, 2002.

Van Den Beukel, J. and Wortel, R.: Thermo-mechanical modelling of arc-trench regions, Tectonophysics, 154, 177–193, 1988.

Warren, C.: Exhumation of (ultra-) high-pressure terranes: concepts and mechanisms, Solid Earth, 4, 75–92, 2013.

Warren, C. J., Beaumont, C., and Jamieson, R. A.: Deep subduction and rapid exhumation: Role of crustal strength and strain weakening in continental subduction and ultrahigh-pressure rock exhumation, Tectonics, 27, 2008.

Wassmann, S. and Stöckhert, B.: Matrix deformation mechanisms in HP-LT tectonic mèlanges – microstructural record of jadeite blueschist from the Franciscan, Tectonophysics, 568, 135–153, 2012.

Seismogenic frictional melting in the magmatic column

J. E. Kendrick[1,2,*], Y. Lavallée[1], K.-U. Hess[2], S. De Angelis[1], A. Ferk[2,3], H. E. Gaunt[4], P. G. Meredith[4], D. B. Dingwell[2], and R. Leonhardt[3]

[1]School of Earth, Ocean and Ecological Sciences, University of Liverpool, Liverpool, L69 3GP, UK
[2]Department of Earth and Environmental Sciences, Ludwig-Maximilians-Universität, Theresienstr. 41, 80333 Munich, Germany
[3]Central Institute for Meteorology and Geodynamics, 1190 Vienna, Austria
[4]Rock & Ice Physics Laboratory, Department of Earth Sciences, University College London, Gower Street, London, UK

Correspondence to: J. E. Kendrick (jackie.kendrick@liverpool.ac.uk)

Abstract. Lava dome eruptions subjected to high extrusion rates commonly evolve from endogenous to exogenous growth and limits to their structural stability hold catastrophic potential as explosive eruption triggers. In the conduit, strain localisation in magma, accompanied by seismogenic failure, marks the onset of brittle magma ascent dynamics. The rock record of exogenous dome structures preserves vestiges of cataclastic processes and thermal anomalies, key to unravelling subsurface processes. Here, a combined structural, thermal and magnetic investigation of a shear band crosscutting a large block erupted in 2010 at Soufrière Hills volcano (SHV) reveals evidence of faulting and frictional melting within the magmatic column. The mineralogy of this pseudotachylyte vein offers confirmation of complete recrystallisation, altering the structure, porosity and permeability of the material, and the magnetic signature typifies local electric currents in faults. Such melting events may be linked to the step-wise extrusion of magma accompanied by repetitive long-period (LP) drumbeat seismicity at SHV. Frictional melting of Soufrière Hills andesite in a high velocity rotary shear apparatus highlights the small slip distances (< 15 cm) thought to be required to bring 800 °C magma to melting point at upper conduit stress conditions (10 MPa). We conclude that frictional melting is a common consequence of seismogenic magma fracture during dome building eruptions and that it may govern the ascent of magma in the upper conduit.

1 Background

Dome-building eruptions hold potential for volcanic catastrophes, with dome collapse leading to devastating pyroclastic flows with almost no warning (Voight and Elsworth, 2000; Carn et al., 2004; Herd et al., 2005). Extrusion of high-viscosity magma at arc volcanoes is frequently accompanied by seismic activity in the form of repetitive drumbeat, LP (long period) events (Iverson et al., 2006; Neuberg et al., 2006; De Angelis, 2009). At Soufrière Hills volcano (SHV) this seismicity has been attributed to magma fracture (De Angelis and Henton, 2011) and cyclic plug extrusion (Costa et al., 2012), which is dictated by magma supply rate, conduit geometry (Rowe et al., 2004), overpressure build-up (Edmonds and Herd, 2007; Lensky et al., 2008) and rheological stiffening (Voight et al., 1999). At SHV crystal growth during ascent suppresses the thermal runaway due to viscous heating and so the magma temperature remains close to that of the magma chamber (Hale et al., 2007), which has been estimated at 830–858 °C (Melnik and Sparks, 2002; Devine et al., 1998; Murphy et al., 2000) up to 880 °C when heated by an incoming mafic intrusion (Devine et al., 1998; Barclay et al., 1998). At these temperatures heterogeneity develops in crystalline magmas in the upper conduit; areas of dense magma are surrounded by shear zones at the conduit margin (Kendrick et al., 2013; Kendrick et al., 2012; Hale and Wadge, 2008; Lavallée et al., 2013), controlling the development of the degassing network (Carn et al., 2004; Watts et al., 2002; Plail et al., 2014; Laumonier et al., 2011).

Fig. 1. Field photo of the shear band. In the shear band up to seven interlayered bands of aphanitic pseudotachylyte and cataclasite layers and lenses are observed. The photograph illustrates the lateral variation in thickness and its morphology, which suggests formation in viscous magma in the conduit.

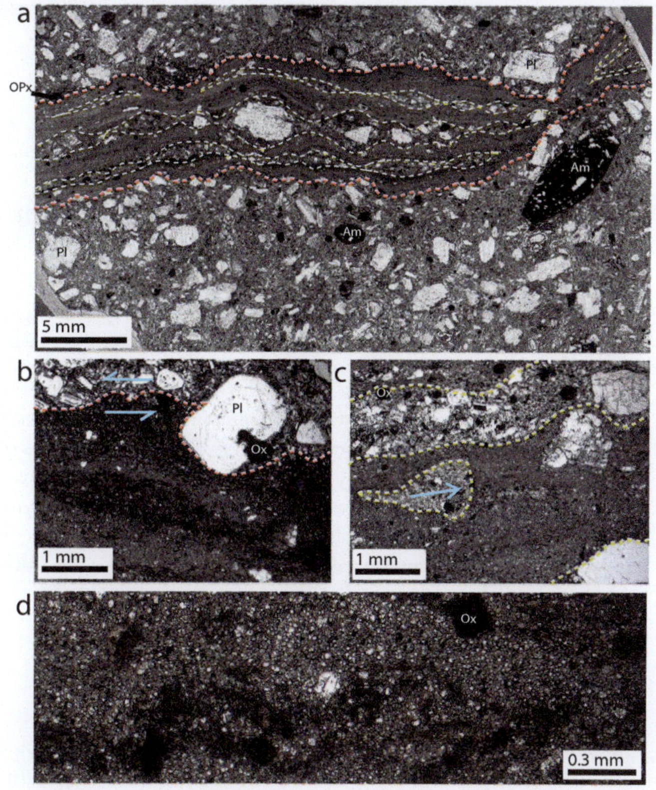

Fig. 2. Photomicrographs of the shear band. **(a)**, a wide view of pseudotachylyte (dark grey) and cataclasite (grey) layers in the shear band cutting through the andesite host rock, containing plagioclase (Pl), amphibole pseudomorphs (Am, replaced by reaction products), pyroxene (here OPx) and FeTi oxides (Ox). Morphology of the vein shows the shear band pinching out and lenses of cataclasites between pseudotachylyte veins of more constant thickness, which show flow textures. **(b)**, inferred shear direction from flow morphology of the pseudotachylyte around a host-rock phenocryst at the vein boundary. **(c)**, cataclastic lens hosted in a pseudotachylyte layer with bulbous onset and elongate tail indicating flow direction of the band. **(d)**, internal structure in the pseudotachylyte–cataclasite interface indicating turbulent flow. Red dashed lines mark the boundary of the shear band, yellow dashed lines show boundaries between pseudotachylyte and cataclasites, arrows represent inferred flow/shear directions.

Beyond this point brittle fracture and sliding can lead to formation of gouge, cataclasite (Kennedy et al., 2009; Cashman et al., 2008; Kennedy and Russell, 2012) and pseudotachylyte (Kendrick et al., 2012) akin to tectonic fault zones (Lin, 1996; Curewitz and Karson, 1999; Kirkpatrick and Rowe, 2013; Sibson, 1975). This may be linked to repetitive seismicity at SHV (Neuberg et al., 2006), hence, the fracture and slip properties of ascending magma (e.g. Lavallée et al., 2008; Tuffen and Dingwell, 2005; Kendrick et al., 2013; Cordonnier et al., 2012; De Angelis and Henton, 2011) are of critical importance to understanding the frequent transitions from effusive to explosive behaviour during volcanic eruptions (Okumura et al., 2010; Castro et al., 2012; Lavallée et al., 2013).

2 Analysis and interpretation

2.1 Sample Description

Metre-scale blocks from block-and-ash flow deposits at SHV present the opportunity to study textural and structural information from conduit and dome material that would otherwise remain inaccessible due to the current volcanic unrest. Of specific interest is a ~ 2 m long shear band located in an andesitic block erupted in 2010 due to its vitreous appearance, lateral extent and structural signature. The shear band cross-cuts the centre of its metre-scale host rock, which confirms its origin within the magmatic column and contrasts to frictional marks on the surface of blocks formed during turbulent flow of pyroclastics at SHV (Grunewald et al., 2000). A multi-parametric approach is employed to assess its structural, mineralogical, kinetic and magnetic character (see detailed methodology in the Appendix).

2.2 Petrography

The shear band consists of interlayered aphanitic (interpreted herein as) pseudotachylyte and granular cataclasite up to 3 cm thick which pinches out, widens and bends along length (~ 2 m) and breadth (Figs. 1, 2). The shear band was distinguished from an injection vein by these features, in contrast to the centimetre-scale half-bell morphology (tapering rapidly from the vein base and less rapidly toward the tip) envisaged for pseudotachylyte injection veins (Griffith et al., 2012). The morphology also demonstrates post-formation ductile deformation in the still-flowing, conduit-dwelling, magma. The interlayered pseudotachylyte and cataclasite indicate repeat slip events along the fault surface as is frequently observed in tectonic fault zones (e.g. Kirkpatrick and Rowe, 2013; Rowe et al., 2005; Kim et al., 2010; Sibson, 1977; Swanson, 1992). Intermittently along the length of the shear band up to seven layers can be identified (Figs. 1, 2), and this can be envisaged in the 3-D reconstruction of the vein and pore space in Fig. 3. Permeability was measured

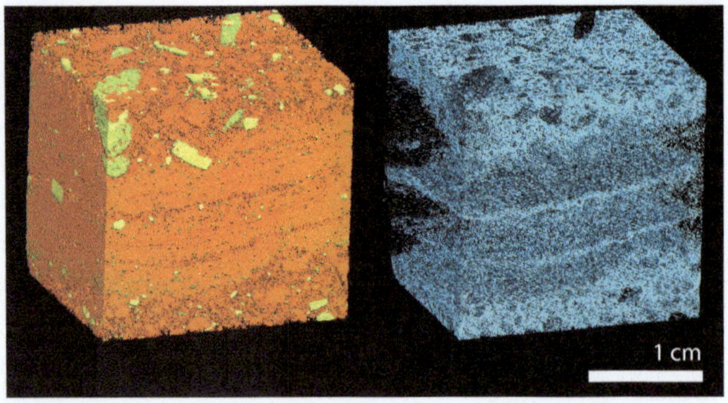

Fig. 3. Tomography of the shear band in the host rock. Solid fraction of the sample is shown in yellow (oxides, amphibole and pyroxenes) and orange (plagioclase and groundmass). A 3-D reconstruction of the pore space is shown in blue, highlighting the negligible porosity of the pseudotachylyte layers. Pores were selected as areas with grey level of 0 (black only), hence the figure may under-represent the porous fraction. Images have a pixel/voxel size of 20.0 μm, 1000 images for 360°(average of 3 single images, one image skipped), with exposure time of 1 s.

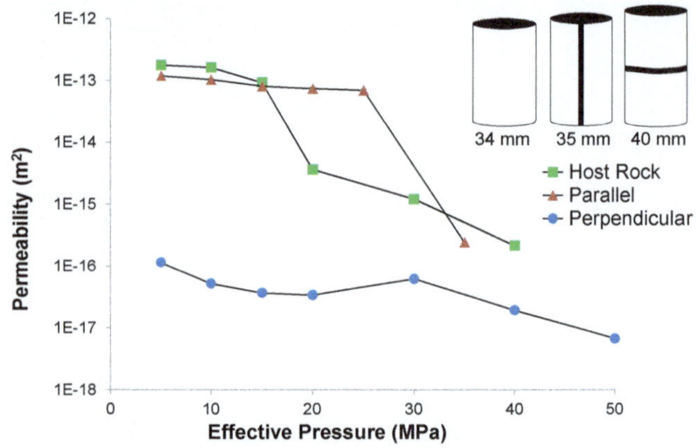

Fig. 4. Shear zone permeability. Steady-state water permeability measured at effective pressures of 5–50 MPa for sample cores (2.5 cm diameter) with no shear band (host rock), and drilled parallel and perpendicular to the pseudotachylyte-bearing shear band (inset). When the pseudotachylyte is parallel to flow, the flow rate is still controlled by the host rock, but when the vein runs perpendicular to flow it controls the permeability, which is reduced. Additionally permeability is reduced as effective pressure is increased and cracks are accordingly closed, this effect is less apparent in the perpendicular-cut sample as the porosity of the pseudotachylyte is already negligible.

on the host rock, and both parallel and perpendicular to the shear band on 34–40 (length) by 25 mm cylindrical samples at 5–10 MPa pressure intervals from 5–50 MPa (for more information on methodology see Heap et al., 2014). Results indicate that at low pressures the shear band may be up to three orders of magnitude less permeable than the host rock, although this contrast diminishes at higher pressure (Fig. 4).

The host rock, a porphyritic andesite with 23 % porosity (determined via helium pycnometry), has a crystal assemblage (total crystallinity < 60 %) of plagioclase, amphibole and orthopyroxene phenocrysts, with some quartz grains and FeTi oxide (Ti-magnetite to ilmenite) set in a groundmass of orthopyroxene and plagioclase with trace amounts of zircon, apatite and chlorite (Fig. 6). There are two types of plagioclase, categorised as having < An80 (type 1) or > An80 (type 2). Type 1 is more common and is present across the sample whereas type 2 tends to form as rims and is only present in the host rock and cataclasite bands more than ~ 1 mm thick (Fig. 6). Amphiboles are broken down into pseudomorphs containing plagioclase, pyroxene, opacite and abundant FeTi oxides, as has been reported from earlier in the eruption (Zellmer et al., 2003). Phenocrysts are for the most part euhedral and have few fractures, and there is no apparent orientation or foliation in phenocrysts or groundmass. What interstitial glass may have existed is fully devitrified; the glass has been replaced by silica residue, including feldspar and cristobalite.

In contrast to the host rock groundmass (Fig. 5a) the pseudotachylyte consists of fine grained (10–40 μm), equant and well sorted quartz, plagioclase, FeTi oxides, pyroxene, feldspar, cristobalite and cordierite (Fig. 5b), and occasional larger sieve-textured plagioclase phenocrysts (Fig. 5d) as well as higher levels of chlorite (Fig. 6e). The pseudotachylyte has a porosity of 1 % (measured by helium pycnometry), has flow textures (Fig. 2b, c) and has sharp but undulating contacts with the host rock. The transitions into the cataclasites are more gradual but are distinguishable by the change in porosity and grain size. The cataclasites appear to be granular aggregates of both starting material and pseudotachylyte, with grain sizes up to several millimetres in diameter for occasional phenocrysts, although more typically 20–400 μm

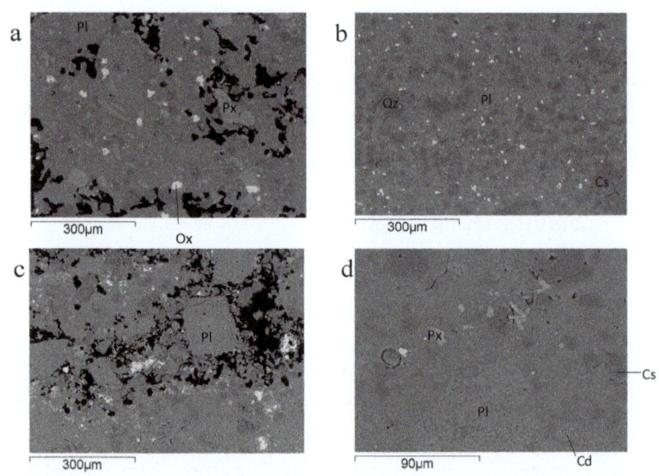

Fig. 5. Backscatter electron (BSE) images of the sample. **(a)** Host rock, **(b)** pseudotachylyte, **(c)** cataclasite and **(d)** sieve textured plagioclase in the pseudotachylyte. Here Qz is quartz, Pl is plagioclase, Px is pyroxene, Ox is FeTi oxide, Cd is cordierite and Cs is cristobalite.

(Figs. 5, 6). Porosity is localised in patches and averages 15 % (by helium pycnometry) while grains are equant but angular and are often fragments. The mineral assemblage is made up of plagioclase, quartz, altered amphibole (which appears in bands > 1 mm thick), FeTi oxides (which appear as granular clusters) and cristobalite. In earlier stages of the eruption silica polymorphs, including cristobalite, have comprised up to 15 wt % in the magma, and its presence here was verified using differential scanning calorimetry (Sect. 2.3).

2.3 Kinetics

High sensitivity differential scanning calorimetry (HS-DSC) measurements on the host rock groundmass revealed a reversible endothermic peak at 190 °C at a heating/cooling rate of 10 °C s^{-1}, which recurred after cooling and reheating, verifying the α–β phase transition of cristobalite. While the ideal phase transition occurs at 272 °C, highly distorted crystals originating from a gel or glass can have a lower transition temperature, and inversion temperatures of 120–272 °C have been recorded (Sosman, 1965). The shear band reveals no such endothermic peak as a result of cristobalite (despite it being visible in thin section (Fig. 5b) in both the pseudotachylyte and cataclasite), however, a repeatable endothermic peak at 572 °C (Fig. 7) can be attributed to the α–β phase transition of quartz (Sosman, 1965). The detection of the phase transitions reflects the relative abundance of the silica polymorphs in the vein versus host rock (Fig. 5).

The melting temperature of the shear zone also differs from the host rock; low sensitivity (LS)-DSC measurements on the groundmass of the host rock show two broad melting peaks between 1050–1250 °C and 1350–1450 °C (post optical analysis of samples after runs to 1300 and 1500 °C indicate partial and complete melting of all phases respectively), whereas the aphanitic pseudotachylyte vein shows a broad melting peak between 1200 and 1400 °C (post optical analysis indicates complete melting of all phases). This difference arises from the distinct mineralogy (Fig. 6) formed during disparate crystallisation histories.

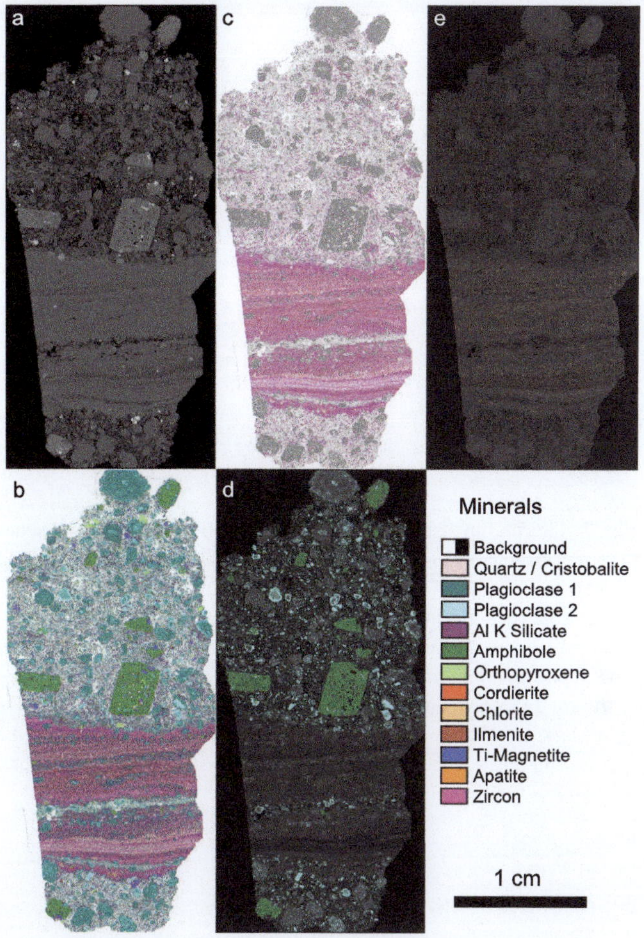

Fig. 6. Distribution of minerals using QEMSCAN (Quantitative Evaluation of Minerals by SCANning electron microscopy). **(a)** BSE image of the thin section used for analysis, including host rock, cataclasite and pseudotachylyte **(b)** composite of all resolved minerals and porosity in white **(c)** distribution of Al K silicate (potentially feldspar) and cordierite with other minerals in greyscale and porosity in white and **(d)** distribution of plagioclase type 2 and amphibole and **(e)** distribution of chlorite with other minerals in greyscale and porosity in black.

2.4 Magnetic anomalies

Crystallisation in the volcanic pseudotachylyte vein recorded a distinct magnetic signature, as is often noted in tectonic and impact generated pseudotachylytes (Ferré et al., 2005; Freund et al., 2007), which supports its status as a frictionally generated melt. The remanence of the pseudotachylyte is carried by a low coercive material with a Curie temperature of 320 °C, whereas the host rock, which also shows low coercive behaviour, has two Curie temperatures at 400 and 540 °C. Alternating field demagnetisation of different remanent magnetisations was used to identify further differences between the two samples (Fig. 8). The demagnetisation pattern of the natural remanence magnetisation (NRM) of the pseudotachylyte is similar to an isothermal remanent magnetisation (IRM) pattern, whereas the demagnetisation patterns of the host rock display similarities between the NRM and anhysteretic remanent magnetisation (ARM). Thus, the pseudotachylyte has seen a strong magnetic field that overwrote the previous thermoremanent magnetisation of the magma. Due to the in situ proximity of the two samples (within 2 cm) this IRM cannot originate from a lightning strike (known to enforce magnetisation). Instead, it demonstrates simultaneous high local electric currents and frictional melt occurring on a fault, known to produce a high IRM (Ferré et al., 2005; Freund et al., 2007; Kendrick et al., 2012).

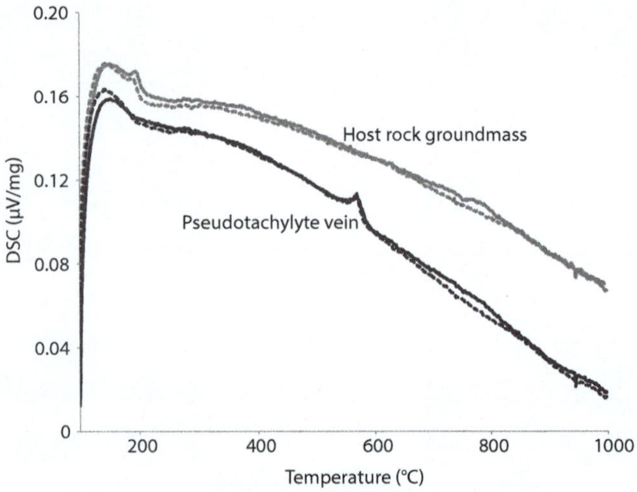

Fig. 7. Thermal measurements by high sensitivity differential scanning calorimetry (HS-DSC). Measurements were carried out using a Netzsch DSC 404 C. A sharp endothermic peak at about 190 °C in the host rock (light grey trace) corresponds to a cristobalite α–β phase transition and the trace shows no evidence of a potential glassy phase. An endothermic peak at about 570 °C corresponding to the α–β phase transition in quartz is seen in the pseudotachylyte vein (dark grey trace). The same exothermic peaks are visible on the second run of the experiments with the same sample (dashed lines), confirming their respective phase transitions.

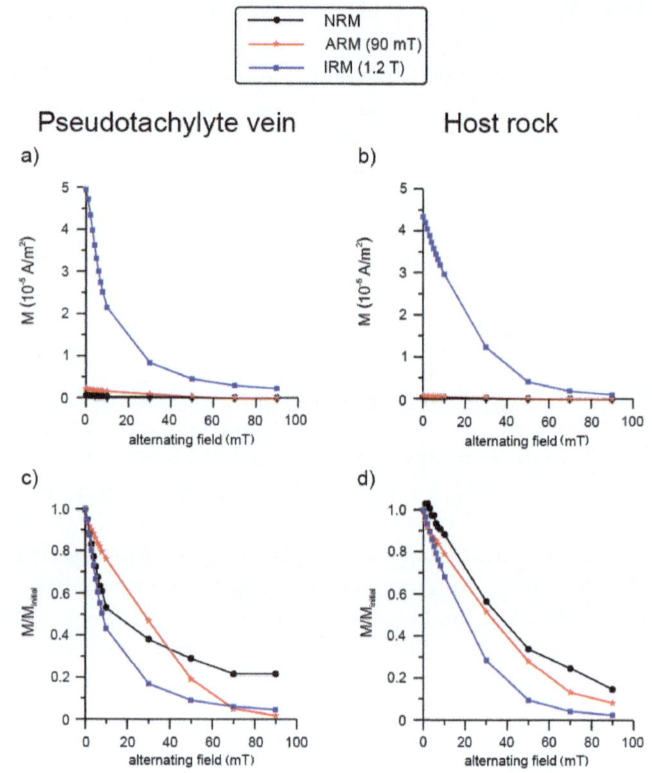

Fig. 8. Rock magnetic measurements. Alternating field demagnetisations of NRM, ARM and IRM of pseudotachylyte and host rock. Shown are first **(a, b)** magnetisation versus alternating field, then **(c, d)** normalised magnetisation (normalisation using initial magnetisation at 0 mT alternating field) versus alternating field. NRM of the pseudotachylyte is analogue to an IRM while the NRM of the host rock is analogue to an ARM.

3 A model

Pseudotachylyte has previously been linked to seismogenic ruptures (Magloughlin and Spray, 1992) at slip velocities over 0.1 m s^{-1} (Spray, 2010), and at SHV may be linked to the recorded repetitive drumbeat seismicity (Rowe et al., 2004; Watts et al., 2002; Luckett et al., 2008; Neuberg et al., 1998) inferred to result from magma failure and stick-slip events along conduit margins (Neuberg et al., 2006; De Angelis and Henton, 2011; Harrington and Brodsky, 2007). Throughout the eruption LP events occurred at $\sim$50 s intervals with a P-wave pulse duration of approximately 0.15 s (Fig. 9), which translates to a source duration of the same time frame (Harrington and Brodsky, 2009). In order to calculate the displacement during these events we average the cumulative source displacement over 24 hr, for example, the 235 m preceding the 29 July 2008 dome collapse (De Angelis, 2009) divided by the 1,588 recorded events (of similar size) during that period to provide a mean slip distance of 15 cm per event. Together, this provides a slip velocity of 1.0 m s^{-1} for 15 cm in 0.15 s.

To assess the effect of such slip conditions on the SHV andesite, we performed a high-velocity rotary shear experiment using a normal stress (10 MPa) representative of approximately 500 m depth in the upper conduit. In this experiment two 25 mm diameter hollow cores (9.5 mm holes) were brought into contact and the axial stress was applied. Next, one side was rotated at a velocity of 1.0 m s^{-1} as the simulated fault was recorded by optical and thermographic videos to track the temperature profile and observe melting (see Hirose and Shimamoto (2005) for more details of methodology; see supplementary video). When considering a magmatic temperature of 800 °C in the conduit (Devine et al., 1998; Hale et al., 2007; Murphy et al., 2000; Barclay et al., 1998; Melnik and Sparks, 2002) at the onset of the process, these slip conditions are more than enough to force the andesite through the amphibole stability field at 855 °C (Rutherford and Devine, 2003) and on into frictional melting above 950 °C (Fig. 10a and supplementary video). At the onset of displacement the shear stress rapidly increases as isolated pockets of melt begin to form, shear resistance to slip is at its maximum immediately prior to a continuous melt layer forming on the slip surface (Fig. 10b). As this melt zone thickens resistance reduces and achieves a stable value of $\sim$3 MPa as the melt is produced and expelled (shortening, Fig. 10b) at a constant rate. Once melt forms, the shear resistance is controlled by melt viscosity, and hence the behaviour may contrast significantly to the frictional behaviour

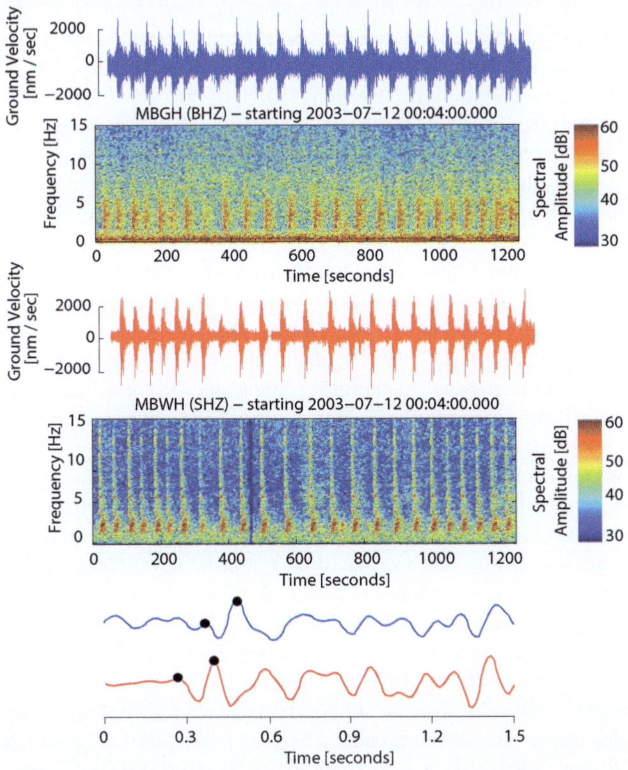

Fig. 9. Repetitive seismicity at Soufrière Hills volcano. Seismograms for 18 min of repeating earthquakes at stations MBGH and MBWH recorded on 12 July 2003 approximately 12 hr before the largest dome collapse in SHV history. Bottom panel: The black dots highlight the width of a single P-wave pulse from the seismograms above, which provide an estimate of source duration.

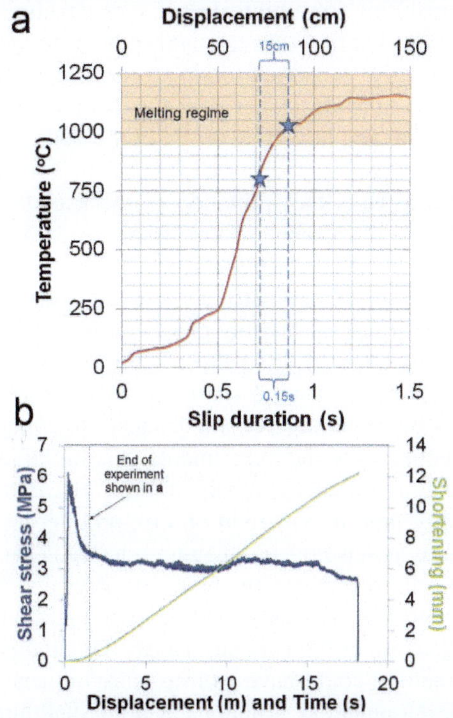

Fig. 10. Frictional melting of SHV andesite using a high velocity rotary shear apparatus. **(a)** Increasing temperature with time and slip distance at 10 MPa normal load and $1.0\,\mathrm{m\,s^{-1}}$ slip velocity. Temperature was recorded using H2640 NEC\Avio at a resolution of $90\,\mu\mathrm{m} \times 90\,\mu\mathrm{m}$ per pixel with no correction for melt cooling. Starting from room temperature melting is achieved in 0.75 s (75 cm). The blue stars represent the change in temperature that results from a slip episode of 15 cm in 0.15 s (displacement parameters calculated from the seismic events at SHV in Fig. 9) from a starting magmatic temperature of 800 °C, where the magma temperature is increased by friction to 1025 °C in that interval, with the onset of melting occurring at approximately 950 °C (see also supplementary video). **(b)** Shear resistance (MPa) with slip duration and displacement during the experiment showing rapid increase in resistance as melting begins, followed by a peak as a full melt layer forms and attainment of steady state as melt supply rate and melt ejection (shortening) equilibrate.

of rock–rock sliding or gouge-hosting shear zones (Spray, 2010; Magloughlin and Spray, 1992; Lin and Shimamoto, 1998).

4 Discussion and conclusions

This study documents evidence of frictional melting in ascending, conduit-dwelling magma. The presence of pseudotachylyte is verified using a multi-parametric approach that assesses its structural and mineralogical character as well as its distinct thermal and magnetic signatures.

The drastic contrast in mineralogical assemblage and petrographic textures between host rock and shear band suggest that it formed due to comminution and localised melting of the host rock during faulting and frictional slip of the viscous magma. The metastable melt was then able to crystallise slowly due to the high volcanic geotherm, forming the crystalline pseudotachylyte vein, with fine-grained, equant crystals (Figs. 2d, 5b, 6). This accounts for the lower porosity and permeability of the pseudotachylyte layers as compared to the cataclasite or the host rock (see Fig. 4) and indicates that when pseudotachylyte forms as part of a shear band this will alter the development of the permeable porous network. Indeed, the presence of chlorite (Fig. 6d), a distinct geochemical signature, and other evidence suggests that the shear zone was involved in degassing in the magma column (Plail et al., 2014).

The melting and recrystallisation of the magma locked-in a distinct thermal history, unravelled here using DSC. Contrasting melting temperatures of the pseudotachylyte versus the host rock, and the occurrence of different phase changes during heating and cooling of the sample subsets reveal their divergent petrogenesis, confirming the thermal anomaly. The disparate magnetic signature of the vein and host rock, resulting from high electric currents during faulting and frictional slip, serves to confirm the pseudotachylyte status. Indeed the demagnetisation experiments performed provide some of the clearest evidence to distinguish between pseudotachylyte and host rock (Ferré et al., 2005; Freund et al., 2007).

At Soufrière Hills the formation of the pseudotachylyte is considered to have been contemporaneous with repetitive LP seismicity, caused by stick-slip extrusion that persisted throughout the eruption. Finally, we tested this hypothesis by recreating the frictional melting process using a high velocity rotary shear apparatus and slip conditions relevant to the volcanic scenario; measured source duration of LP events, combined with calculations of slip distances provided the parameters required for the experiment. This highlighted that andesite magma, subjected to slip at $1\,\mathrm{m\,s^{-1}}$, can be forced to frictionally melt in $< 0.15\,\mathrm{m}$ of slip, proving to be a viable response to seismogenic magma fracture. This may be a common occurrence; causing buoyant forces acting on the magma at depth to be superseded by frictional and viscosity controlled slip at the conduit margin. This indicates that frictional melting could have an important influence on the ongoing development of eruptions at dome-building volcanoes.

Acknowledgements. Thanks to A. Biggin and E. Hurst from the Geomagnetism Group at the University of Liverpool for performing the VFTB measurements. Thanks also to Alan R. Butcher (FEI) and Gavyn K. Rollinson (CSM, Exeter) for performing the QEMSCAN study and to S. Wiesmaier and D. Mueller for technical assistance. Y. Lavallée wishes to acknowledge the support of the Deutsche Forschungsgemeinschaft grant LA2191/3-1 as well as the ERC Starting Grant Strain Localisation in Magma (SLiM, #306488). D. B. Dingwell wishes to acknowledge the support of a research professorship of the Bundesexzellenzinitiative (LMUexcellent), the EU funded FP7 activity 6.1 VUELCO consoritum and ERC Advanced Grant Explosive Volcanism in the Earth System (EVOKES, #247076). A. Ferk and R. Leonhardt acknowledge funding by the Austrian Science foundation FWF (Grant P21221-N14).

Edited by: M. Heap

References

Barclay, J., Rutherford, M. J., Carroll, M. R., Murphy, M. D., Devine, J. D., Gardner, J., and Sparks, R. S. J.: Experimental phase equilibria constraints on pre-eruptive storage conditions of the Soufriere Hills magma, Geophys. Res. Lett., 25, 3437–3440, doi:10.1029/98GL00856, 1998.

Carn, S. A., Watts, R. B., Thompson, G., and Norton, G. E.: Anatomy of a lava dome collapse: The 20 march 2000 event at Soufrière Hills volcano, Montserrat, J. Volcanol. Geoth. Res., 131, 241–264, doi:10.1016/S0377-0273(03)00364-0, 2004.

Cashman, K. V., Thornber, C. R., and Pallister, J. S.: From dome to dust: Shallow crystallization and fragmentation of conduit magma during the 2004-2006 dome extrusion of Mount St. Helens, Washington, in: A volcano rekindled: A volcano rekindled: The renewed eruption of mount St. Helens, 2004–2006. Professionnal paper 1750., edited by: Sherrod, D. R., Scott, W. E., and Stauffer, P. H., U.S. Geological Survey, 387–413, 2008.

Castro, J. M., Cordonnier, B., Tuffen, H., Tobin, M. J., Puskar, L., Martin, M. C., and Bechtel, H. A.: The role of melt-fracture degassing in defusing explosive rhyolite eruptions at volcán Chaitén, Earth Planet. Sci. Lett., 333–334, 63–69, doi:10.1016/j.epsl.2012.04.024, 2012.

Cordonnier, B., Caricchi, L., Pistone, M., Castro, J., Hess, K.-U., Gottschaller, S., Manga, M., Dingwell, D. B., and Burlini, L.: The viscous-brittle transition of crystal-bearing silicic melt: Direct observation of magma rupture and healing, Geology, 40, 611–614, doi:10.1130/g3914.1, 2012.

Costa, A., Wadge, G., and Melnik, O.: Cyclic extrusion of a lava dome based on a stick-slip mechanism, Earth Planet. Sci. Lett., 337–338, 39–46, doi:10.1016/j.epsl.2012.05.011, 2012.

Curewitz, D. and Karson, J. A.: Ultracataclasis, sintering, and frictional melting in pseudotachylytes from east Greenland, J. Struct. Geol., 21, doi:10.1016/S0191-8141(99)00119-4, 1693–1713, 1999.

De Angelis, S.: Seismic source displacement by coda wave interferometry at Soufrière Hills volcano, Montserrat, W.I., Nat. Hazards Earth Syst. Sci., 9, 1341–1347, doi:10.5194/nhess-9-1341-2009, 2009.

De Angelis, S. and Henton, S. M.: On the feasibility of magma fracture within volcanic conduits: Constraints from earthquake data and empirical modelling of magma viscosity, Geophys. Res. Lett., 38, L19310, doi:10.1029/2011gl049297, 2011.

Devine, J. D., Murphy, M. D., Rutherford, M. J., Barclay, J., Sparks, R. S. J., Carrol, M. R., Young, S. R., and Gardner, J. E.: Petrologic evidence for pre-erptive pressure-temperature conditions, and recent reheating, of andesitic magma erupting at the Soufrière Hills volcano, Montserrat, W.I, Geophys. Res. Lett., 25, doi:10.1029/98GL01330, 3669–3672, 1998.

Edmonds, M., and Herd, R. A.: A volcanic degassing event at the explosive-effusive transition, Geophys. Res. Lett., 34, L21310, doi:10.1029/2007GL031379, 2007.

Ferré, E. C., Zechmeister, M. S., Geissman, J. W., MathanaSekaran, N., and Kocak, K.: The origin of high magnetic remanence in fault pseudotachylites: Theoretical considerations and implication for coseismic electrical currents, Tectonophysics, 402, 125–139, doi:10.1016/j.tecto.2005.01.008, 2005.

Freund, F., Salgueiro da Silva, M. A., Lau, B. W. S., Takeuchi, A., and Jones, H. H.: Electric currents along earthquake faults and the magnetization of pseudotachylite veins, Tectonophysics, 431, 131–141, doi:10.1016/j.tecto.2006.05.039, 2007.

Gottlieb, P., Wilkie, G., Sutherland, D., Ho-Tun, E., Suthers, S., Perera, K., Jenkins, B., Spencer, S., Butcher, A., and Rayner, J.: Using quantitative electron microscopy for process mineralogy applications, JOM, 24–25, 2000.

Griffith, W. A., Mitchell, T. M., Renner, J., and Di Toro, G.: Coseismic damage and softening of fault rocks at seismic depths, Earth and Planetary Science Letters, 353, 219-230, doi:10.1016/j.epsl.2012.08.013, 2012.

Grunewald, U., Sparks, R. S. J., Kearns, S., and Komorowski, J. C.: Friction marks on blocks from pyroclastic flows at the Soufrière Hills volcano, Montserrat: Implications for flow mechanisms, Geology, 28, 827–830, doi:10.1130/0091-7613(2000)28<827:fmobfp>2.0.co;2, 2000.

Hale, A. J., Wadge, G., and Muhlhaus, H. B.: The influence of viscous and latent heating on crystal-rich magma flow in a conduit, Geophys. J. Int., 171, 1406–1429, doi:10.1111/j.1365-246X.2007.03593.x, 2007.

Hale, A. J. and Wadge, G.: The transition from endogenous to exogenous growth of lava domes with the development of shear bands, J. Volcanol. Geoth. Res., 171, 237–257, doi:10.1016/j.jvolgeores.2007.12.016, 2008.

Harrington, R. M. and Brodsky, E. E.: Volcanic hybrid earthquakes that are brittle-failure events, Geophys. Res. Lett., 34, doi:10.1029/2006GL028714, 2007.

Harrington, R. M. and Brodsky, E. E.: Source duration scales with magnitude differently for earthquakes on the San Andreas fault and on secondary faults in Parkfield, California, Bull. Seismol. Soc. Am., 99, 2323–2334, doi:10.1785/0120080216, 2009.

Heap, M. J., Baud, P., Meredith, P. G., Vinciguerra, S., and Reuschlé, T.: The permeability and elastic moduli of tuff from Campi Flegrei, Italy: Implications for ground deformation modelling, Solid Earth, 5, 25–44, doi:10.5194/se-5-25-2014, 2014.

Herd, R. A., Edmonds, M., and Bass, V. A.: Catastrophic lava dome failure at Soufrière Hills volcano, Montserrat, 12–13 july 2003, J. Volcanol. Geoth. Res., 148, 234–252, doi:10.1016/j.jvolgeores.2005.05.003, 2005.

Hirose, T. and Shimamoto, T.: Slip-weakening distance of faults during frictional melting as inferred from experimental and natural pseudotachylytes, Bull. Seis. Soc. Am., 95, 1666–1673, doi:10.1785/0120040131, 2005.

Iverson, R. M., Dzurisin, D., Gardner, C. A., Gerlach, T. M., LaHusen, R. G., Lisowski, M., Major, J. J., Malone, S. D., Messerich, J. A., Moran, S. C., Pallister, J. S., Qamar, A. I., Schilling, S. P., and Vallance, J. W.: Dynamics of seismogenic volcanic extrusion at Mount St. Helens in 2004–05, Nature, 444, 439–443, doi:10.1038/nature05322, 2006.

Kendrick, J. E., Lavallée, Y., Ferk, A., Perugini, D., Leonhardt, R., and Dingwell, D. B.: Extreme frictional processes in the volcanic conduit of Mount St. Helens (USA) during the 2004–2008 eruption, J. Struct. Geol., 38, 61–76, doi:10.1016/j.jsg.2011.10.003, 2012.

Kendrick, J. E., Lavallée, Y., Hess, K.-U., Heap, M. J., Gaunt, H. E., Meredith, P. G., and Dingwell, D. B.: Tracking the permeable porous network during strain-dependent magmatic flow, J. Volcanol. Geoth. Res., doi:10.1016/j.jvolgeores.2013.05.012, 2013.

Kennedy, L. A., Russell, J. K., and Nelles, E.: Origins of Mount St. Helens cataclasites: Experimental insights, American Mineralogist, 94, 995–1004, doi:10.2138/am.2009.3129, 2009.

Kennedy, L. A. and Russell, J. K.: Cataclastic production of volcanic ash at Mount St. Helens, Physics and Chemistry of the Earth, Parts A/B/C, 45–46, 40–49, doi:10.1016/j.pce.2011.07.052, 2012.

Kim, J. W., Ree, J. H., Han, R., and Shimamoto, T.: Experimental evidence for the simultaneous formation of pseudotachylyte and mylonite in the brittle regime, Geology, 38, 1143–1146, doi:10.1130/g31593.1, 2010.

Kirkpatrick, J. D. and Rowe, C. D.: Disappearing ink: How pseudotachylytes are lost from the rock record, J. Struct. Geol., 52, 183–198, doi:10.1016/j.jsg.2013.03.003, 2013.

Laumonier, M., Arbaret, L., Burgisser, A., and Champallier, R.: Porosity redistribution enhanced by strain localization in crystal-rich magmas, Geology, 39, 715–718, doi:10.1130/g31803.1, 2011.

Lavallée, Y., Meredith, P. G., Dingwell, D. B., Hess, K. U., Wassermann, J., Cordonnier, B., Gerik, A., and Kruhl, J. H.: Seismogenic lavas and explosive eruption forecasting, Nature, 453, 507–510, doi:10.1038/nature06980, 2008.

Lavallée, Y., Benson, P., Hess, K.-U., Flaws, A., Schillinger, B., Meredith, P. G., and Dingwell, D. B.: Reconstructing magma failure and the permeable degassing network, Geology, 41, 515–518, doi:10.1130/G33948.1, 2013.

Lensky, N. G., Sparks, R. S. J., Navon, O., and Lyakhovsky, V.: Cyclic activity at Soufrière Hills volcano, Montserrat: Degassing-induced pressurization and stick-slip extrusion, Geol. Soc. Lond. Spec. Publ., 307, 169–188, doi:10.1144/sp307.10, 2008.

Lin, A. and Shimamoto, T.: Selective melting processes as inferred from experimentally generated pseudotachylytes, J. Asian Earth Sci., 16, 533–545, doi:10.1016/s0743-9547(98)00040-3, 1998.

Lin, A. M.: Injection veins of crushing-originated pseudotachylyte and fault gouge formed during seismic faulting, Engineer. Geol., 43, 213–224, doi:10.1016/0013-7952(96)00062-2, 1996.

Luckett, R., Loughlin, S., De Angelis, S., and Ryan, G.: Volcanic seismicity at Montserrat, a comparison between the 2005 dome growth episode and earlier dome growth, J. Volcanol. Geoth. Res., 177, 894–902, doi:10.1016/j.jvolgeores.2008.07.006, 2008.

Magloughlin, J. F. and Spray, J. G.: Frictional melting processes and products in geological materials, Geological society of america annual meeting, Science Direct, 141 pp., doi:10.1016/0040-1951(92)90307-R, 1992.

Melnik, O. E. and Sparks, R. S. J.: Dynamics of magma ascent and lava extrusion at the Soufrière Hills volcano, Montserrat, in: The eruption of the Soufrière Hills volcano, Montserrat from 1995 to 1999, edited by: Kokelaar, T. D. B., Geol. Soc. Lond., 153–172, 2002.

Murphy, M. D., Sparks, R. S. J., Barclay, J., Carroll, M. R., and Brewer, T. S.: Remobilization of andesite magma by intrusion of mafic magma at the Soufrière Hills volcano, Montserrat, West Indies, J. Petrol., 41, 21–42, doi:10.1093/petrology/41.1.21, 2000.

Neuberg, J., Baptie, B., Luckett, R., and Stewart, R.: Results from the broadband seismic network on Montserrat, Geophys. Res. Lett., 25, 3661–3664, doi:10.1029/98GL01441, 1998.

Neuberg, J., Tuffen, H., Collier, L., Green, D., Powell, T., and Dingwell, D.: The trigger mechanism of low-frequency earthquakes on Montserrat, J. Volcanol. Geoth. Res., 153, 37–50, doi:10.1016/j.jvolgeores.2005.08.008, 2006.

Okumura, S., Nakamura, M., Nakano, T., Uesugi, K., and Tsuchiyama, A.: Shear deformation experiments on vesicular rhyolite: Implications for brittle fracturing, degassing, and compaction of magmas in volcanic conduits, J. Geophys. Res.: Solid Earth, 115, n/a-n/a, doi:10.1029/2009JB006904, 2010.

Pirrie, D., Butcher, A. R., Power, M. R., Gottlieb, P., and Miller, G. L.: Rapid quantitative mineral and phase analysis using automated scanning electron microscopy (qemscan); potential applications in forensic geoscience, Forensic geoscience, principles, techniques and applications, edited by: Pye, K., Croft, D. J., Geol. Soc. Spec. Publ., London, 2004.

Plail, M., Edmonds, M., Barclay, J., Humphreys, M. C. S., and Herd, R.: Geochemical evidence for relict degassing pathways preserved in andesite, Earth Planet. Sci. Lett., 386, 21–33, doi:10.1016/j.epsl.2013.10.044, 2014.

Rowe, C. A., Thurber, C. H., and White, R. A.: Dome growth behavior at Soufrière Hills volcano, Montserrat, revealed by relocation of volcanic event swarms, 1995–1996, J. Volcanol. Geoth. Res., 134, 199–221, doi:10.1016/j.jvolgeores.2004.01.008, 2004.

Rowe, C. D., Moore, J. C., Meneghini, F., and McKeirnan, A. W.: Large-scale pseudotachylytes and fluidized cataclasites from an ancient subduction thrust fault, Geology, 33, 937–940, doi:10.1130/g21856.1, 2005.

Rutherford, M. J. and Devine, J. D.: Magmatic conditions and magma ascent as indicated by hornblende phase equilibria and reactions in the 1995–2002 Soufrière Hills magma, J. Petrol., 44, 1433–1453, doi:10.1093/petrology/44.8.1433, 2003.

Sibson, R. H.: Generation of pseudotachylyte by ancient seismic faulting, Royal Astronomical Society of London Journal, 43, doi:10.1111/j.1365-246X.1975.tb06195.x, 1975.

Sibson, R. H.: Fault rocks and fault mechanisms, J. Geolog. Soc., 133, doi:10.1144/gsjgs.133.3.0191, 1977.

Sosman, R. B.: The phases of silica, Rutgers University Press, New Brunswick, N. J., 389 pp., 1965.

Spray, J. G.: Frictional melting processes in planetary materials: From hypervelocity impact to earthquakes, Annu. Rev. Earth Planet. Sci., 38, 221–254, doi:10.1146/annurev.earth.031208.100045, 2010.

Swanson, M. T.: Fault structure, wear mechanisms and rupture processes in pseudotachylyte generation, Tectonophysics, 204, 223–242, doi:10.1016/0040-1951(92)90309-T, 1992.

Tuffen, H. and Dingwell, D.: Fault textures in volcanic conduits: Evidence for seismic trigger mechanisms during silicic eruptions, Bull. Volcanol., 67, 370–387, doi:10.1007/s00445-004-0383-5, 2005.

Voight, B., Sparks, R. S., Miller, A. D., Stewart, R. C., Hoblitt, R. P., Clarke, A., Ewart, J., Aspinall, W. P., Baptie, B., Calder, E. S., Cole, P., Druitt, T. H., Hartford, C., Herd, R. A., Jackson, P., Lejeune, A. M., Lockhart, A. B., Loughlin, S. C., Luckett, R., Lynch, L., Norton, G. E., Robertson, R., Watson, I. M., Watts, R., and Young, S. R.: Magma flow instability and cyclic activity at Soufrière Hills volcano, Montserrat, British West Indies, Science, 283, 1138–1142, 1138-1142, doi:10.1126/science.283.5405.1138, 1999.

Voight, B. and Elsworth, D.: Instability and collapse of hazardous gas-pressurized lava domes, Geophys. Res. Lett., 27, 1–4, doi:10.1029/1999GL008389, 2000.

Watts, R. B., Herd, R. A., Sparks, R. S., and Young, S. R.: Growth patterns and emplacement of the andesitic lava dome at soufriere hills volcano, in: The eruption of Soufrière Hills volcano, Montserrat, from 1995 to 1999, edited by: Druitt, T. H., and Kokelaar, B. P., Geol. Soc. Lond. Memoirs, 115–152, 2002.

Zellmer, G. F., Hawkesworth, C. J., Sparks, R. S. J., Thomas, L. E., Harford, C. L., Brewer, T. S., and Loughlin, S. C.: Geochemical evolution of the Soufrière Hills volcano, Montserrat, Lesser Antilles volcanic arc, J. Petrol., 44, 1349–1374, doi:10.1093/petrology/44.8.1349, 2003.

Appendix A

Materials and methods

A1 Thermal measurements

Thermogravimetric measurements (TG) and low sensitivity scanning calorimetry (LS-DSC) were carried out using a Netzsch STA 449 C simultaneous thermal analysis equipment. Small chips of about 50 mg were heated in a Pt crucible (with lid) with a heating rate of $10\,\mathrm{K\,min^{-1}}$ up to 1500 °C in air. High sensitivity scanning calorimetry measurements (HS-DSC) were carried out using a Netzsch DSC 404 C. Small chips of about 25 mg were heated in a Pt crucible (with lid) with a heating rate of $10\,°\mathrm{C\,min^{-1}}$ up to 1000 °C in air, cooled with $10\,\mathrm{K\,min^{-1}}$ to 100 °C and then reheated to 1000 °C.

A2 Magnetic measurements

Rock magnetic measurements were made on the variable field translation balance by Petersen Instruments at the University of Liverpool. Remanence carriers defined as low coercive materials (saturation of an isothermal remanence, IRM, below 200 mT). A second experiment used alternating field demagnetisation of different remanent magnetisations. This experiment was run in a magnetically shielded room at LMU Munich, using the SushiBar – an automated system for palaeomagnetic investigations (Wack and Gilder, 2012). First, the natural remanence NRM was measured and then demagnetised using 14 steps of increasingly higher alternating fields. Then an anhysteretic remanent magnetisation (ARM) was implied. An ARM is produced by the combination of a slowly decaying alternating field and a steady unidirectional field. For the ARM a maximum field of 90 mT was applied. This ARM was measured and then demagnetised in the same manner as the NRM. Finally an IRM using a 1.2 T magnetic field was implied, measured and again stepwise demagnetised.

A3 Scanning electron microprobe

Uncovered thin sections of each sample were carbon coated for analysis in a CAMECA SX100 scanning electron microprobe. Backscatter electron (BSE) images (Fig. 5) highlight grain-size and density differences between the pseudotachylyte, cataclasite and host rock. Specific minerals within the samples were analysed using wavelength dispersive analysis (WDA) to verify the mineralogy identified by optical analysis.

A4 QEMSCAN

QEMSCAN (Quantitative Evaluation of Minerals by SCANning electron microscopy) analysis was completed at the Camborne School of Mines laboratory facility, University of Exeter, UK. The uncovered, polished thin section was carbon coated and then scanned using the Fieldscan measurement mode, which in this case was programmed to collect an X-ray analysis every 10 μm across the sample surface in a grid (for further details see Gottlieb et al., 2000; Pirrie et al., 2004). System settings were 25 kV with a 5 nA beam, and X-rays were acquired at 1000 total X-ray counts per spectrum. Resulting data (approximately 5 million data points) were processed using iDiscover software to produce customised mineral data appropriate for the sample type. This involved modifying the SIP (database) to be accurate for the sample type, assisted by the geological context of the sample. Data output included quantitative modal mineralogy, a BSE map, a full false colour mineral map and individual mineral maps for each mineral type.

A5 Tomography

3-D high-resolution tomography images were acquired through v/tome/x s 240 micro-CT scanner from GE phoenix using a high-power X-ray tube and a drx-250 rt (real time) detector system (experimental conditions: pixel/voxel size: 20.0 μm, 1000 images for 360° (average of three single images, one image skipped), exposure time of 1 s, voltage of 80 kV and current of 250 μA).

A6 Porosity and permeability

Porosity was measured on 5 mm by 5 mm cores using an AccuPyc 1330 helium pycnometer from Micromeritics. Water permeability was measured in a servo-controlled steady-state-flow permeameter at effective pressures of 5–50 MPa (Fig. 6). The effective pressure is taken as the confining pressure minus the pore fluid pressure (this assumes that the poroelastic constant α is equal to one, see Guéguen and Palciauskas, 1994). Upstream and downstream pore fluid pressures were 9.5 and 10.5 MPa, respectively (a 1 MPa pressure differential across the sample). Water permeability (κ water) was calculated during steady-state flow using Darcy's law:

$$Q/A = \kappa_{\mathrm{water}}/\eta L(P_{\mathrm{up}} - P_{\mathrm{down}}) \quad \text{(A1)}$$

where Q is the volume of fluid measured per unit time, A is the cross-sectional area of the sample, η is the viscosity of the pore fluid, and L is the length of the sample.

Focal mechanisms in the southern Aegean from temporary seismic networks – implications for the regional stress field and ongoing deformation processes

W. Friederich[1], A. Brüstle[1], L. Küperkoch[3], T. Meier[2], S. Lamara[1], and Egelados Working Group[1]

[1]Ruhr-University Bochum, Institute of Geology, Mineralogy and Geophysics, Bochum, Germany
[2]Christian-Albrechts-University Kiel, Institute of Geosciences, Kiel, Germany
[3]BESTEC GmbH, Landau, Germany

Correspondence to: W. Friederich (wolfgang.friederich@rub.de)

Abstract. The lateral variation of the stress field in the southern Aegean plate and the subducting Hellenic slab is determined from recordings of seismicity obtained with the CYCNET and EGELADOS networks in the years from 2002 to 2007. First motions from 7000 well-located microearthquakes were analysed to produce 540 well-constrained focal mechanisms. They were complemented by another 140 derived by waveform matching of records from larger events. Most of these earthquakes fall into 16 distinct spatial clusters distributed over the southern Aegean region. For each cluster, a stress inversion could be carried out yielding consistent estimates of the stress field and its spatial variation. At crustal levels, the stress field is generally dominated by a steeply dipping compressional principal stress direction except in places where coupling of the subducting slab and overlying plate come into play. Tensional principal stresses are generally subhorizontal. Just behind the forearc, the crust is under arc-parallel tension whereas in the volcanic areas around Kos, Columbo and Astypalea tensional and intermediate stresses are nearly degenerate. Further west and north, in the Santorini–Amorgos graben and in the area of the islands of Mykonos, Andros and Tinos, tensional stresses are significant and point around the NW–SE direction. Very similar stress fields are observed in western Turkey with the tensional axis rotated to NNE–SSW. Intermediate-depth earthquakes below 100 km in the Nisyros region indicate that the Hellenic slab experiences slab-parallel tension at these depths. The direction of tension is close to east–west and thus deviates from the local NW-oriented slab dip presumably owing to the segmentation of the slab. Beneath the Cretan sea, at shallower levels, the slab is under NW–SE compression. Tensional principal stresses in the crust exhibit very good alignment with extensional strain rate principal axes derived from GPS velocities except in volcanic areas, where both appear to be unrelated, and in the forearc where compressional principal stresses are very well aligned with compressional principal strain rates. This finding indicates that, except for volcanic areas, microseismic activity in the southern Aegean is not controlled by small-scale local stresses but rather reflects the regional stress field. The lateral and depth variations of the stress field reflect the various agents that influence tectonics in the Aegean: subduction of the Hellenic slab, incipient collision with continental African lithosphere, roll back of the slab in the southeast, segmentation of the slab, arc volcanism and extension of the Aegean crust.

1 Introduction

The Hellenic subduction zone in the southern Aegean belongs to the seismically most active regions in Europe and has therefore been the target of many geoscientific research efforts. Seismicity is and was observed by the global seismic network, by permanent seismic observatories in Greece (National Observatory Athens (NOA), Thessaloniki, Patras and Chania universities, now the Hellenic Unified Seismic Network (HUSN)) and Turkey (Kandilli Observatory) as well as by temporary seismic deployments in the entire southern

Aegean (Hatzfeld et al., 1993; Friederich and Meier, 2008), on Crete (Bohnhoff et al., 2005; Jost et al., 2002; Meier et al., 2004; Becker et al., 2009), in the Cyclades (Bohnhoff et al., 2004; Dimitriadis et al., 2005, 2009) and in the western Hellenic subduction zone (Haslinger, 1998; Rigo et al., 1996; Papadimitriou et al., 2010). In view of the severe seismic hazard of the region, a central aim of all research efforts is and was to promote the understanding of the current seismotectonics of the Aegean.

Seismicity observations contributed to the understanding of Aegean seismotectonics via earthquake locations which delineate active faults (Comninakis and Papazachos, 1980; Papazachos et al., 1984, 2009; Papazachos, 1990), via focal mechanisms of earthquakes which allow an (ambiguous) determination of fault planes and slip direction (e.g. Papazachos et al., 1991; Taymaz et al., 1991; Benetatos et al., 2004; Bohnhoff et al., 2005; Kiratzi et al., 2007), via observation of aftershocks which permit inferences on the size of the fault plane and the amount of slip (Drakatos and Latoussakis, 2001) and finally via stress field determinations from catalogues of focal mechanisms (Papazachos and Delibasis, 1969; Bohnhoff et al., 2005; Rontogianni et al., 2011). Besides seismicity, direct observations of plate motions by geodetic measurements (e.g. Kahle et al., 2000; McClusky et al., 2000; Reilinger et al., 2006; Hollenstein et al., 2006, 2008; Aktug et al., 2009; Floyd et al., 2010) provide constraints on today's stress field. These observations are complemented by the study of surface fault populations (e.g. Angelier, 1978), sea bathymetry measurements and reflection seismic profiles (Sachpazi et al., 1997; Bohnhoff et al., 2001).

The focus of this paper is on the lateral and depth variations of the stress field in the southeastern Aegean which we determine from focal mechanisms of shallow and intermediate-depth earthquake clusters recorded during the temporary CYCNET (Bohnhoff et al., 2004) and EGELADOS (Friederich and Meier, 2008) experiments. In the southeastern section of the Aegean, the availability of seismic data suitable for focal mechanism and stress field determinations has been particularly sparse in the past. Most previous studies rely on data from larger events recorded by teleseismic stations (Benetatos et al., 2004; Kiratzi et al., 2007). Focal mechanisms of microseismic events have been obtained by Hatzfeld et al. (1993) using data from a temporary network covering the entire southern Aegean. A notable exception is the work by Rontogianni et al. (2011) who use focal mechanisms of intermediate-depth earthquakes from previous literature and the NOA database to derive the stress field of the subducting slab in four large sectors along the Hellenic arc. Bohnhoff et al. (2005) used a focal mechanism from previous work, from a local network on Crete, from the Harvard CMT catalogue and from web sites operated by INGV and ETH for a determination of the stress field around Crete.

With the CYCNET and EGELADOS temporary networks, a dense observational coverage of the entire southern Aegean could be accomplished. A sufficient number of high-quality focal mechanisms could be determined for several clusters of microseismicity allowing us to obtain estimates of the local stress field for each cluster region separately. The inherent assumption of homogeneity of the stress field commonly made in stress tensor determinations from focal mechanisms should hence be much better fulfilled than in studies where widely distributed earthquakes are used. In this way, similar to work done in California by e.g. Hardebeck and Michael (2004), more reliable constraints on the lateral variation of the stress field can be gained. Since it is not a priori clear that focal mechanisms of microearthquakes reflect the regional stress field but may rather be controlled by small-scale local stresses, we compare the principal stress directions obtained for the shallow earthquake clusters with principal directions of the crustal strain rate field calculated from a smooth interpolation of the GPS velocity measurements of Floyd et al. (2010).

2 Data

The results of this study are derived from two passive, temporary seismic deployments in the Aegean Sea, the CYCNET (Bohnhoff et al., 2004) and the EGELADOS (Friederich and Meier, 2008) experiments. The CYCNET was deployed for two years beginning in autumn 2002 and covered the Hellenic volcanic arc. It consisted of in total 22 seismic stations and comprised both short-period (1 Hz) and 120 s Streckeisen broadband sensors. The EGELADOS network was a follow-up project started in autumn 2005 which covered the entire Hellenic subduction zone from the Peloponnese in the west to western Turkey in the east. EGELADOS was an amphibic network encompassing on-shore 45 Güralp 60 s broadband seismometers, 4 Streckeisen 120 s sensors, 7 Mark 1 Hz seismometers and 24 ocean-bottom stations equipped with 60 s Güralp broadband sensors and long-period hydrophones. In addition, data of eight permanent stations of the GEOFON network equipped with Streckeisen broadband sensors could be used.

The analysis of the seismicity recorded by the two networks is described by Bohnhoff et al. (2006), Meier et al. (2004), Becker et al. (2006) and Brüstle (2012). First onsets for P and S waves were manually picked for all events. During the picking, the analysts also determined first motions of the P arrival for later use in focal mechanism determination. Brüstle (2012) focused her attention on the southeastern part of the Aegean and derived a minimum 1-D model for that region using the VELEST program (Kissling et al., 1994). They used this new reference model to relocate all earthquakes recorded by CYCNET and those recorded in the southeastern Aegean by EGELADOS using the nonlinear location program NonLinLoc (Lomax and Curtis, 2001; Lomax et al., 2009). In this way, a catalogue of about 7000 earthquakes with location uncertainty of less than 20 km was compiled which forms the observational basis of this study.

Events from this catalogue with magnitudes larger than 3.8 were checked for suitability of focal mechanism determination by waveform matching. For about 140 of them, a reliable focal mechanisms with an acceptable waveform fit could be obtained. For all other events of the catalogue, focal mechanisms were determined from first motions using the HASH-method of Hardebeck and Shearer (2002).

3 Methodology

3.1 Moment tensors from waveform matching

For about 140 earthquakes of magnitude larger than 3.8 recorded during the EGELADOS experiment and providing at least eight good-quality traces, we were able to obtain focal mechanism solutions using waveform fitting. The signal-to-noise ratio of each trace was checked automatically during extraction from the data archive and checked visually later. Instead of performing a direct inversion for the six elements of the moment tensor, we applied a grid search. Assuming a double-couple mechanism, we searched through regularly spaced values of strike and dip of the fault plane, rake of the slip vector in the fault plane and, in addition, source depth. Epicentral source coordinates, which can be determined with much less uncertainty than source depth, were adopted from the locations previously obtained using first arrivals. For each set of angles and depths, a moment tensor with unit seismic moment was calculated and synthetic seismograms were computed for each seismic station and available component using the GEMINI code (Friederich and Dalkolmo, 1995). We took a minimum 1-D reference model that was derived from local earthquake first arrivals (Brüstle, 2012) using the program VELEST (Kissling et al., 1994). The synthetic and data traces were low-pass filtered with corner frequency of 0.1 Hz and the synthetic time series were adjusted to the same length and sampling rate as those of the data.

A misfit between instrument-corrected synthetic seismograms and observed seismic records was then determined as follows: first, a normalized cross-correlation function of data and synthetic traces with time lags in the range of −60 s to +60 s was calculated; second, the synthetic seismogram was shifted by the time lag associated with the maximum cross-correlation; third, an amplitude scaling factor was calculated from the ratio of the rms–amplitude of the highest amplitude data trace and the rms–amplitude of the corresponding synthetic trace, and each synthetic trace was scaled by this factor; fourth, a misfit was calculated from the summed squared difference of the corresponding time samples of data and scaled and time-shifted synthetic trace weighted by the inverse energy of the maximum amplitude data trace and by the inverse square of the maximum cross-correlation. In this way, traces with high cross-correlation yield smaller misfits. The misfits were summed for all traces of the event. In addition, we determined the seismic moment from the square root of the ratio of the average energy of the data traces and the average energy of the synthetic traces. Misfits for each set of fault angles and depth were calculated providing a list of ranked moment tensors and source depths. The resulting best fitting focal mechanisms are depicted in Fig. 1.

With the minimum 1-D reference velocity model based on VELEST, this procedure typically provided source mechanisms by which the observed seismograms could be well reproduced in amplitude and phase by the time-shifted synthetic seismograms (Fig. 2). For some events, time shifts were large but did not vary much across different traces, indicating an error in origin time or source location or both. To obtain an agreement in phase of less than half a dominant period we added a relocation step for these events which was also done by waveform fitting. Synthetic seismograms for epicentral locations on a small rectangular grid around the original location were calculated and a rms-misfit between the traces was computed. The grid point associated with the smallest misfit was taken as the improved location.

3.2 Focal mechanisms from first motions

Since waveform inversion for moment tensors was only feasible for larger events with magnitudes greater than 3.8, we turned to determination of focal mechanisms from first motions to include smaller but well-recorded events. We chose an approach developed by Hardebeck and Shearer (2002) (HASH) which performs a search through a grid of fault plane normals and slip directions to find focal mechanisms which are compatible with the observed first motions. Since first-motion readings, locations and the velocity model may be erroneous, the HASH method attempts to not only determine solutions exactly fitting the first motions but also solutions which become acceptable when errors of first motions, station azimuth and take-off angles are taken into account. From the entire set of solutions the method calculates statistical properties and provides a preferred focal mechanism and estimates of the quality of the solution. The latter include the rms angular difference between the acceptable nodal planes and the preferred one and the fraction of solutions that differ by less than a given angle from the preferred one. In addition, the station distribution is ranked according to the distance of the stations from the nodal planes. Together with the number of unexplained polarities, these quality criteria allow a ranking of each obtained focal mechanism.

To use HASH with EGELADOS and CYCNET data, we searched our database of 7000 events in the southeastern Aegean established by Brüstle (2012) for suitable readings of P wave first motions. At least 10 first-motion readings were required. The errors of azimuth and take-off angle were estimated from the location error determined during localization with NonLinLoc (Lomax and Curtis, 2001). We rejected all focal mechanism solutions with a relative polarity error of

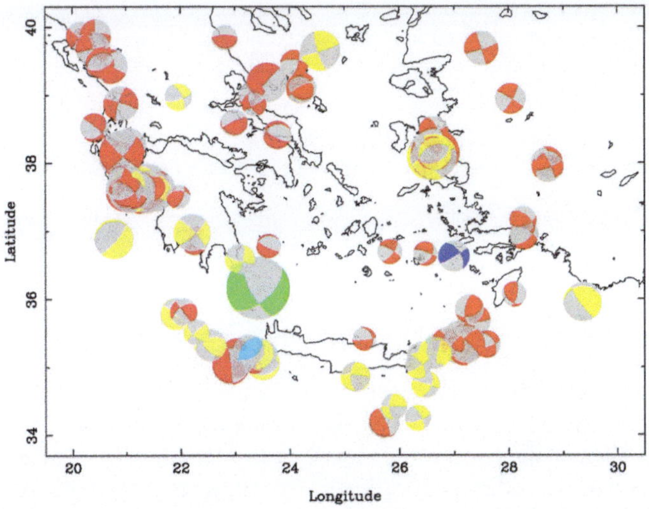

Fig. 1. Map of earthquakes for which focal mechanisms could be determined by waveform inversion. Colour of beach balls indicates source depth. Red: 0–20 km, yellow: 20–40 km, green: 40–80 km, blue: 80–100 km, deep blue: 100–150 km, magenta: > 150 km. Size of beach ball indicates magnitude.

greater than 0.3, a variation of the nodal planes of more than 45°, a fraction of less than 50 % of the solutions within 30° of the preferred one and a station distribution rank of less than 0.3. From the about 7000 considered events 540 passed the quality criteria (Fig. 3). The average location error of these events is 7.5 km. Adding the moment tensor solutions from waveform inversion provides us with a database of 680 good-quality focal mechanisms.

3.3 Stress tensor inversion

Focal mechanisms are typically highly variable. One reason is the variations owing to the previously mentioned uncertainties in focal mechanism determination. A further reason is the fact that the condition for compatibility of a focal mechanism with the local stress field, namely that the slip direction be parallel to the projection of the traction vector onto the fault plane (i.e. the resolved shear stress) (Gephart and Forsyth, 1984; McKenzie, 1969), is not very restrictive. Third, local stresses may exhibit small-scale variations and differ from the large-scale regional stress field.

It can be shown that the angle between the slip vector and the resolved shear stress on the fault plane only depends on four quantities (McKenzie, 1969): the directions of the principal stress axes and a dimensionless number defined by

$$R = \frac{\lambda_1 - \lambda_2}{\lambda_1 - \lambda_3}, \tag{1}$$

where the λ_i are the principal values of the stress deviator with $\sum_i \lambda_i = 0$. We follow here the convention used by engineers and physicists that tensional stresses are positive and compressional stresses negative. Thus, λ_1 as the largest (positive) eigenvalue is associated with the least compressive and λ_3 with the most compressive stress. $R = 0$ implies $\lambda_1 = \lambda_2$ and $\lambda_3 = -2\lambda_1$ and indicates a biaxial compressive stress regime. For $R = 0.5$, $\lambda_2 = 0$ and $\lambda_3 = -\lambda_1$ indicating a plane deviatoric stress state. In the case of $R = 1$, $\lambda_2 = \lambda_3$ and $\lambda_3 = -0.5\lambda_1$ signifying a biaxial tensional stress regime.

We implemented an approach that combines the grid search proposed by Gephart and Forsyth (1984) and an approach by Michael (1984). For principal stress directions and R value varying on a regular grid, we determine the angle between slip vector and resolved shear stress on the fault plane for each focal mechanism entering the stress tensor determination. The angles are summed to yield a misfit for each specific stress tensor. The stress tensors are ranked according to this misfit and the best fitting solution is the preferred stress tensor. However, since focal mechanisms contain substantial uncertainty, we need meaningful confidence regions of the preferred solution. They can be obtained by a bootstrap procedure (Efron and Tibshirani, 1986) which was applied to stress tensor determination by Michael (1987). In principle, one should repeat the experiment with new earthquakes in the same region many times and then observe the variation of the preferred solution. Since this is impossible, one resorts to a random resampling of the available data set by randomly picking N focal mechanisms out of the original N ones. Some focal mechanisms will occur more than one time and some will be absent. By finding the preferred solution for many of these randomly picked data sets of focal mechanism we can obtain an estimate of the variability of the preferred solution. The p % confidence limit is obtained by finding the p % of the stress tensors which are closest to the one determined from the original data set. To calculate the similarity of two tensors, we follow Michael (1984) who uses a normalized scalar product of two tensors defined as follows:

$$S = \frac{\sum_{i=1,3} \sum_{j=1,3} M_{ij} N_{ij}}{\sqrt{\sum_{i=1,3} \sum_{j=1,3} M_{ij}^2} \sqrt{\sum_{i=1,3} \sum_{j=1,3} N_{ij}^2}}. \tag{2}$$

The confidence limits of the principal axes are visualized by plotting the directions of the principal axes of the p % closest solutions into binned lower-hemisphere Schmidt projections and colour coding each bin according to its relative frequency of occurrence in the set of solutions. In addition, we display the frequency distribution of R values occurring in the p % closest solutions.

The stress tensor itself is graphically represented by visualizing the dependency of deviatoric tangential and normal stress on the orientation of the surface they act on. Surface normals are discretized as bins on a sphere, and the values of tangential and normal stresses are represented by assigning a colour code to each bin. A graphical display is finally realized by a Schmidt projection of the lower half of the binned and colour-shaded sphere onto the earth's surface. We use separate Schmidt projections for the absolute value of tangential stress and the signed value of deviatoric normal

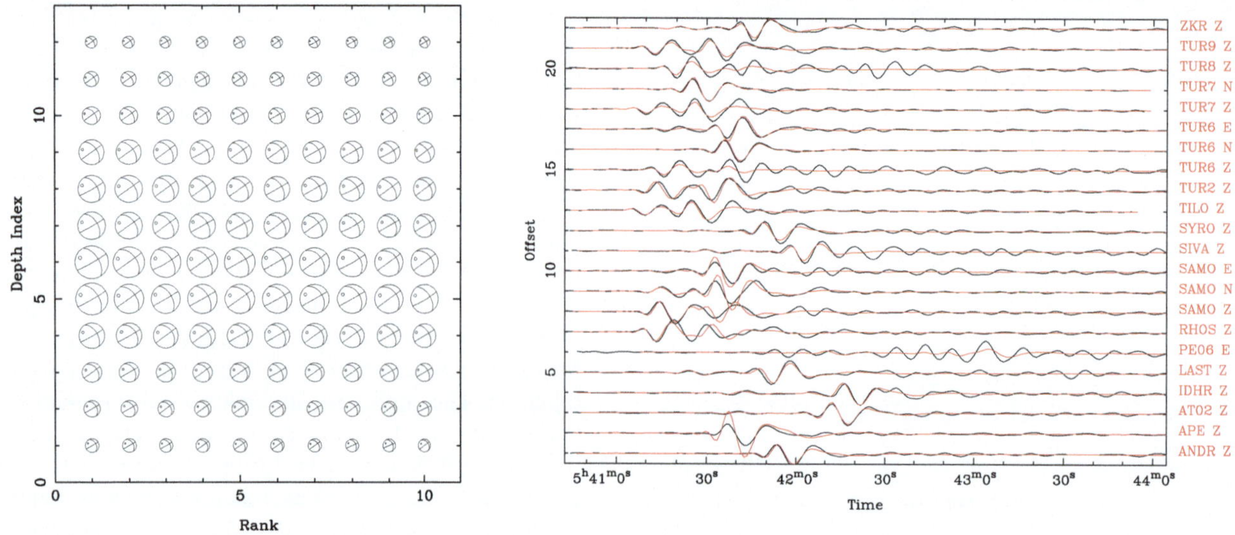

Fig. 2. Left: example of ranking of source mechanisms for an intermediate-depth earthquake at 135 km depth. Each row of beach balls displays the 10 best solutions for a given source depth. The size of the beach ball is inversely proportional to the waveform misfit. Depths range from 160 km at the bottom to 100 km at the top. The small circle in the beach balls indicates the location of the tension axis. Right: example of the waveform fit for the intermediate-depth earthquake. Data are shown as black lines, shifted synthetic traces as red lines. Relative amplitudes of data and synthetic traces are correctly reproduced.

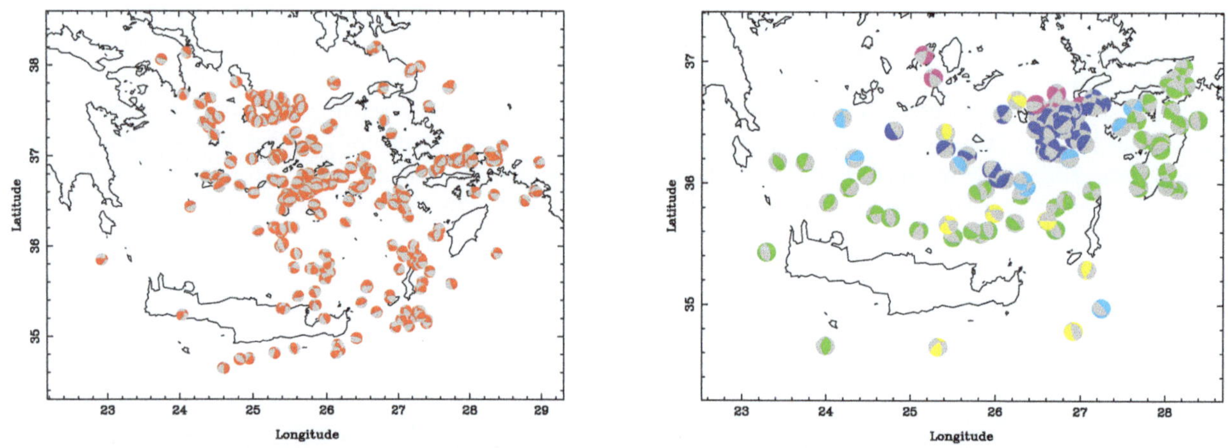

Fig. 3. Map of earthquakes for which focal mechanisms could be determined from first motions. Left: shallow earthquakes. Right: earthquakes deeper than 20 km. Colour of beach balls indicates source depth. Red: 0–20 km, yellow: 20–40 km, green: 40–80 km, blue: 80–100 km, deep blue: 100–150 km, magenta: $>$ 150 km.

stress. In both cases the deviatoric stress tensor is normalized by setting the largest eigenvalue λ_1 to 1. For tangential stress, the principal stress axes show up as places of zero tangential stress whereas for normal deviatoric stress, the principal stress directions lie at the extrema of normal stress. According to our sign convention, positive (negative) deviatoric normal stress indicates tension (compression). While tangential stress only depends on deviatoric stress, the true normal stress differs from deviatoric normal stress by the isotropic pressure which cannot be determined by stress inversion from focal mechanisms.

4 Results and discussion

Stress inversion from focal mechanisms builds on the assumption of homogeneous stress within the region spanned by the associated earthquakes. For this reason, stress inversion studies attempt to divide the region of interest into small subregions under the constraint that they still contain a sufficient number of focal mechanisms. Because of the limited number of focal mechanisms available in many studies, these subregions have to be chosen fairly large, making the validity of the homogeneity assumption debatable. In this study, the large number of high-quality focal mechanisms and the clustering of seismicity allows a determination of the stress field separately for each cluster. Homogeneity of the stress field is much better fulfilled in this case than for regionally distributed seismicity. In the following we will separately present and discuss results obtained for shallow seismicity that provide constraints on the crustal stress field and results obtained for intermediate-depth earthquakes from which we get information on the stress field in the Hellenic slab.

4.1 Stress state of the Aegean crust from shallow seismicity

We have grouped the shallow seismicity (depth less than 20 km) into 12 clusters as depicted in Fig. 4: Iraklion basin, Kamilonisi basin, Gulf of Gökova, Kos, Astypalea, Mykonos, Amorgos, Columbo volcano, South and North Karpathos, Sigacik and Zante (not shown on map). In this way, average deviations of the slip direction from the resolved shear stress on the fault plane did not exceed 28°. The resulting stress solutions allow a fair image of the distribution of stress in the southeastern Aegean at the crustal level (Table 1). In general, the compressional principal axes dip steeply while the tensional and intermediate axes are subhorizontal. Notable exceptions are the clusters at Zante, S-Karpathos and Gökova. In the majority of cases, the stress ratio R is less than 0.5 indicating a larger magnitude of compressive principal stress compared to tensional principal stresses. Unfortunately, the stress ratio R does not allow any inferences on the ratio of compressive to tensional stress because it is independent of the isotropic part of the stress tensor. Hence, for example, a dominant vertical compressive principal stress could be caused either by gravity stresses inside a vertically layered elastic crust or by additional compressive stress created by a magma upwelling in the mantle. In both cases, we would obtain the same stress ratio of $R = 0$. However, there are also clusters with $R \neq 0$. If $R < 0.5$, the compressive principal stress dominates but tensional and intermediate principal stresses differ, indicating a deviation of the stress field from that of a vertically layered elastic crust. Is $R > 0.5$, the tensional principal stress is larger than the compressive ones. Both cases indicate the action of some tectonic force on the crust.

In the following we discuss the results of focal mechanism determination and stress inversion for each cluster. They are visualized by a composite figure displaying the distribution of T and P axes of the focal mechanisms, the 80 % confidence regions of the three principal stress axes, the frequency distribution of R values among the 80 % stress solutions closest to the preferred one and a graphical representation of the stress tensor.

4.1.1 Amorgos and Andros–Mykonos

The biggest number of focal mechanisms in this study is available for the Amorgos fault zone region. The 72 focal

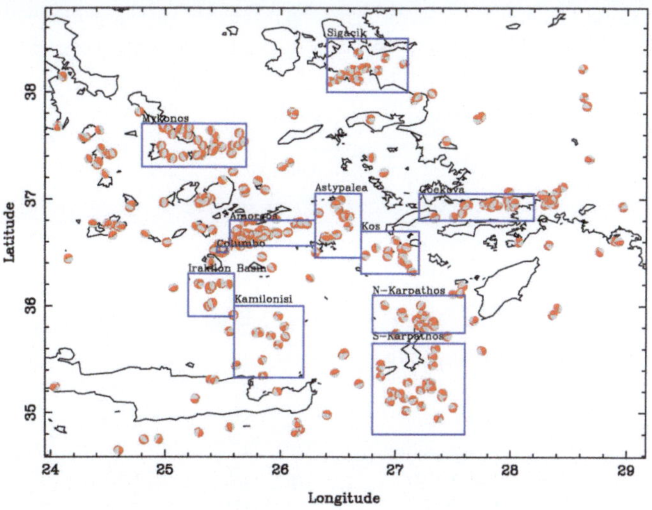

Fig. 4. Shallow earthquake locations with focal mechanisms and definition of cluster regions.

mechanisms can be explained by a single stress solution with a small average angular misfit of 23.6° (Fig. 5). The P axes of the focal mechanisms tend to dip steeply while the T axes are close to horizontal with an accumulation in the N–S direction. The tensional and compressional stress axes are well determined with concentrated confidence regions while the intermediate stress axis is less well constrained. Compressional stress slightly dominates and is directed subvertical with a dip of 62°. The tensional axis is close to horizontal with a dip of 12°. Its orientation is about NNW–SSE with an azimuth of 155°. The distribution of stress ratio values also indicates a well-constrained value of around 0.3. Stress directions and stress ratio indicate a normal faulting regime. Maximum tangential stress is reached either on rather steeply SE-dipping planes striking SW–NE or W–E-striking planes dipping at about 45° towards north. These results are consistent with the general SW–NE trend of the Amorgos fault zone and the preferred occurrence of normal faulting earthquakes. A similar result with a tensional axis slightly rotated to the west was found for the Andros–Mykonos cluster described in the Appendix (Fig. A6).

4.1.2 Columbo volcano, Kos–Nisyros and Astypalea

At the southwestern tip of the Amorgos fault zone and northeast of Santorini sits Columbo volcano which exhibited high seismic activity during the observation period of the EGELADOS and CYCNET deployments (Fig. 6). As for the Amorgos cluster the P axes concentrate in the centre of the focal sphere while the T axes tend to appear in the outer parts with a large spread of directions. As a consequence, only the compressional stress axis can be well determined. The 80 % closest bootstrap solutions exhibit a large spread of directions for both tensional and intermediate stress axis. In contrast, the R values concentrate at very low values with

Table 1. Summary of stress inversion results for shallow earthquake clusters. Clusters are ordered according to the number of available focal mechanisms. Azimuth is measured in degrees from north over east.

Cluster	Number of focal mechanisms	Average misfit/deg	Tensional axis Dip/deg	Tensional axis Azimuth/deg	Interm. axis Dip/deg	Interm. axis Azimuth/deg	Compr. axis Dip/deg	Compr. axis Azimuth/deg	R
Amorgos	72	23.6	12	155	25	59	62	268	0.3
Columbo	40	26.9	13	0	20	95	65	239	0.0
Zante	29	28.0	43	90	34	320	27	209	0.1
Mykonos	23	7.4	10	309	11	217	75	82	0.2
S-Karpathos	19	27.7	43	311	8	48	46	147	0.1
Sigacik	21	14.0	0	25	11	295	79	115	0.9
Gökova	16	23.9	38	20	5	286	52	190	0.3
N-Karpathos	15	11.6	0	47	26	317	64	137	0.5
Astypalea	14	22.1	29	40	14	302	58	189	0.0
Iraklion	10	11.8	31	270	10	174	58	69	0.7
Kos	10	17.6	17	149	25	248	59	29	0.0
Kamilonisi	9	20.8	6	72	36	165	54	332	0.8

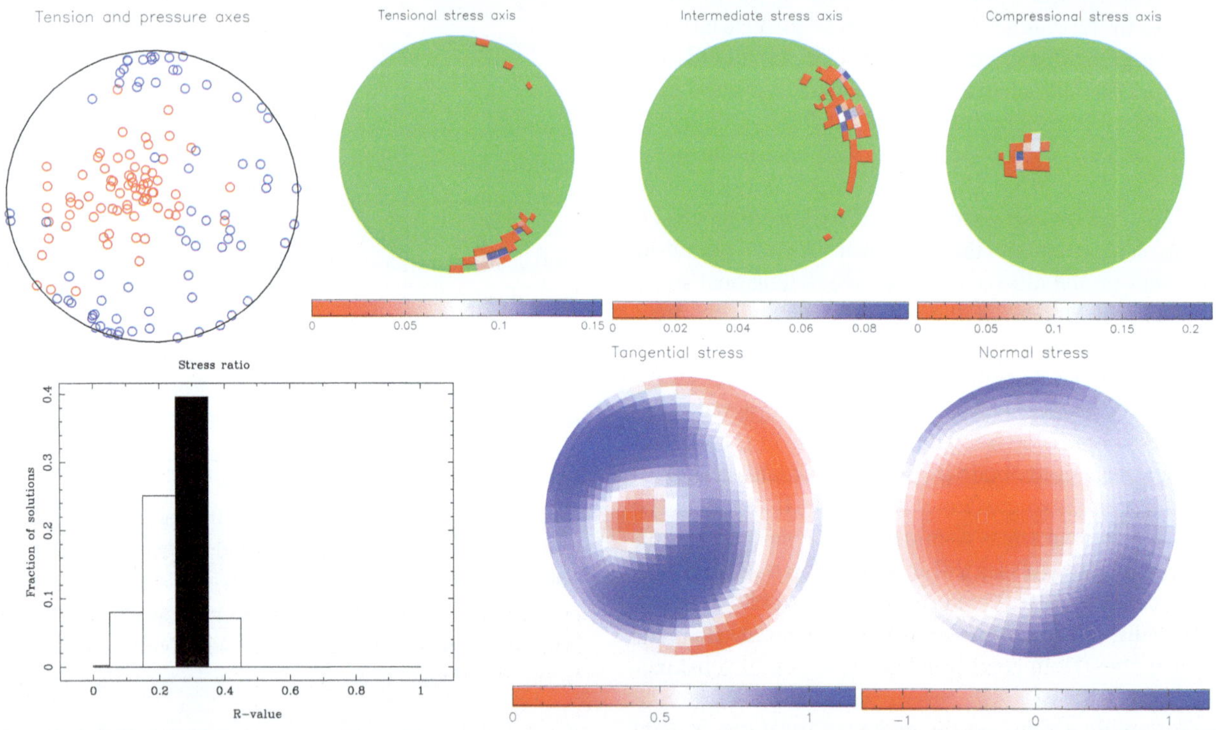

Fig. 5. Results of stress inversion for cluster of shallow earthquakes near Amorgos. Top row from left to right: location of tension (blue circles) and pressure axes (red circles) of focal mechanisms, 80 % confidence region for least compressive (deviatoric tensional), intermediate and most compressive (deviatoric compressive) principal stress axes. Bottom row from left to right: distribution of R values for best 80 % of stress tensor solutions, absolute value of tangential stress and signed normal stress versus surface normal.

a sharp maximum at $R = 0$. The best fitting stress solution with an average angular misfit of 26.9° is characterized by a steeply dipping compressive axis and degenerate tensional principal stresses. Hence, there is no preferred tensional direction. Planes of maximum shear stress can strike in any direction. One could speculate that magma upwelling from the mantle puts the crust under vertical compressive stress without any preference for near-horizontal tensional stress. However, as discussed before, the stress solution does not enable us to make any statements about the ratio of true compressive to tensional principal stresses. Very similar solutions

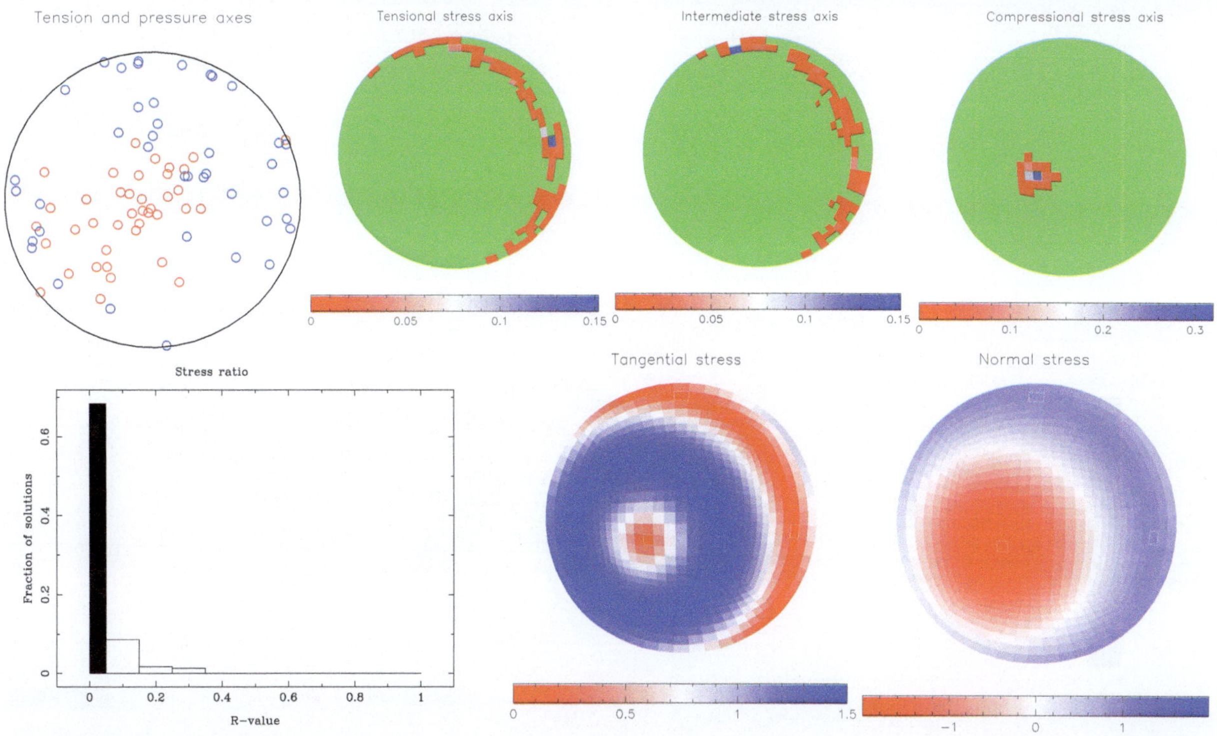

Fig. 6. Results of stress inversion for cluster of shallow earthquakes near Columbo volcano. Explanation of subfigures is as in Fig. 5.

were obtained for the clusters near Kos–Nisyros and Astypalea which are described in the Appendix.

4.1.3 Karpathos, Iraklion and Kamilonisi basin

The focal mechanisms of the earthquakes around Karpathos can only be well explained by a stress tensor if they are split into two subregions, one encompassing the events north of Karpathos and one containing the events surrounding the southern part of Karpathos. Results for the two regions are very different. For the events south of Karpathos (Fig. 7), the compressional stress axis is well constrained while the bootstrap solutions exhibit a strong spread for the other two axes. The distribution of R values is very narrow with a clear maximum at $R = 0.1$ indicating a degeneracy of the tensional and intermediate principal stresses. The compressional axis dips at only 46° which is the second smallest value among all clusters considered. The direction of the compressional axis is about SE–NW (azimuth 147°) and nicely coincides with the local direction of subduction. This finding suggests that the stress field in the Aegean plate south of Karpathos is modified by the downgoing African lithosphere which exerts some horizontal pressure on the upper plate and hence rotates the compressional stress axis from its subvertical orientation found for the other event clusters. Due to the degeneracy of the tensional principal stresses planes of maximum shear stress may strongly vary in strike and dip.

The situation is very different for the earthquakes north of Karpathos (Fig. 8) where the compressional axis moves back to a dip of 64° with about the same azimuth as south of Karpathos. Tensional and intermediate principal stresses are clearly different. All principal axes have small 80 % confidence regions. The distribution of R values is narrow with a maximum at 0.5. The tensional axis is exactly horizontal with an NE–SW orientation parallel to the Hellenic arc. Planes of maximum shear stress have intermediate dip and strike either roughly N–S or E–W. Apparently, the stress state in the Aegean plate significantly changes with distance from the plate contact from dominating subduction parallel compression to dominating tension perpendicular to the subduction direction. Arc parallel tension is also found for the Iraklion and Kamilonisi basins (Figs. A3 and A4) as described in more detail in the Appendix.

4.1.4 Gökova graben and Sigacik basin

The Gökova graben is located at the easternmost tip of the Hellenic subduction zone on the transition from the Aegean microplate to the Anatolian plate. GPS observations in this area (Reilinger et al., 2010) indicate a significant relative motion between the Aegean and western Turkey which may influence the regional stress field. Our stress analysis (Fig. 9) exhibits a well-defined compressional axis, a less well defined tensional axis and a variable intermediate axis. The stress ratio R shows a fairly concentrated distribution around

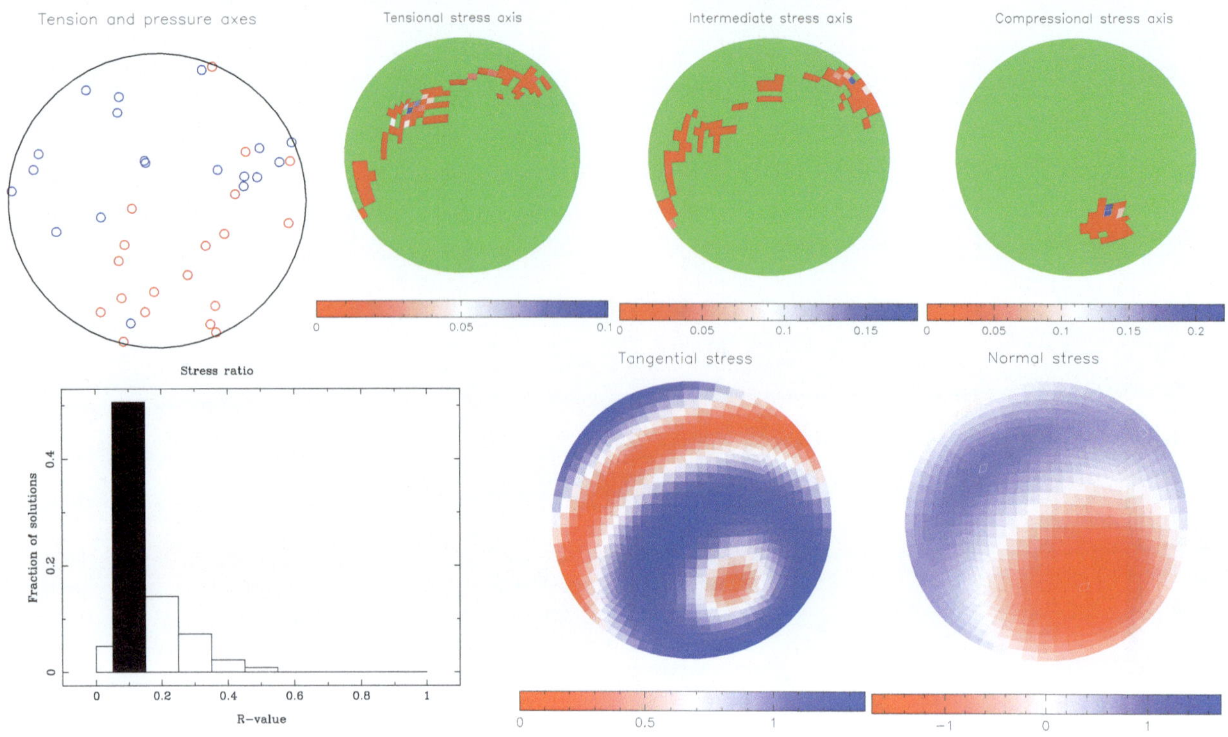

Fig. 7. Results of stress inversion for cluster of shallow earthquakes south of Karpathos. Explanation of subfigures is as in Fig. 5.

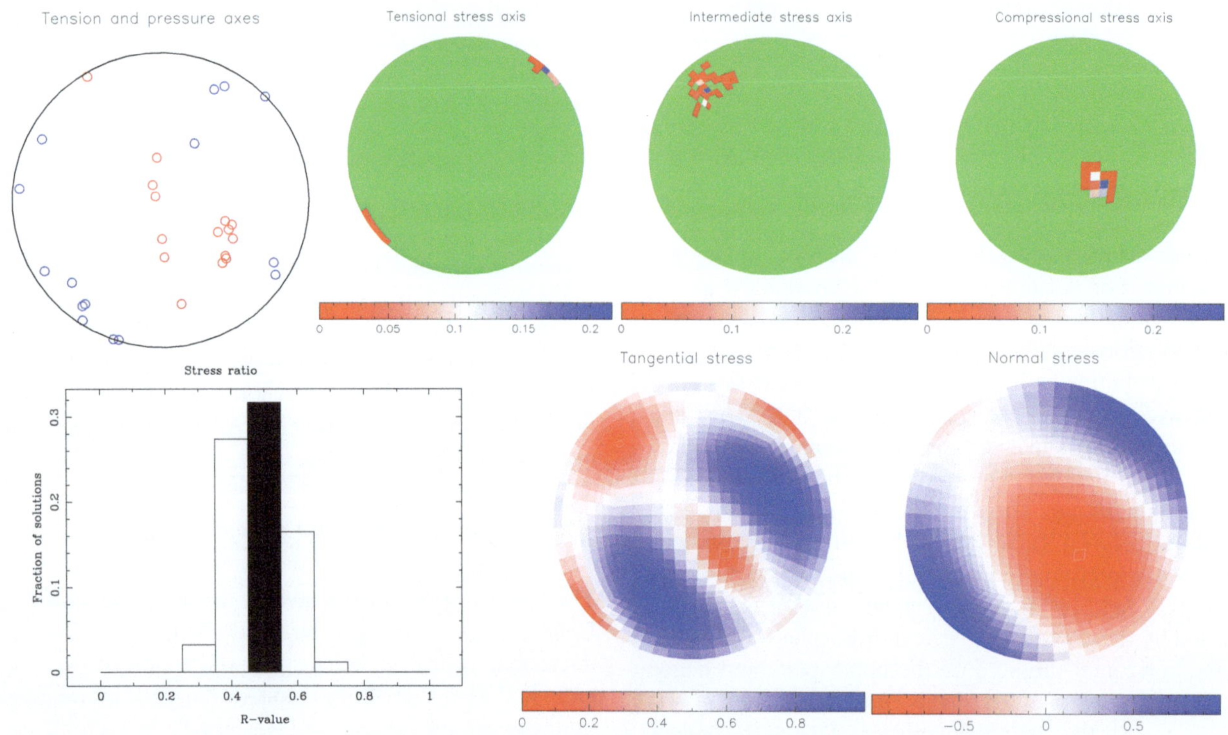

Fig. 8. Results of stress inversion for cluster of shallow earthquakes north of Karpathos. Explanation of subfigures is as in Fig. 5.

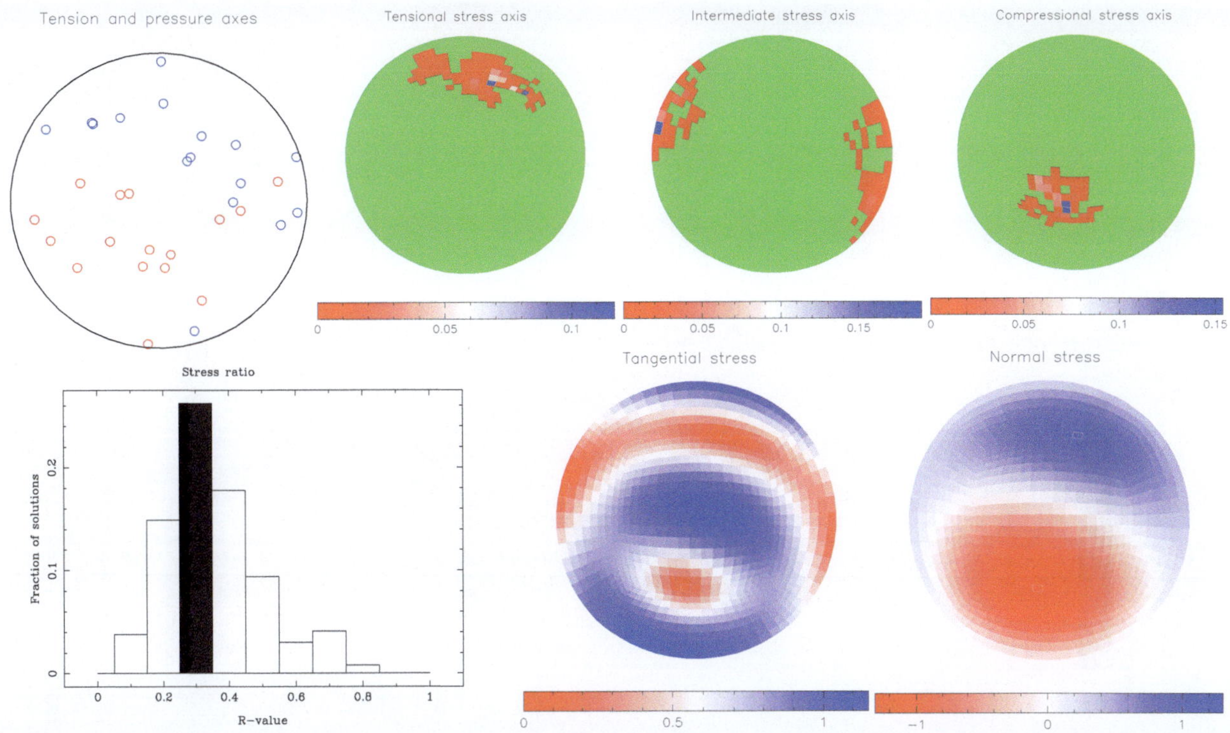

Fig. 9. Results of stress inversion for cluster of shallow earthquakes near Gökova. Explanation of subfigures is as in Fig. 5.

0.3. Thus, tensional and intermediate principal stresses significantly differ. The dip of the compressional axis is relatively small (52°) with a N–S orientation and suggests some influence of subduction on the stress field. The tensional axis also dips by 38° with a NNE–SSW orientation. Planes of maximum shear stress are either horizontal or steeply dipping striking roughly E–W. One may speculate that the superposition of stresses related to subduction and to the northward motion of western Turkey relative to the central Aegean (Reilinger et al., 2010) controls the regional stress field. The stress field in the Sigacik basin (Fig. A5) that exhibits a very clearly constrained horizontally oriented tensional axis is described in the Appendix.

4.1.5 Zante

The cluster in Zante is very close to the western Hellenic subduction front along western Peloponnese. The stress determination for these events (Fig. 10) results in a well-constrained compressional stress axis dipping subhorizontal at an angle of 27° and less well defined tensional and intermediate axes. A small stress ratio $R = 0.1$ indicates nearly degenerate tensional and intermediate principal stresses. The azimuth of the compressional axis is 209° which is close to the SSW–NNE direction. Planes of maximum shear stress strike WNW–ESE with either very small or very large dip. The findings suggest a strong influence of subduction on the stress field. The subduction direction is roughly parallel to the azimuth of the compressional axis. Its small dip also indicates the action of horizontal pressure exerted by the subducting slab at the plate contact. Otherwise, we would expect a steeply dipping compressional axis.

4.1.6 Overview of the stress field in the Aegean micro-plate

In order to provide a comprehensive picture of the variation of the stress field in the Aegean micro-plate we plot the stress tensor visualizations showing normal stress into a map of the southeastern Aegean region (Fig. 11). Dark blue colours indicate tensional normal stress and dark red colours signify compressional normal stress. We observe a tensional stress parallel to the Hellenic arc from the Iraklion basin in the west over the Kamilonisi basin to north Karpathos in the east. This behaviour can be explained by the fan-like spread of velocity vectors in the southern part of the Aegean microplate which is mainly caused by the rollback of the Hellenic slab in the southeastern Aegean (Reilinger, 2010). A similar observation was made by Benetatos et al. (2004) based on averaged focal mechanisms of larger-magnitude earthquakes ($M > 5$). The stress fields at Zante, South Karpathos and also Gökova appear to be influenced by nearby subduction processes which exert a subhorizontal pressure on the upper plate leading to a rotation of the compressional stress axis towards smaller dips. This effect is most pronounced for the Zante cluster. The stress field in the volcanic arc could be

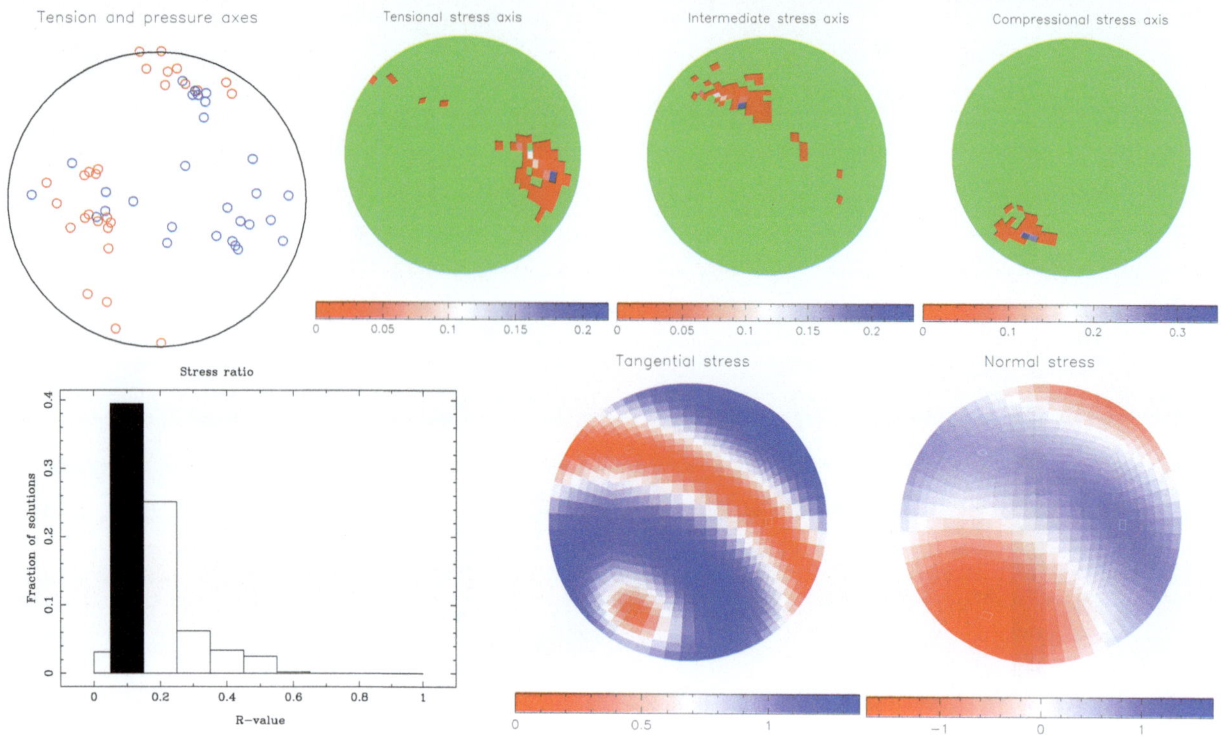

Fig. 10. Results of stress inversion for cluster of shallow earthquakes near Zante. Explanation of subfigures is as in Fig. 5.

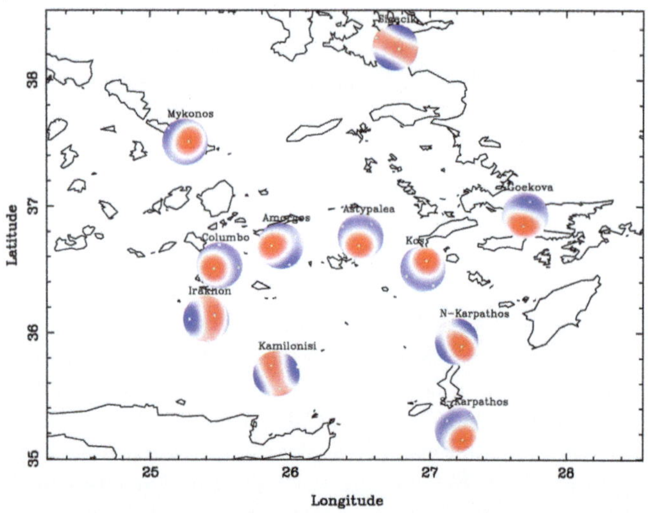

Fig. 11. Stress solutions visualized by distribution of normal stress versus fault plane normal.

influenced by magmatic processes in the mantle because the stress solutions exhibit dominant subvertical compressional axes with nearly degenerate tensional and intermediate principal stresses. The exception is the Amorgos fault zone where a NNW–SSE oriented tensional stress direction was identified. Nearly the same trend (NW–SE) of the tensional axis is observed for the Andros–Tinos–Mykonos cluster in the northwest of Amorgos while tensional stresses are oriented NNE–SSW in the Sigacik and Gökova area. A similar observation is made by Benetatos et al. (2004) who find focal mechanisms with generally N–S directed T axes in the region north of the volcanic arc. However, our results indicate a systematic rotation of the tensional stress axis which is reflected in the curved shape of graben systems extending from Mykonos to Sigacik in the north and from Amorgos to Gökova further in the south.

4.1.7 Comparison with the GPS-derived regional strain rate field

Since in this study stress tensors have been obtained for local clusters of microearthquakes, it is not a priori clear that they represent the regional stress field. The fault area ruptured by microearthquakes analysed in this study may be well below 1 km^2. For example, according to Wells and Coppersmith (1994), the rupture area of a magnitude 3 earthquake is about 0.2 km^2. Hence, our stress solutions might well reflect local stresses expected to exhibit small-scale spatial variations and potentially being unrelated to the large-scale regional stress field.

There are two major arguments against this proposition: first, if microearthquakes were controlled by local stresses on a scale of 1 km or less, we would not have been able to find stress solutions for an entire cluster with such small angular misfits (Table 1). Second, the principal stress axes obtained in this study align very well with the principal directions of the regional strain rate field as will be shown in the following.

For comparison of stress and strain rates, we start out with GPS velocities obtained by Floyd et al. (2010) and compute a continuous velocity field using a method and associated computer program developed by Tape et al. (2009). The method uses a representation of the velocity field by multiscale spherical wavelets and determines wavelet coefficients by a least-squares fit of theoretical to observed velocities. We use wavelets of order 3 to 8 with half-widths ranging from 1400 km down to 40 km. The GPS velocities (Fig. 12, left) can be fit by the continuous velocity field in the southeastern Aegean with residuals of less than 5 % (Fig. 12, right). There appears to be no point in defining blocks, as potentially discontinuous motion by a finite number of blocks does not result in a better fit of the observed velocities (Floyd et al., 2010). The program of Tape et al. (2009) also computes the components of the symmetric strain rate tensor by differentiating the velocity field. Since neither vertical velocities nor vertical derivatives of velocity are available, the program assumes a stress-free spherical Earth's surface and a linear viscous rheology to obtain constraints on the unavailable tensor components. With these assumptions, two eigenvectors of the strain rate tensor are parallel to the Earth's surface and one points perpendicular to it. If, in addition, vertical velocity is negligible, the surface-parallel eigenvectors can be calculated from the horizontal velocity components only. Figure 13 shows the resulting extensional and compressional principal directions of strain rate in the southeastern Aegean. The former ones compare very well with a similar result obtained by Floyd et al. (2010).

Comparison of principal stress and strain rate axes yields the following results: for cluster regions with distinct tensional stress axes and near-vertical compressional stress axes, azimuths of principal directions of tensional stress and extensional strain rate are very similar (Table 2). The largest deviation is 37° for the cluster north of Karpathos. Moreover, for clusters with inclined or subhorizontal compressional stress axes (Zante and S-Karpathos), the azimuths of the compressional principal stress axes compare well with those of compressional strain rate with deviations of about 30°. In view of the errors of the determined stress field and the fact that the strain rate field in the southern Aegean is an interpolation of rather sparse velocity observations, these deviations appear acceptable. The only regions where strain rate differs significantly from deviatoric stress are the volcanic areas around Columbo, Astypalea and Kos–Nisyros. They exhibit a well-constrained near-vertical compressional stress axis but no distinct tensional axes. In contrast, extensional strain rate in these regions is much greater than compressional strain rate. Apparently, volcanic activity creates local stresses that also control the character of seismicity.

Table 2. Comparison of azimuths of tensional/compressional principal stress axes with extensional/compressional principal axes of strain rate for shallow earthquake clusters.

Cluster	Azimuth/deg tensional stress	Azimuth/deg extensional strain rate
Amorgos	155	162
Mykonos	309	296
Sigacik	25	17
Gökova	20	−4
N-Karpathos	47	84
Iraklion	270	300
Kamilonisi	72	100
Cluster	**Azimuth/deg compressional stress**	**Azimuth/deg compressional strain rate**
S-Karpathos	147	173
Zante	209	240

4.2 Stress state of the Hellenic slab from intermediate depth events

Events located at depths greater than 50 km could be grouped into four different clusters (Fig. 14): Cretan Sea, Rhodos and surroundings, Cyclades and a big cluster in the Nisyros–Kos–Astypalea region. Table 3 summarizes the stress tensor solutions obtained for these four clusters. These clusters differ very much from the shallow ones. The compressional stress axis dips much less and is closer to horizontal than vertical. The tensional axis varies strongly and can dip very steeply. The angular misfit for the Rhodos and Cyclades events is quite large and thus reflects a violation of the assumption of a homogeneous stress field. Hence, the results for the Cyclades cluster and to a lesser degree those of the Rhodos cluster should be regarded with caution.

4.2.1 Nisyros–Astypalea

We were able to collect focal mechanisms for 42 earthquakes deeper than 100 km clustering in the Nisyros–Astypalea area. T and P axes of the focal mechanisms nicely cluster on different halves of the focal sphere (Fig. 15). The 80 % closest bootstrap stress solutions exhibit a small scatter for all three principal axes. The stress ratio concentrates at a value of $R = 0.4$ indicating nearly equal magnitudes for the tensional and compressional principal stresses. The compressional axis dips at an angle of 32° indicating near slab-normal compression. The azimuth is rotated by 12° towards the south from SE–NW. This direction only slightly deviates from the SW–NE direction of subduction. The dip of the tensional axis is 47°, about parallel to the expected dip of the eastern part of the Hellenic slab. Its azimuth, however, only deviates by 7° from straight E–W and thus differs by 38° from the presumed

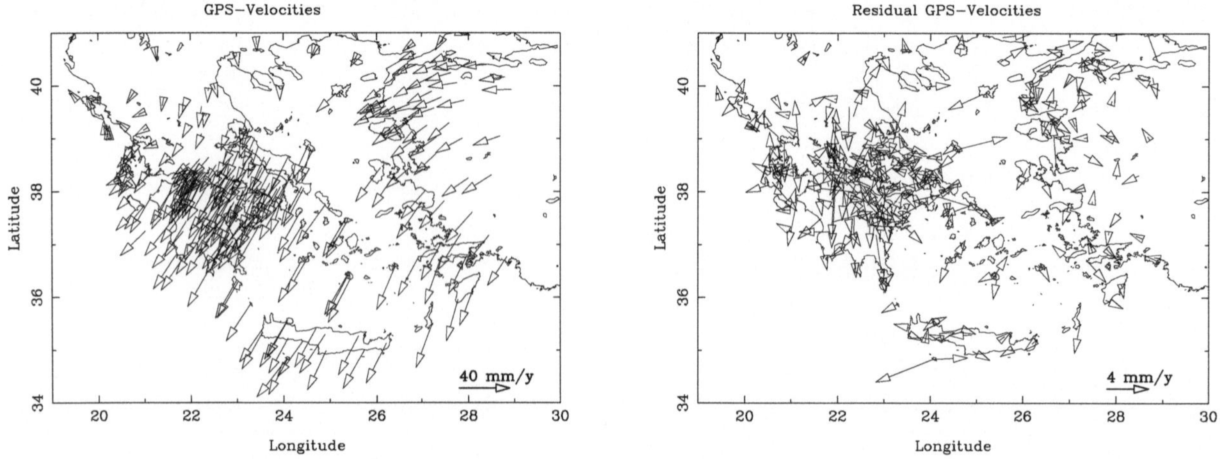

Fig. 12. Observed GPS velocities of Floyd et al. (2010) (left) and residuals between observed and interpolated continuous velocity field (right).

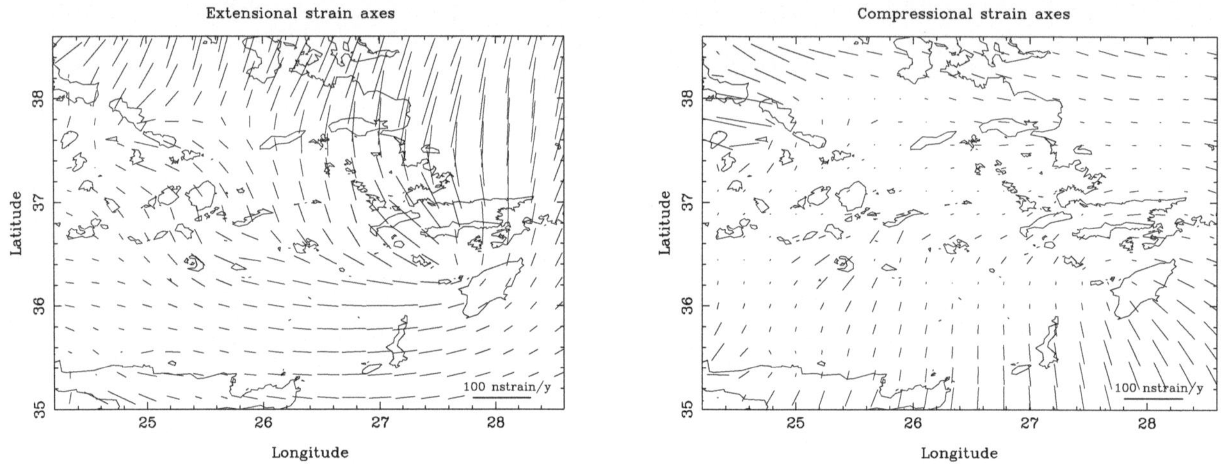

Fig. 13. Extensional (left) and compressional (right) principal directions of strain rate derived from a continuous interpolation of observed GPS velocities in the Aegean (Floyd et al., 2010).

SE–NW subduction direction. Maximum shear stress occurs at either nearly horizontal planes or very steeply dipping planes striking about SW–NE, favouring normal faulting inside the slab.

The solution for the stress field suggests that the slab experiences slab-parallel tension that is however not exactly down-dip but slightly rotated to the west. Down-dip tension has been found by Fujita and Kanamori (1981) for old slabs with low convergence velocities at the trench. Due to their increased density caused by thermal contraction they tend to sink faster than they converge at the trench. This mechanism will put the slab under down-dip tension. We speculate that the observed rotation of tensional stress relative to the down-dip direction is related to the segmentation of the Hellenic slab which dips more strongly in the east than in the west. Beneath Astypalea, hypocentres reach depths of about 180 km while further in the west beneath the Cyclades they only reach depths of about 140 km. Hence, the slab is either curved in the case it is still continuous or it is torn somewhere between the Cyclades and Astypalea. This slab deformation or tearing could explain a rotation of the tensional axis to the west because the counterforce exerted by western slab parts is missing.

Rontogianni et al. (2011) give a stress tensor solution for the Hellenic slab in this region derived from focal mechanisms of 19 earthquakes at depths between 90 km and 180 km. They find a N–S-oriented compressional axis dipping at 46° and a slab-parallel tensional axes directed along NW–SE. Their solution is roughly consistent with ours but does not exhibit the rotation of the tensional axis towards the W–E direction. However, their solution exhibits a high angular misfit (51°) interpreted by them as indication of the heterogeneity of the stress field. In contrast, our stress solution with the tensional axis rotated towards W–E obtained from

Table 3. Summary of stress inversion results for shallow earthquake clusters. Clusters are ordered according to the number of available focal mechanisms. Azimuth is measured in degrees from north over east.

Cluster	Number of focal mechanisms	Average misfit/deg	Dip/deg	Azimuth/deg	Dip/deg	Azimuth/deg	Dip/deg	Azimuth/deg	R
			Tensional axis		Interm. axis		Compr. axis		
Nisyros	42	20.1	47	277	27	39	32	147	0.4
Rhodos	16	28.6	75	297	2	32	15	123	0.8
Cyclades	11	34.6	64	284	17	52	19	148	0.2
Cretan Sea	10	13.4	16	35	52	283	33	136	0.3

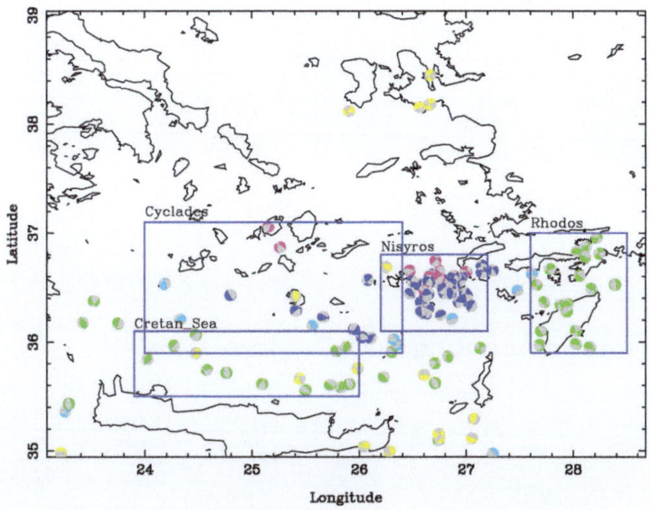

Fig. 14. Intermediate-depth earthquake locations with focal mechanisms and definition of cluster regions. The meaning of the colours of the beach balls are as given in Fig. 3.

42 densely clustered focal mechanisms allows a match with an angular misfit of 20° only.

4.2.2 Cretan Sea

The cluster in the Cretan Sea comprises 10 events at depths ranging from 50 to 80 km. Their focal mechanisms can be fit very well by a single stress tensor (Fig. 16). As for the Nisyros events, the scatter of the bootstrap solutions and for the stress ratio is very small. The compressional principal axis dips at 33° with an exact SE–NW orientation. The tensional axis dips at a low angle of 16° pointing roughly SW–NE (35° azimuth). The intermediate axis dips more strongly at 52°. The stress ratio of 0.3 indicates a slight dominance of compression and also a dominance of horizontal over vertical tension. Nevertheless, the stress field allows both strike slip and thrusting mechanisms.

The slab appears to be under down-dip tension and lateral compression, although the compressional axis deviates by 33° from the horizontal. In addition, the SE–NW orientation of compression is close to but not exactly parallel to the Hellenic arc in this area which rather trends along ESE–WNW. Lateral compression could be caused by reduced curvature of the slab in the segment along the island of Crete. According to the ping-pong model of Frank (1968), a slab of reduced curvature would become too "wide" laterally during subduction and thus develop compressive lateral membrane stresses. We speculate that either the rather straight shape of the island of Crete or the incipient collision with continental African lithosphere imposes this curvature reduction of the arc.

Rontogianni et al. (2011) provide a stress solution for a region encompassing Crete, the Cretan Sea and the Cyclades area based on 12 earthquakes in a depth range of 50–80 km. They find a W–E-oriented compressional axis roughly consistent with our result. Their tensional axis is, however, near-vertical while ours is nearly horizontal and pointing SW–NE. This discrepancy illustrates the potential variability of stress solutions that are derived from only a few focal mechanisms and based on different collections of earthquakes.

The stress field in the slab beneath Rhodos (Fig. A7) and the Cyclades (Fig. A8) is much less well constrained and therefore described in the Appendix.

5 Conclusions

Analysis of first motions of 7000 well-located earthquakes in the southeastern Aegean recorded by the CYCNET and EGELAOS networks produced 540 well-constrained focal mechanisms. An additional 140 focal mechanism of larger events could be determined by waveform matching. Many of these earthquakes fall into spatial clusters of several tens of events permitting a local determination of the stress field from the focal mechanisms. In this way, the lateral variation of the stress field in the southern Aegean could be mapped. In addition, the intrinsic assumption of a homogeneous stress field in stress inversions is much better fulfilled for the individual clusters than for spatially distributed earthquakes.

At crustal levels the stress inversion yields steeply dipping compressional principal stress axes for nearly all clusters. Stress ratios indicate a dominance of compressional over

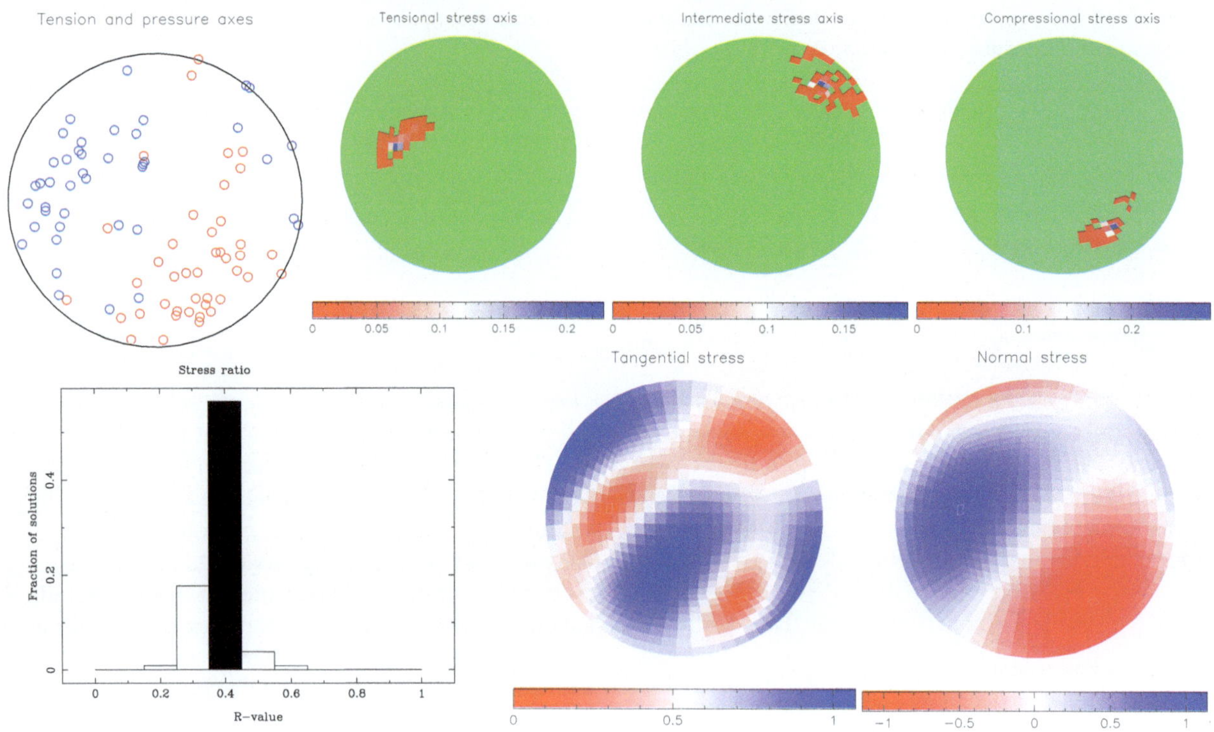

Fig. 15. Results of stress inversion for cluster of intermediate-depth earthquakes near Nisyros. Explanation of subfigures is as in Fig. 5.

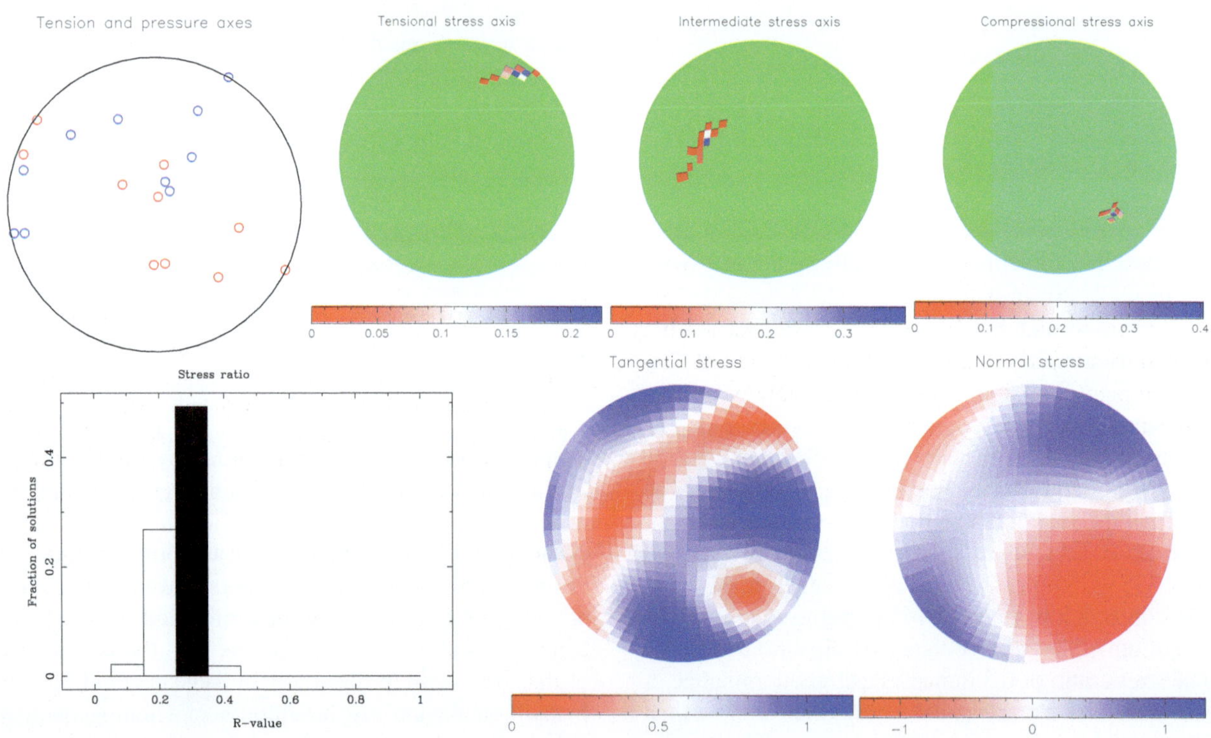

Fig. 16. Results of stress inversion for cluster of intermediate-depth earthquakes beneath the Cretan Sea. Explanation of subfigures is as in Fig. 5.

tensional stresses. According to the behaviour of the tensional stress axis the clusters fall into three groups. For the clusters around Kos, Astypalea and Columbo in the southern volcanic arc, the tensional and intermediate principal stresses are nearly degenerate, indicating a possible magmatic source for the near-vertical compressional stresses. A very different stress regime is deduced for the Santorini–Amorgos, Gökova and Sigacik grabens where the tensional stress shows a clearly identifiable direction rotating from NW–SE in the west to NNE–SSW in the east. The tensional stress appears to stay perpendicular to the curved graben systems extending from Gökova to Amorgos and from Sigacik to Mykonos. In the area stretching from the Cretan Sea to north of Karpathos the tensional stress axes apparently follow the curvature of the Hellenic arc, indicating a control of slab rollback over the stress field there. Much less dipping compressional stress axes such as for the South Karpathos and Zante cluster reflect the tectonic influence of the subducting Hellenic slab which exerts a subhorizontal stress on the Aegean plate close to the trench.

Azimuths of principal stress axes agree very well with azimuths of principal axes of a continuous strain rate field derived by interpolation from observed GPS velocities, indicating a dominating influence of regional stress on microseismic activity. Exceptions occur in the volcanic areas around Columbo, Astypalea and Nisyros–Kos where local stresses associated with volcanic activity appear to control seismicity.

Two major clusters of intermediate-depth events allowed a determination of the stress field in the subducting plate. Beneath the Dodecanese islands the slab is roughly under down-dip tension and slab-normal compression. But the tensional stress axis is rotated by about 35° to the west relative to the presumed subduction direction. This behaviour could be caused by the missing support of the western part of the slab due to segmentation or even vertical tearing along a N–S line west of Karpathos. Beneath the Cretan Sea the Hellenic slab is found to be under NW–SE compression which may be explained by a reduced curvature of the arc south of Crete owing to incipient collision of the Aegean plate with continental African lithosphere.

Acknowledgements. This study was supported by the German Research Foundation through the Collaborative Research Centre 526 "Rheology of the Earth – from the upper crust to the subduction zone". The EGELADOS experiment greatly benefited from seismic stations kindly provided by the Geophysical Instrument Pool Potsdam (GIPP). Waveform data of the CYCNET and EGELADOS experiments are openly available at the GEOFON data centre. An inspiring review by N. Houlie is acknowledged.

Special Issue: "Rheology of the Earth – observations, laboratory experiments and numerical modelling from the micro- to the macro-scale"
Edited by: H. Thybo and W. Friederich

References

Aktug, B., Nocquet, J. M., Cingöz, A., Parsons, B., Erkan, Y., England, P., Lenk, O., Gürdal, M. A., Kilicoglu, A., Akdeniz, H., and Tekgül, A.: Deformation of Western Turkey from a combination of permanent and campaign GPS data: Limits to block-like behaviour, J. Geophys. Res., 114, B10404, doi:10.1029/2008JB006000, 2009.

Angelier, J.: Tectonic evolution of the Hellenic arc since the late Miocene, Tectonophysics, 49, 23–36, 1978.

Becker, D., Meier, T., Rische, M., Bohnhoff, M., and Harjes, H.-P.: Spatio-temporal microseismicity clustering in the Cretan region, Tectonophysics, 423, 3–16, 2006.

Becker, D., Meier, T., Bohnhoff, M., and Harjes, H.-P.: Seismicity at the convergent plate boundary offshore Crete, Greece, observed by an amphibian network, J. Seismol., 14, 369–392, 2009.

Benetatos, C., Kiratzi, A., Papazachos, C., and Karakaisis, G.: Focal mechanisms of shallow and intermediate depth earthquakes along the Hellenic Arc, J. Geodyn., 37, 253–296, 2004.

Bohnhoff, M., Markis, J., Papanikolaou, D., and Stavrakakis, G.: Crustal investigations of the Hellenic subduction zone using wide aperture seismic data, Tectonophysics, 343, 239–262, 2001.

Bohnhoff, M., Rische, M., Meier, T., Endrun, B., Harjes, H.-P., and Stavrakakis, G.: CYC-NET: a temporary seismic network on the Cyclades (Aegean Sea, Greece), Seismol. Res. Lett., 75, 352–357, 2004.

Bohnhoff, M., Harjes, H.-P., and Meier, T.: Deformation and stress regimes in the Hellenic subduction zone from focal Mechanisms, J. Seismol., 9, 341–366, 2005.

Bohnhoff, M., Rische, M., Meier, T., Becker, D., Stavrakakis, G., and Harjes, H.-P.: Microseismic activity in the Hellenic Volcanic Arc, Greece, with emphasis on the seismotectonic setting of the Santorini-Amorgos zone, Tectonophysics, 423, 17–33, 2006.

Brüstle, A.: Seismicity of the eastern Hellenic Subduction Zone, Ph.D. thesis, Ruhr-University, Bochum, 2012.

Comninakis, P. and Papazachos, B.: Space and time distribution of the intermediate focal depth earthquakes in the Hellenic arc, Tectonophysics, 70, 35–47, 1980.

Dimitriadis, I., Panagiotopoulos, D., Papazachos, C., Hatzidimitrou, P., Karagianni, E., and Kane, I.: Recent seismic activity (1994–2002) of the Santorini volcano using data from local seismological network, Dev. Volcano., 7, 185–203, 2005.

Dimitriadis, I., Karagianni, E., Panagiotopoulos, D., Papazachos, C., Hatzdimitriou, P., Bohnhoff, M., Rische, M., and Meier, T.: Seismicity and active tectonics at Coloumbo Reef (Aegean Sea, Greece): monitoring an active volcano at Santorini Volcanic Center using a temporary seismic network, Tectonophysics, 465, 136–149, 2009.

Drakatos, G. and Latoussakis, J.: A catalog of aftershock sequences in Greece (1971–1997): their spatial and temporal characteristics, J. Seismol., 5, 137–145, 2001.

Efron, B. and Tibshirani, R.: Bootstrap methods for standard errors, confidence intervals, and other measures of statistical accuracy, Stat. Sci., 1, 54–75, 1986.

Frank, F.: Curvature of island arcs, Nature, 220, 363–363, 1968.

Friederich, W. and Dalkolmo, J.: Complete synthetic seismograms for a spherically symmetric earth by a numerical computation of the Green's function in the frequency domain, Geophys. J. Int., 122, 537–550, 1995.

Friederich, W. and Meier, T.: Temporary seismic broadband network acquired data on Hellenic subduction zone, EOS T. Am. Geophys. Un., 89, 378–378, 2008.

Floyd, M. A., Billiris, H., Paradissis, D., Veis, G., Avallone, A. Briole, P., McClusky, S., Nocquet, J.-M., Palamartchouk, K., Parsons, B., and England, P. C.: A new velocity field for Greece: Implications for the kinematics and dynamics of the Aegean, J. Geophys. Res. ,115, B10403, doi:10.1029/2009JB007040, 2009.

Fujita, K. and Kanamori, H.: Double seismic zones and stresses of intermediate depth earthquakes, Geophys. J. Roy. Astr. S., 66, 131–156, 1981.

Gephart, J. and Forsyth, D.: Improved method for determining the regional stress tensor using earthquake focal mechanism data: application to the San Fernando earthquake sequence, J. Geophys. Res., 89, 9305–9320, 1984.

Hardebeck, J. and Michael, A. L.: Stress orientations at intermediate angles to the San Andreas Fault, California, J. Geophys. Res., 109, B11303, doi:10.1029/2004JB003239, 2004.

Hardebeck, J. and Shearer, P.: A new method for determining first-motion focal mechanisms, B. Seismol. Soc. Am., 92, 2264–2276, 2002.

Haslinger, F.: Velocity structure, seismicity and seismotectonics of northwestern Greece between the Gulf of Arta and Zakynthos, Ph.D. thesis, Eidgenössische Technische Hochschule, Zürich, No. 12966, 1998.

Hatzfeld, D., Besnard, M., Makropoulos, K., and Hatzidimitriou, P.: Microearthquake seismicity and fault-plane solutions in the southern Aegean ant its geodynamic implications, Geophys. J. Int., 115, 799–818, 1993.

Hollenstein, C., Geiger, A., Kahle, H. G., and Veis, G.: CGPS time-series and trajectories of crustal motion along the West Hellenic Arc, Geophys. J. Int., 164, 182–191, 2006.

Hollenstein, C., Müller, M. D., Geiger, A., and Kahle, H. G.: Crustal motion and deformation in Greece from a decade of GPS measurements, 1993-2003, Tectonophysics, 449, 17–40, 2008.

Jost, M., Knabenbauer, O., Cheng, J., and Harjes, H.-P.: Fault plane solutions of microearthquakes and small events in the Hellenic arc, Tectonophysics, 356, 87–114, 2002.

Kahle, H.-G., Concord, M., Peter, Y., Geiger, A., Reilinger, R., Barka, A., and Veis, G.: GPS derived strain rate field within the boundary zones of the Eurasian, African, and Arabian plates, Geophys. J. Int., 105, 23353–23370, 2000.

Kiratzi, A. and Papazachos, C. B.: Active deformation of the shallow part of the subducting lithospheric slab in the southern Aegean, J. Geodyn., 19, 65–78, 1995.

Kiratzi, A., Benetatos, C., and Roumelioti, Z.: Distributed earthquake focal mechanisms in the Aegean Sea, Bull. Geol. Soc. Greece, 40, 1125–1137, 2007.

Kissling, E., Ellsworth, W. L. Eberhart-Phillips, D., and Kradolfer, U.: Initial reference models in local earthquake tomography, J. Geophys. Res., 99, 19635–19646, 1994.

Lomax, A. and Curtis, A.: Fast, probabilistic earthquake location in 3D models using Oct-Tree Importance sampling, Geophys. Res. Abstr., 3, 2001.

Lomax, A., Michelini, A., and Curtis, A.: Earthquake location, direct, global-search methods, in: Encyclopedia of Complexity and System Science, Part 5, edited by: Meyers, R., Springer, New York, 2449–2473, 2009.

McClusky, S., Balassanian, S., Barka, A., Demir, C., Ergintav, S., Georgiev, I., Gurkan, O., Hamburger, M., Hurs, K., Kahle, H., Kastens, K., Kekelidze, G., King, R., Kotzev, V., Lenk, O., Mahmoud, S., Mishin, A., Nadariya, M., Ouzounis, A., Paradissis, D., Peter, Y., Prilepin, M., Reilinger, R., Sanli, I., Seeger, H., Tealeb, A., Toksoez, M., and Veis, G.: Global Positioning System constraints on plate kinematics and dynamics in the eastern Mediterranean and Caucasus, J. Geophys. Res., 105, 5695–5719, 2000.

McKenzie, D.: The relation between fault plane solutions for earthquakes and the directions of the principal stresses, B. Seismol. Soc. Am., 59, 591–601, 1969.

Meier, T., Dietrich, K., Stöckert, B., and Harjes, H.-P.: One-dimensional models of shear-wave velocity for eastern Mediterranean obtained from the inversion of Rayleigh wave phase velocities and tectonic implications, Geophys. J. Int., 156, 45–58, 2004.

Michael, A.: Determination of stress from slip data: faults and folds, J. Geophys. Res., 89, 11517–11526, 1984.

Michael, A. J.: Use of focal mechanisms to determine stress: a control study, J. Geophys. Res., 92, 357–368, 1987.

Papadimitriou, P., Kaviris, G., Karakonstantis, A., and Makropoulos, K.: The Cornet seismological network: 10 years of operation,recorded seismicity and significant applications, Hellenic J. Geosci., 45, 193–208, 2010.

Papazachos, B.: Seismicity of the Aegean and surrounding area, Tectonophysics, 178, 287–308, 1990.

Papazachos, B. and Delibasis, N.: Tectonic stress field and seismic faulting in the area of Greece, Tectonophysics, 7, 231–255, 1969.

Papazachos, B., Kiratzi, A., Hatzidimitriou, P., and Rocca, A.: Seismic faults in the Aegean area, Tectonophysics, 106, 71–85, 1984.

Papazachos, B., Kiratzi, A., and Papadimitriou, E.: Regional focal mechanisms for earthquakes in the Aegean area, Pageoph, 136, 405–420, 1991.

Papazachos, B., Comninakis, P., Scordilis, E., Karakaisis, G., and Papazachos, C.: A catalogue of earthquakes in the Mediterranean and surrounding area for the period 1901 – Sept 2009, Publ. Geophys. Laboratory, University of Thessaloniki, 2009.

Reilinger, R., McClusky, S., Vernant, P., Lawrence, S., Ergintav, S., Cakmak, R., Ozener, H., Kadirov, F., Guliev, I., Stepanyan, R., Nadyriya, M., Hahubia, G., Mahmoud, S., Sakr, K., ArRajehi, A., Paradissis, D., Al-Aydrus, A., Prilepin, M., Guseva, T., Evren, E., Dmitrotsa, A. Filikov, S., Gomez, F., Al-Ghazzi, R., and Karam, G.: GPS constraints on continental deformation in the African-Arabian-Eurasia continental collision zone and implications for the dynamics of plate interactions, J. Geophys. Res., 111, B05411, doi:10.1029/2005JB004051, 2006.

Reilinger, R., McClusky, S., Paradissis, D., Ergintav, S., and Vernant, P.: Geodetic constraints on the tectonic evolution of the Aegean region and strain accumulation along the Hellenic subduction zone, Tectonophysics, 488, 22–30, 2010.

Rigo, A., Lyon-Caen, H., Armijo, R., Deschamps, A., Hatzfeld, D., Makropoulos, K., Papadimitriou, P., and Kassaras, I.: A microseismic study in the western part of the Gulf of Corinth (Greece): implications for large-scale normal faulting mechanisms, Geophys. J. Int., 126, 663–688, 1996.

Rontogianni, S., Konstantinou, K. I., Melis, N. S., and Evangelidis, C. P.: Slab stress field in the Hellenic subduction zone as inferred from intermediate-depth earthquakes, Earth Planets Space, 63, 139–144, 2011.

Sachpazi, M., Hirn, A., Nercessian, A., Avedik, F., Mc Bride, J., Loucoyannakis, M., Nicolich, R., and the STREAMERS-PROFILES group: A first coincident normal-incidence and wide-angle approach to studying the extending Aegean crust, Tectonophysics, 270, 301–312, 1997.

Tape, C., Musé, P., Simons, M., Dong, D., and Webb, F.: Multiscale estimation of GPS velocity fields, Geophys. J. Int., 179, 945–971, 2009.

Taymaz, T., Jackson, J., and McKenzie, D.: Active tectonics of the north and central Aegean Sea, Geophys. J. Int., 106, 433–490, 1991.

Wells, D. L., and Coppersmith, K. J.: New empirical relationships among magnitude, rupture length, rupture width, rupture area, and surface displacement, Bull. Seis. Soc. Am., 84, 974–1002, 1994.

Appendix A

Stress solutions for further cluster regions

A1 Kos and Astypalea

Similar results as for the cluster at Columbo volcano were obtained for the shallow clusters near Kos and Astypalea (Figs. A1 and A2). The compressional stress axis dips steeply and the distribution of the bootstrap solutions is well focused. With respect to the other two axes, the bootstrap solutions exhibit a wide spread. In addition, and contrary to the Columbo case, there is a spread of R values as well. The best fitting solutions yield $R = 0$. With the same restrictions as for the Columbo case, one could speculate here as well about a magmatic contribution to the earthquake activity because the Kos–Astypalea region is located right above the intermediate-depth seismicity associated with the eastern part of the Hellenic slab.

A2 Iraklion and Kamilonisi basin

The stress solutions for these two regions (Figs. A3 and A4) show dominating subhorizontal tensional principal stress and again nearly subvertical compressional stress. The stress ratio is around 0.7 to 0.8. For the Iraklion basin, all stress axes are fairly well determined but R values show a large spread with the majority of values greater than 0.5. The tensional stress axis is oriented W–E indicating horizontal extension and again a normal faulting regime. Planes of maximum shear stress strike N–S and are either near-vertical or near-horizontal. For the Kamilonisi region, only the tensional axis is well determined. Its direction is rotated counterclockwise by about 20° to WSW–ENE. The strike of planes of maximum shear stress is rotated in the same way and the dip is around 45°. Interpretation of these results should be done with caution because of the small number of available focal mechanisms.

A3 Sigacik basin

The stress field in the Sigacik basin exhibits nearly degenerate intermediate and compressional principal stresses (Fig. A5) and a dominating and very clearly constrained horizontally oriented tensional stress axis striking NNE–SSW. This is the same direction that was found for the tensional axis in the Gökova graben. Planes of maximum shear stress have intermediate dip and are oriented about WNW–ESE. The reason for the tensional stress could again be the northward motion of western Turkey relative to the central Aegean as observed by recent GPS measurements (Reilinger et al., 2010).

A4 Andros–Mykonos

A cluster of shallow earthquakes was observed along the islands of Andros, Tinos and Mykonos characterized mainly by normal faulting events. The stress determination works extraordinarily well with an average misfit of only 7.4° (Fig. A6). All three principal stress axes have small 80 % confidence regions. The stress ratio is also well constrained around values of 0.3. Thus, a meaningful tensional stress axis can also be defined pointing about NW–SE. The compressional axis dips very steeply at 75°. Planes of maximum shear stress strike around SW–NE with intermediate to strong dip. The stress solution indicates a tensional tectonic stress in that part of the Aegean microplate in the NW–SE direction.

A5 Rhodos

For the intermediate-depth events of the Rhodos cluster (Fig. A7), the stress field is less well constrained as the angular misfit is rather large (28.6°) and the distribution of R values of the bootstrap solutions is very broad. In addition, only the tensional axis appears to be well constrained at a dip of 75°. This value is much greater than the gentle dip of the slab in the Rhodos region as indicated by depth profiles of micro-seismicity (Brüstle, 2012). A similar observation was made by Kiratzi and Papazachos (1995) who analysed focal mechanisms of large intermediate-depth earthquakes along the Hellenic arc. The compressional and intermediate principal stresses are close to degenerate ($R = 0.8$) and exhibit a very small dip. With an azimuth of 42° the intermediate stress axis which is also compressional points approximately arc-parallel and thus lies in the slab plane.

A6 Cyclades

The stress solution for the intermediate-depth Cyclades cluster (Fig. A8) even more suggests that the stress homogeneity condition is violated. The area covered by the selected earthquakes is too large and stress in the subducting slab in this region is potentially heterogeneous. The angular misfit is rather large in spite of the small number of events, neither axis is well constrained and the distribution of stress ratio values is very wide. We therefore refrain from further analysis and interpretation of the stress solution for this cluster. More fault plane solutions are needed to allow a further subdivision of this region.

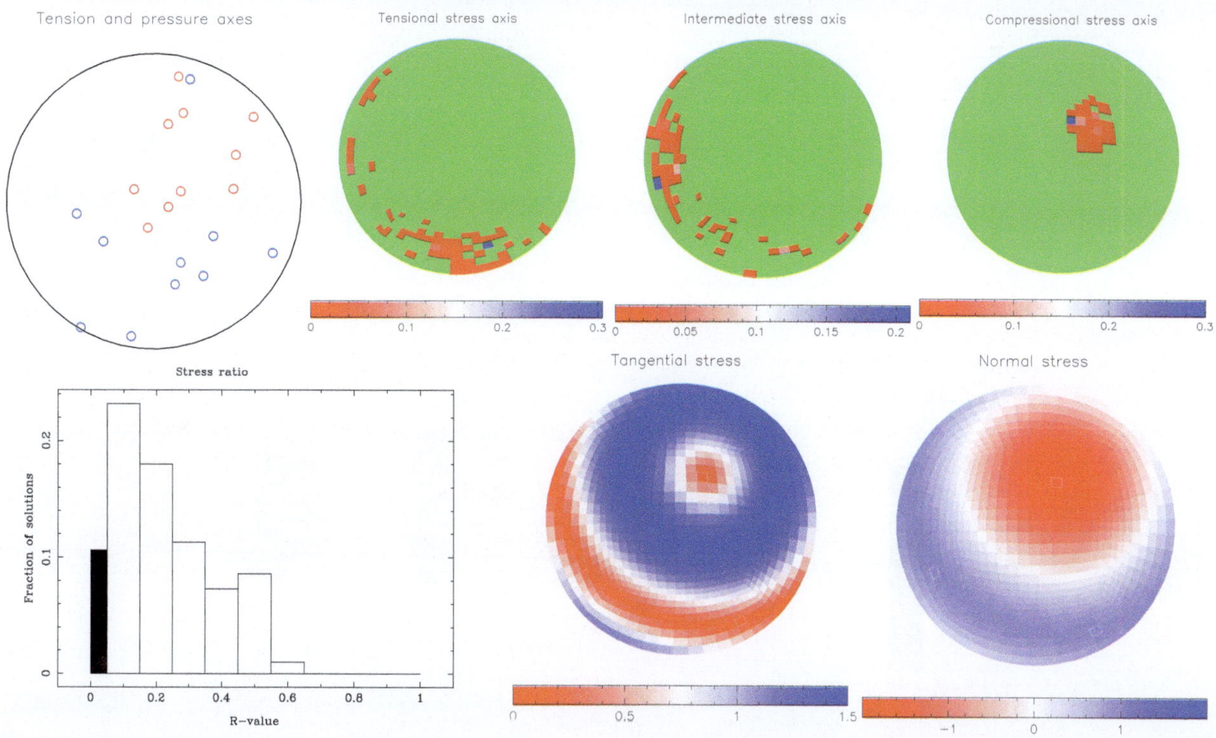

Fig. A1. Results of stress inversion for cluster of shallow earthquakes near Kos. Explanation of subfigures is as in Fig. 5.

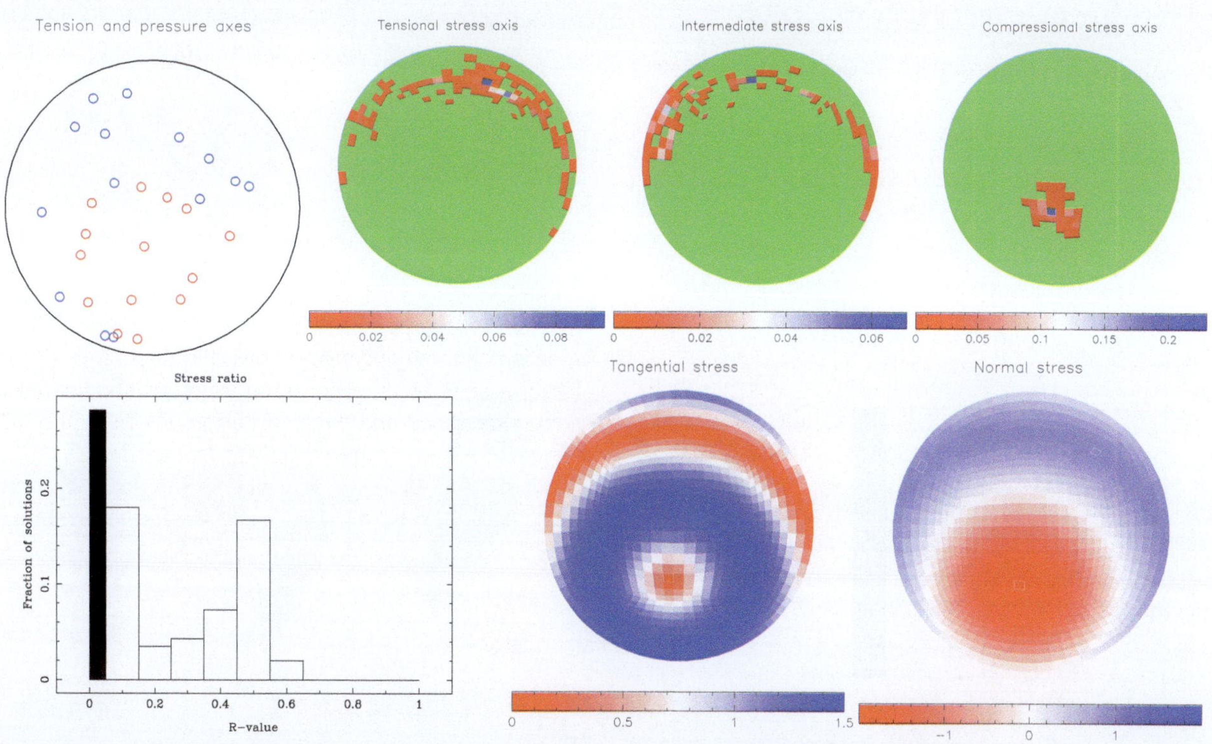

Fig. A2. Results of stress inversion for cluster of shallow earthquakes near Astypalea. Explanation of subfigures is as in Fig. 5.

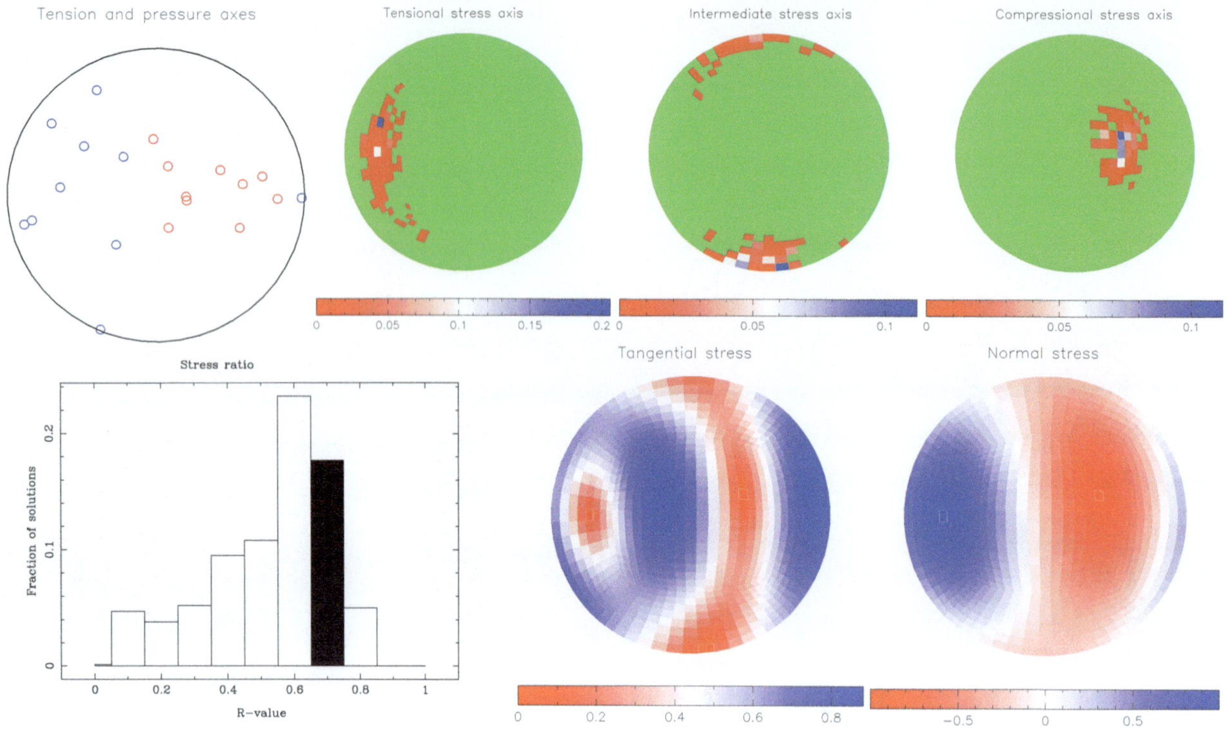

Fig. A3. Results of stress inversion for cluster of shallow earthquakes in the Iraklion basin. Explanation of subfigures is as in Fig. 5.

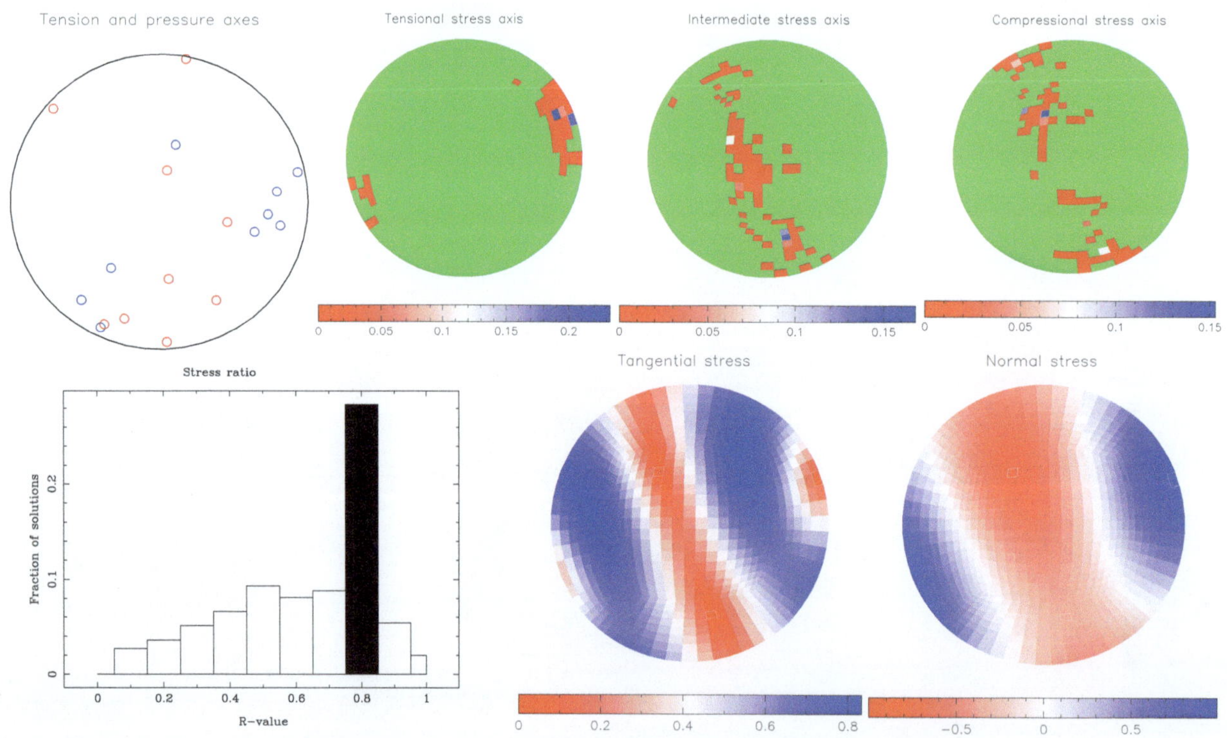

Fig. A4. Results of stress inversion for cluster of shallow earthquakes in the Kamilonisi basin. Explanation of subfigures is as in Fig. 5.

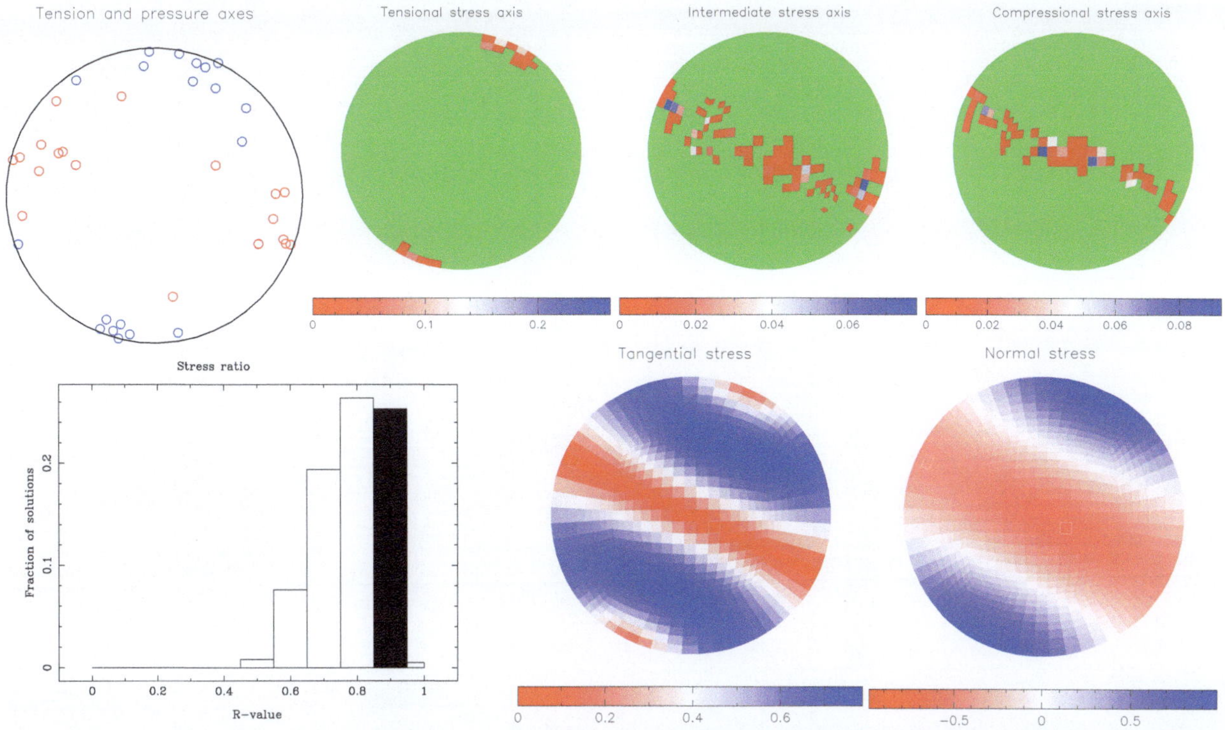

Fig. A5. Results of stress inversion for cluster of shallow earthquakes near Sigacik. Explanation of subfigures is as in Fig. 5.

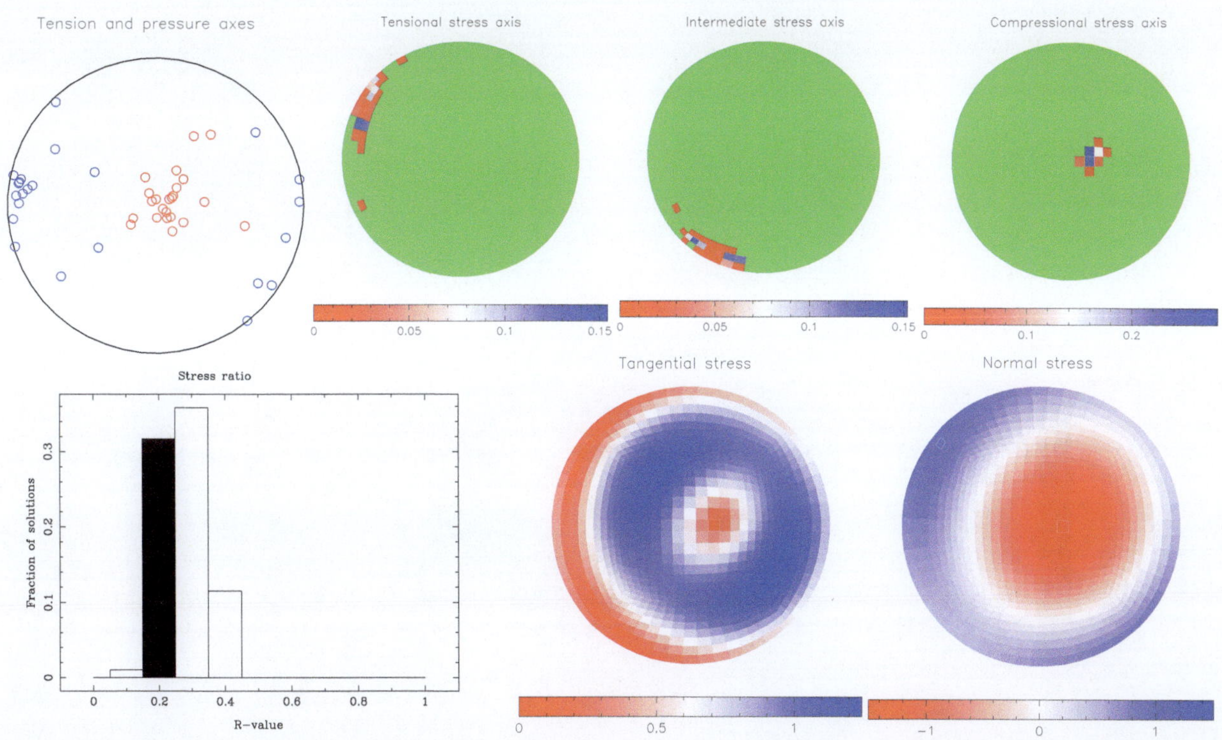

Fig. A6. Results of stress inversion for cluster of shallow earthquakes near Mykonos. Explanation of subfigures is as in Fig. 5.

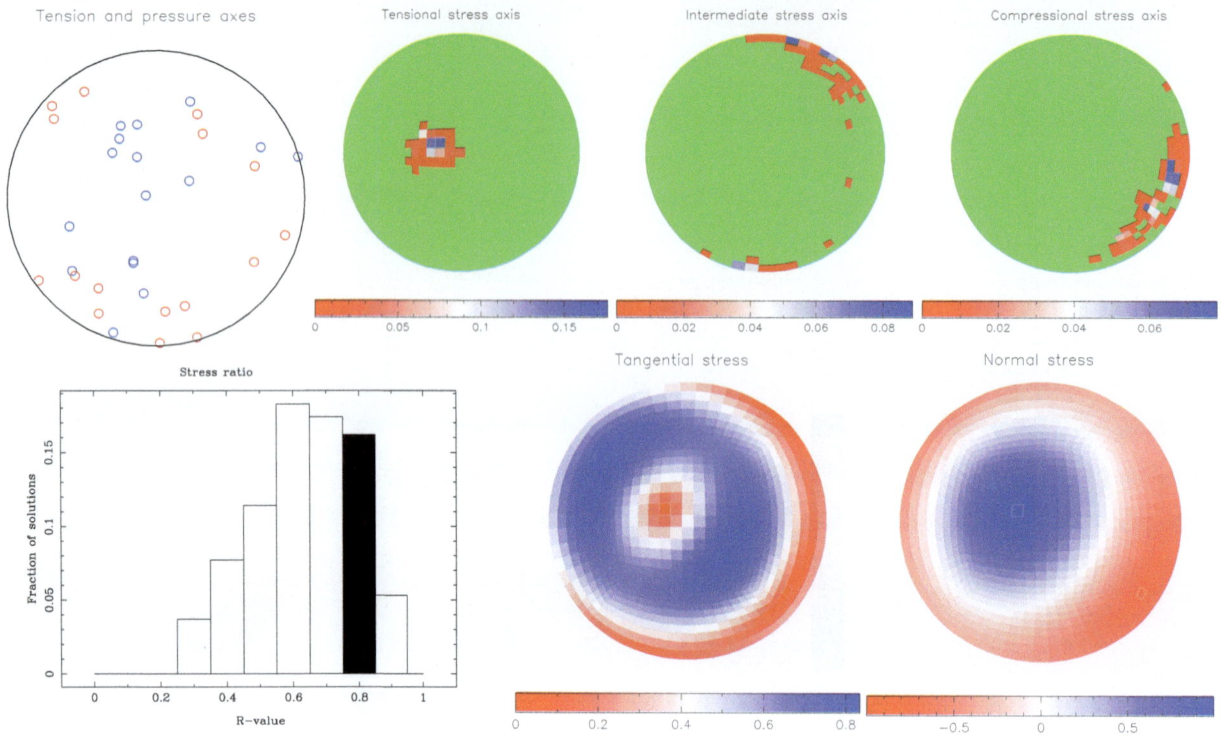

Fig. A7. Results of stress inversion for cluster of intermediate-depth earthquakes near Rhodos. Explanation of subfigures is as in Fig. 5.

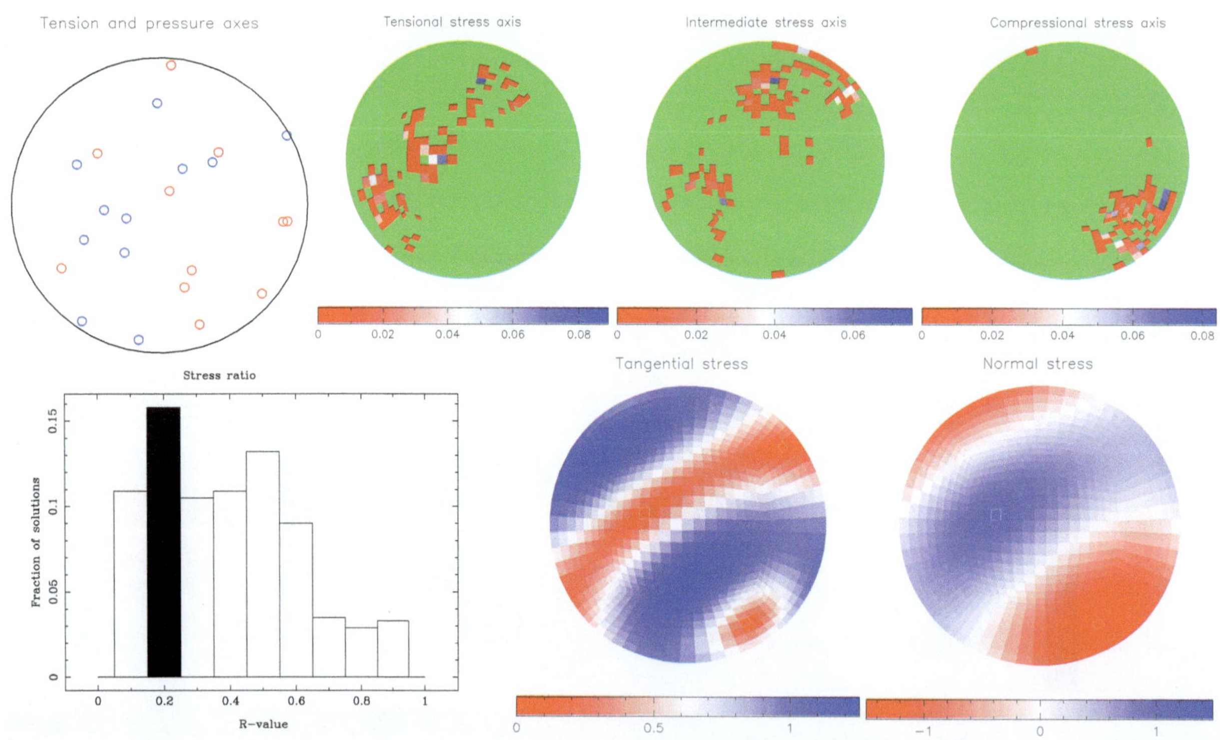

Fig. A8. Results of stress inversion for cluster of intermediate-depth earthquakes beneath the Cyclades. Explanation of subfigures is as in Fig. 5.

13

Seismic structure of the lithosphere and upper mantle beneath the ocean islands near mid-oceanic ridges

C. Haldar[1], **P. Kumar**[1], **and M. Ravi Kumar**[1]

[1]National Geophysical Research Institute (CSIR), Uppal Road, Hyderabad-500007, India

Correspondence to: P. Kumar (prakashk@ngri.res.in)

Abstract. Deciphering the seismic character of the young lithosphere near mid-oceanic ridges (MORs) is a challenging endeavor. In this study, we determine the seismic structure of the oceanic plate near the MORs using the P-to-S conversions isolated from quality data recorded at five broadband seismological stations situated on ocean islands in their vicinity. Estimates of the crustal and lithospheric thickness values from waveform inversion of the P-receiver function stacks at individual stations reveal that the Moho depth varies between $\sim 10 \pm 1$ km and $\sim 20 \pm 1$ km with the depths of the lithosphere–asthenosphere boundary (LAB) varying between $\sim 40 \pm 4$ and $\sim 65 \pm 7$ km. We found evidence for an additional low-velocity layer below the expected LAB depths at stations on Ascension, São Jorge and Easter islands. The layer probably relates to the presence of a hot spot corresponding to a magma chamber. Further, thinning of the upper mantle transition zone suggests a hotter mantle transition zone due to the possible presence of plumes in the mantle beneath the stations.

1 Introduction

Mid-oceanic ridges (MORs) are the largest sources of magma on the earth. Being the locus of new lithosphere generation and accretion to the existing ones, their study assumes importance in understanding the dynamics of plate tectonics. These linear features on the ocean floor control the rheology of oceanic lithosphere, ridge topography, the style of oceanic crustal accretion and also affect the earth's deeper discontinuities. Created at the mid-oceanic ridges, new oceanic lithosphere undergoes cooling and thickening as it subsequently spreads away and again plunges downward into the mantle along the trenches by the process of subduction. In a broader perspective, the thermal contraction of lithosphere created at the ridge also defines the shape of the mid-oceanic ridge (Parsons and Sclater, 1977). The lithosphere is the rigid outer shell, made up of the crust and the uppermost part of the mantle that floats over a lower-viscosity asthenosphere that is seismically characterized by a low-velocity zone. The first seismic observation of the asthenospheric low-velocity layer was reported by Gutenberg, at a depth range of ~ 50–200 km (Gutenberg, 1959). Therefore, any velocity reduction in the uppermost mantle, similar to the one at the lithosphere–asthenosphere boundary (LAB), has been sometimes referred to as the Gutenberg (G) discontinuity (Revenaugh and Jordan, 1991), which may define the top of the low-velocity layer.

Most of the seismological knowledge of the oceanic plates, particularly near the ridges, comes from high-resolution 3-D seismic tomography studies using active source data (Evangelidis et al., 2004). The results reveal fine layering within the crust in the volcanic edifices of Ascension. Seismic reflection/refraction studies have also been carried out near the ridges (Bjarnson and Menke, 1993; Evangelidis et al., 2004; Rodgers and Harben, 1999) in and around the Atlantic. All these studies reveal a highly variable crustal structure. Surface-wave dispersion studies first elucidated the nature of the oceanic plate and LAB (Kanamori and Press, 1970; Leeds, 1975; Zhang and Tanimoto, 1993; Nettles and Dziewonski, 2008). However, the depth resolution of surface-wave and body-wave tomography studies is limited to >40 km. Shear-wave anisotropy studies suggest horizontally connected melts that characterize the LAB near the MOR and cause radial anisotropy at these depths (Nowacki et al., 2012). Attempts to understand the structure of the oceanic plates have been made through a number of seismological studies at different locations beneath the Pacific Ocean (Gaherty, 1999; Tan and Helmberger, 2007; Kawakatsu, 2009; Kumar and Kawakatsu, 2011; Kumar et al,. 2012; Rychert and Shearer, 2011; Schmerr, 2012, etc.). Logistic problems and the extreme costs compared to land surveys impose severe constraints on obtaining similar information about the nature of other MORs. Another factor which hampers body-wave observations from the oceanic data is the water reverberations that mostly contaminate the vertical components of the seismograms of the ocean bottom seismometers (Kumar et al., 2012). In this study, we attempt to investigate the seismic structure using the available data from five seismological stations situated on the islands located close to the mid-oceanic ridges. Although the crustal structure may not truly represent the nature of the oceanic plate due to possible influence of the islands, the deeper structure is devoid of such effects.

2 Data and methods

We used all the available teleseismic data recorded at stations located on five oceanic islands near MORs. The locations of the stations are shown in Fig. 1, together with the major plate boundaries and mid-oceanic ridges. In the present study, we use converted wave techniques, namely P-receiver functions (PRFs) (Burdick and Langston, 1977; Langston, 1977; Vinnik, 1977). Receiver functions are time series, computed from three component seismograms, which show the relative response of the earth structure near the receiver. The events with magnitude $\geq$5.5 Mb in the epicentral distance range of 30–90° are selected from all available back azimuths. We visually picked the waveforms having clear P-waves in the vertical component for further analysis. First, the Z, N and E components are rotated into Z (vertical), R (radial), and T (transverse) system using back azimuths derived from the waveforms. We further rotate the ZRT components into the LQT system using the theoretical angle of incidence determined by the global velocity model IASP91 and also with the angle of maximum polarization direction of P-wave. We chose only those events whose differences in theoretical and waveform back-azimuth and incidence angle are less than 30° and 5°, respectively. Further, we derive the P-receiver functions in time domain by deconvolving the radial components by their respective vertical ones. The deconvolution makes the equalization of source and propagation path effects. All of the above steps involved in computation of P-receiver functions are performed using the codes written in Seismic Handler (http://www.seismic-handler.org, K. Stammler, 1992), a seismological analysis tool. The P-receiver functions thus obtained are then moveout corrected to a reference slowness of 6.4 s/° (Yuan, 1997) in order to remove the dependence of the travel times of the converted phases on the source receiver distance. In addition, we stack several P-receiver functions after moveout correction, to enhance the signal to noise ratio.

3 Modeling of receiver functions

At each of the stations, the stacked receiver function traces along with the 2 sigma error limits, computed utilizing the bootstrap resampling technique, are shown in Fig. 2. In this bootstrap analysis, a single receiver function stack is generated by summation of those traces indicated by a random number generator where the number of traces for summation remains the same as the total number of P-receiver functions at a given station. This procedure is repeated 500 times, by stacking a totally different sequence of traces corresponding to randomly generated trace numbers. The mean and standard deviations of the amplitudes from 500 realizations at each point on the time axis are then calculated. In Fig. 2, we marked the prominent phases, namely Moho, LAB and another low-velocity layer (L). All these phases have their amplitudes above the 2 sigma error limit. Although the positive conversions from the Moho and a negative one from the LAB are prominent in this figure, the multiples from either of these are not obvious. This could be because of to their annihilation due to interference from other phases.

In order to constrain the seismic structure of the lithosphere, the stacks of the moveout-corrected traces at each station are inverted using a linearized time domain iterative waveform inversion scheme (Ammon et al., 1990), where for each model, the synthetic seismograms for the slowness corresponding to the stacked trace are computed using the reflectivity method (Fuchs and Müller, 1971; Kind, 1978). We allowed the generation of all primary and multiple phases. The synthetic traces are rotated and deconvolved in the same manner as the observed traces, for a meaningful comparison.

The receiver function results are summarized in Table 1 and the stacked traces observed are shown in Fig. 2. Since the islands are volcanic in nature, the velocities and crustal thickness values from other studies are used as constraints in the inversion. While modeling the P-receiver functions at various stations, the velocity of the basaltic layer was adopted from the available published values (Rodgers and Harben,1999; Christensen, 1973). Also, the velocities in the crustal part are kept close to the estimates obtained regionally and globally from active seismic experiments (Shinohara, 2008; White, 1992), except for the station ROSA, where the data required a top layer to have very low shear velocity, probably due to the presence of sediments near the surface.

The piercing points of the P-to-S conversions at a depth of 100 km, plotted in the inset maps for each station in Fig. 1, reveal a maximum offset of 0.5°. Since we model the stacked traces, any lateral heterogeneity within 0.5° is averaged. Figure 3 shows the corresponding L-components of our stack traces, which essentially contain only the P-waves as first arrivals. Figure 4 shows the velocity models obtained at each

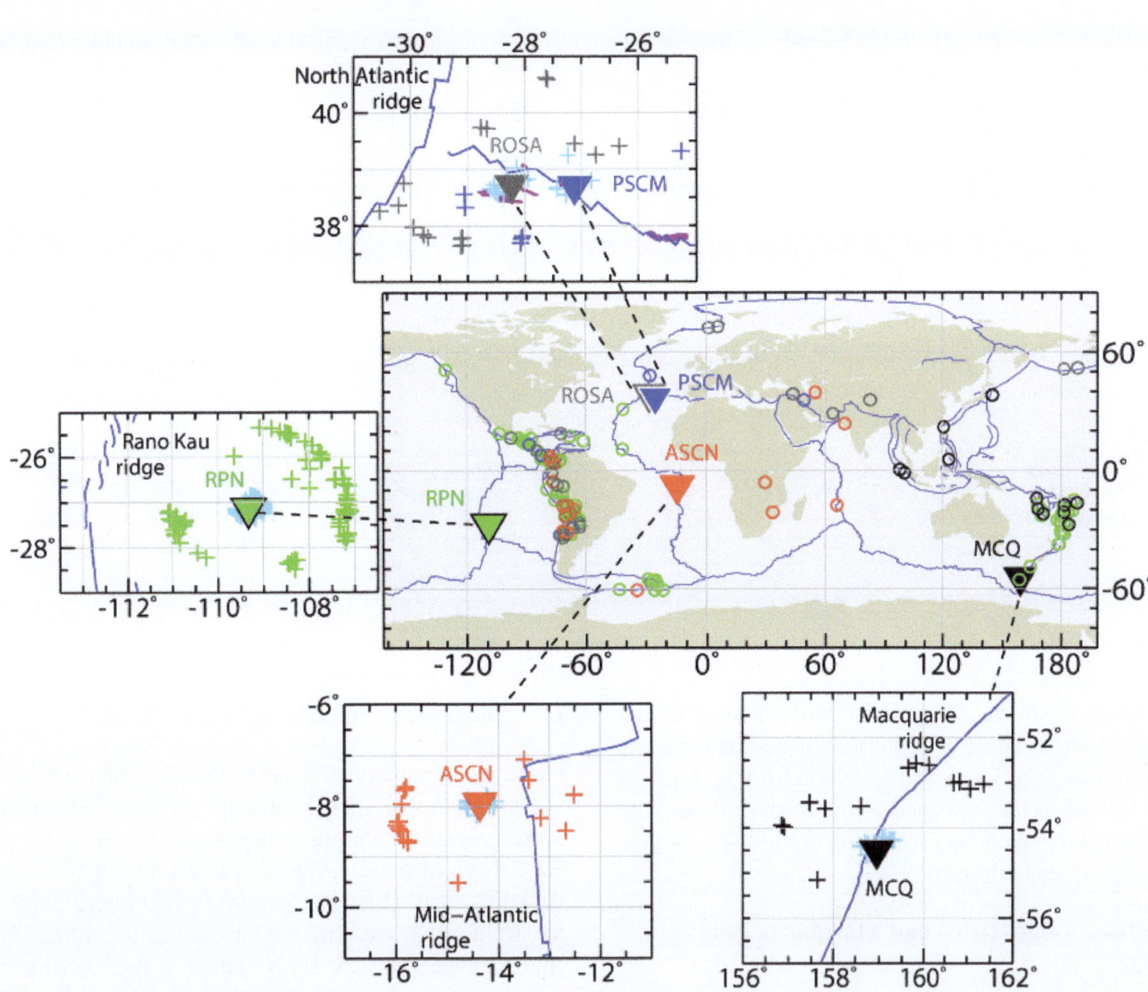

Figure 1. Location of the seismic stations (inverted triangles) used in the present study. Mid-oceanic ridge (MOR) boundaries are also shown (blue lines) with their names in the inset maps. The color code of the stations corresponds to the colors of the lines in Figs. 2 and 4. Open circles indicate the geographical distribution of the events corresponding to the stations of the same color. Insets corresponding to each station show their relative position with respect to the mid-oceanic ridges. The crosses in the insets are the P-to-S conversion points corresponding to the depths at 100 km (cyan, nearest to the station) and at 520 km (same color as the corresponding station, farther away).

station from inversion, together with the waveform matching of the observed and synthetic traces. Apart from the Moho and LAB, the models reveal additional low-velocity layers (L) in the sub-lithospheric mantle below the three stations RPN, ROSA and ASCN.

4 Results and discussions

4.1 Mid-Atlantic Ridge (PSCM, ROSA and ASCN)

Our modeling results indicate that the crust–mantle boundary is located at depths of $\sim 12 \pm 1$, 16 ± 1 and 10 ± 1 km for ROSA, PSCM and ASCN respectively. Both the stations PSCM (in Terceira) and ROSA (in São Jorge), in the middle of the North Atlantic Ocean (see Fig. 1), are volcanic in origin and about 10 km away from the MOR, making them some of the closest landmasses to an active, oceanic spreading center. ASCN is located in the South Atlantic Ocean on Ascension, a volcanic island located 100 km west of the MOR.

Near the station ASCN, wide-angle reflection seismic tomography studies indicate the Moho at a depth of ~ 10.2 km (Evangelidis et al., 2004). The depth to the LAB (see Table 1) beneath the stations PSCM, ROSA and ASCN is 65 ± 4, 40 ± 7 and 50 ± 4 km, respectively. Here we define the LAB as the depth corresponding to the middle part of the velocity decrease below the Moho (see Fig. 4). The result for the station ASCN is consistent with observations by Li, et al. (2003). The depths of the LAB beneath stations ROSA ($\sim 40 \pm 4$ km) and PSCM ($\sim 65 \pm 7$ km) are quite different, although they are separated by only ~ 100 km. The older crust (19 My) at PSCM shows a larger plate thickness compared to a younger crust (13 My) at ROSA (http://www.ngdc.noaa.gov/mgg/ocean_age/data/2008/grids/age/). The inset of Fig. 1 shows that station PSCM is located on Terceira, which

Table 1. The Moho, LAB and LVL depths beneath the stations used in this study, together with the *P*-to-*S* conversion times from the upper mantle discontinuities.

Station	Region	No. of traces	Lat	Lon	Period	Elv (m)	Moho (km)	LAB (km)	L (km)	410 (s)	660 (s)	Age (Ma)
ASCN	Ascension, Atlantic Ocean	20	−7.93	−14.36	1994–2010	173	10±1	50±4	75±4	46.0 ±0.2	68.4 ±0.3	5
MCQ1	Macquarie Island, Pacific Ocean	7	−54.49	158.95	2005–2012	14	19±1	42±2	–	48.6 ±0.3	70.6 ±0.3	3
MCQ2		8								46.8 ±0.2	–	
PSCM	Terceira, Atlantic Ocean	7	38.70	−27.11	2000–2002	400	16±1	65±7	–	47.8 ±0.3	71.6 ±0.2	19
ROSA	São Jorge, Atlantic Ocean	18	38.72	−28.24	2008–2012	310	12±1	40±4	106±4	46.8 ±0.3	70.6 ±0.3	13
RPN1	Easter Island, Pacific Ocean	29	−27.12	−109.33	1988–2012	110	12±1	46±2	75±2	45.2 ±0.3	–	6
RPN2		58								46.8 ±0.3	69.4 ±0.3	

± indicates the ±2 SE in these values.

is much further from the main north Mid-Atlantic Ridge compared to ROSA. These two stations are situated on a complex triple junction. The possible reason for this discrepancy might be due to the influence of the compositional nature of the oceanic islands and structural complexity as evident from the surface bathymetry.

4.2 Rano Kau Ridge (RPN) and Macquarie Ridge (MCQ)

Station RPN is situated on Easter Island ~350 km east from the Rano Kau Ridge in the Pacific, while station MCQ is located on Macquarie Island, which lies in the middle of the Macquarie Ridge (Duncan et al., 1998; Tapley et al., 2004). Both these volcanic islands are dominated by the presence of basalt and gabbro (Duncan and Varne, 1988; Jorg, 2005). The results of receiver function modeling (Fig. 5) suggest a crustal thickness of ~12±1 km and a plate thickness of 46±2 km at station RPN. On the other hand, the crust is much thicker (~19±1 km) below the station MCQ and the observation of LAB is dubious. The negative phases seen close to 6 s and 10 s are barely above the error limit (Fig. 2). Station MCQ is directly located on the MOR, but the retrieved Moho depth is larger than expected. This might be due to the fact that the observations are more representative of the islands. The LAB depth found below RPN is consistent with the results obtained earlier from *S* receiver functions (Heit et al., 2007; Li, et al., 2003). This study suggests that the numerical value corresponding to the negative discontinuity below RPN might correspond to an oceanic LAB.

4.3 Shear-wave velocity drop across LAB

There are a number of models that describe the generation and migration of melt in a ridge system (Spiegelman and McKenzie, 1987; Phipps-Morgan, 1987; Sparks and Parmentier, 1993; Garmany, 1989; Buck and Su, 1989; Kelemem et al., 1997, etc.). However, geophysical studies, which help in constraining the existing models, are quite limited. The most important geophysical study aimed at deciphering the melt generation was the MELT experiment (Forsyth, 1992). Results from inversion of Love wave dispersion data from the East Pacific Rise revealed an asymmetric velocity structure across the ridge (Dunn and Forsyth, 2003), where a crustal magma chamber is confined.

Our modeling results suggest variable shear-wave velocity contrasts across the LAB for the five stations on the oceanic islands close to the mid-oceanic ridges. The inversion results reveal that the first-order discontinuities are not characterized by an abrupt change in shear velocity. The need for incorporating a gradational nature of the velocity interfaces in order to fit the observed and computed waveforms is due to lack of multiples. The velocity contrast across the LAB varies from ~10±4 % to ~12±3 %, except for station ROSA, where an abnormal reduction of 22 % is observed. This value may not be so reliable in view of the large error associated with the value (22±9 %). Hence, in the present discussion we will not take into account the velocity jump for the station ROSA. Here, we estimated the velocity contrast by taking the average lid velocity and minimum asthenospheric shear velocity. Also, the drop across LAB is sharp in nature, implying the presence of partial melt in the asthenosphere (Holtzman and Kohlsedt, 2007; Mierdel et al., 2007; Kawakatsu et al., 2009). The alteration of seismic velocities or attenuation is often used to interpret the amount and spatial disposition

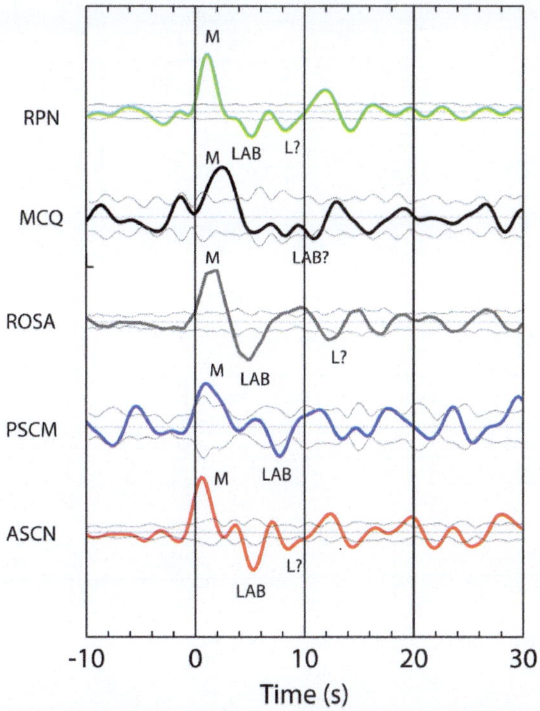

Figure 2. Observed P-receiver function stacks (solid lines) at the five stations. Before stacking, the receiver functions are low-pass filtered with a corner frequency of 3 s and moveout corrected to a reference slowness of 6.4 s/° using the IASP91 Earth model. The two gray lines on both sides of the mean lines represent the 2 σ standard error estimated using bootstrap resampling technique. The color codes correspond to different stations: Black, MCQ; Red, ASCN; Blue, PSCM; Green, RPN and Grey, ROSA. Different discontinuities are marked as M, Moho; LAB, lithosphere–asthenosphere boundary and L. sub-lithospheric low-velocity layer.

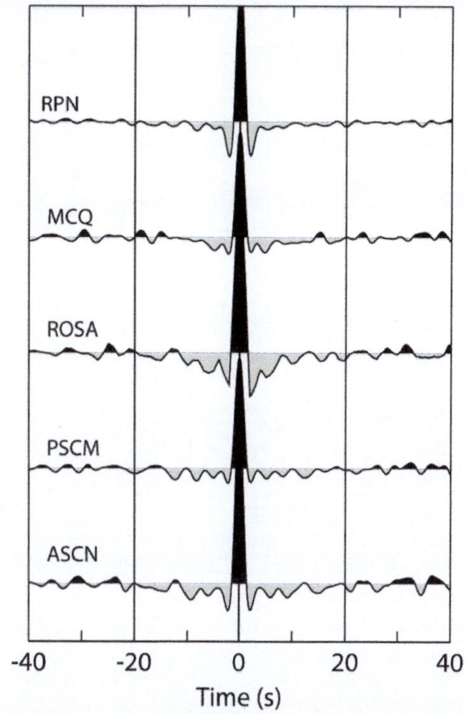

Figure 3. Stacked L-component receiver functions at different stations that essentially indicate the P-wave at zero time.

of partial melts in the uppermost mantle (i.e., Humphreys and Dueker, 1994; Zhao et al., 1992; Sobolev et al., 1996; Xu and Wiens, 1997; Dunn and Toomey, 1997; Kawakatsu et al., 2009). The sharp and abrupt drop in shear-wave velocity can be reconciled by the fact that the hot mantle upwelling near the ridges possibly enhances the reflectivity of the LAB (Schmerr, 2012). Moreover, the fraction of melt retention in the asthenosphere may cause such a large drop in shear velocity. Inclusion of little more than 1 % melt in the asthenosphere, in addition to the temperature contrast, provides a realistic explanation of this velocity drop (Hammond and Humphreys, 2000; Rychert et al., 2012). Karato (1990) indicated that the presence of H_2O in olivine weakens the material. Recently, Karato (2013) argued against the various partial melt models to explain the seismically observed anomalies and emphasized the important role played by the geometry of melt. However, the presence of hydrogen in the uppermost mantle may present an alternate model (Dai and Karato, 2009). Anelastic relaxation caused by the elastically accommodated grain-boundary sliding may also be a cause of large velocity drops of more than 5 % (Karato, 2012). In the normal oceanic (not ocean island) plate, there is an indication of an abrupt increase in electrical conductivity (Shankland and Waff, 1977; Honkura, 1975; Utada and Baba, 2013), often interpreted as the electrical asthenosphere, favoring the idea of partial melts. Interestingly this layer corresponds to the seismologically identified layer.

Figure 6 illustrates the variation of plate thickness with age. It is clearly seen that the seismologically estimated lithospheric thicknesses in oceanic environment generally show an asymptotic increase with age, in agreement with the thermally predicted models (e.g., Stein and Stein, 1992; Kawakatsu et al., 2009). The presence of a significant scatter in the data requires incorporation of additional controlling factors. The present values obtained from the receiver function analysis (Table 1) seem to fill the existing gap in the lower end of the age spectrum and are consistent with the already observed values using the receiver function analysis.

5 Sub-lithospheric low-velocity layer

The observed P-receiver function traces (Figs. 2, 4 and 5) show an additional negative phase for stations RPN, ROSA and ASCN. We interpret the shallower LVLs as conversions from the LAB, since they are consistent with previous receiver function studies (Li et al., 2003; Heit et al., 2007; Rychert and Shearer, 2009 etc.). These values are close to

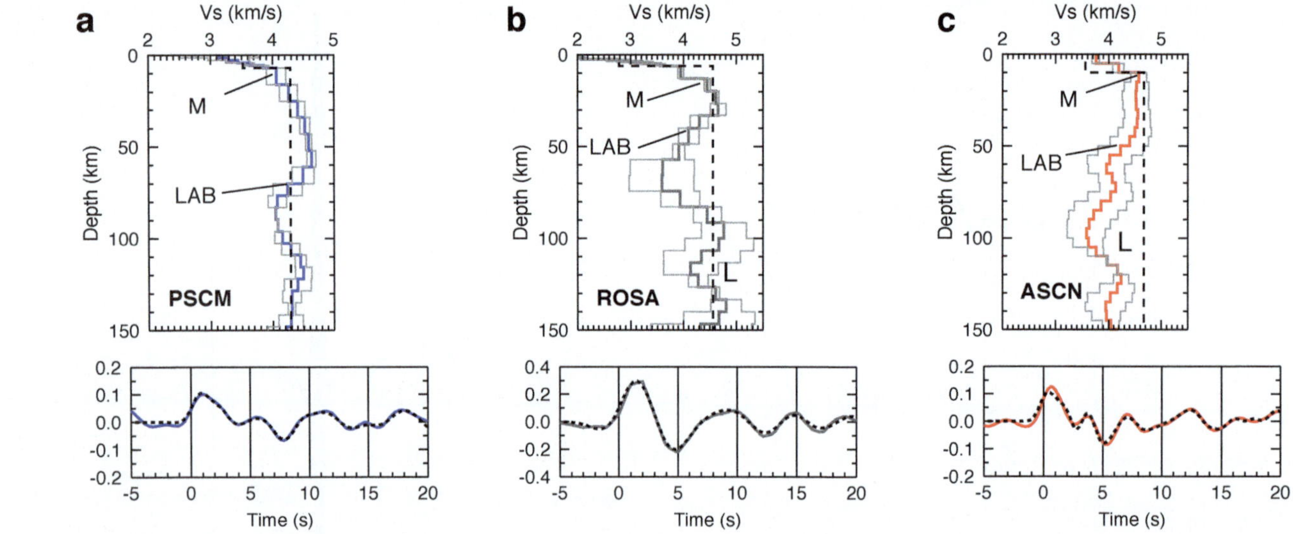

Figure 4. Models obtained by inversion of stacked traces in the bottom panel (colored as in Fig. 1) of each subplot. Synthetic traces (black dashed lines) computed from these models fit the observed traces well for the stations PSCM, ROSA and ASCN. The two gray lines that bound the thick colored lines model in the upper panel of each subplot are the ± 2 SE in the velocity models derived from inversion. The dashed black line shows the 1-D starting velocity model used to invert the data.

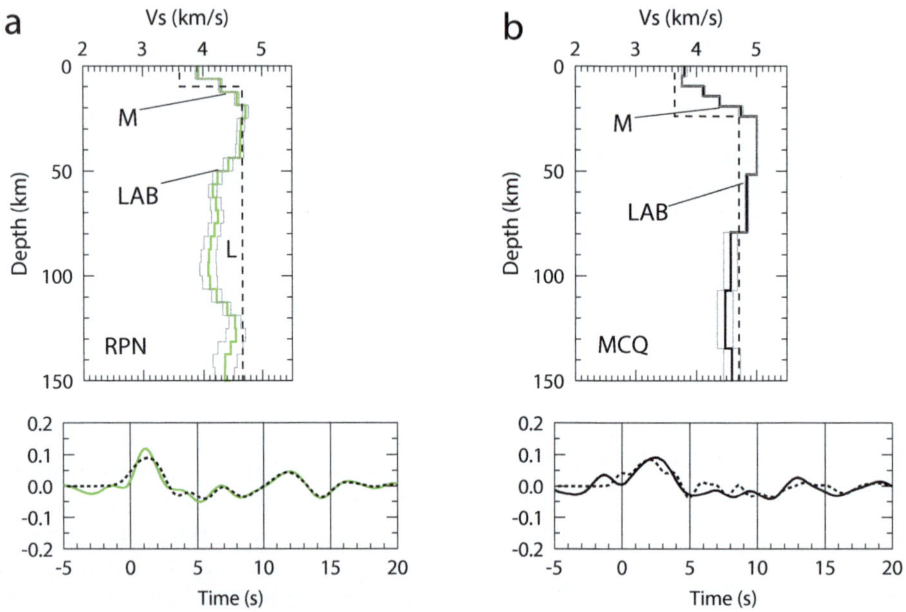

Figure 5. Same as Fig. 4 for the stations RPN and MCQ.

those predicted by the cooling model for the ocean lithosphere. The converted phases related to the deeper low-velocity layers (Fig. 2) below RPN, ROSA and ASCN are observed much beyond the error limits. Owing to the proximity of these stations to the mid-oceanic ridges, the possible source for the generation of these LVLs may be related to the source of magma at a depth range of ~75–106 km, from where the magma ascends along narrow lithospheric openings. For the region close to the ridges, occurrence of horizontal layers of connected melt pockets has been proposed based on the nature of shear-wave anisotropy (Nowacki et al, 2012). The mid-oceanic ridges receive magma from the plume that feeds the hot magma from below. The most volatile magmas reside at deeper depths. For example, Ascension is supposed to be a product of flow from an off-axis plume to the spreading axis (Brozena, 1986). These locations are where the MORs interact with the mantle plume (Montagner and Ritsema, 2001) and possibly produce the second

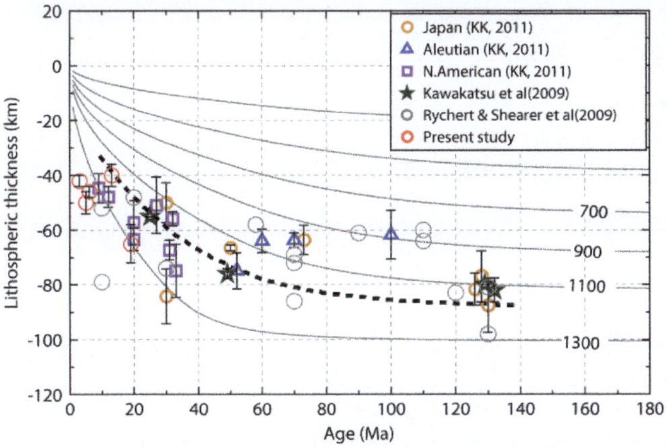

Figure 6. Observed lithospheric thickness values plotted against their age. Different color symbols denote data from different sources and red open circles denote the observations from the present study. Isotherms (gray contour lines with a 200 °C interval) and the estimated top of partial melting region (the black, dashed contour line) are taken from Kawakatsu et al. (2009). KK: Kumar and Kawakatsu et al. (2011).

low-velocity layers (L), as observed along the Mid-Atlantic Ridge, where regions affected by the shallow mantle plumes are detected.

An interesting review by Sinton and Detrick (1992) presented geophysical and petrological evidence for the distribution and behavior of magma bodies beneath the ridges. When these magma bodies assume a crystalline form greater than about 50 %, rheologically they behave much like a solid. The presence of melts in the lower crust and also in the upper mantle is suggested by active source tomography (Toomey et al., 1990; Dunn and Toomey, 1997) and compliance techniques (Crawford et al., 1991; Crawford et al., 1998; Crawford and Webb, 2002). In the latter technique, the pressure field caused by the long-period ocean waves that deform the sea floor is studied. The relation between pressure and deformation is called compliance and used to investigate the structure under MOR locations.

6 Upper mantle discontinuities

Figure 7 shows the *P*-receiver function stacks at individual stations after the application of a low-pass filter with a corner frequency of 6 s in the time domain. The stacks clearly reveal the global upper mantle seismic discontinuities typically detected at 44.1 and 68.1 s. The lateral extent of the *P*-to-*S* conversion points at a depth of 520 km (insets in Fig. 1) suggests that they sample regions far away from the ridges. Stacked traces, in fact, represent an average character of the upper mantle below the station. Due to the large sampling of upper mantle *P*-to-*S* conversions, we grouped the traces in such a way that one group is far from the ridge and other nearer to it. Station RPN (in Fig. 7) has two groups labeled as RPN1 and RPN2. Similarly, we made two clusters for MCQ. Other stations like ASCN, ROSA and PSCM have only one cluster, since the other clusters had much fewer traces. The *P*-to-*S* conversion times from the 410 and 660 km discontinuities with their errors are depicted in Table 1. The oceanic stations reveal a delay in both the phases (see Table 1). These seismic discontinuities are globally observed and generally interpreted to be formed due to a phase transformation in the mantle from olivine to spinel and from spinel-structured gamma phase to perovskite-structured magnesiowüstite (Duffy and Anderson, 1989) respectively. These phases are in a state of equilibrium governed by the temperature and pressure at those depths, described by the Clapeyron slope. The Clapeyron slope is positive and negative for the 410 and 660 km discontinuities respectively (e.g., Bina and Helffrich, 1994). Thus, the respective locations of these phases can be interpreted in terms of the temperature in the upper mantle. For temperatures in excess of ~ 200 °C, the separation between these discontinuities can be reduced by ~ 20–30 km (Helffrich, 2000). The *P*-to-*S* conversion times of the upper mantle discontinuities also depend on the average shear velocity in the upper mantle, implying that any discrepancies in these timings are related to the velocity perturbation in the uppermost mantle. The delays in both the phases might also indicate the lower average velocity in the upper mantle implying a thinner mantle lid and/or higher temperature of the oceanic lithosphere as compared to the continental mantle, corroborated by the LAB at shallower depths.

In the present study, the conversion times from the 410 km global discontinuity vary from 46 to 48.6 s, while those from the 660 km discontinuity range between 68.4 s and 71.6 s. All the stations show a delay in both the discontinuities with respect to the global average values predicted by the IASP91 model. These observed delays might be related to elevated upper-mantle temperatures just below the mid-oceanic ridges where hot magmas continuously pour out to make new lithosphere. These values are in general agreement with the earlier results in oceanic environments obtained from receiver function analysis (Li et al., 2003) and SS precursors (Gu and Dziewonski, 2002). The observed values suggest that the thickness of the upper mantle transition zone varies between 22 s (~ 220 km) and 24 s (~ 240 km) compared to the global average value of 24 s (~ 240 km). This thinning of the transition zone favors a hotter mantle (e.g., Agee, 1997; Shearer, 2000; Deuss, 2007). For the station ASCN, the transition zone is ~ 16 km (1.6 s) thinner on the western side of the ridge compared to the global average value (Fig. 7). For the eastern side, we do not have enough data to arrive at a conclusive result. For the station MCQ, the transition zone on the western side of ridge is ~ 20 km thinner (2 s). For stations PSCM and ROSA, our conversion point plots show that the data sample a similar upper mantle region (Fig. 7) and the upper mantle transition zone (within the error limits) is normal compared to the global average value. On the other

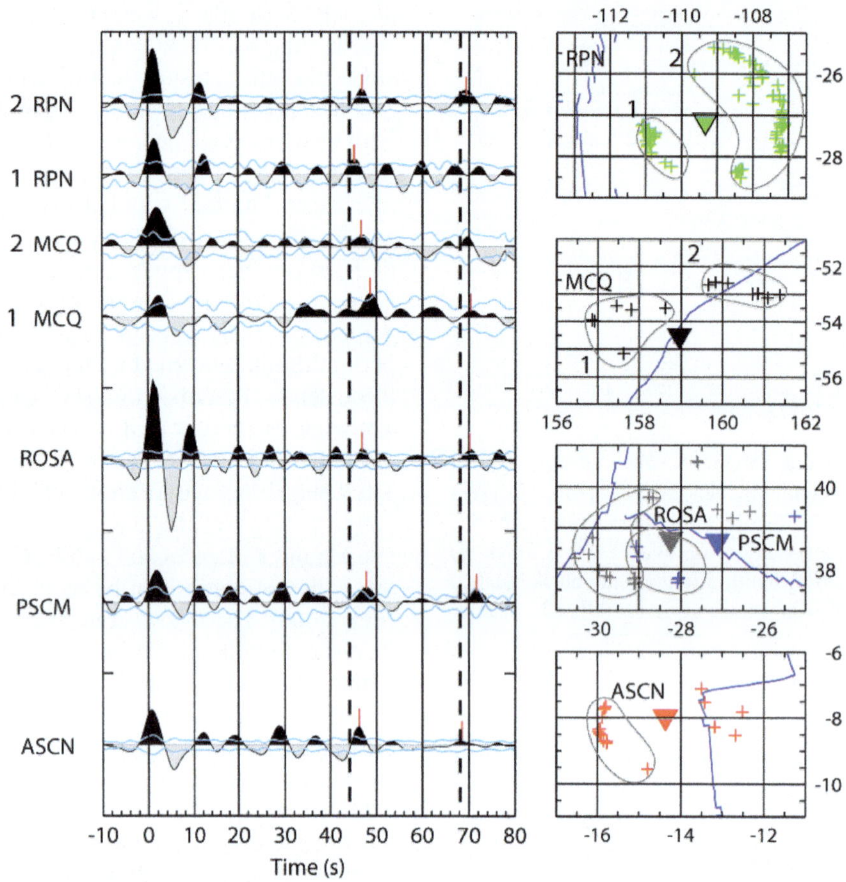

Figure 7. Stacked receiver functions similar to Fig. 2 but shown to deeper depths in order to display the upper mantle discontinuities. The traces are low-pass filtered with corner frequency of 6 s and moveout corrected to a reference slowness of 6.4 s/° using the IASP91 Earth model prior to stacking. Two bounding blue lines parallel to the mean represent 2 sigma error bounds. The stack traces are based on the geographical locations at a depth of 520 km for *P*-to-*S* conversions. We grouped those data that fall near and far away from the mid-oceanic ridge in order to avoid any structural complexities. For the station ASCN we made only one cluster since the data near the ridge are too scant to make a meaningful stack. All stations show clear arrivals of the Ps conversions from the 410 and 660 km discontinuities (marked by short red lines) and they are delayed with respect to the predicted times from the IASP91 global Earth models (dashed lines).

hand, for the station RPN, the transition zone is thinner on the eastern side of the ridge by ~ 14 km (1.4 s) .

7 Conclusions

P-receiver function analysis at the five stations located near the mid-ocean ridges reveals significant converted energy from the crust–mantle boundary, LAB and an additional deeper low-velocity layer (L). Modeling results reveal that the Moho depths vary from ~ 10 km to ~ 20 km. Hence, we conclude that the crustal thickness beneath the MORs is ~ 15 km. The LAB depths beneath the MORs vary from ~ 40 km to ~ 65 km. The observed LAB depths are consistent with the thermal cooling model (Stein and Stein, 1992); however, the depth estimates may not represent a pure oceanic plate in view of the contamination from the composition of the ocean islands on which the stations are located. Further, we observe sufficient converted energy from the upper mantle discontinuities, namely 410 km and 660 km. These phases are delayed (see Table 1) with respect to the global average. Also, the thickness of the transition zone is found to be small, due to a deepening of the 410 km and shallowing of the 660 km discontinuities. This thinning of the transition zone indicates the existence of mantle plumes beneath the stations (e.g., ASCN and RPN). These findings are very much coherent with the fact that the present observations are in close vicinity of the mid-oceanic ridges. However, it must be borne in mind that the data are from the stations located on oceanic islands, and hence the uppermost mantle observations may not be an exact proxy for the true oceanic plate.

Acknowledgements. Seismic data are from IRIS, DMC. Seismic data analysis was performed in Seismic Handler (K. Stammler). Plots are generated using Generic Mapping Tool (Wessel and Smith, 1995). Xiaohui Yaun (GFZ) provided the moveout script. Thanks to Rainer Kind for his valuable suggestions on the initial version of the manuscript. M. K. Sen, the director of NGRI has kindly permitted the publication of this work. Chinmay Haldar is a PhD student funded by the University Grants Commission (UGC). This work has been performed under the GENIAS Project PSC0104 (PK) of CSIR-NGRI. We thank the Topical Editor, J. Plomerova, for constructive suggestions and excellent handling of the manuscript. We thank an anonymous reviewer and W. Geissler for their thoughtful and constructive comments and suggestions that improved the original manuscript.

Edited by: J. Plomerova

References

Agee, C. B.: Phase transformations and seismic structure in the upper mantle and transition zone, in: Ultrahigh-Pressure Mineralogy: Physics And Chemistry of the Earth's Deep Interior, edited by: Hemley, R. J., Reviews in Mineralogy, Mineralogical Society of America, Washington DC, 37, 165–201, 1997.

Atwater, T. and Mudie, J. D.: Block faulting on the Gorda Rise, Science, 159, 729–731, 1968.

Backus, G. E.: Long-wave elastic anisotropy produced by horizontal layering, J. Geophys. Res., 67, 4427–4440, 1962.

Ballard, R. D. and Van Andel, T. H.: Morphology and tectonics of the inner rift valley at lat. 36°50′ N on the Mid-Atlantic Ridge, Geol. Soc. Am. Bull., 88, 507–530, 1977.

Barclay, A. H. and Toomey, D. R.: Shear wave splitting and crustal anisotropy at the Mid-Atlantic Ridge, 35° N, J. Geophys. Res, 108, 2378, doi:10.1029/2001JB000918, 2003.

Bina, C. R. and Helffrich, G.: Phase transition Clapeyron slopes and transition zone seismic discontinuity topography, J. Geophys. Res., 99, 15853–15860, 1994.

Bjarnason, I., Menke, W., Flovenz, O., and Caress, D.: Tomographic image of the spreading center in southern Iceland, J. Geophys. Res., 98, 6607–6622, 1993.

Brozena, J. M.: Temporal and spatial variability of sea-floor spreading processes in the northern South-Atlantic, J. Geophys. Res., 91, 497–510, 1986.

Buck, W. R. and Su, W.: Focused mantle upwelling below mid-ocean ridges due to feedback between viscosity and melting, Geophys. Res. Lett., 16, 641–644, 1989.

Burdick, L. J. and Langston, C. A.: Modelling crustal structure through the use of converted phases in teleseismic body wave forms, B. Seismol. Soc. Am., 67, 677–691, 1977.

Christensen, N. I.: Compressional and shear wave velocities in basaltic rocks: Deep Sea Drilling Project, Leg 16, in Initial Reports of the Deep Sea Drilling Project, XVI, 647–650, 1973.

Crawford, W. C., Webb, S. C., and Hildebrand, J. A.: Seafoor compliance observed by long-period pressure and displacement measurements, J. Geophys. Res., 96, 16151–16160, 1991.

Crawford, W. C., Webb, S. C., and Hildebrand, J. A.: Estimating shear velocities in the oceanic crust from compliance measurements by two-dimensional finite difference modeling, J. Geophys. Res., 103, 9895–9916, 1998.

Crawford, W. C. and Webb, S. C.: Variations in the distribution of magma in the lower crust and at the Moho beneath the East Pacific Rise at 9°–10° N, Earth Planet. Sci. Lett., 203, 117–130, 2002.

Deuss, A.: Seismic observations of transition zone discontinuities beneath hotspot locations, Geol. Soc. Am., 430, 121–136, 2007.

Du, Z. J. and Foulger, G. R.: The crustal structure beneath the northwest fjords, Iceland, from receiver functions and surface waves, Geophys. J. Int., 139, 419–432, 1999.

Duffy, T. S. and Anderson, D. L.: Seismic velocities in mantle minerals and the mineralogy of the upper mantle, J. Geophys. Res., 94, 1895–1912, 1989.

Duncan, R. A. and Varne, R.: The age and distribution of the igneous rocks of Macquarie Island: Papers and Proceedings of the Royal Society of Tasmania, 122, 45–50, 1988.

Dunn, R. A. and Toomey, D. R.: Seismological evidence for three-dimensional melt migration beneath the East Pacific Rise, Nature, 388, 259–262, 1997.

Dunn, R. A. and Forsyth, D. W.: Imaging the transition between the region of mantle melt generation and the crustal magma chamber with short-period Love wave propagation along the southern East Pacific Rise, J. Geophys. Res., 108, 2352, doi:10.1029/2002JB002217, 2003.

Efron, B. and Tibshirani, R.: An Introduction to the Bootstrap, Stanford University, California, USA, 1993.

Evangelidis, C. P., Minshull, T. A., and Henstock, T. J.: Three-dimensional crustal structure of Ascension Island from active source seismic tomography, Geophys. J. Int., 159, 311–325, 2004.

Forsyth, D. W.: Geophysical constraints on mantle flow and melt generation beneath mid-ocean ridges, In: mantle flow and melt generation at mid-ocean ridges, edited by: Morgan, J. P., Blackman, D. K., and Sinton, J. M., Geophys. Monogr. Ser., 71, 1–65, AGU, Washington, 1992.

Frederiksen, A. W. and Bostock, M. G.: Modelling teleseismic waves in dipping anisotropic structures, Geophys. J. Int., 141, 401–412, 2000.

Fuchs, K. and Müller, G.: Computation of synthetic seismograms with the reflectivity method and comparison with observations, Geophys. J. R. Astr. Soc., 23, 417–433, 1971.

Gaherty, J. B., Kato, M., and Jordan, T. H.: Seismological structure of the upper mantle: a regional comparison of seismic layering, Phys. Earth Planet. In., 110, 21–41, 1999.

Garmany, J.: Accumulations of melt at the base of young oceanic crust, Nature, 340, 628–632, 1989.

Gu, Y. J. and Dziewonski, A. M.: Global variability of transition zone thickness, J. Geophys. Res., 107, 2135, doi:10.1029/2001JB000489, 2002.

Gutenberg, B.: Physics of the Earth's Interior, Academic Press, New York, 1959.

Hammond, W. C. and Humphreys, E. D.: Upper mantle seismic wave velocity: Effects of realistic partial melt geometries, J. Geophys. Res., 105, 10975–10986, 2000.

Heit, B., Sodoudi, F., Yuan, X., Bianchi, M., and Kind, R.: An S receiver function analysis of the lithospheric structure in South America, Geophys. Res. Lett., 34, L14307, doi:10.1029/2007GL030317, 2007.

Helffrich, G.: Topography of the transition zone seismic discontinuities, Rev. Geophys., 38, 141–158, 2000.

Holtzman, B. K.: Melt segregation and strain partitioning: implications for seismic anisotropy and mantle flow, Science, 301, 1227–1230, 2003.

Holtzman, B. K. and Kohlstedt, D. L.: Stress-driven melt segregation and strain partitioning in partially molten rocks: Effects of stress and strain, J. Petrology, 48, 2379–2406, 2007.

Honkura, Y.: Partial melting and electrical conductivity anomalies beneath the Japan and Philippine seas, Phys. Earth Planet. Inter., 10, 128–134, 1975.

Humphreys, E. D. and K. G.: Dueker, Physicalstate of the western U.S. upper mantle, J. Geophys. Res., 99, 9635–9650, 1994.

Jorg, W. E.: Fassbinder, Easter Island (Chile): Magnetometry of archaeological structures basaltic, J. Org. Geology, 2005.

Kanamori, H. and Press, F.: How thick is the lithosphere?, Nature, 226, 330–331, 1970.

Karato, S.: The role of hydrogen in the electrical conductivity of the upper mantle, Nature, 347, 272–273, 1990.

Katz, R. F., Spiegelman, M., and Holtzman, B.: The dynamics of melt and shear localization in partially molten aggregates, Nature, 442, 676–679, 2006.

Kawakatsu, H., Kumar, P., Takei, Y., Shinohara, M., Kanazawa, T., Araki, E., and Suyehiro, K.: Seismic evidence for sharp lithosphere-asthenosphere boundaries of oceanic plates, Science, 324, 499–502, 2009.

Kelemen, P. B., Hirth, G., Shimizu, N., Spiegelman, M., and Dick, H. J.: A review of melt migration processes in the adiabatically upwelling mantle beneath oceanic spreading ridges, Phil. Trans. Roy. Soc. Lond., A, 355, 1–35, 1997.

Kendall, J. M.: Teleseismic arrivals at a mid-ocean ridge: effects of mantle melt and anisotropy, Geophys. Res. Lett., 21, 301–304, 1994.

Kind, R.: The reflectivity method for a buried source, J. Geophys., 44, 511–523, 1978.

Kind, R., Yuan, X., and Kumar, P.: Seismic receiver functions and lithosphere-asthenosphere Boundary, Tectonophysics, 536, 25–43, 2012.

Kumar, P. and Kawakatsu, H.: Imaging the seismic lithosphere-asthenosphere boundary of the oceanic plate, Geochem. Geophys. Geosyst., 12, Q01006, doi:10.1029/2010GC003358, 2011.

Kumar, P., Kawakatsu, H., Shinohara, M., Kanazawa, T., Araki, E., and Suyehiro, K.: P and S receiver function analysis of seafloor borehole broadband seismic data, J. Geophys. Res., 116, B12308, doi:10.1029/2011JB008506, 2011.

Langmuir, J., Klein, E. M., and Plank, T.: in Mantle Flow and Melt Generation at Mid-Ocean Ridges:American Geophysical Union, Washington, DC, 183–200, 1992.

Langston, C. A.: Corvallis, Oregon, crustal and upper mantle structure from teleseismic P and S waves, B. Seismol. Soc. Am., 67, 713–724, 1977.

Leeds, A. R.: Lithospheric thickness in the western Pacific, Phys. Earth Planet. In., 11, 61–64, 1975.

Li, X., Kind, R., and Yuan, X.: Seismic study of upper mantle and transition zone beneath hotspots, Phys. Earth Planet Int., 136, 79–92, 2003.

Menke, W., Brandsdottir, B., Jakobsdottir, S., and Stefansson, R.: Seismic anisotropy in the crust at the Mid-Atlantic plate boundary in south-west Iceland, Geophys. J. Int., 119, 783–790, 1994.

Mierdel, K., Keppler, H., Smyth, J. R., and Langenhorst, F.: Water Solubility in Aluminous Orthopyroxene and underline Science and the Origin of Earth's Asthenosphere, Science, 315, 364–368, 2007.

Montagner, J. P. and Ritsema, J.: Interactions between ridges and plumes, Science, 294, 1472–1473, 2001.

Nettles, M. and Dziewo'nski, A.: Radially anisotropic shear velocity structure of the upper mantle globally and beneath North America, J. Geophys. Res.-Sol. Ea., 113, B02303, doi:10.1029/2006JB004819, 2008.

Nowacki, A., Kendall, J. M., and Wookey, J.: Mantle anisotropy beneath the Earth's mid-ocean ridges, Earth Planet. Sc. Lett., 317–318, 56–67, 2012.

Parsons, B. and Sclater, J. G.: An analysis of the variation of ocean floor bathymetry and heat flow with age, J. Geophys. Res., 82, 803–827, 1977.

Phipps Morgan, J.: Melt migration beneath mid-ocean spreading centers, Geophys. Res. Lett., 145, 1238–1241, 1987.

Revenaugh, J. and Jordan, T. H.: Mantle layering from reverberations: 2. The transition zone, J. Geophys. Res., 96, 19763–19780, 1991.

Rodgers, A. and Harben, P.: Modeling the Conversion of Hydroacoustic to Seismic Energy at Islands and Continental Margins: Preliminary Analysis of Ascension Island Data, 21st Seismic Research Symposium, Technologies for Monitoring the Comprehensive Nuclear-Test-Ban Treaty Las Vegas, Nevada, 1999.

Rychert, C. A. and Shearer, P. M.: A global view of the lithosphere–asthenosphere boundary: Science, 324, 495–498, 2009.

Rychert, C. A. and Shearer, P. M.: Imaging the lithosphere-asthenosphere boundary beneath the Pacific using SS waveform modeling, J. Geophys. Res., 116, B07307, doi:10.1029/2010JB008070, 2011.

Schmerr, N.: The Gutenberg discontinuity: melt at the lithosphere-asthenosphere boundary, Science, 335, 1480–1483, 2012.

Shankland, T. J. and Waff, H. S.: Partial melting and electrical conductivity anomalies in the upper mantle, J. Geophys. Res., 82, 5409–5417, 1977.

Shearer, P. M.: Upper Mantle Seismic Discontinuities, American Geophysical Union, 115–131, 2000.

Shinohara, M., Fukano, T., Kanazawa, T., Araki, E., Suyehiro, K., Mochizuki, M., Nakahigashi, K., Yamada, T., and Mochizuki, K.: Upper mantle and crustal seismic structure beneath the Northwestern Pacific Basin using seafloor borehole broadband seismometer and ocean bottom seismometers, Phys. Earth Planet. In., 170, 95–106, 2008.

Sinton, J. M. and Detrick, R. S.: Mid-Ocean Ridge Magma Chambers, J. Geophys. Res., 97, 197–216, 1992.

Sobolev, S. V., Zeyen, H., Stoll, G., Werling, F., R. Althern, and K. Fuchs, Upper mantle temperatures from teleseismic tomography of French Massif Central including effects of composition, mineral reactions, anharmonicity, anelasticity and partial melt. Earth Planet. Sci. Left., v. 139, p. 147-163, 1996.

Sparks, D. W., Parmentier, E. M., and Morgan, J. P.: Three-Dimensional Mantle Convection Beneath a Segmented Spreading Center: Implications for Along-Axis Variations in Crustal Thickness and Gravity, J. Geophys. Res., 98, 21977–21995, 1993.

Spiegelman, M. and McKenzie, D.: Simple 2-D models for melt extraction at mid-ocean ridges and island arcs, Earth Planet. Sci. Lett., 83, 137–152, 1987.

Stammler, K.: Seismic Handler: programmable multichannel data handler for interactive and automatic processing of seismological analysis: Comput. Geosci., 19, 135–140, 1993.

Stein, C. A. and Stein, S.: A model for the global variation in oceanic depth and heat flow with lithospheric age, Nature, 359, 123–129, 1992.

Tan, Y. and Helmberger, D. V.: Trans-Pacific upper mantle shear velocity structure, J. Geophys. Res., 112, B08301, doi:10.1029/2006JB004853, 2007.

Tapley, I., Dijkstra, A. H., and Brolsma, H.: The Geology, Landforms and Topography of Sub-Antarctic Macquarie Island, Australia, as revealed by AIRSAR, Geosci. Remote Sens. Symposium, 1, 2004.

Toomey, D. R., Purdy, G. M., Solomon, S. C., and Wilcock, W. S. D.: The three-dimensional seismic velocity structure of the East Pacific Rise, near latitude 9°30′ N, Nature, 347, 639–645, 1990.

Utada, H. and Baba, K.: Estimating the electrical conductivity of the melt phase of a partially molten asthenosphere from seafloor magnetotelluric sounding data, Phys. Earth Planet. Int., doi:10.1016/j.pepi.2013.12.004, 2013.

Vinnik, L. P.: Detection of waves converted from P to SV in the mantle, Phys. Earth Planet. In., 15, 294–303, 1977.

White, R. S., McKenzie, D., and Onions, R. K.: Oceanic crustal thickness from seismic measurements and rare earth element inversions, J. Geophys. Res., 97, 19683–19715, 1992.

Xu, Y. and Wiens, D. A.: Upper mantle structure of the southwest Pacific from regional waveform inversion, J. Geophys.Res., 102, 27439–27451, 1997.

Yaun, X., Ni, J., Kind, R., Mechie, J., and Sandvol, E.: Lithospheric and upper mantle structure of southern Tibet from a seismological passive source experiment, J. Geophys. Res., 102, 27491–27500, 1977.

Zhang, Y. and Tanimoto, T.: High resolution global upper mantle structure and plate tectonics, J. Geophys. Res, 98, 148–227, 1993.

Zhao, D., Hasegawa, A., and Horiuchi, S.: Tomographic imaging of P and S wave velocity structure beneath northeastern Japan, J. Geophys. Res., 97, 19909–19928, 1992.

14

On the complexity of surface ruptures during normal faulting earthquakes: excerpts from the 6 April 2009 L'Aquila (central Italy) earthquake (M_w 6.3)

L. Bonini[1], D. Di Bucci[2], G. Toscani[1], S. Seno[1], and G. Valensise[3]

[1]Dipartimento di Scienze della Terra e dell'Ambiente, Università di Pavia, Pavia, Italy
[2]Dipartimento della Protezione Civile, Rome, Italy
[3]Istituto Nazionale di Geofisica e Vulcanologia, Rome, Italy

Correspondence to: D. Di Bucci (daniela.dibucci@protezionecivile.it)

Abstract. Over the past few years the assessment of the earthquake potential of large continental faults has increasingly relied on field investigations. State-of-the-art seismic hazard models are progressively complementing the information derived from earthquake catalogs with geological observations of active faulting. Using these observations, however, requires full understanding of the relationships between seismogenic slip at depth and surface deformation, such that the evidence indicating the presence of a large, potentially seismogenic fault can be singled out effectively and unambiguously.

We used observations and models of the 6 April 2009, M_w 6.3, L'Aquila, normal faulting earthquake to explore the relationships between the activity of a large fault at seismogenic depth and its surface evidence. This very well-documented earthquake is representative of mid-size yet damaging earthquakes that are frequent around the Mediterranean basin, and was chosen as a paradigm of the nature of the associated geological evidence, along with observational difficulties and ambiguities.

Thanks to the available high-resolution geologic, geodetic and seismological data aided by analog modeling, we reconstructed the full geometry of the seismogenic source in relation to surface and sub-surface faults. We maintain that the earthquake was caused by seismogenic slip in the range 3–10 km depth, and that the slip distribution was strongly controlled by inherited discontinuities. We also contend that faulting was expressed at the surface by pseudo-primary breaks resulting from coseismic crustal bending and by sympathetic slip on secondary faults.

Based on our results we propose a scheme of normal fault hierarchization through which all surface occurrences related to faulting at various depths can be interpreted in the framework of a single, mechanically coherent model. We stress that appreciating such complexity is crucial to avoiding severe over- or under-estimation of the local seismogenic potential.

1 Introduction

Starting in the 1980s many workers have attempted to infer the earthquake potential of large continental faults from their surface length and displacement (see Kim and Sanderson, 2005 for a review). As a result, over the past two decades several empirical relationships between earthquake magnitude and the extent of surface ruptures have been developed (e.g., Wells and Coppersmith, 1994; Wesnousky, 2008) and have since become the most popular analytical tool in earthquake geology.

On the one hand, these relationships have certainly helped in quantifying field geological observations, ultimately allowing them to be incorporated in analytical earthquake and seismic hazard assessment models (e.g., Stirling et al., 2013). In fact, the assessment of the hazard posed by large continental faults is increasingly relying on field investigations on all scales, and even a worldwide initiative for seismic hazard assessment such as the Global Earthquake Model (GEM) has "Gathering all knowledge on active faults worldwide" as one of its founding pillars (http://www.globalquakemodel.org/what/seismic-hazard/active-faults-database/). On the other hand, however, these empirical relationships have also contributed to put in the background the complexity of the relationships between the seismogenic source at depth and its surface expression. As a reminder to geologists, nature has recently spawned a number of damaging earthquakes that turned out to have been generated by blind, hidden, or otherwise hard-to-identify faults (e.g., Jan 2010, Haiti, M_w 7.0; Sep 2010–Feb 2011, Darfield–Christchurch, New Zealand, M_w 7.1–6.3; Oct 2011, Van, eastern Turkey, M_w 7.1; July–August 2013).

The M_w 6.3, 6 April 2009 L'Aquila (Abruzzi, central Italy) earthquake belongs to this category. Although the L'Aquila region had long been known for its high seismicity level along with most of Abruzzi, the 2009 earthquake challenged the standard approach for surface active fault identification because (1) clearly visible surface faults, that prior to 2009 were presumed to be active, showed no genetic relationships with the seismogenic source, and only some showed negligible reactivation; and (2) post-earthquake field analyses documented only limited and discontinuous coseismic surface ruptures.

The 2009 event is currently one of the best documented continental extensional earthquakes worldwide, and hence it makes a first-rate case for exploring the relationships between the activity of a seismogenic normal fault and its surface evidence in a structurally complex area such as the Apennines.

Notwithstanding the high quality of available data, the nearly 200 papers published to date about the 2009 L'Aquila earthquake have produced widely divergent seismotectonic interpretations and models (see Table 5 in Vannoli et al., 2012, for a summary).

We used a wealth of high-resolution geological, geodetic and seismological data combined with analog modeling to reconstruct the geometry of the seismogenic rupture in relation to sub-surface and surface faults. We aimed at devising a scheme for active fault hierarchization that explains all surface outcomes of shallow crustal seismogenic faulting in the framework of a single, mechanically coherent interpretative model. Proper appreciation of such complexity forms the basis for a correct assessment of the local earthquake potential.

2 Tectonic setting

The Italian Apennines exhibit a remarkably complex structure resulting from the overprinting of a number of subsequent tectonic phases (Fig. 1). In the early Mesozoic the region was part of the African passive margin of the Tethys Ocean; it hosted large carbonate platforms and intervening pelagic basins that were subsequently broken up by E–W Triassic–Lower Jurassic extension (Calamita et al., 2011; Di Domenica et al., 2014). Since the Cretaceous, the region evolved within the framework of the convergent motion between the African and European plates; east- to northeast-verging thrusts along with their associated foredeep/thrust-top basins progressed toward the Adriatic foreland up to the Middle Pleistocene (Patacca and Scandone, 1989) and were subsequently dissected by strike-slip and normal faulting.

Following a major geodynamic change at $\sim$800 ka, SW–NE extension became the dominant tectonic style over the core of the Apennines (e.g., Hyppolite et al., 1994; Galadini, 1999), as demonstrated also by breakout, seismicity and crustal strain data (e.g., Montone et al., 2012; Carafa and Barba, 2013). Extension is definitely a youthful process in the Apennines, however, and proceeds at the relatively slow rate of 2–3 mm yr^{-1} (D'Agostino et al., 2011). In contrast, the core of the Apennines is undergoing vigorous regional-scale uplift at 1–2 mm yr^{-1} (D'Anastasio et al., 2006); this process induces fast exhumation and widespread differential erosion of the Meso–Cenozoic rocks comprising it, being by far more effective than extension in building and modifying the landscape. As a result of these competing processes, the Apennines landscape is largely dominated by the older compressional structures (Fig. 1), which tend to be emphasized by erosion despite their being inactive: some even simulate the typical basin-and-range landforms associated with the action of a mature normal fault, according to a process of geomorphological convergence referred to as "mimicking" (Valensise and Pantosti, 2001). In addition to that, and due to the combined effect of tectonic stress and gravity, extended terrains often exhibit a level of complexity that makes the correct hierarchization of active normal faults – or even their mere identification – extremely challenging. On the outcrop scale, a large normal fault of crustal significance, a shallow reactivated normal fault on the backlimb of older thrust sheets, or an even shallower sackung scarp may appear equally evident and similarly convincing as to the existence of an underlying major seismogenic source (Gori et al., 2014).

3 The 6 April 2009 earthquake

The 6 April 2009 L'Aquila earthquake struck a seismically very active portion of the Apennines chain. It came as the culmination of a long foreshock/aftershock sequence recorded by permanent and temporary INGV seismometers (Chiarabba et al., 2009; Chiaraluce et al., 2011). The earthquake caused intensity up to IX-X MCS effects in a small number of villages located southeast of L'Aquila, but the largest number of collapsed buildings and casualties was reported in L'Aquila itself.

Extensive albeit limited coseismic surface breaks were reported by several workers over a 100 km^2 region elongated in the NW–SE direction between L'Aquila and Monticchio,

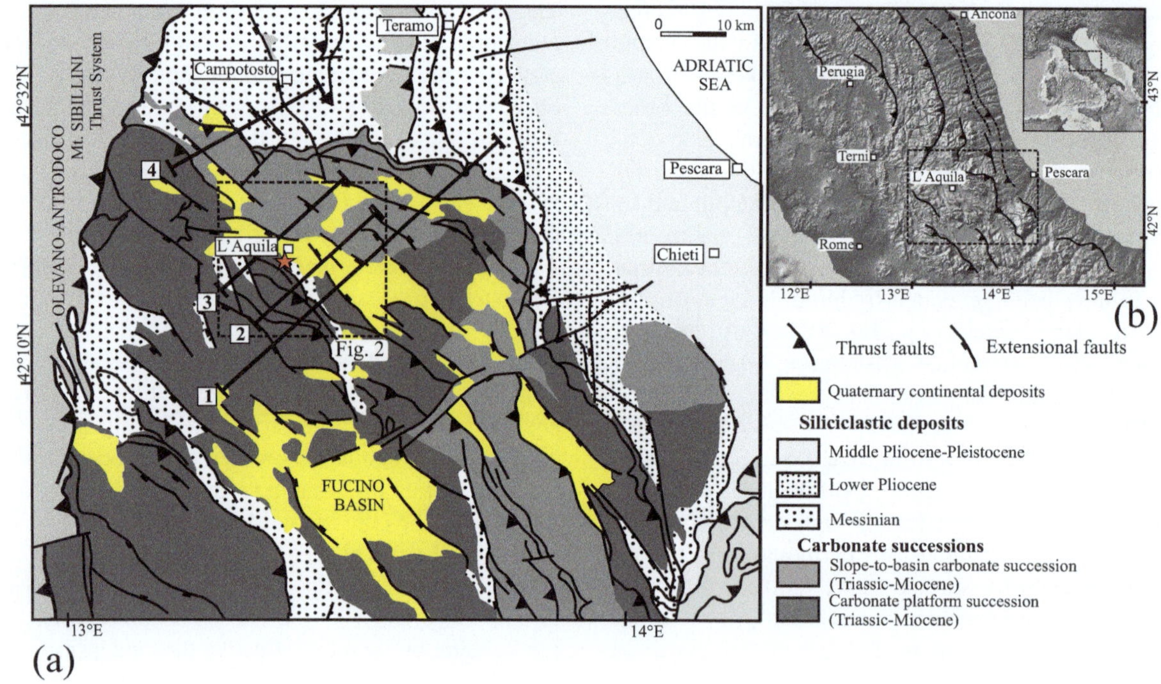

Figure 1. **(a)** Tectonic map of the Abruzzi region (modified from Vezzani et al., 2009). The present morphology of the axial portion of the Apennines is still dominated by the folding and thrusting through which the chain was built up over a long time interval between the upper Miocene and the middle Pleistocene. Subsequent extension and intervening sedimentation in intermontane basins has so far only slightly modified the compressional architecture of the Apennines. A dashed box indicates the exact location of the study area. 1, 2, 3, and 4 indicate cross sections. **(b)** Digital elevation model of the central Apennines, showing the trend of the main thrust fronts. A number of secondary fronts are also clearly delineated by the topography.

about 15 km to the southeast (see detailed descriptions in Boncio et al., 2010; the Emergeo Working Group, 2010, and Vittori et al., 2011). The most continuous surface breaks were extensional cracks, some with a maximum net throw of a few centimeters, often seen at the top of a ~10 m-high scarp formed by Quaternary continental deposits and running above the village of Paganica (Fig. 3).

DInSAR measurements based on extensive Envisat and COSMO-SkyMed data sets constrained by scattered GPS observations revealed sizable coseismic crustal deformation resulting in bowl-shaped, gently asymmetric surface subsidence peaking at 15–20 cm (Atzori et al., 2009; D'Agostino et al., 2012; Cheloni et al., 2014; Fig. 4). All published coseismic slip models agree that most coseismic slip – up to 1.0 m – occurred between 9–10 and 3–4 km in depth, whereas slip in the shallowest crust was found to be 0.1 m or less over most of the fault length (e.g., Atzori et al., 2009; Cirella et al., 2009; Cheloni et al., 2010; D'Agostino et al., 2012; Cheloni et al., 2014). Shallower slip was documented mainly near the northern end of the fault (D'Agostino et al., 2012), about 3 km to the northeast of L'Aquila (Fig. 4).

P and *S* wave analyses pointed out that the main rupture corresponded to an asperity located in a high-velocity layer (Vp > 6 km s^{-1} and Vs ~ 4.2 km s^{-1}; from Bianchi et al., 2010 and Di Stefano et al., 2011, respectively). This high-velocity body has been interpreted as being composed of partially hydrated mafic rocks between 3–4 and 10 km in depth (Fig. 4c), although in nearby areas the observed Vp and Vs values correspond to the characteristic velocities of the Meso–Cenozoic limestones and terrigenous rocks (Chiarabba et al., 2010; Di Stefano et al., 2011).

The sequence is well described by a detailed double-difference catalog of relocated events (3000 events with $M \geq 1.9$ from Chiaraluce et al., 2011; ~64 000 events of all magnitudes from Valoroso et al., 2013). When coupled with moment tensor solutions for the largest shocks ($M_w \geq 2.7$; Herrmann et al., 2011), these data allow imaging of individual faults activated during this complex sequence in great detail (Fig. 4). In the depth interval 3–4 to 10 km the aftershocks align rather regularly along a ~9 km wide, single planar surface dipping 45–50° to the SW over the entire fault length and extending for ~16 km in the NW–SE direction. Some shallower aftershocks (i.e., above 3–4 km in depth) are seen around the northern end of the fault; even though they are rather scattered, their locations highlight the activation of minor high-angle faults. In contrast, no shallow aftershocks are seen along the central and southern portions of the fault.

While the models based on seismological and geodetic data agree with fault parameters at depth (for both fault geometry and coseismic slip distribution), those based on

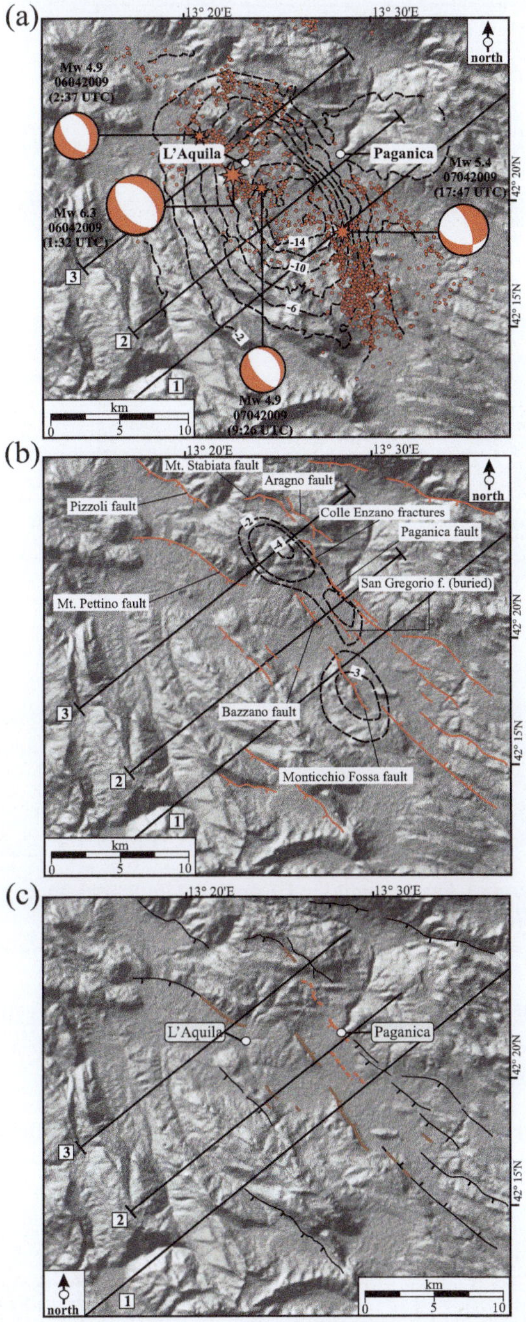

Figure 2. **(a)** Map view of the L'Aquila area showing the location (red stars) and focal mechanisms of the largest events of the 2009 sequence (Scognamiglio et al., 2010). Dashed lines are contours of elevation changes observed between 4 and 12 April 2009 (D'Agostino et al., 2012). Red dots indicate relocated aftershocks (Chiaraluce et al., 2011). **(b)** Presumed active faults, shown by red lines. S1 and S2 indicate the traces of cross sections 1 and 2. Dashed lines are contours of the elevation changes observed between 12 April and 5 October 2009 (D'Agostino et al., 2012). **(c)** Faults reported as activated during the 2009 earthquake. Those shown in red are discontinuous fractures mapped close to the Paganica fault; black lines are presumed active faults and brown segments represent the portion of the faults reported as partially reactivated in 2009 (Emergeo Working Group, 2010).

coseismic surface ruptures showed widely divergent results, especially regarding the net throw and the lateral extent of surface ruptures, respectively, ranging from none to about 10 cm and from 2.6 to 19 km (see Fig. 14 in Vittori et al., 2011, and Table 5 in Vannoli et al., 2012).

4 Buried tectonic structures in the L'Aquila area from geological and seismological data

To investigate the buried tectonic setting of the Apennines in the L'Aquila area we combined (1) surface geological data from various sources (e.g., Vezzani et al., 2009; see Fig. 1), (2) a three-dimensional distribution of geological units at depths derived from seismic tomography (Di Stefano et al., 2011), and (3) a reconstruction of the extensional active fault systems obtained from aftershock locations and moment tensor solutions (Herrmann et al., 2011; Valoroso et al., 2013).

Cross- and down-dip sections of the L'Aquila region (Fig. 5) show its intricate structural framework. Such complexity is not surprising in view of the overall structure of the Abruzzi Apennines, which were built by progressive stacking of heterogeneous thrust sheets. Strong lateral variations occur within each tectonic unit, mainly due to the architecture inherited from the Mesozoic extensional tectonics, and most tectonic structures of this part of the Apennines thrust belt are not cylindrical but exhibit a rather complex shape. These circumstances result in a complex lateral distribution of rock volumes that may have significant geometric and rheological implications for contemporary deformation.

The extensional faults activated during the 2009 seismic sequence lie in the hanging wall of the large Gran Sasso thrust. The upper tip of the main fault surface (at $\sim$3–4 km) coincides with a distinct lithological change. Aftershock alignments, focal mechanisms, and receiver function analyses highlighted the presence of a discontinuity striking N 334° and dipping 20° towards SW near the upper tip of the fault. When coupled with the distribution of geological units, such a discontinuity can be interpreted as a thrust plane that appears to play a primary role in the development of the newly formed extensional faults. In the northwestern part of the study area just above this plane there is a change in fault geometry, whereas in the central (cross sections 2 and 3 in Fig. 5) and southern parts (cross section 1) of the study area the aftershocks are confined below it. In our view this observation alone prevents the coseismic ruptures mapped at the surface (e.g., the Paganica ruptures) from being simply connected with the seismogenic fault identified at depth.

To shed light on the genetic relationships between the master fault at depth and surface structures we simulated the evolution of a normal fault using analog models.

Figure 3. Landscape view of the faults in the L'Aquila area (2x vertical exaggeration). On the right the Paganica fault (central image) with two details of the surface ruptures (upper and lower panels) recognized after the 2009 earthquake (Emergeo Working Group, 2010).

5 Deep and surface fault patterns: an analog experiment perspective

To analyze the mid- to long-term evolution of an upward propagating blind normal fault and the associated secondary structures related to the bending of the overlying rocks we constructed a series of analog models. We chose wet clay and dry sand as our preferred analog materials; this allowed us to reproduce different aspects of the faulting process and to make the most out of the characteristics of both these materials. Recent studies compared results obtained using wet clay and dry sand (e.g., Eisenstadt and Sims, 2005; Withjack and Schlische, 2006; Withjack et al., 2007), highlighting differences and similarities. On the one hand, wet clay has been used extensively to analyze brittle deformation related to folding (e.g., Cloos, 1968; Withjack and Jamison, 1986; Withjack and Schlische, 2006; Henza et al., 2010; Miller and Mitra, 2011), and its effectiveness as analog material has been stressed recently by new rheological tests (Cooke and van der Elst, 2012). On the other hand, dry sand is the most commonly used material for modeling tectonic deformation (see Graveleau et al., 2012 for a review). Using these two different materials allowed us to explore if and how the rheological differences between the two materials, and primarily the cohesion, affect our observations. All the experiments took place in a normal gravity field.

5.1 Pre-existing mechanical discontinuities and fault evolution: wet kaolin experiments

Our first goal was to understand if and how low-angle mechanical discontinuities affect the upward propagation of a normal fault. As a clay type we used kaolin with 65 % water content. Wet kaolin has a shear strength in the range 40 to 100 Pa (Eisenstadt and Sims, 2005; Cooke and van der Elst, 2012: Cooke et al., 2013) and its frictional coefficient is about 0.6. Following well-established scaling laws (e.g., Hubbert, 1937; Schellart, 2000), these parameters imply that 1 cm in the experiment represents about 1 km in nature.

The deformation device is composed of two plates (Fig. 6); one remains fixed throughout the experiment, thus representing the footwall, whereas the other one is pulled by a stepper-motor simulating hanging-wall subsidence. The displacement rate was fixed at 0.005 mm s^{-1}, a strain rate that is considered appropriate for wet clay experiments (e.g., Cooke and van der Elst, 2012). The apparatus has the top and both sides free so as to prevent undesired boundary effects.

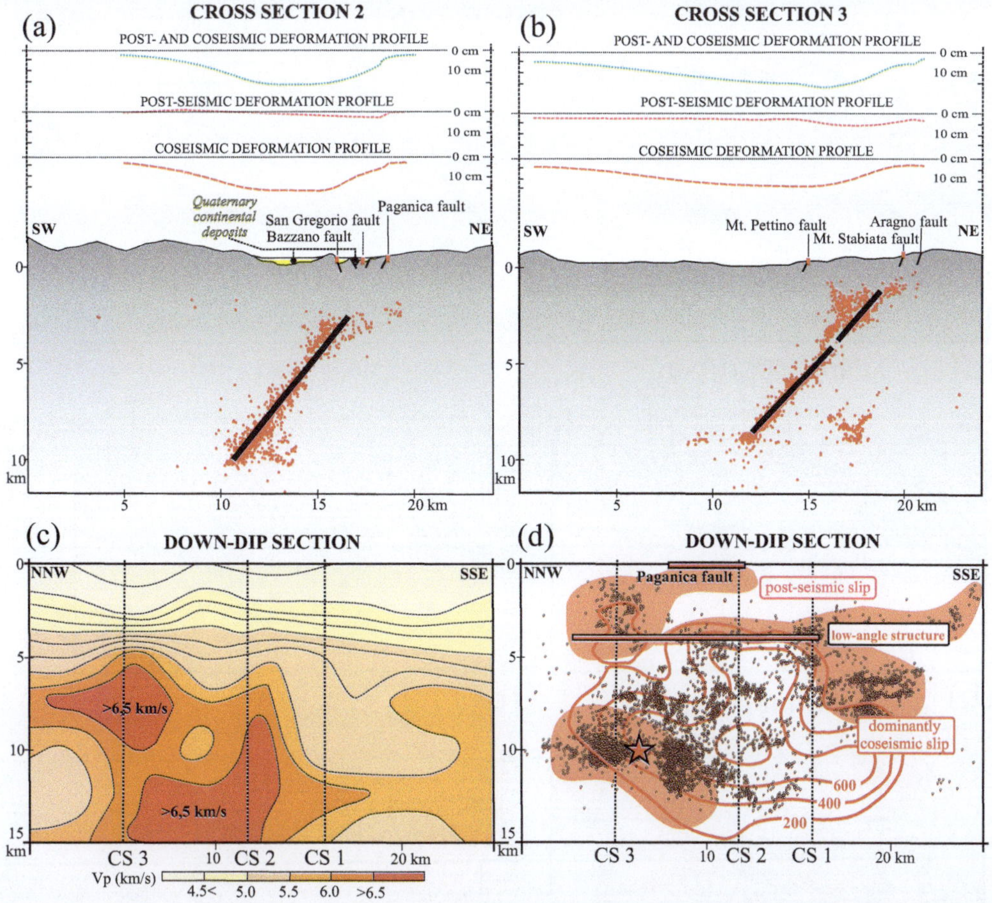

Figure 4. **(a)** and **(b)**: cross sections, showing dominantly coseismic (4 April 2009–12 April 2009; D'Agostino et al., 2012), post-seismic (12 April–5 October 2009; D'Agostino et al., 2012) and cumulative elevation changes measured along the section trace. In all figures red dots indicate relocated aftershocks (Valoroso et al., 2013). Thick black lines represent the fault plane derived by geodetic inversions with coseismic slip (Atzori et al., 2009). Section traces are shown in Figs. 1 and 2. **(c)** and **(d)**: down-dip sections along the L'Aquila master fault (strike N135E; dip 50° to the WSW); **(c)** shows Vp values imaged along the fault plane (from Di Stefano et al., 2011); **(d)** shows the mainshock location (star). Red dots are aftershocks located within 0.5 km of the fault plane (Chiaraluce et al., 2011). Coseismic slip pattern, shown by red lines (contour interval is 200 mm; Atzori et al., 2009), and location of the inferred inherited thrust faults (orange thick lines). Red areas mark portions of the fault that experienced post-seismic slip (modified from D'Agostino et al., 2012).

An experiment is terminated when the upward-propagating faults reach the surface of the model. To monitor and quantify the deformation we took lateral photographs every 1 mm of displacement. To this end we used displacement image correlation (DIC) analyses, a non-destructive, non-invasive, high-resolution optical technique that has the capability to detect any subtle motion of the model material during the experiments.

The clay experiments were carried out based on two experimental configurations: the first, Wet Kaolin #1 (WK1), representing a buried fault growing in a homogeneous setting (Fig. 6a); the second, Wet Kaolin #2 (WK2), similar to the previous one but with a low-angle discontinuity placed along the presumed propagating trajectory of the master fault (Fig. 6b). Such a pre-existing discontinuity has been reproduced by sliding an electrified blade through the analog material. This innovative procedure, which has been successfully tested by Cooke et al. (2013), allowed us to introduce in the model very thin discontinuities that simulate pre-existing fault systems. Figures 7 and 8 show significant steps selected from the two experiments, though both experiments are photographed at every 1 mm of displacement. The initial thickness of the WK models is 5 cm, representing the shallower 5 km of a generic continental crust.

At the beginning of both experiments (0.1 mm), the displacement fields showed a triangular deformation zone (Figs. 7a and 8a), corresponding to the trishear zone described and modeled by several investigators (e.g., Erslev, 1991; Hardy and Ford, 1997; Jin and Groshong, 2006). The displacement field showed by the two experiments during the very first deformation step is very similar, confirming that the

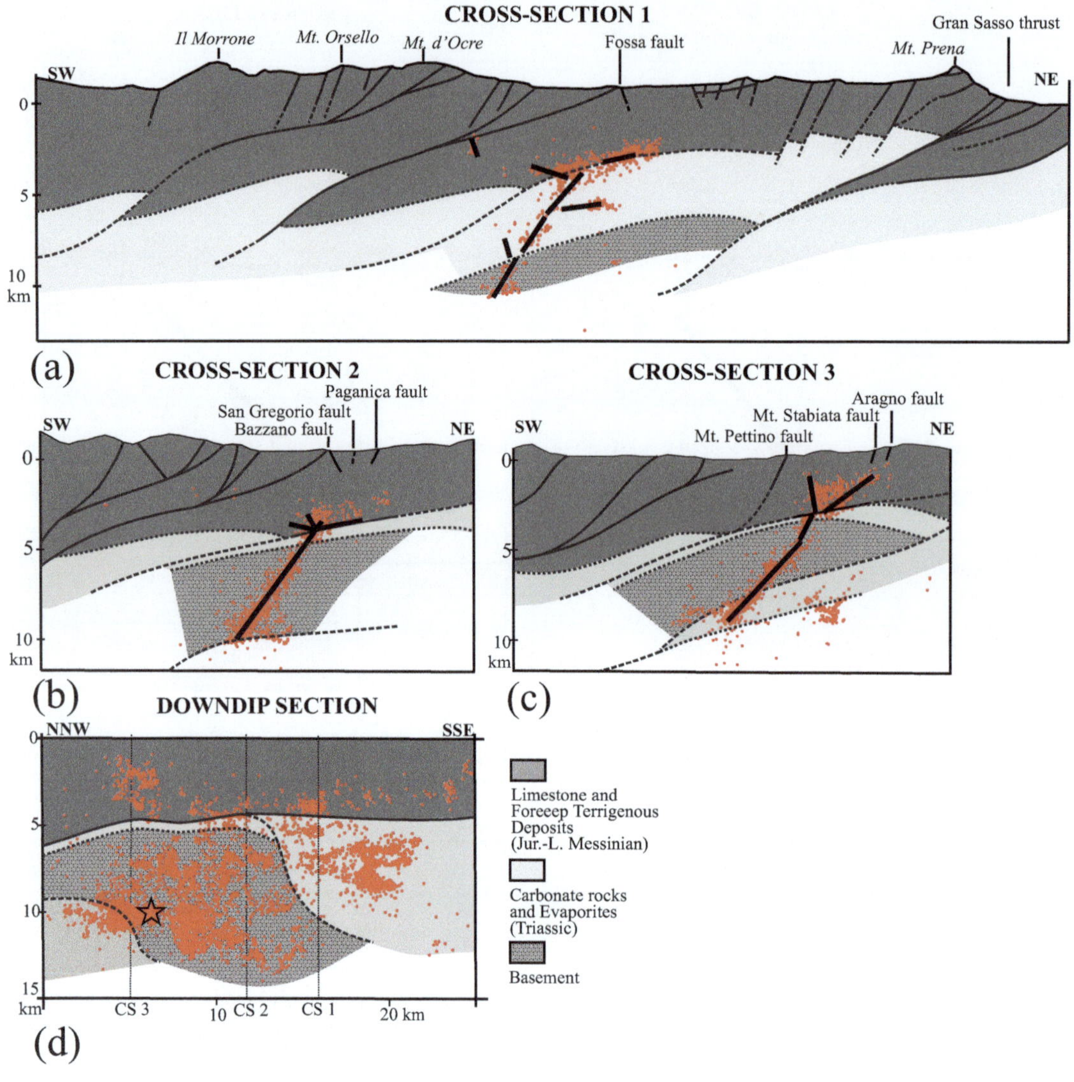

Figure 5. Geological sections across the L'Aquila area. (**a**) Regional geological section showing the main tectonic features of the area (modified after Satolli and Calamita, 2008 and Vezzani et al., 2009). (**b**) and (**c**): geological sections across the central (**b**) and northern (**c**) parts of the L'Aquila seismogenic source. (**d**) Down-dip section (strike N135E; dip 50° to the WSW). In all sections red dots indicate relocated aftershocks (Valoroso et al., 2013). In all figures thick black lines represent fault segments from Valoroso et al. (2013) based on the analysis of the aftershock locations and focal mechanisms. Red star is the hypocenter of the April 6 mainshock.

mechanical properties of the analog material are the same in both wet kaolin experiments.

As displacement proceeds, brittle and ductile deformation taking place in the experiments begins to be visible. A monoclinal structure developed during both experiments after a displacement of 10 mm (Figs. 7b and 8b). In WK1, three synthetic faults and one antithetic extensional fault are formed at the tip of the master fault with a Mode II mechanism (Scholz, 1990). As a result of bending, two crestal fractures formed with a Mode I mechanism at the surface of the model (Fig. 7b), where tensile stress is maximal. A strain distribution analysis corroborates our observations showing two high-strain areas, a lower one and an upper one, spatially corresponding to the previously described brittle structures. At the same displacement level (10 mm) WK2 showed a different development of brittle structures (Fig. 8b). An individual upward-propagating fault stemming from the buried tip of the master fault appeared more developed than those seen in WK1 (Fig. 9a) and with a steeper dip angle. When this fault encountered the pre-existing discontinuity, it stopped its propagation, splitting into two minor elements. Meanwhile, minor faults were seen to displace the pre-existing low-angle plane. Also in this experiment, some downward-propagating faults related to the bending developed at the surface of the model (Fig. 8b). Strain analyses confirmed the previous observations highlighting a narrow zone where strain is highest; significantly, there is a sudden decrease in the strain value

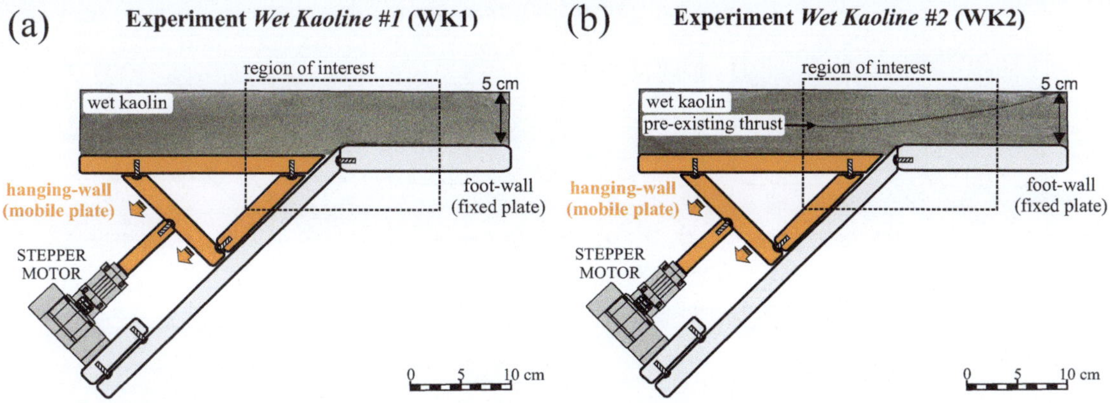

Figure 6. Experimental setup used to model **(a)** a fault evolving into a medium configuration and **(b)** with a low-angle discontinuity.

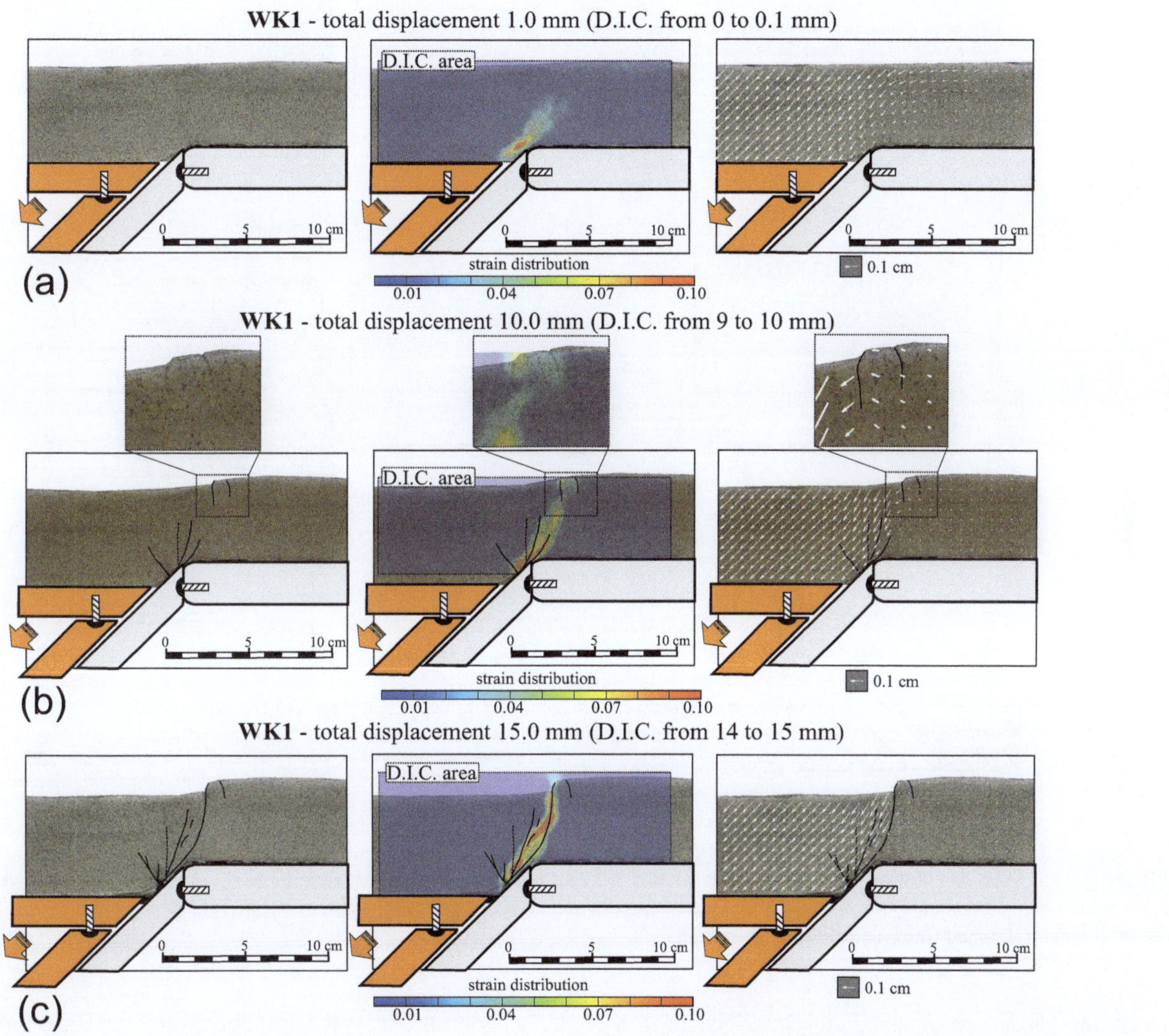

Figure 7. Progressive evolution of the fault patterns in the WK1 wet clay experiment (left panels). For each step the brittle and ductile deformations have been analyzed using the DIC technique, which returned the displacement field (right panels) and the strain distribution (central panels). Dashed lines indicate precursor faults (see text).

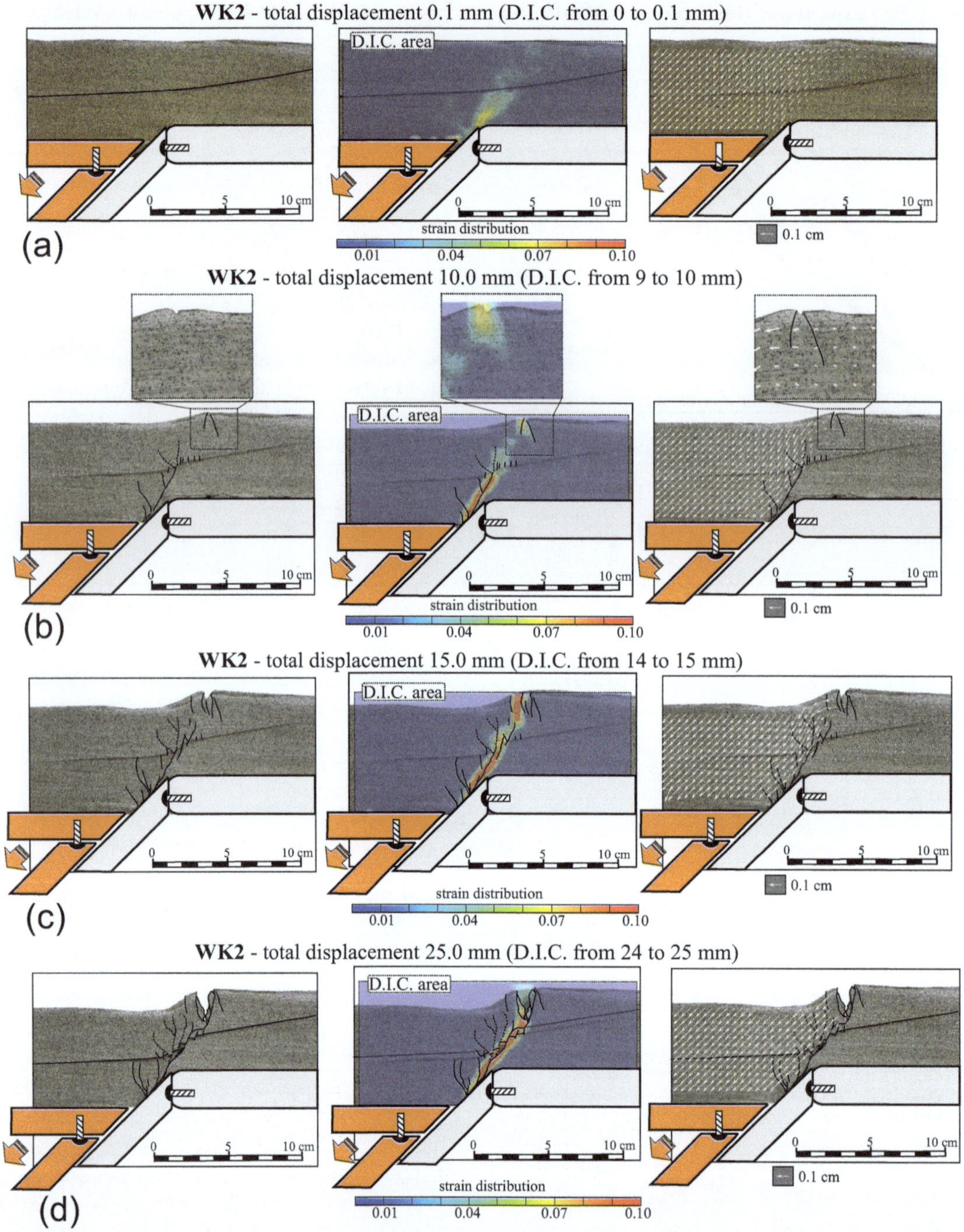

Figure 8. Progressive evolution of the fault patterns in the WK2 wet clay experiment (left panels). For each step the brittle and ductile deformations have been analyzed using the DIC technique, which returned the displacement field (right panels) and the strain distribution (central panels). Dashed lines indicate precursor faults.

where the upward-propagating fault intersects the low-angle plane.

After 15 mm of displacement, in WK1 the upward-propagating fault reached the surface, connecting with one of the downward-propagating faults and showing an overall listric geometry (Fig. 7c); the connection of the two structures is well predicted by the strain distribution analysis. In contrast, in WK2 the downward- and upward-propagating faults do not appear to have connected at the same displacement step (Fig. 8c). This different behavior must necessarily be related to the role played by the low-angle discontinuity; a circumstance confirmed by the strain analysis, which shows that most of the deformation remains confined below the discontinuity. The rate of upward propagation of the

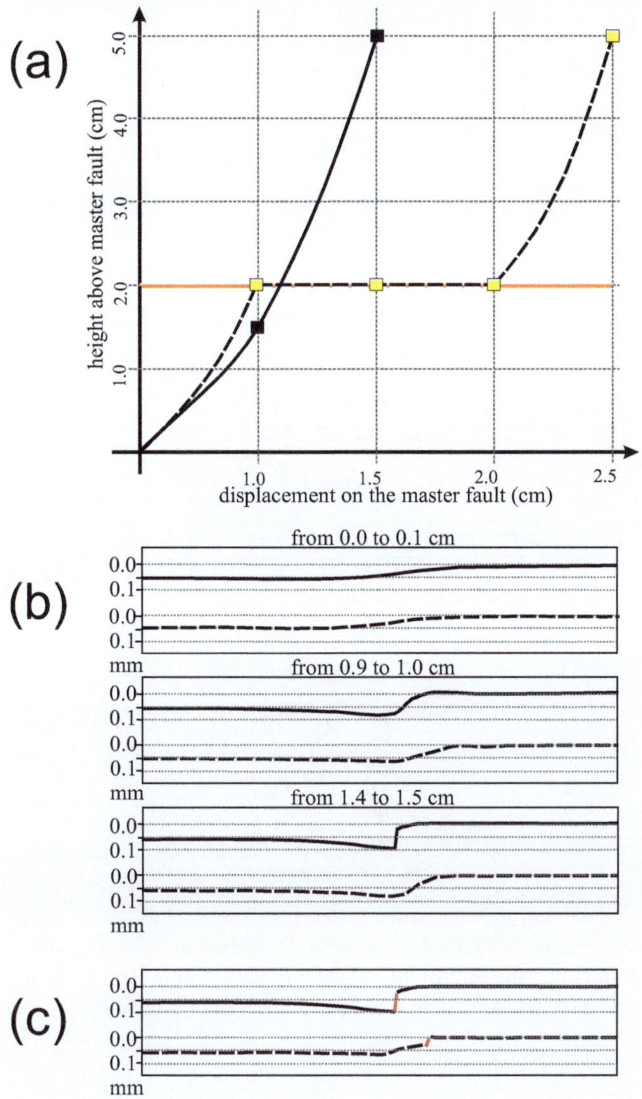

Figure 9. (a) Diagram showing the vertical location of the upper tip of the upward-propagating faults against the amount of displacement on the master fault. Thick black and dashed lines represent results from WK1 and WK2, respectively. Orange lines represent the location of the pre-existing discontinuity in WK2. **(b)** Topography profiles of the WK1 (thick black lines) and WK2 (dashed lines) experiments at different intervals. **(c)** Final topographic profile of the wet clay experiments, showing the location of the surface-breaking faults (red lines). Data in diagrams **(b)** and **(c)** come from the DIC analyses shown in Figs. 7 and 8 (right panels).

fault in WK2 is much slower than in WK1, and downward-propagating faults in WK2 are more developed than in WK1 (Fig. 9a).

Notice that in WK2 a direct link between upward- and downward-propagating faults was reached only after 25 mm of total displacement (Fig. 8d), 10 mm more than was needed for WK1 (Fig. 7c), and with a peculiar ramp-flat-ramp configuration.

A comparison of the two models at this final development stage, i.e., when the upward- and downward-propagating faults are connected and the displacement produced by the two rigid blocks has reached the surface, shows significant differences. Regarding fault geometry, we observed a listric profile in WK1 and a change in dip in WK2 coincident with the pre-existing low-angle discontinuity. Such a ramp-flat-ramp profile in WK2 further stresses the role of the fault-bend folding mechanism connected to the growth of the upward-propagating fault; as a result, the bending-moment faults are more developed in this experiment. Despite the rigidity of the blocks that simulate the hanging wall and the footwall, the faulting evolution in WK1 and WK2 produced subsiding basins having different shapes (Figs. 9b and c); while in WK1 the basin depocenter falls close to the surface-breaking fault, the depocenter of the bowl-shaped basin that developed in WK2 lies 40 mm away from the surface-breaking fault.

In summary, the results of the experiments show how strongly the mechanical discontinuity cut in the uniform material of the model affected the growth of the simulated master blind fault, delaying the upward propagation of the synthetic faults that developed from its tip (Fig. 9). Both experiments displayed that sub-vertical bending-moment faults are the first structures to form at the surface, as shown also by other investigators (e.g., van Gent et al., 2010). In our wet kaolin experiments, such secondary structures mainly developed with a Mode I mechanism showing an excessive width (see for example the final step of WK2; Fig. 5g), probably due to the quite high cohesion of the wet clay itself.

5.2 Quartz sand experiments

Although wet kaolin is very effective as an analog material, its high cohesion may result in some undesirable modeling effects. To overcome this limitation we reproduced some key features of the models made with clay using dry sand.

We performed two different experiments. The first one, Quartz Sand #1 (QS1), used the same experimental apparatus of the wet kaolin experiments, formed by two rigid blocks, one of which is mobile and acts as a subsiding hanging-wall fault block (Fig. 10a). This experiment was conceived to evaluate how much the rheological differences between wet kaolin and dry sand affect deformation patterns in an isotropic material. In the second experiment, Quartz Sand #2 (QS2), the hanging-wall fault block was replaced by a flexing plate connected to the stepper motor. The aim of this experiment was strictly to reproduce a bending of the hanging wall caused by a slip on a blind fault (Fig. 10b). This configuration is reminiscent of the plateau phase of the upward propagation in the WK2 experiment, namely the phase when the deformation in the block on top of the low-angle discontinuity was dominated by bending rather than simple shearing (Fig. 9a). The low cohesion of the quartz sand was expected

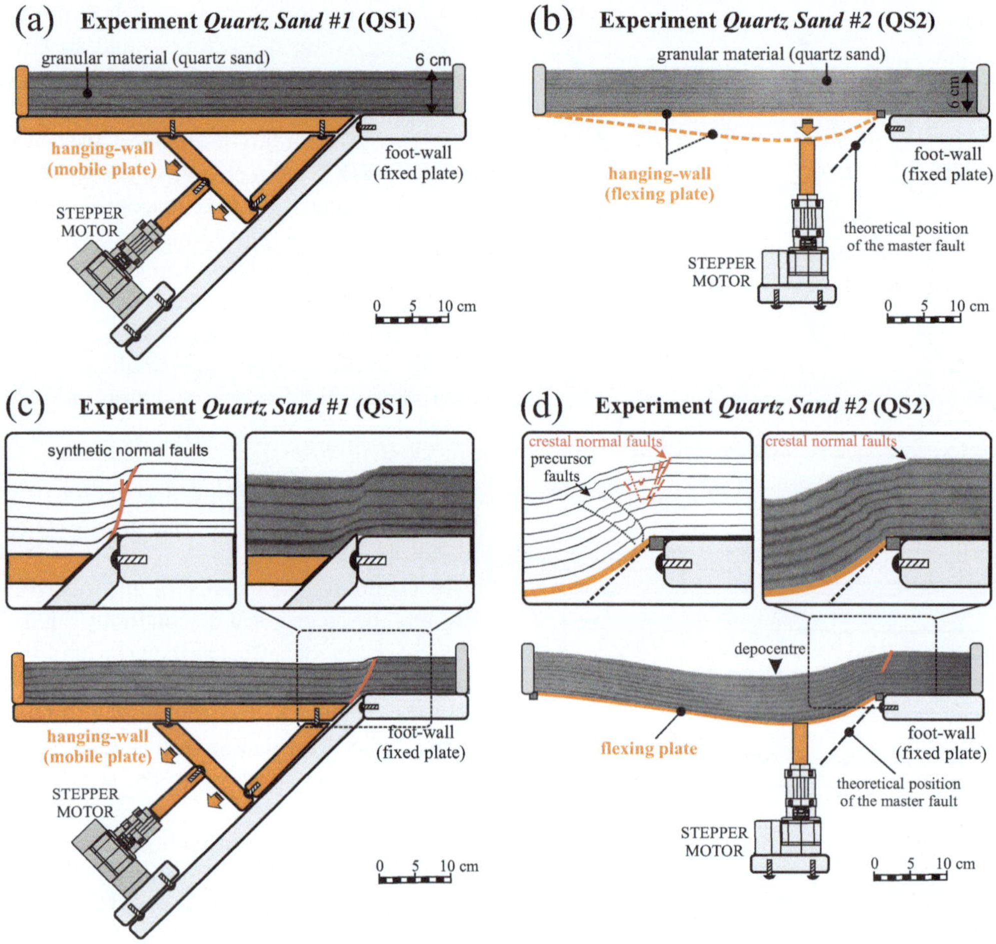

Figure 10. Analog modeling setup and results of dry sand experiments QS1 and QS2. (**a**) and (**b**) are sketches of the experimental modeling apparatus. (**c**) and (**d**) show the results of the experiment. Red lines represent newly formed faults and dashed black lines indicate precursor faults.

to prevent the unnatural generation of open cracks that we observed in WK2.

The experiment used quartz sand having the following mechanical properties: cohesion 30 Pa, coefficient of internal friction 0.88, angle of internal friction 41°. Using standard scaling rules (Hubbert, 1937), under our experimental conditions 1 cm represents 0.5 km in nature, implying that our model aimed at reproducing the shallowest 3 km of a continental crust. In both experiments, two lateral glass walls confined the quartz sand. To prevent undesired boundary effects, we reduced the friction by polishing the glass walls with graphite powder.

The results of QS1 did not show remarkable differences with respect to WK1; the upward-propagating fault nucleated at the upper tip of the simulated master fault and rapidly reached the surface (Fig. 10c), as already observed in several previous sand box models (e.g., Bonini et al., 2011, and references therein). The basin associated with slip on the master fault was quite similar to its equivalent in WK1.

In QS2, a forced fold developed above the tip of the hypothetical fault as a result of the imposed bending (Fig. 10d). A series of crestal fractures, some showing small but measurable vertical throw due to low cohesion of the dry sand, formed where the tensile stress was maximum, that is to say along the hypothetical up-dip prolongation of the master fault. Overall, these structures are reminiscent of the bending-moment faults seen in the wet kaolin experiments. Additionally, some upward-convex faults formed at the boundary between rigid and flexing blocks. Such structures had also been recognized in the WK models (dashed lines in Figs. 7 and 8); they are known as "precursor faults" (Mandl, 2000) and have been described as one of the first steps of brittle deformation connected to a buried fault, both in analog models and in natural examples (Mandl, 2000).

As observed in the WK models, the QS experiments also show that the relative position between the initial tip point of the master fault at depth and the surface-breaking faults (both bending-moment faults and secondary synthetic faults) is not much different. Bending-moment faults, however, are located in a more external position, i.e., towards the footwall, so that they appear to be aligned with the deep master fault plane even though they are not connected with it.

6 Discussion

Our analog models, especially WK2 and QS2, show the mechanical feasibility of a long-lived buried extensional master fault, accompanied by disconnected secondary normal faults lying at the surface roughly along its projection. Model QS2 also demonstrates that bending-moment faults developed in a low-cohesion material may display some vertical displacement, mimicking a genuine upward-propagating normal fault, and that to some extent the size and the shape of the basin related to the growth of upward-propagating faults depends on the growth rate of the faults (Fig. 9).

Based on the available data about the L'Aquila sequence and on our modeling results, we conclude that the low-angle discontinuity that exists around 3 km depth played an active role in preventing coseismic slip from propagating to the surface, ultimately affecting the mid- to long-term evolution of the shallow portion of the entire fault system.

We believe that our results cast doubts on the nature of the surface fractures detected after the 2009 L'Aquila earthquake, in particular on the interpretation of the Paganica fault as the primary expression of seismogenic faulting at depth. In the following we discuss this point analyzing the available seismological and geological evidence in the light of our modeling results. We will address in detail what is seen in different portions of the fault system.

6.1 The Paganica fault: upward or downward propagating?

To address this question we move from the evidence supplied by the aftershock pattern. In this respect we wish to recall that, although very shallow aftershocks were imaged between 1 and 3 km depth near the northern end of the seismogenic source, no aftershocks where observed between the surface and 2–3 km depth in its central portion, i.e., near Paganica (Chiarabba et al., 2009; Chiaraluce et al., 2011; Chiaraluce, 2012; Valoroso et al., 2013). This could be well explained by the velocity strengthening behavior of faults at shallow crustal depth, that is to say, in the upper stability transition (UST) zone defined by Scholz (1988), but two additional explanations are equally likely: (1) the fault normal stress, which controls the effective coefficient of friction of the rupture, might be especially low in the shallowest portion of the fault plane, thus generating stable sliding (e.g., Brace and Byerlee, 1970); or, more simply, (2) there is no fault plane continuity in the shallowest 2–3 km of the crust in the Paganica area. It is well known that the UST zone is not always found in seismogenic areas; for instance, Marone and Scholz (1988) investigated continental faults and concluded that the UST occurs only in mature fault zones. As a result, young faults and faults with long recurrence intervals or negligible gouge zones do not exhibit a UST zone.

Unfortunately, neither field nor trenching observations allow the nature of the Paganica fault gouge to be assessed, as the only available information consists of extensional and shear joints nearby the village of Paganica. Paleoseismological investigations across the Paganica fault used standard trenches dug in loose Quaternary deposits up to 4 m in depth (e.g., Cinti et al., 2011) that did not reach the bedrock (i.e., the pre-Quaternary carbonate rocks). A seismic refraction and resistivity survey run perpendicular to the Aterno Valley south of L'Aquila (Improta et al., 2012) did constrain the shape of the basin at depth; the characteristics and effective penetration depth (around 300 m) of the selected methodology, however, do not allow one to draw firm conclusions on whether or how the surface ruptures extend at depth, or to understand the mechanical properties of the hypothetical fault gouge.

In summary, there is no surface exposure of the Paganica fault that could be used to state whether or not this fault is well developed at depth. Moreover, even if we considered the Paganica extensional and shear joints as secondary branches of the seismogenic source reaching the surface, we would expect afterslip along these features to result from velocity strengthening hours or days after the mainshock (e.g., Perfettini and Ampuero, 2008). Luckily, we know exactly the timing of the surface breakage near Paganica, as the Paganica–Tempera aqueduct high-pressure pipe, which crosses the Paganica fault not far from the village center, was reported broken during the mainshock (Vittori et al., 2011). Hence this effect is incompatible with afterslip along a secondary branch of the main seismogenic fault.

As stated earlier, our experiments demonstrated the mechanical feasibility of a long-lived buried extensional master fault with disconnected secondary normal faults lying roughly along the same hypothetical plane: a configuration that reproduces satisfactorily all information available on the current setting and on the presumed evolution of the Paganica surface rupture. We hence interpret the Paganica breaks as tensional and shear joints that accommodate the deformation within the surface fold induced by the buried master fault.

6.2 Northern end of the L'Aquila fault system

With regard to the northern end of the fault system activated in 2009, the aftershock distribution indeed appears to delineate a master fault plane and its associated shallow secondary structures (Fig. 3a). These may resemble a set of secondary splays of a master fault approaching the surface. As shown by the results of the WK2 experiment, however, this would imply that this portion of the fault system is more developed than the central portion.

To test this hypothesis we compared the central and northern portions of the L'Aquila fault system (Figs. 5b and c) in the light of distinct subsequent steps of the wet clay experiments; one could visually relate the central portion to the step shown in Fig. 3d, and the northern portion to the step of Fig. 3g. Several observations, however, make this view questionable.

Seismological data, observations of surface co- and post-seismic deformation, and inferences on the morphotectonic evolution of the Aterno River Valley, all indicate that the L'Aquila fault system generated the maximum displacement in its central portion, as seen along most normal fault systems worldwide (Fig. 4). As a consequence, the secondary structures directly generated by the master fault are also expected to be more developed in its central portion, in contrast to all available observations.

A plausible explanation for this ambiguous configuration involves the reactivation of inherited extensional fault systems located close to the buried upper tip of the seismogenic master fault. As we mentioned describing the local geological setting, the Abruzzi Apennines are the result of three subsequent deformation phases: Mesozoic extension, Cenozoic compression and shortening, and Plio–Quaternary extension. The high-angle faults that we see today at the surface may be equally well related to any of these deformation events: for instance, they could be Mesozoic faults incorporated and translated within the Cenozoic thrust sheets, or normal faults that developed in the backlimbs of thrust-related anticlines during Cenozoic compression (see Scisciani et al., 2002 for a review on this topic).

We have shown that a discontinuity – specifically a thrust plane – occurs at 3 km depth in the crustal volume above the master fault; therefore, some of the surface faults that slipped during the 2009 sequence and that are in the hanging wall of that thrust may also be related to previous deformation phases. A feasible hypothesis for the nature of the fault system seen near the northern end of the L'Aquila fault system is the reactivation of inherited extensional faults fortuitously located near the upper tip of the seismogenic blind master fault.

We conclude by suggesting that, although most coseismic slip was confined below 3 km depth during the mainshock, it did trigger sympathetic slip on inherited extensional faults such as the buried portion of the Mt. Stabiata and Aragno faults. It is well known that blind earthquake ruptures impart stress on the overlying crust, possibly triggering pre-existing faults (e.g., Lin and Stein, 2004). This hypothesis is supported by the post-seismic strain recovery observed just above this shallow structure (D'Agostino et al., 2012).

6.3 A down-dip segmentation scheme of the seismogenic source

Based on the evidence discussed above we propose (a) that the causative rupture of the 2009 L'Aquila earthquake was confined at a depth > 3 km, and (b) that the complex interaction with inherited low-angle faults (thrusts) not only prevented coseismic slip from propagating upwards, but also reduced the natural mid- to long-term tendency of the master fault to reach the surface. Accordingly, we interpret the Paganica breaks as tensional and shear joints that accommodate the deformation within the fold induced by the normal fault propagation. This outcome explains both the hidden nature of the earthquake causative fault prior to 2009 and the wide scatter of rupture models described in the literature (see Vittori et al., 2011, and Vannoli et al., 2012, for a review). We also suggest that in the depth interval 3–10 km the causative fault of the 2009 L'Aquila earthquake was laterally segmented due to the interaction with inherited structures. We believe that these limitations, which are due to permanent characteristics of the local geological setting, contributed to limit the fault size and hence the magnitude of the 2009 earthquake, and we expect it to have done so also through previous seismic cycles.

6.4 Categorization of active faults

Faulting is a complex phenomenon on all spatial and temporal scales. In the Abruzzi Apennines this complexity arises largely from (a) the superposition of a young (< 1 My) extensional tectonic regime over the outcomes of a compressional regime that built the chain by progressive thrusting over a timescale of a few million years, (b) the similarity between surface tectonic features that represent the primary outcome of the ongoing extensional regime and secondary, inherited extensional features that are commonly associated with thrusting, and (c) the progressive rejuvenation of landforms by regional uplift, a fast landscape-building process that largely outpaces the ongoing regional extension and that is also responsible for generating fault-like gravity features. An extremely well-documented earthquake such as the 2009 event, occurring in a relatively isolated intermontane depression such as the middle Aterno Valley, affords the rare possibility to investigate the interplay between these competing processes without being tricked by appearances or – even more importantly – without being ruled by dogmas.

Based on previous observations we devised a hierarchization scheme of active normal faults in the L'Aquila area (Fig. 11). Our scheme includes four categories:

I. Seismogenic fault system (i.e., the L'Aquila master fault system): fault segments that generated the mainshock. They are the main players in the assessment of ground shaking hazards.

II. Inherited subsurface faults (i.e., the low-angle thrusts located above the main planar seismogenic plane) are faults generated during previous deformation phases and in this case acting passively as segment boundaries, both horizontally and vertically, effectively limiting the size and hence the magnitude of the earthquake generated by the master fault (mainshock). As such they

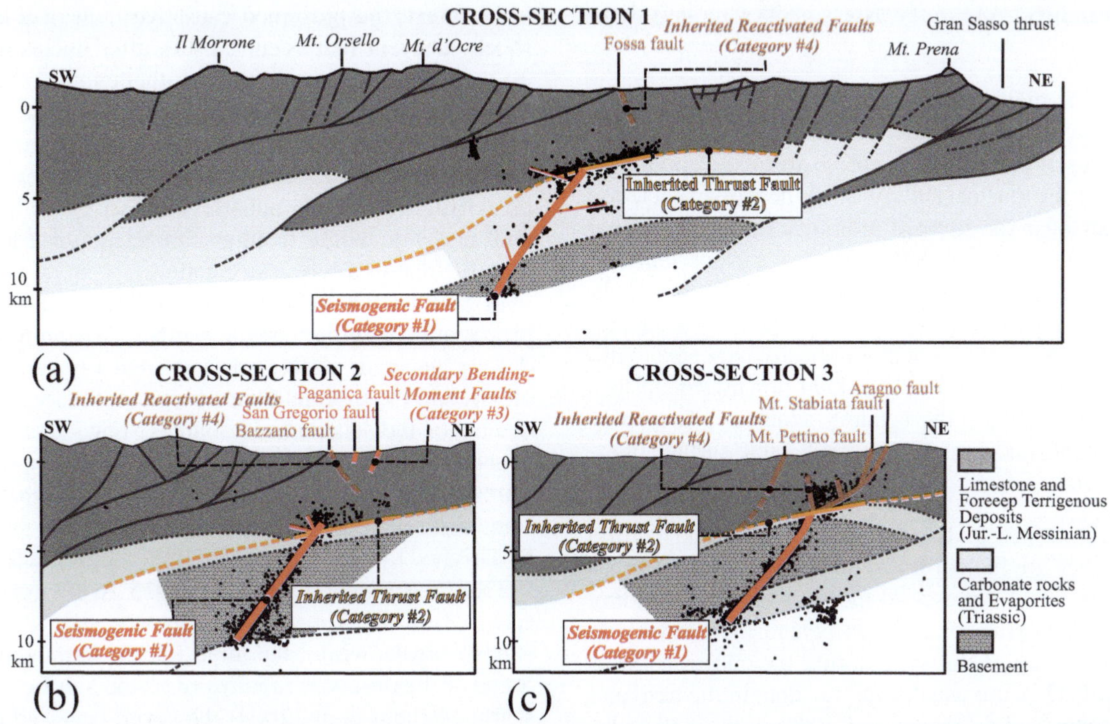

Figure 11. Geological cross sections showing different fault categories recognized in the L'Aquila area. Section traces are shown in Fig. 1.

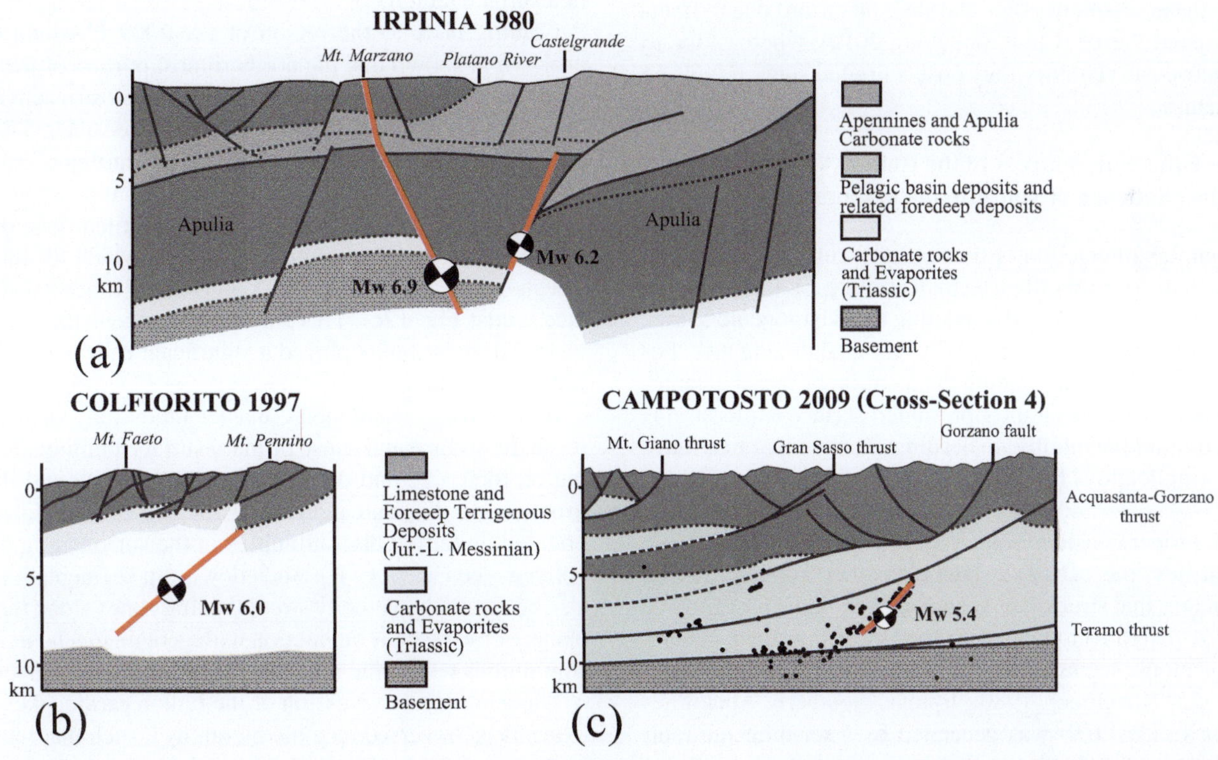

Figure 12. Geological cross sections across the causative faults of the 1980 Irpinia (modified from Mazzotti et al., 2007; faults from Pantosti and Valensise, 1990), 1997 Colfiorito (modified from Chiaraluce et al., 2005), and 2009 Campotosto (from Bigi et al., 2013) earthquakes. Thick red lines represent the seismogenic sources (Category I).

play an indirect role in the assessment of ground shaking hazards.

III. Bending-moment secondary surface faults (i.e., Paganica ruptures): they represent breaks that are generated by the crustal bending taking place above the uppermost portion of the master fault. As such they are expected to (a) occur near the upward prolongation of the deeper master fault, thus simulating primary surface faulting, and (b) be restricted to the middle of the master fault, where slip is usually larger and bending is consequently tightest. Bending moment faults are also expected to nucleate at the surface and to extend downward up to a depth controlled by the bending geometry. They may cause sizable and somehow unpredictable surface faulting hazards.

IV. Inherited surface faults (e.g., the Bazzano and Pettino faults): they are faults formed during previous deformation phases and correspond mostly to faults bounding piggy-back basins, or more in general accompanying the progression of thrusting during the construction of the chain. They are usually very evident in the field and may or may not be reactivated, depending on their location and geometry relative to the coseismic strain pattern imposed by the master fault. They may be relevant to the assessment of ground shaking hazard due to fault-trapped waves (Calderoni et al., 2012), whereas due to their clear visibility they pose a limited surface faulting hazard.

6.5 Seismogenic sources in the Italian Apennines: the challenge of youthful normal faulting

The complex interaction of the recent to currently active extension with the inherited tectonic setting of the L'Aquila area once again suggests that relating the seismogenic source to its surface expressions may be a very challenging task. For the same reasons, anticipating the location of major seismogenic sources based on their presumed surface expression is a difficult and potentially misleading exercise that may have crucial implications for the assessment of local seismic hazards.

With proper consideration of local geological and tectonic peculiarities, the causative faults of other normal faulting earthquakes that struck the Apennines during the past century can be revisited following our observations on the L'Aquila seismotectonic scenario.

The 23 November 1980, Irpinia, Southern Apennines earthquake (M_W 6.9) was generated by a seismogenic fault (Category I) that clearly reached the surface over a distance of nearly 40 km (Pantosti and Valensise, 1990: Fig. 12a). This causative fault is directly visible, but it is worth recalling that the observed ruptures did not bound any intermontane basin, but rather crossed the landscape, showing a tendency to reverse the existing topographic setting. In the same geological context, the presumed causative fault of a large mainshock subevent that occurred 40 s after nucleation with an estimated M_W 6.2 was a blind antithetic fault confined in the Apulia carbonate rocks (Pantosti and Valensise, 1990). The observed lateral fragmentation of the 1980 earthquake source can be interpreted as related to the widespread occurrence of preexisting tectonic boundaries interacting with the master fault at depth. These features can be assigned to our Category II (Inherited subsurface faults).

In contrast, the 26 September 1997 Umbria–Marche complex sequence in the central–northern Apennines (M_W 6.0) did not generate sizable surface breaks (e.g., Basili et al., 1998), and the available knowledge on the location and geometry of the seismogenic sources shows that they do not connect with any of the known range-bounding faults (e.g., Barba and Basili, 2000). The surface faults that displayed some rejuvenation of the bedrock fault plane should hence be assigned to our Category IV (Inherited surface faults).

The catastrophic 13 January 1915 Avezzano earthquake (M_W 6.8) in the Abruzzi Apennines caused large coseismic surface breaks within the large Fucino intramontane basin. Some of the observed ruptures (e.g., the San Benedetto fault scarp, Michetti et al., 2004), however, occurred away from the basin boundaries, suggesting that some of the rejuvenated bedrock faults should be assigned to our Category IV (Inherited surface faults).

Coming back to the region of the 2009 L'Aquila earthquake, it is known that the northernmost portion of the aftershock sequence, located near Lake Campotosto, activated a further seismogenic fault system that released a M_W 5.4 event on 9 April 2009 (Bigi et al., 2013). This earthquake and its aftershocks occurred in a crustal volume confined both at the bottom and at the top by thrust faults inherited from the extinct compressional tectonic regime, and which for this reason should be assigned to our Category II (Inherited subsurface faults: Fig. 12c). This further strengthens the view that inherited thrust faults played a significant role as well in the L'Aquila earthquake.

In summary, seismogenic master faults may or may not reach the surface and cause primary surface faulting, depending on their size and depth, the amount of slip and its distribution on the fault plane, and the presence of favorably oriented, inherited discontinuities in the surrounding crustal volume (see Category II). Such down-dip segmentation may also be caused by velocity strengthening zones along the fault plane, or by the fault interaction with generic mechanical discontinuities within the host rocks (e.g., weak rock layers).

The geological expression of the Italian earthquakes leads to a sort of "reversed tectonic hierarchy", such that the main seismogenic fault (Category I) may be expressed at the surface only by rather subdued bending-moment faults (Category III), may be constrained in length and width by preexisting faults and discontinuities (Category II), and may lie hidden beneath a blanket of clearly visible yet substantially harmless faults (Category IV).

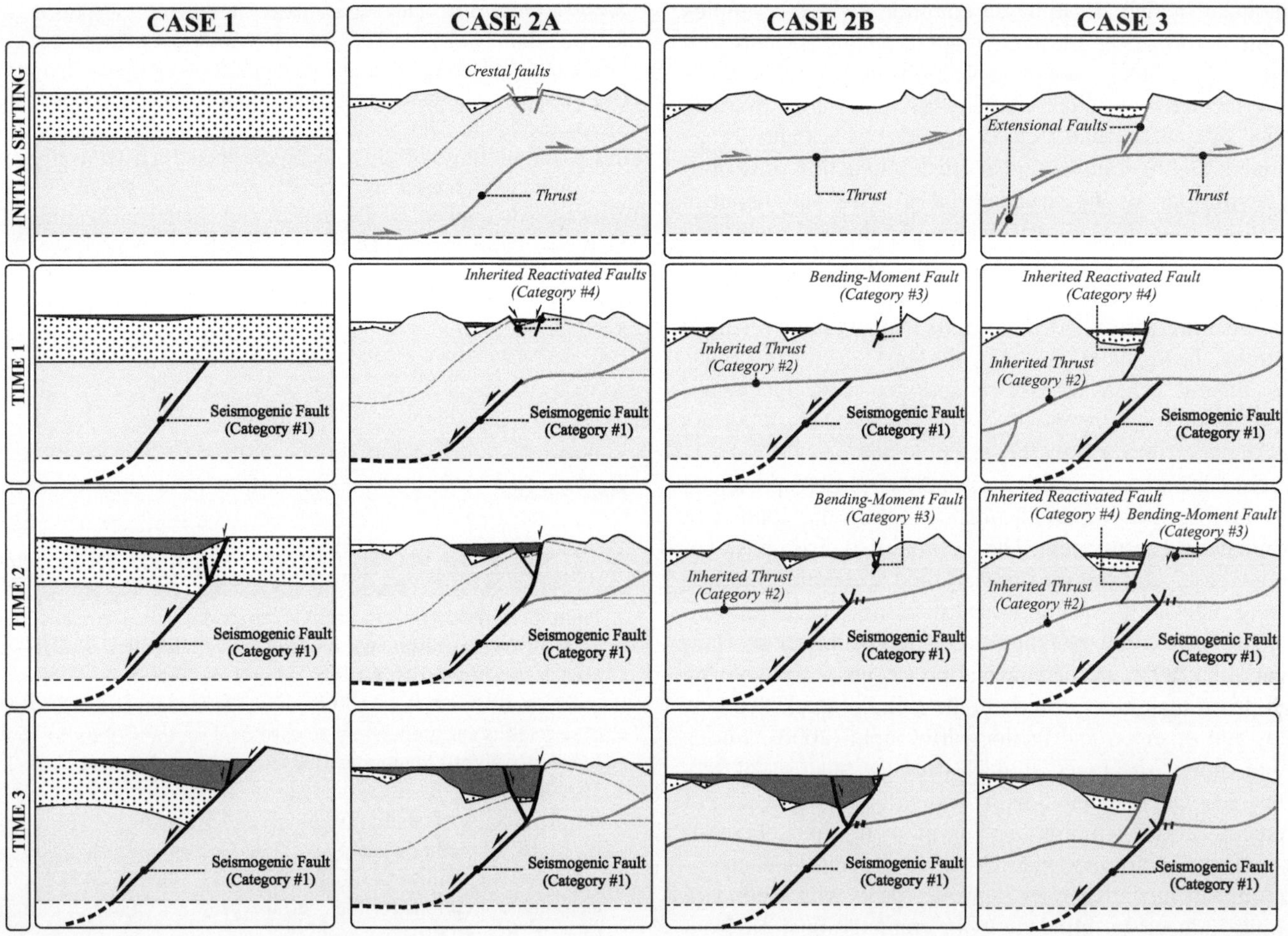

Figure 13. Images showing an inferred sequential development of a seismogenic master fault (Category #1) and its ancillary structures (Category #2, Category #3, and Category #4) in four different tectonic scenarios. Gray and black lines represent inherited structures and active faults, respectively. The panels labeled TIME 1, TIME 2 and TIME 3 show successive steps of fault evolution. In CASE 1 the initial setting does not exhibit significant inherited mechanical discontinuities; the seismogenic fault easily reaches the surface. CASE 2 (A and B) comprises two scenarios of geodynamic change, i.e., from compression to extension. In CASE 2A the seismogenic master fault (Category #1) is an inherited thrust ramp (Category #2) reactivated since the early stages of extension (TIME 1); crestal faults that developed along the extrados of the thrust-related fold may be reactivated (Category #4), mimicking a surface-breaking master fault. In CASE 2B an inherited thrust plays a segmentation role (Category #2) and slows down the upward propagation of the master fault, causing extreme bending of the surface rather than faulting and promoting the creation of bending-moment faults (Category #3). CASE 3 is the most complex, as it shows a scenario of both positive and negative inversion. The seismogenic fault interacts with an inherited thrust at shallower crustal levels, generating bending-moment faults (Category #3) and remobilizing an inherited extensional fault plane that simulates a fully primary surface rupture.

7 Conclusions

Our analysis of seismogenic faulting in the L'Aquila area reveals an unprecedented complexity in the interaction between coseismic slip and inherited structural features. In particular, it suggests that the main surface coseismic rupture related to the 2009 L'Aquila earthquake – the Paganica fault – can be better explained as a result of surface bending rather than as the direct prolongation of the seismogenic master fault. Under these circumstances, the length of the faulted zone and the extent of surface slip would be controlled more by the rheology of the shallow rocks than by the fault slip at depth. This would ultimately prevent the surface rupture parameters from being used to obtain the earthquake magnitude using empirical relationships (e.g., Wells and Coppersmith, 1994), both for the considered event and for previous earthquakes detected through paleoseismological trenching.

We believe that the 2009 L'Aquila earthquake illustrates well the nature of the interaction between the seismogenic fault and other inherited structures at depth (e.g., buried thrust planes), the partial reactivation of inherited surface faults (e.g., the Bazzano, Mt. Stabiata and Pettino faults), and the occurrence of pseudo-primary surface ruptures (the Paganica fault). We contend that the 6 April 2009 L'Aquila earthquake in fact illustrates a common style of complex tectonic deformation, implying that the number of unrecognized, blind or hidden seismogenic faults in Italy – and probably elsewhere – could be larger than previously thought. Driven by the current tectonic regime such faults break through a highly complex upper crust, interacting in various ways with the existing structural fabric. This may result in limitations in their size, reactivation of inherited faults and generation of new surface breaks (Fig. 13). What is absolutely crucial for the geological reconnaissance work is that in the field these highly diverse faults may exhibit a reversed hierarchy, the most obvious being the least relevant in a fault-based seismic hazard assessment and vice versa. Proper appreciation of such complexity forms the basis for a correct assessment of the local earthquake potential.

From this perspective, the work and the ideas we presented have obvious and important implications in the context of seismic hazard assessment. The "...ability to distinguish between tectonically induced primary and secondary faulting, faulting induced by strong ground motions, and faulting induced by nontectonic phenomena ..." has been seen by Hanson et al. (1999) as a fundamental pre-requisite for devising appropriate regulatory criteria in the siting of nuclear power plants and other critical facilties. McCalpin (2000), among several others, discussed strategies for the analysis of secondary features associated with the distributed expression of reverse faulting, including bending-moment faults, flexural-slip faults and folds, and placed them in a hierarchical classification. Much of their work, however, deals with secondary faulting induced by blind faults in compressional environments, because this is the dominating tectonic style in the countries where these studies were initiated. So far there has been little appreciation for the fact that the same features and interpretative problems may also be seen in extensional environments: our study aims to fill this gap using the unique evidence from the 2009 earthquake.

Italy no longer has nuclear facilities in operation, but it is a highly seismic country that hosts a large fraction of the worldwide cultural heritage. Not all areas are equally at risk, hence any building retrofitting strategy requires priorities to be set based on careful examination of the true seismogenic potential of each individual seismogenic area. Much work has already been carried out on the potential of Italian faults, and several quiescent areas are being watched closely following the identification of presumed active tectonic features. The results of our analysis suggest that an objective evaluation of the hazard posed by any recognized active fault requires full and proper appreciation of its true hierarchy level. This is to be achieved by blending surface, subsurface, geomorphic, structural and seismological data and, more importantly, by avoiding preconceptions and overly simplified models.

Acknowledgements. This work was financially supported by Regione Lombardia, by the Italian Ministry of Research and Education (MIUR) in the framework of INGV's Progetto Abruzzo, and by Presidenza del Consiglio dei Ministri – Dipartimento della Protezione Civile (DPC). The views and conclusions contained in this paper are those of the authors and should not be interpreted as necessarily representing official opinions and policies, either expressed or implied, of the Italian Government. The editor F. Storti and the referee K. Decker are gratefully acknowledged for the review and the constructive criticisms that improved this paper.

Edited by: F. Storti

References

Atzori, S., Hunstad, I., Chini, M., Salvi, S., Tolomei, C., Bignami, C., Stramondo, S., Trasatti, E., Antonioli, A., and Boschi, E.: Finite fault inversion of DInSAR coseismic displacement of the 2009 L'Aquila earthquake (central Italy), Geophys. Res. Lett., 36, L15305, doi:10.1029/2009GL039293, 2009.

Barba, S. and Basili, R.: Analysis of seismological and geological observations for moderate-size earthquakes: the Colfiorito Fault System (Central Apennines, Italy), Geophys. J. Int., 141, 241–252, 2000.

Basili, R., Bosi, V., Galadini, F., Galli, P., Meghraoui, M., Messina, P., Moro, M., and Sposato, A.: The Colfiorito earthquake sequence of September-October 1997: Surface breaks and seismotectonic implications for the central Apennines (Italy), J. Earthq. Eng., 2, 291–302, 1998.

Bianchi, I., Chiarabba, C., and Piana Agostinetti, N.: Control of the 2009 L'Aquila earthquake, central Italy, by a high-velocity structure: A receiver function study, J. Geophys. Res., B12326, doi:10.1029/2009JB007087, 2010..

Bigi, S., Casero, P., Chiarabba, C., and Di Bucci, D.: Contrasting surface active faults and deep seismogenic sources unveiled by the 2009 L'Aquila earthquake sequence (Italy), Terra Nova, 25, 1, 21–29, doi:10.1111/ter.12000, 2013.

Boncio, P., Pizzi, A., Brozzetti, F., Pomposo, G., Lavecchia, G., Di Naccio, D., and Ferrarini, F. : Coseismic ground deformation of the 6 April 2009 L'Aquila earthquake (central Italy, M_W 6.3), Geophys. Res. Lett., 37, L06308, doi:10.1029/2010GL042807, 2010.

Bonini, L., Di Bucci, D., Toscani, G., Seno, S., and Valensise, G.: Reconciling deep seismogenic and shallow active faults through analogue modeling; the case of the Messina Straits (southern Italy), J. Geol. Soc. Lond., 168, 191–199. 2011.

Brace, W. F. and Byerlee, J. D.: California earthquakes: Why only shallow focus?, Science, 168, 1573–1575, 1970.

Calamita, F., Satolli, S., Scisciani, V., Esestime, P., and Pace, P.: Contrasting styles of fault reactivation in curved orogenic belts: Examples from the Central Apennines (Italy), Geol. Soc. Am. Bull., 123, 1097–1111, 2011.

Calderoni, G., Di Giovambattista, R., Vannoli, P., Pucillo, S., and Rovelli, A.: Fault-trapped waves depict continuity of the fault system responsible for the 6 April 2009 M_W 6.3 L'Aquila earthquake, central Italy, Earth Planet. Sci. Lett., 323–324, 1–8, 2012.

Carafa, M. M. C. and Barba, S.: The stress field in Europe: optimal orientations with confidence limits, Geophys. J. Int., 193, 531–548, doi:10.1093/gji/ggt024, 2013.

Chiarabba, C., Amato, A., Anselmi, M., Baccheschi, P., Bianchi, I., Cattaneo, M., Cecere, G., Chiaraluce, L., Ciaccio, M.G., De Gori, P., De Luca, G., Di Bona, M., Di Stefano, R., Faenza, L., Govoni, A., Improta, L., Lucente, F. P., Marchetti, A., Margheriti, L., Mele, F., Michelini, A., Monachesi, G., Moretti, M., Pastori, M., Piana Agostinetti, N., Piccinini, D., Roselli, P., Seccia, D., and Valoroso, L.: The 2009 L'Aquila (central Italy) M_W 6.3 earthquake: main shock and aftershocks, Geophys. Res. Lett., 36, L18308, doi:10.1029/2009GL039627, 2009.

Chiarabba, C., Bagh, S., Bianchi, I., De Gori, P., and Barchi, M.: Deep structural heterogeneities and the tectonic evolution of the Abruzzi region (Central Apennines, Italy) revealed by microseismicity, seismic tomography, and teleseismic receiver functions, Earth Planet. Sci. Lett., 295, 3, 462–476, 2010.

Cheloni, D., D'Agostino, N., D'Anastasio, E., Avallone, A., Mantenuto, S., Giuliani, R., Mattone, M., Calcaterra, S., Gambino, P., Dominici, D., Radicioni, F., and Fastellini, G.: Coseismic and initial post-seismic slip of the 2009 M_W 6.3 L'Aquila earthquake, Italy, from GPS measurements, Geophys. J. Int., 181, 1539–1546, 2010.

Cheloni, D., Giuliani, R., D'Anastasio, E., Atzori, S., Walters, R. J., Bonci, L., D'Agostino, N., Mattone, M., Calcaterra, P., Deninno, R., Maseroli, R., and Stefanelli, G.: Coseismic and post-seismic slip of the 2009 L'Aquila (central Italy) M_W 6.3 earthquake and implications for seismic potential along the Campotosto fault from joint inversion of high-precision levelling, InSAR and GPS data, Tectonophysics, doi:10.1016/j.tecto.2014.03.009, 2014.

Chiaraluce, L.: Unravelling the complexity of Apenninic extensional fault systems: A review of the 2009 L'Aquila earthquake (Central Apennines, Italy), J. Struct. Geol., 42, 2–18, 2012.

Chiaraluce, L., Barchi, M., Collettini, C., Mirabella, F., and Pucci, S.: Connecting seismically active normal faults with Quaternary geological structures in a complex extensional environment: The Colfiorito 1997 case history (northern Apennines, Italy), Tectonics, 24, TC1002, doi:10.1029/2004TC001627, 2005.

Chiaraluce, L., Valoroso, L., Piccinini, D., Di Stefano, R., and De Gori, P.: The anatomy of the 2009 L'Aquila normal fault system (central Italy) imaged by high resolution foreshock and aftershock locations, J. Geophys. Res., 116, B12311, doi:10.1029/2011JB008352, 2011.

Cinti, F. R, Pantosti, D., De Martini, P. M., Pucci, S., Civico, R., Pierdominici, S., Cucci, L., Brunori, C. A., Pinzi, S., and Patera, A.: Evidence for surface faulting events along the Paganica fault prior to the 6 April 2009 L'Aquila earthquake (central Italy), J. Geophys. Res., 116, B07308, doi:10.1029/2010JB007988, 2011.

Cirella, A., Piantanesi, A., Cocco, M., Tinti, E., Scognamiglio, L., Michelini, A., Lomax, A., and Boschi, E.: Rupture history of the 2009 L'Aquila (Italy) earthquake from non-linear joint inversion of strong motion and GPS data, Geophys. Res. Lett., 36, L19304, doi:10.1029/2009GL039795, 2009.

Cloos, E.: Experimental analysis of Gulf Coast fracture patterns, AAPG Bulletin, 52, 420–444, 1968.

Cooke, M. L., Schottenfeld, M. T., and Buchanan, S. W.: Evolution of fault efficiency at restraining bends within wet kaolin analog experiments, J. Struct. Geol., 51, 180–192, 2013.

Cooke, M. L. and van der Elst, N. J.: Rheologic testing of wet kaolin reveals frictional and bi-viscous behaviour typical of crustal materials, Geophys. Res. Lett., 39, L01308, doi:10.1029/2011GL050186, 2012.

D'Agostino, N., Cheloni, D., Fornaro, G., Giuliani, R., and Reale, D.: Space-time distribution of afterslip following the 2009 L'Aquila earthquake, J. Geophys. Res., 117, B02402, doi:10.1029/2011JB008523, 2012.

D'Agostino, N., Mantenuto, S., D'Anastasio, E., Giuliani, R., Mattone, M., Calcaterra, S., Gambino, P., and Bonci, L.: Evidence for localized active extension in the central Apennines (Italy) from Global Positioning System observation, Geology, 39, 291–294, 2011.

D'Anastasio, E., De Martini, P. M., Selvaggi, G., Pantosti, D., Marchioni, A., and Maseroli, R.: Short-term vertical velocity field in the Apennines (Italy) revealed by geodetic levelling data, Tectonophysics, 418, 219–234, 2006.

Di Domenica, A., Bonini, L., Fernando, C., Toscani, G., Galuppo, C., and Seno, S.: Analogue modeling of positive inversion tectonics along differently oriented pre-thrusting normal faults: An application to the Central-Northern Apennines of Italy, Geol. Soc. Am. Bull., doi:10.1130/B31001.1, 2014.

Di Stefano, R., Chiarabba, C., Chiaraluce, L., Cocco, M., De Gori, P., Piccinini, D., and Valoroso, L.: Geophys. Res. Lett. 38, L10310, doi:10.1029/2011GL047365, 2011.

Eisenstadt, G. and Sims, D.: valuating sand and clay models; do rheological differences matter? J. Struct. Geol., 27, 1399–1412, 2005.

Emergeo Working Group: Evidence for surface rupture associated with the M_W 6.3 L'Aquila earthquake sequence of April 2009 (central Italy), Terra Nova, 22, 43–51, 2010.

Erslev, E. A.: Trishear fault-propagation folding, Geology, 19, 617–620, doi:10.1130/0091-7613(1991)019<0617:TFPF>2.3.CO;2,1991.

Galadini, F.: Pleistocene changes in the central Apennine fault kinematics: a key to decipher active tectonics in central Italy, Tectonics, 18, 877–894, 1999.

Geological Map of Italy: L'Aquila sheet, n. 359, 1 : 50000 scale, available at: http://www.isprambiente.gov.it/MEDIA/carg/359_LAQUILA/Foglio.html, 2009.

Graveleau, F., Malavieille, J., and Dominguez, S.: Experimental modelling of orogenic wedges: a review, Tectonophysics, 538–540, 1–66, 2012.

Gori, S., Falcucci, E., Dramis, F., Galadini, F., Galli, P., Giaccio, B., Messina, P., Pizzi, A. Sposato, A., and Cosentino, D.: Deep-seated gravitational slope deformation, large-scale rock failure, and active normal faulting along Mt. Morrone (Sulmona basin, Central Italy): Geomorphological and paleoseismological analyses, Geomorfology, 208, 88–101, 2014.

Hanson, K. L., Kelson, K. I., Angell, M. A., and Lettis, W. R.: Techniques for identifying faults and determining their origins, NUREG/CR-5503, U.S. Nuclear Regulatory Commission, 1999.

Hardy, S. and Ford, M.: Numerical modeling of trishear fault propagation folding, Tectonics, 16, 841–854, doi:10.1029/97TC01171, 1997.

Henza, A., Withjack, M. O., and Schlische, R. W.: Normal-fault development during two phases of non-coaxial extension: An experimental study, J. Struct. Geol., 32, 1656–1667, doi:10.1016/j.jsg.2009.07.007, 2010.

Herrmann, R. B., Malagnini, L., and Munafò, I.: Regional moment tensors of the 2009 L'Aquila earthquake sequence, Bull. Seismol. Soc. Am., 101, 975–993, 2011.

Hyppolite, J.-C., Angelier, J., and Roure, F.: A major geodynamic change revealed by Quaternary stress patterns in the Southern Apennines (Italy), Tectonophysics, 230, 199–210, 1994.

Hubbert, M. K.: Theory of scale models as applied to the study of geologic structures, Geol. Soc. Am. Bull., 48, 1459–1520, 1937.

Improta, L., Villani, F., Bruno, P. P., Castiello, A., De Rosa, D., Varriale, F., Punzo, M., Brunori, C. A., Civico, R., Pierdominici, S., Berlusconi, A., and Giacomuzzi, G.: High-resolution controlled-source seismic tomography across the Middle Aterno basin in the epicentral area of the 2009, M_W 6.3, L'Aquila earthquake (central Apennines, Italy), Ital. J. Geosci., 131, 373–388, 2012.

Jin, G. and Groshong Jr., R. H.: Trishear kinematic modeling of extensional fault-propagation folding, J. Struct. Geol., 28 , 170–183, 2006.

Kim, Y.-S. and Sanderson, D. J.: The relationship between displacement and length of faults: a review, Earth-Sci. Rev., 68, 317–334, 2005.

Lin, J. and Stein, R. S.: Stress triggering in thrust and subduction earthquakes and stress interaction between the southern San Andreas and nearby thrust and strike-slip faults, J. Geophys. Res., 109, B02303, doi:10.1029/2003JB002607, 2004.

Mandl, G.: Faulting in brittle rocks: an introduction to the mechanics of tectonic faults, Springer Berlin, 2000.

Mazzotti, A., Stucchi, E., Fradelizio, G. L., Zanzi, L., and Scandone, P.: Re-processing of the CROP-04 seismic data: Bollettino della Società Geologica Italiana, 7, 141–153, 2007.

McCalpin, J. P.: Paleoseismology, Academic Press, 2 Edn., 647 pp, 2009.

Marone, C. and Scholz, C. H.: The depth of seismic faulting and the upper transition from stable to unstable slip regimes, Geophys. Res. Lett., 15, 621–624, 1988.

Michetti, A. M., Brunamonte, F., Serva, L., and Vittori, E.: Trench investigations of the 1915 Fucino earthquake fault scarp (Abruzzo, central Italy): geological evidence of large historical events, J. Geophys. Res., 101, 5921–5936, 1996.

Miller, J. F. and Mitra, S.: Deformation and secondary faulting associated with basement-involved compressional and extensional structures, AAPG Bull., 95, 675–689, 2011.

Montone, P., Mariucci, M. T., and Pierdominici, S.: The Italian present-day stress map, Geophys. J. Int., 189, 705–716, 2012.

Pantosti, D. and Valensise, G.: Faulting mechanism and complexity of the 23 November, 1980, Campania-Lucania earthquake inferred from surface observations, J. Geophys. Res., 95, 15319–15341, 2000.

Patacca, E. and Scandone, P.: Post-Tortonian mountain building in the Apennines. The role of the passive sinking of a relic lithospheric slab, in: The Lithosphere in Italy, edited by: Boriani, A., Bonafede, M., Piccardo, G. B., and Vai, G. G., Accademia dei Lincei, Roma, 157–176, 1989.

Perfettini, H. and Ampuero, J.-P.: Dynamics of a velocity strengthening fault region: Implications for slow earthquakes and post-seismic slip, J. Geophys. Res., doi:10.1029/2007JB005398, 11, 901-11, B09411, 912, 2008.

Satolli, S. and Calamita, F.: Differences and similarities between the Central and the Southern Apennines (Italy): examining the Gran Sasso vs. the Matese-Frosolone salients using paleomagnetic, Geological and Structural Data, J. Geophys. Res., 113, B10101, doi:10.1029/2008JB005699, 2008.

Schellart, W. P.: Shear test results for cohesion and friction coefficients for different granular materials: scaling implications for their usage in analogue modeling, Tectonophysics, 324, 1–16, 2000.

Scholz, C. H.: The brittle-plastic transition and the depth of seismic faulting, Geol. Rundsch. 77, 319–328, 1988.

Scholz, C. H.: The mechanics of earthquakes and faulting, Cambridge University Press, 439 pp, 1990.

Scisciani, V., Tavarnelli, E., and Calamita, F.: The interaction of extensional and contractional deformations in the outer zones of the Central Apennines, Italy, J. Struct. Geol., 24, 1647–1658.

Scognamiglio, L., Tinti, E., Michelini, A., Dreger, D. S., Cirella, A., Cocco, M., Mazza, S., and Piatanesi, A.: Fast determination of moment tensors and rupture history: What has been learned from the 6 April 2009 L'Aquila earthquake sequence, Seismol. Res. Lett., 81, 892–906. 2010.

Stirling, M., Goded, T., Berryman, K., and Litchfield, N.: Selection of Earthquake Scaling Relationships for Seismic-Hazard Analysis, Bull. Seismol. Soc. Am., 103, 1–19, 2013.

Valensise, G. and Pantosti, D.: The investigation of potential earthquake sources in peninsular Italy: a review, J. Seismol., 5, 287–306, 2001.

Valoroso, L., Chiaraluce, L., Piccinini, D., Di Stefano, R., Schaff, D., and Waldhauser, F.: Radiography of a normal fault system by 64,000 high-precision earthquake locations: the 2009? L'Aquila (central Italy) case study., J. Geophys. Res., doi:10.1002/jgrb.50130, 2013.

van Gent, H. W., Holland, M., Urai, J. L., Loosveld, R.: Evolution of fault zones in carbonates with mechanical stratigraphy – Insights from scale models using layered cohesive powder, J. Struct. Geol., 32, 1375–1391, 2010.

Vannoli, P., Burrato, P., Fracassi, U., and Valensise, G.: A fresh look at the seismotectonics of the Abruzzi (Central Apennines) following the 6 April 2009 L'Aquila earthquake (M_W 6.3), Ital. J. Geosci., 131, 309–329, 2012.

Vezzani, L., Festa, A., and Ghisetti, F.: Geological-structural map of the Central-Southern Apennines (Italy), 1:250,000 scale (S.E.L.CA. ed., Firenze), 2009.

Vittori, E., Di Manna, P., Blumetti, A. M., Comerci, V., Guerrieri, L., Esposito, E., Michetti, A. M., Porfido, S., Piccardi, L., Roberts, G. P., Berlusconi, A., Livio, F., Sileo, G., Wilkinson, M., McCaffrey, K. J. W., Phillips, R. J., and Cowie P. A.: Surface Faulting of the 6 April 2009 M_W 6.3 L'Aquila Earthquake in Central Italy, Bull. Seismol. Soc. Am., 101, 1507–1530, 2011.

Wells, D. L. and Coppersmith, K. J.: New empirical relationships among magnitude, rupture length, rupture width, rupture area, and surface displacement, Bull. Seismol. Soc. Am., 84, 974–1002, 1994.

Wesnousky, S. G.: Displacement and geometrical characteristics of earthquake surface ruptures: Issues and implications for seismic hazard analysis and the earthquake rupture process, Bull. Seismol. Soc. Am., 98, 1609–1632, 2008.

Withjack, M. O. and Jamison, W. R.: Deformation produced by oblique rifting, Tectonophysics, 126, 99–124, doi:10.1016/0040-1951(86)90222-2, 1986.

Withjack, M. O. and Schlische, R. W.: Geometric and experimental models of extensional fault-bend folds, Geol. Soc. Spec. Pub., 253, 285–305, 2006.

Withjack, M. O., Schlische, R. W, and Henza, A. A.: Scaled experimental models of extension: dry sand vs. wet clay, Houston Geol. Soc. Bull., 49, 31–49, 2007.

15

BrO / SO_2 molar ratios from scanning DOAS measurements in the NOVAC network

P. Lübcke[1,2], N. Bobrowski[1], S. Arellano[3], B. Galle[3], G. Garzón[4], L. Vogel[1,*], and U. Platt[1]

[1]Institute of Environmental Physics, University of Heidelberg, Heidelberg, Germany
[2]Max Planck Institute for Chemistry, Mainz, Germany
[3]Department of Earth and Space Sciences, Chalmers University of Technology, Gothenburg, Sweden
[4]FISQUIM Research Group, Laboratory Division, Colombian Geological Survey, Cali, Colombia
[*]now at: Earth Observation Science Group, Space Research Centre, Department of Physics and Astronomy, University of Leicester, Leicester, UK

Correspondence to: P. Lübcke (pluebcke@iup.uni-heidelberg.de)

Abstract. The molar ratio of BrO to SO_2 is, like other halogen / sulfur ratios, a possible precursor for dynamic changes in the shallow part of a volcanic system. While the predictive significance of the BrO / SO_2 ratio has not been well constrained yet, it has the major advantage that this ratio can be readily measured using the remote-sensing technique differential optical absorption spectroscopy (DOAS) in the UV. While BrO / SO_2 ratios have been measured during several short-term field campaigns, this article presents an algorithm that can be used to obtain long-term time series of BrO / SO_2 ratios from the scanning DOAS instruments of the Network for Observation of Volcanic and Atmospheric Change (NOVAC) or comparable networks. Parameters of the DOAS retrieval of both trace gases are given. The influence of co-adding spectra on the retrieval error and influences of radiative transfer will be investigated. Difficulties in the evaluation of spectroscopic data from monitoring instruments in volcanic environments and possible solutions are discussed. The new algorithm is demonstrated by evaluating data from the NOVAC scanning DOAS systems at Nevado del Ruiz, Colombia, encompassing almost 4 years of measurements between November 2009 and end of June 2013. This data set shows variations of the BrO / SO_2 ratio several weeks prior to the eruption on 30 June 2012.

1 Introduction

The molar ratio of halogen (mainly HCl, HF) to sulfur species (mainly SO_2) in volcanic plumes has long been known as a possible tracer for volcanic activity (e.g. Noguchi and Kamiya, 1963; Menyailov, 1975; Pennisi and Le Cloarec, 1998). This is due to the different solubility of volatiles in magma (e.g. Carroll and Holloway, 1994) and its dependence on pressure. As a consequence, different gaseous species are exsolved in equilibrium at different magmastatic pressures, i.e. at different depths in the magma column. The most commonly used example of halogen / sulfur ratios has been the measurement of the HCl/SO_2 molar ratio (e.g. Noguchi and Kamiya, 1963). Another halogen species is BrO, which has now been detected at several volcanoes (e.g. Bobrowski et al., 2003; Bobrowski and Platt, 2007; Boichu et al., 2011; Kelly et al., 2012; Hörmann, 2013) and whose abundance was suggested as an additional indicator for changes in volcanic activity (Bobrowski and Giuffrida, 2012). One major advantage of measuring the BrO / SO_2 molar ratio is that both species can be detected in a single spectroscopic remote sensing measurement using differential optical absorption spectroscopy (DOAS) in the UV (Platt and Stutz, 2008), which is a rather simple-to-implement and sensitive technique.

Scanning DOAS has become the most widespread method for measuring the emission rate of volcanic SO_2 (e.g. Galle et al., 2003). In particular, within the framework of the EU project NOVAC (Network for Observation of Volcanic and

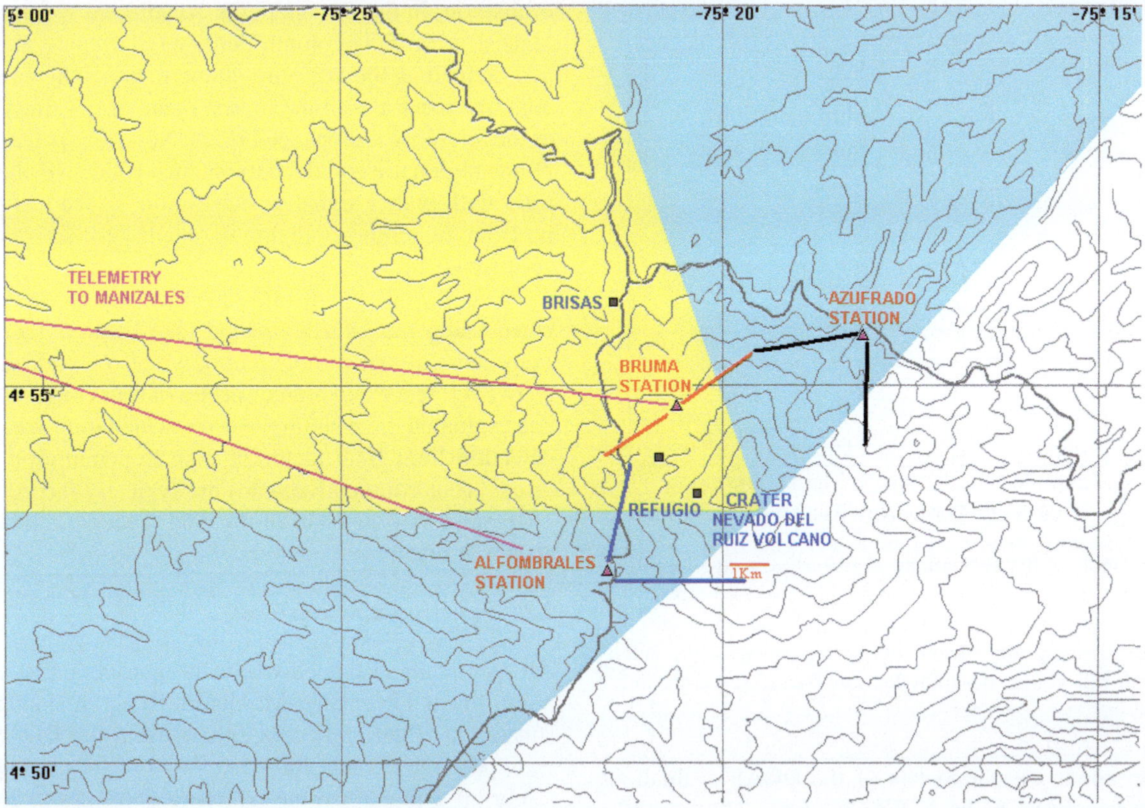

Figure 1. Topographic map (contour levels every 200 m, the upper contour level is at 5200 m, Bruma station is at 4800 m a.s.l.) showing the location of Nevado del Ruiz volcano (Colombia) and the three NOVAC stations used for gas monitoring. The blue, red and black lines depict the viewing directions of the scanning DOAS systems. The two original instruments at Alfombrales Station and Bruma Station were installed in November 2009 and transmit data in real time to the volcanological observatory located at Manizales city. The third instrument at Azufrado Station was installed in May 2012; its data are not included in this paper. The yellow background colour shows the wind direction for approximately 70 % of the time, and the light blue shaded area shows the wind direction for the remaining 30 %.

Atmospheric Change, Galle et al., 2010) to date about 70 scanning DOAS systems have been installed at 26 volcanoes worldwide, with the goal of performing automated spectroscopic gas emission rate measurements. These instruments measure the volcanic SO_2 emission rates by acquiring scattered UV radiation spectra over a scan through a surface intercepting the volcanic plume and combining the DOAS-derived gas column densities with geometrical factors and the component of transport speed normal to the scanned surface. Typically, about 40 individual gas emission rate measurements are obtained daily at each volcano by NOVAC instruments, which have allowed operational volcanic gas monitoring to approach temporal resolutions comparable to other methods of geophysical surveillance.

While the spectra are routinely evaluated for SO_2 by the local observatories, this is not the case for BrO. The BrO absorption cross section (in this work the one of Fleischmann et al., 2004, is used) is roughly a factor of 150 larger than that of SO_2 (in this work the one of Vandaele et al., 2009, is used) in the wavelength ranges typically used for DOAS evaluations (Vogel, 2011), but the abundance of BrO is significantly lower. BrO / SO_2 ratios between 8×10^{-5} and 1×10^{-3} have been found in volcanic plumes (Bobrowski and Platt, 2007), thus leading to one to two orders of magnitude weaker BrO absorption structures compared to SO_2. The smaller optical density makes it technically more challenging to detect and accurately measure BrO. In particular, small instrumental imperfections like changes in the instrument's slit function and the wavelength-channel alignment due to, for example, changes of the instrument's temperature (e.g. Chapter 3 in Kern, 2009), which have only a negligible effect on the SO_2 evaluation, pose a challenge for the BrO evaluation. Additionally, the noise due to photon statistics has to be lower in order to obtain a good signal to noise ratio for the BrO evaluation. One way to achieve this is by co-adding several spectra obtained by the same instrument under similar measurement conditions (e.g. within the same horizon-to-horizon scan or from consecutive scans with close time proximity).

This work presents an algorithm to evaluate the spectra recorded by the NOVAC instruments in order to obtain BrO column densities and BrO / SO_2 ratios. We give detailed

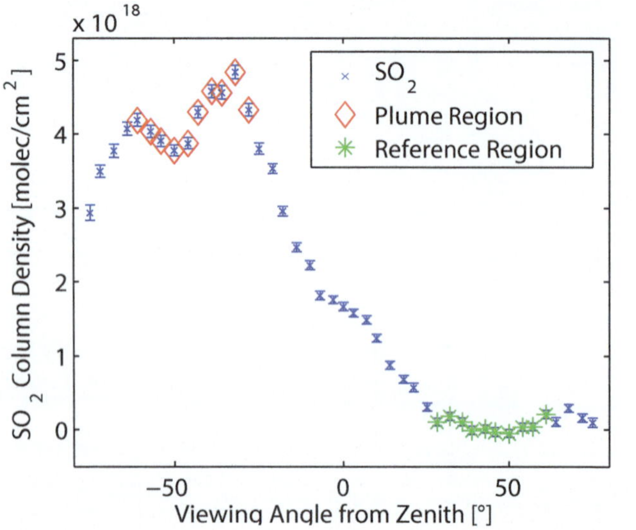

Figure 2. Example of the SO_2 column densities from one complete scan (−75° – zenith – +75° elevation angle, measured clockwise when looking towards the volcano). The red diamonds show the column densities defining the volcanic plume region, while the green asterisks show the column densities defining the reference region.

information about the parameters of the DOAS evaluation used for the analysis of the NOVAC spectra. The feasibility of the BrO evaluation will be demonstrated as well as the magnitude of DOAS fit errors. As an example, a first long-term time series from the Colombian volcano Nevado del Ruiz (see Fig. 1) is presented.

2 Method

2.1 Evaluation procedure

The NOVAC Version 1 instruments installed at all volcanoes in NOVAC use a "flat" or "conical" scanning geometry; that is, all viewing directions of a scan lie in a plane or on the surface of a semi-cone open towards the volcano (Galle et al., 2010). The set of consecutive scattered sunlight spectra collected from horizon to horizon (in this work only viewing directions between −75° to +75° elevation angle were evaluated) is referred to as a plume scan in the following. The instruments utilize Ocean Optics S2000 spectrometers (with a wavelength range of 280–425 nm and an optical resolution of ≈ 0.6 nm) that are not thermally stabilized, which results in a robust, energy-efficient design suitable for operation in remote locations that are sometimes difficult to access.

All spectra of each plume scan are pre-evaluated for SO_2 using a zenith spectrum acquired at the beginning of each plume scan as a reference spectrum. At this time underexposed and overexposed spectra (intensity below 15 % or above 85 % of the maximum exposure in the spectral region used for the DOAS evaluation) were removed from further evaluation. In the next step, the 10 adjacent spectra with the highest mean SO_2 column density (CD) value are selected and defined as the volcanic plume region. Then a set of 10 adjacent spectra with the lowest mean SO_2 value is defined as the reference region (see Fig. 2). Next, the spectra from the plume region are co-added to obtain a collective plume spectrum with higher signal to noise ratio, as well as the spectra from the reference region to obtain a collective reference spectrum.

The collective plume and reference spectra are then evaluated using the DOAS method (see below). To obtain an even better signal to noise ratio, we also co-added the collective reference and plume spectra of four consecutive scans and evaluated the resulting spectra, which improved our BrO detection limit by a factor of two. Depending on the time of day, four consecutive scans are typically recorded within 15–60 min. A flow chart of the data processing procedure is shown in Fig. 3.

2.2 Spectroscopic retrieval

All spectra are evaluated with a fitting routine that combines a non-linear Levenberg–Marquardt fit and a standard least-squares fit (Platt and Stutz, 2008) using the DOASIS software package (Kraus, 2006).

For all evaluations, the reference spectrum is first wavelength calibrated using a high-resolution solar spectrum (Chance and Kurucz, 2010) that was convoluted with the line shape of the 334.15 nm line of a low-pressure mercury lamp to match the lower spectral resolution of the spectrometers used in the NOVAC instruments (of about 0.6 nm FWHM – full width at half maximum). This line of the Hg spectrum was also used for the convolution of the high-resolution absorption cross sections; it was selected due to its proximity to the spectral region of interest for our evaluations. The calibration from the reference spectrum is also used for the plume spectra. The fit coefficient of the reference spectrum in the DOAS fit is fixed to −1. Two Ring spectra are included in the DOAS fit to correct for the Ring effect (Grainger and Ring, 1962). The first Ring spectrum is a standard Ring spectrum calculated using the DOASIS software package; the second is created from the first Ring spectrum by multiplying it with a wavelength-dependent term (in this work λ^{-4}) to account for multiple Rayleigh scattering in the atmosphere (Wagner et al., 2009).

A third-order polynomial is included in the fit to account for broadband absorption structures and Rayleigh as well as Mie scattering structures in the spectra. An additional wavelength-independent offset in the intensity space is allowed to correct for stray light inside the spectrometer.

The SO_2 evaluation is performed in the wavelength range between 314.8 nm and 326.8 nm including one SO_2 absorption cross section at 298 K (Vandaele et al., 2009) and one O_3 absorption cross section at 221 K (Burrows et al., 1999).

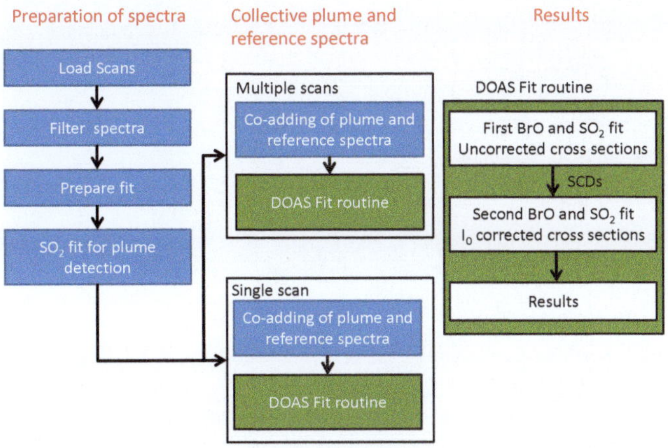

Figure 3. Flow chart of the data evaluation to automatically retrieve BrO/SO_2 ratios.

The BrO evaluation is performed between 330.6 nm and 352.75 nm using the following absorption cross sections: BrO at 298 K (Fleischmann et al., 2004), the same SO_2 and O_3 as above, O_4 (Hermans, 2003), NO_2 at 298 K (Vandaele et al., 1998) and CH_2O at 298 K (Meller and Moortgat, 2000). An example of a BrO fit is shown in Fig. 4. The evaluation ranges and trace gases included in the SO_2 as well as in the BrO fit were chosen according to Vogel (2011) and Vogel et al. (2013), who performed a series of studies on the optimal evaluation wavelength range for BrO and SO_2 in a combination of real measurement data and theoretical studies.

To verify the results, both trace gases were also evaluated in additional wavelength ranges. SO_2 was additionally evaluated between 326.5 and 353.3 nm, a wavelength range already used by Hörmann et al., 2013, as an additional wavelength range for the SO_2 retrieval. BrO was additionally evaluated between 327 and 347 nm. This wavelength range was used by Kelly et al., 2013, for the BrO retrieval.

All trace gas reference cross sections are convoluted to match the instrument resolution using the 334.15 nm line of a mercury lamp, a sample of the instrument line shape, recorded with the respective spectrometer. To account for the I_0 effect (Platt and Stutz, 2008), an iterative approach is used. First, a fit is performed to retrieve uncorrected column densities of the trace gases; these column densities are then used to create I_0-corrected absorption cross sections (Wagner et al., 2002).

In order to correct for small inaccuracies in the wavelength calibration, the reference spectrum and both Ring spectra, as one set, and all trace gas absorption cross sections, as another set, are allowed to be shifted and squeezed against the measurement spectrum. A shift of ± 0.2 nm and a squeeze of 0.98–1.02 were allowed.

2.3 Temperature influences on the DOAS retrieval

The instruments installed at Nevado del Ruiz are not thermally stabilized. While this greatly improved the robustness of these instruments, it should be noted that variations of the spectrometer's temperature can lead to problems in the DOAS retrieval.

Figure 5 shows the range of temperatures the instruments at Nevado del Ruiz are exposed to and that the temperature has an effect on the instrument calibration as well as the instrumental line shape. In general variations of the instrument's wavelength calibration are not a problem as spectra can be calibrated against a high-resolution solar spectrum, an approach also used here.

However, there are two problems if the instrument's optical alignment changes: the first problem appears when co-adding spectra. The spectral intensities are summed up for each channel; if the pixel-wavelength alignment changes between different spectra which are co-added, the resulting spectrum can be erroneous. However, the temperature changes slowly, and most spectra that are co-added were recorded within a time of 15 min up to 1 h. Therefore, this effect should be negligible.

The second effect of the instrument's temperature regards the instrumental line shape. Currently, all spectra are convolved with the 334.15 nm line of a mercury lamp that was recorded at room temperature. Instrumental characterization of S2000 spectrometers performed within the NOVAC project (i.e. Pinardi, 2007, or Kern, 2009) reported changes of up to 10 % for the SO_2 column densities caused by temperature-induced changes of the instrument line shape for temperatures between 5 and 40 °C (comparable to the temperatures occurring at Nevado de Ruiz).

To assess the error due to the second effect, we performed studies using synthetic spectra (similar to Vogel, 2011). A measurement spectrum containing a BrO CD of 1.5×10^{14} molecules cm^{-2} and an SO_2 CD of 1×10^{18} molecules cm^{-2} and a Fraunhofer reference spectrum free of volcanic absorbers were simulated (see Appendix for details). The measurement spectrum and Fraunhofer reference spectrum were convolved to match the lower resolution of a theoretical instrument with a Gaussian profile with varying FWHM (steps of 0.025 nm between 0.525 nm and 0.775 nm). The absorption cross sections for the DOAS retrieval of all spectra were convolved with a Gaussian profile with a FWHM of 0.65 nm. After preparation each set of spectra (measurement and Fraunhofer reference spectrum convolved with the same Gaussian profile) was evaluated using the DOAS retrieval described above. The results of these evaluations are shown in Fig. 6. The BrO/SO_2 ratio deviates up to 15 % if the instrument line shape varies up to ± 0.125 nm (which is a slightly larger range than can be typically observed for the NOVAC instruments).

Since for most NOVAC spectrometers there are no records of the particular instrument's line shape at different

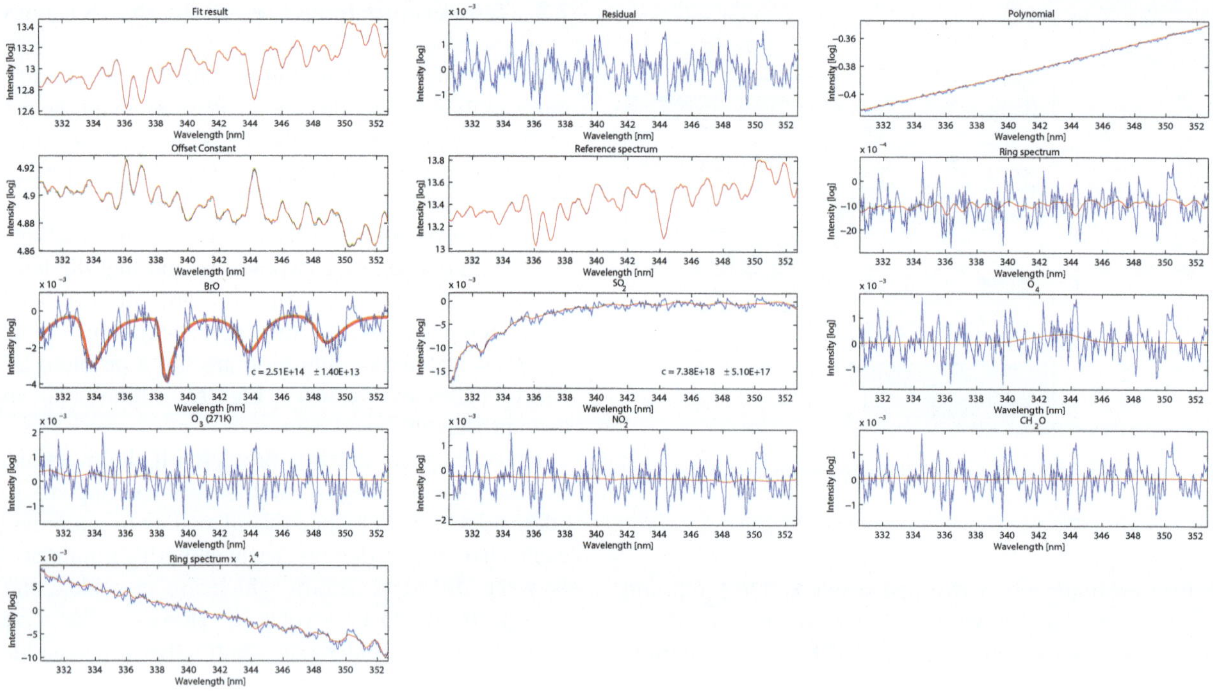

Figure 4. Example of a DOAS BrO evaluation performed in the wavelength range of 330.6–352.75 nm. The red lines show the fit, the blue lines the measurement data. A reference spectrum and two Ring spectra were included in the DOAS fit. The BrO column density is 2.51×10^{14} molecules cm^{-2}. Besides BrO, other trace gases (SO_2, O_4, O_3, NO_2, and CH_2O) were included in the DOAS evaluation. The spectra recorded for this example were recorded on 31 October 2012 between 16:32 and 17:05 GMT.

temperatures available, one possible way to further improve data quality would be to determine the instrument line shape from a comparison between the high-resolution solar reference spectrum and measurement spectra at different temperatures.

2.4 Radiative transfer modelling

The distance between the instrument and the volcanic plume and clouds can both influence the retrieved trace gas CDs. To assess the order of magnitude of errors due to these influences, we performed calculations using the 3-D radiative transfer model McArtim (Deutschmann, 2011). The SO_2 and BrO CDs were approximated by simulations at the centre of the wavelength interval used for the DOAS evaluation (320 nm for SO_2 and 340 nm for BrO).

A volcanic plume with a circular cross-section shape with diameter 1 km in the $x-z$ plane was simulated. The plume centred at 6 km altitude and was assumed to be infinitely extended in the y direction (see Fig. 7). Typical profiles of the trace gases O_3, O_4 and NO_2 and a Rayleigh atmosphere were used to account for non-volcanic absorbers and scattering. Trace-gas concentrations of SO_2 and BrO were set to 1×10^{13} molecules cm^{-3} and 1×10^{9} molecules cm^{-3} respectively, inside the volcanic plume. This leads to column densities of 1×10^{18} molecules cm^{-2} for SO_2 and 1×10^{14} molecules cm^{-2} for BrO, assuming the light path is

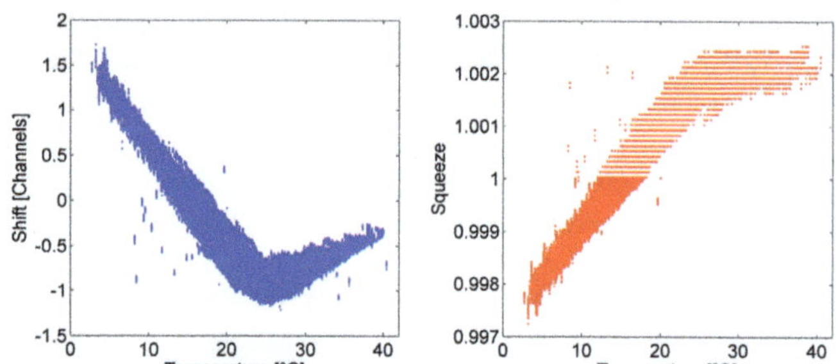

Figure 5. Shift and squeeze for a DOAS fit only containing one reference spectrum with an instrument temperature of 15 °C. For this plot half a year of data were evaluated. It can be clearly seen that the instrument pixel-to-wavelength alignment is influenced by changes in the ambient temperature.

a straight line through the plume centre. The resulting BrO/SO_2 ratio is 1.0×10^{-4}. The instrument was placed below the plume at a height of 4.5 km, and the viewing direction was always directed towards the centre of the volcanic plume. The influence of an increasing distance between instrument and plume has been investigated by varying the lateral distance between 0 km and 6.5 km.

In order to investigate the influence of clouds, we simulated clouds in different heights. Three different cloud layers with heights between 5.5 and 6.5 km (the height of the

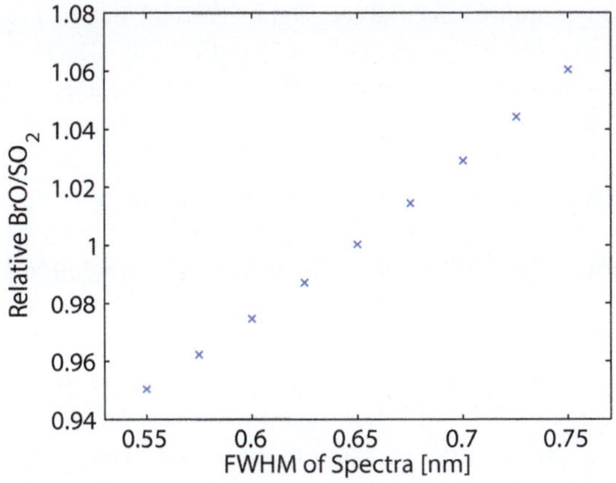

Figure 6. Deviation of the BrO/SO_2 ratio for varying synthetic spectra with different optical resolution. All spectra were evaluated using absorption cross sections that were convolved with a Gaussian profile with a FWHM of 0.65 nm.

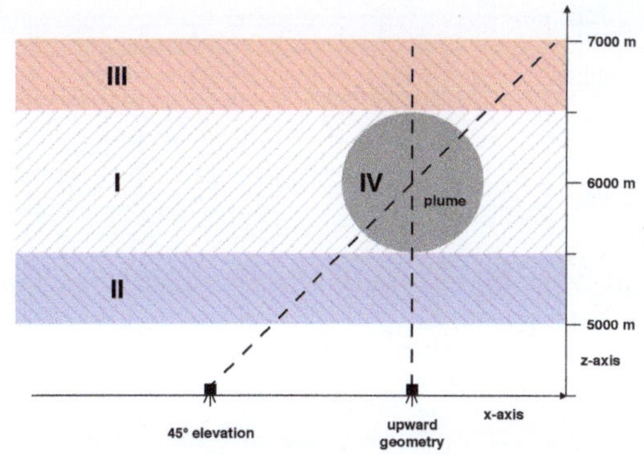

Figure 7. Sketch of the simulated geometries for the McArtim simulations. A round plume with a diameter of 1 km was simulated with the instrument either directly below the volcanic plume or at a horizontal distance. The hatched areas indicate the three different regions in which clouds were simulated. The fourth scenario assumes the absence of clouds but a homogeneous aerosol enhancement inside the volcanic plume (round grey area).

volcanic plume, CS I), 5 and 5.5 km (a cloud layer between the volcanic plume and the instrument, CS 2) and 6.5 and 7 km (a cloud layer above the instrument, CS 3) were simulated. Aerosol with an Ångström exponent of 1, a single scattering albedo of 1 and an asymmetry parameter $g = 0.85$ was used in simulating the cloud (see, e.g. Wagner et al., 2013). Two different aerosol extinction coefficients were simulated to assess variations of the cloud optical depth. The aerosol extinction coefficient was chosen so that the aerosol optical depth for an upward looking instrument was 4.25 and 4 or 12.75 and 12 at 320 and 340 nm, respectively. Additionally as a fourth scenario an enhanced aerosol abundance coinciding with the volcanic plume was simulated. For all cloud simulations only two geometries – an instrument below the volcanic plume with an upward-looking telescope (elevation angle 90°), and an instrument at a lateral distance of 1.5 km with an elevation angle of 45° – were simulated.

3 Results from Nevado del Ruiz

Nevado del Ruiz is a 5400 m high stratovolcano located in the Andes in Colombia, approximately 130 km west of Bogotá. Nevado del Ruiz became infamous in 1985, when on the 13 November approximately 23 000 people were killed by lahars in the aftermath of an eruption. A map showing the location of Nevado del Ruiz and the original two NOVAC instruments whose data are used in this work and their scanning directions are given in Fig. 1.

3.1 Spectroscopy results

Scans between November 2009 and the end of June 2013 were evaluated for two instruments at Nevado del Ruiz, Colombia. Figure 8 shows the number distribution of the BrO DOAS retrieval errors for both instruments (with identification numbers D2J2200 for Bruma and D2J2201 for Alfombrales; see Fig. 1). The fit error is shown for the evaluation of spectra from individual scans as well as for spectra resulting from co-adding four consecutive scans. The DOAS fit errors for the two instruments are essentially equal. For a single scan the fit error is centred around 2×10^{13} molecules cm^{-2}; for the co-added spectra the fit error is centred around 1×10^{13} molecules cm^{-2}. That is, averaging four scans reduces to half the fit error, which indicates that the evaluation is still limited by photon shot noise.

Adding up spectra from different scans is always a trade-off between gaining in signal to noise and losing in absolute signal (for some spectra) and time resolution of the retrieved slant column densities (SCDs). An additional problem may be caused by a drifting spectral instrument response due to changes of the instrument temperature, which would compromise the accuracy of the results when summing up spectra. The variability of the instrument slit function and the wavelength-channel alignment with changing instrument temperature for the spectrometers used in this study is shown in Fig. 5. For this example, one zenith looking spectrum was fitted to spectra from half a year of data with a fit range covering the SO_2 and the BrO evaluation range. Figure 5 shows, as one example, the shift and squeeze of the reference spectrum, which both clearly vary with changes of the ambient temperature.

However, as the signal to noise ratio for four consecutive scans is still close to the ideal photon statistics limited case, the approach to add up spectra from four consecutive scans is thought to not be influenced by these temperature issues and is used in the rest of this manuscript.

SO_2 and BrO column density time series and the SO_2 emission rates are shown in Fig. 9. Both the SO_2 and the BrO column densities start to increase at the beginning of

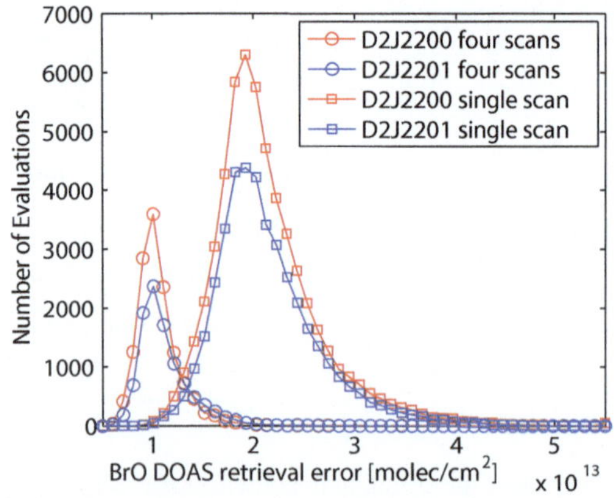

Figure 8. The BrO DOAS fit errors for both instruments whose data are presented in this study. The plot shows the results for spectra from single scans and for spectra co-added from four consecutive scans. The latter have a 50 % lower BrO DOAS fit error than single scans.

2012, as well as the SO_2 emission rates. Towards the beginning of 2013, the SO_2 column densities start to decline while the BrO column densities stay elevated, thus leading to increasing BrO / SO_2 ratios. Unfortunately, both instruments have some periods without transmitted data. D2J2201 has very few scans between July 2011 and November 2011, and the instrument with the serial number D2J2200 has a gap between June 2012 and September 2012. These gaps are due to technical difficulties that can occur when operating the instruments in remote locations. That is, the gap of D2J2200 after June 2012 can be explained by ash from the eruption falling onto the surface of the solar panels, which leads to a lack of power supply.

As there is only a relative small amount of data points ($\sim$ 1–30) per day after the co-addition of scans, the usual approach of calculating the linear regression for a large amount of data points (Bobrowski et al., 2003) to find the BrO / SO_2 ratio cannot be used here. We therefore calculated daily averages of the BrO / SO_2 ratios from the spectra created from co-adding four consecutive scans for the sake of lower noise. Averaging over all ratios can lead to erroneous results, since the average would include many scans without significant volcanic gas and therefore negligible SO_2 and BrO column densities. Taking the ratio of two values (i.e. SO_2 and BrO SCDs) that are close to zero gives unpredictable and unrealistic results. Therefore, SCDs measured outside the volcanic plume have to be excluded. One possibility is setting a BrO threshold equivalent to a certain factor of the retrieval error as a BrO detection limit. An example of a time series created for all scans containing BrO above the detection limit (in this case 4 times the retrieval error was chosen) is shown in Fig. 10. The BrO / SO_2 ratio is almost constant between January 2011 and January 2012 with a value of roughly 6–8×10^{-5}. Starting January 2012, the ratio drops down to values as low as 2×10^{-5}, and increases again up to 4×10^{-5} after an eruption on 30 June 2012 (vertical bar in Fig. 10). However, this approach is problematic since BrO is often close to the detection limit. In this case, setting a BrO threshold would remove low BrO values, and thus lead to elevated BrO / SO_2 ratios.

In this work we therefore chose to set an SO_2 threshold of 7×10^{17} molecules cm^{-2} to select spectra for the evaluation of the BrO / SO_2 ratio. This threshold is a relatively high SO_2 column density. However for the lower values of the BrO / SO_2 ratio in Fig. 10 ($\sim 2 \times 10^{-5}$) this would result in a BrO column density as low as 1.4×10^{13} molecules cm^{-2}, a value only slightly higher than the average DOAS retrieval error for BrO. This approach assures that scans not seeing significant amounts of volcanic gas are filtered out and thus will not significantly influence the BrO / SO_2 ratio. The result is shown in the middle of Fig. 10; the BrO / SO_2 ratio is again constant until January 2012 but with a lower value of 4×10^{-5}. In 2012 there is a drop before the ratio increases again up to 4×10^{-5} after the eruption. Towards the end of 2012, the ratio sinks until the end of June 2013 (no more data were available at the time of this manuscript).

The trends found in the BrO / SO_2 molar ratios are similar irrespective of the thresholds chosen for the calculation of the mean BrO / SO_2 ratio. Interestingly, they show a clear correspondence with the general evolution of the volcanic activity, as expressed by the levels of risks defined by the authorities. The decrease in the measured BrO / SO_2 ratio starts a few months before the main eruptive event that led to a rise of the alert to the highest (red) level.

After the eruption the BrO / SO_2 ratio recovers again, when the level of activity decreases. Many causes can produce the observed pattern in the BrO / SO_2 molar ratio, as for example the injection of an SO_2-rich batch of magma, or atmospheric effects like a different depletion rate of the two species due to scavenging by aerosols and ash in the plume. Garzon et al. (2013) suggested that the time between 30 March 2012 and 30 May 2012 was dominated by magma intrusion processes and had high sulfur emissions as well as first ash emissions (see Fig. 9 for daily averages of the SO_2 emission rates). Although a detailed analysis of this process is beyond the scope of this paper, the correspondence with the overall eruptive activity shows the potential of measuring both gases simultaneously as a precursory signal of volcanic eruptions.

3.2 The influence of radiative transfer

The results of the radiative transfer model calculations for an increasing lateral distance between instrument and plume (without any clouds) are shown in Figs. 11 and 12. Figure 11 shows the decline of the SO_2 and BrO CDs. At a distance of 6.5 km, only 75 % of the SO_2 CD and and 78 % of the BrO CD are retrieved. However, the variation of the BrO/SO_2 ratio is significantly lower (see Fig. 12). Even at a lateral distance of 6.5 km, the BrO/SO_2 ratio is only ≈ 6 % above the

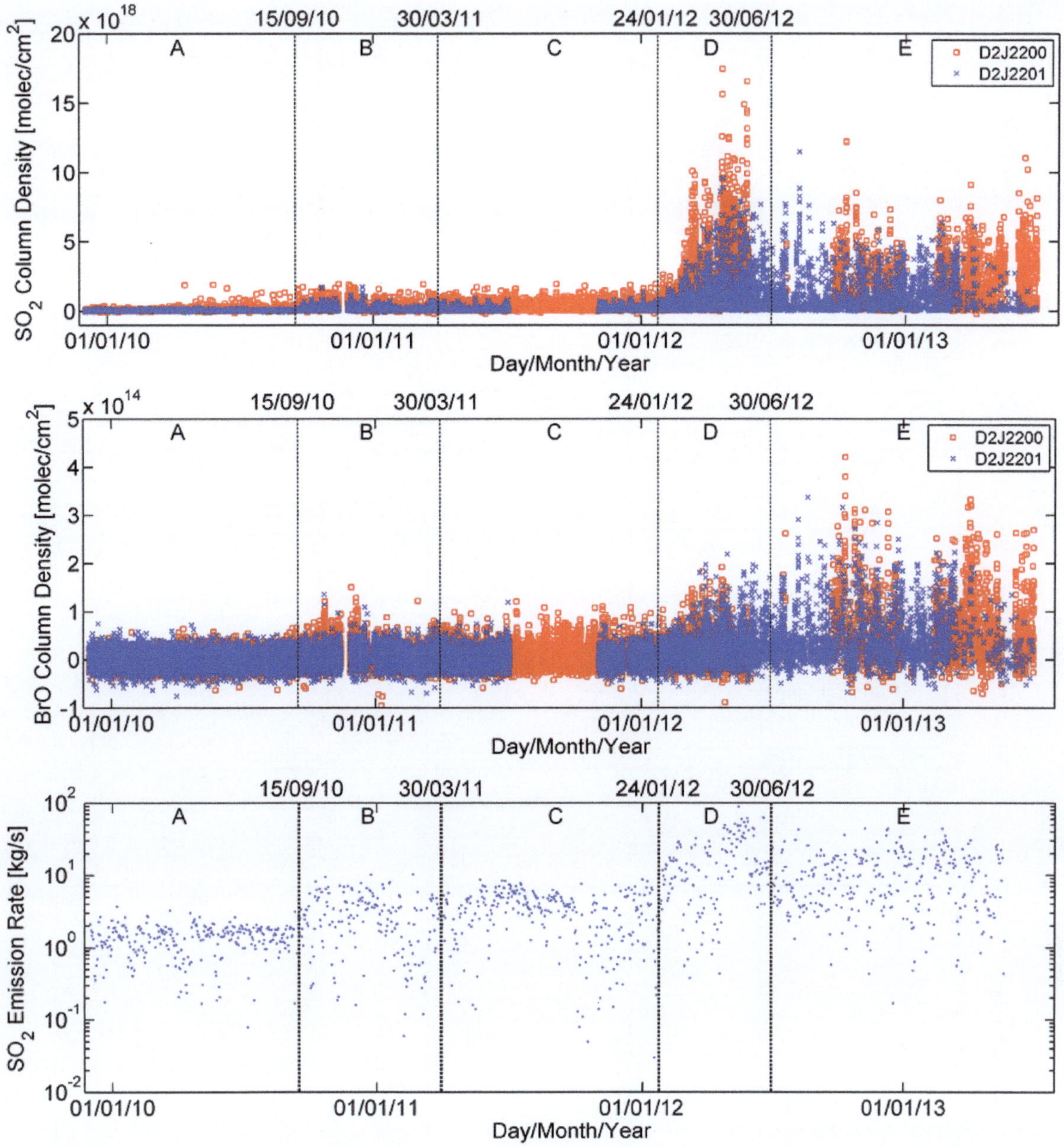

Figure 9. Time series from the two instruments at Nevado del Ruiz. The instrument D2J2200 is located in a way that measures elevated SO_2 and BrO column density values more frequently (see Fig. 1). The spectra evaluated were created from co-adding reference and plume spectra from four consecutive scans. Top: SO_2 CDs; centre: BrO CDs; bottom: Daily averages of the SO_2 emission rates. The wind speed was taken from the ECMWF database. More details are available in Galle et al. (2010) and Arellano (2013).

true value. Generally speaking, the slight overestimation of the BrO/SO_2 ratio arises from the fact that the light dilution effect leads to a decrease of the detected CDs and is more pronounced at smaller wavelengths. The situation is more complex in reality. The light-dilution effect depends on the wavelength as well as on the optical density of the absorber. However, these simulations show that the influence of light dilution is smaller for the ratio of trace gases than for the CDs.

The results for the different clouds scenarios are shown in Table 1 (aerosol optical depth, AOD = 4 at 340 nm for an upward-looking instrument) and Table 2 (AOD = 12 at 340 nm for an upward-looking instrument). We can see that a cloud at the same height as the volcanic plume (CS I and CS IV) leads to larger trace gas column densities (up to 40 % overestimation depending on viewing geometry and AOD). However, the BrO/SO_2 ratio is retrieved correctly within 2.4 %. The largest deviations can be observed for a cloud

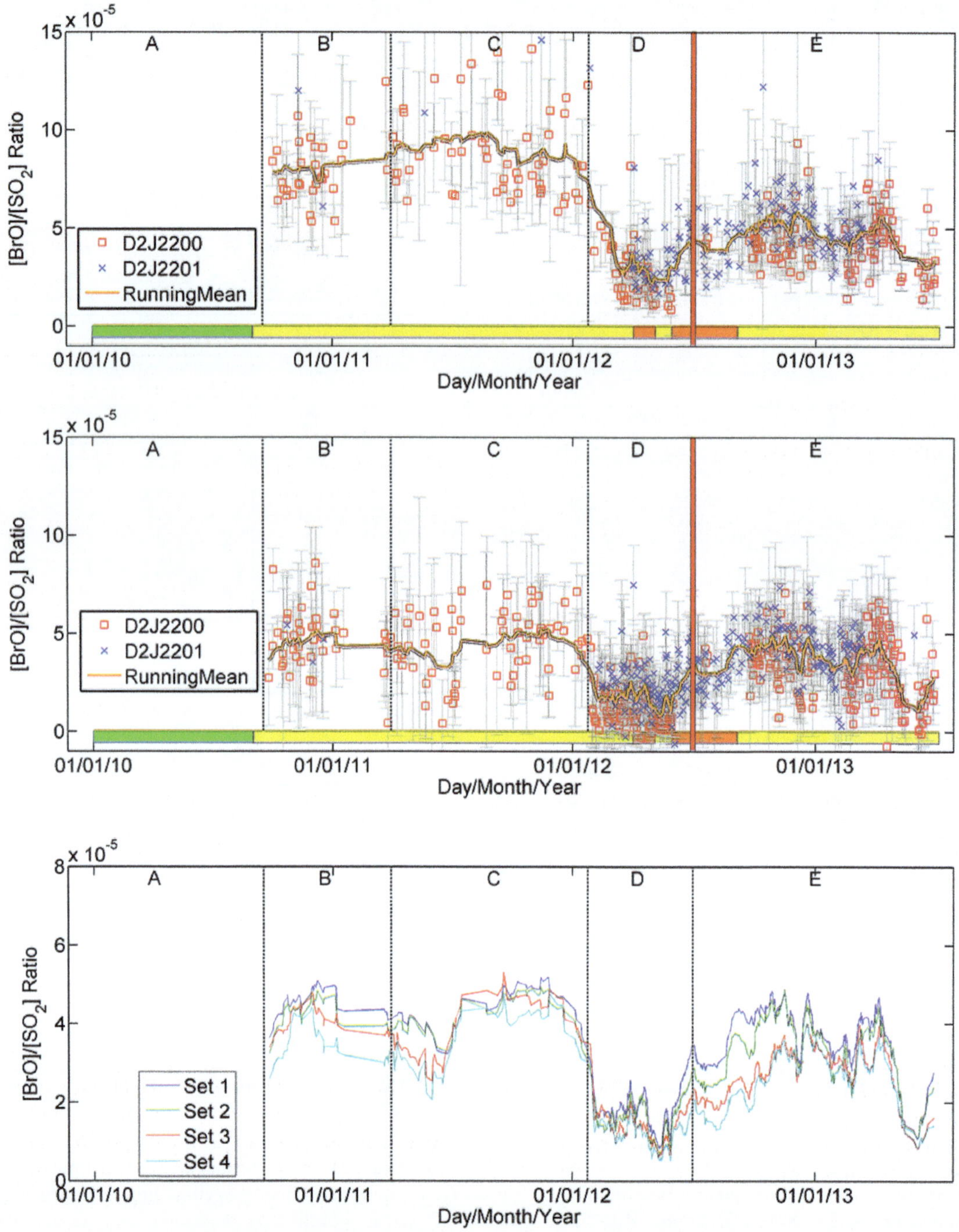

Figure 10. Daily averages of the BrO/SO_2 molar ratios from Nevado del Ruiz from November 2009 to June 2013. The ratio drops down to values below 5×10^{-5}, and increases again after the eruption on 30 June 2012 (red vertical bar). Also shown at the bottom of the image is the volcanic activity risk level defined by the Colombian Geological Survey (coloured horizontal bars). The black-orange line shows a running mean around seven data points (from both instruments). Top: the criterion for a valid measurement is a BrO column density above four times the BrO DOAS retrieval error. Center: for this figure an SO_2 threshold of 7×10^{17} molecules cm^{-2} was chosen as the criterion for valid data points. Bottom: Running mean of the BrO/SO_2 time series retrieved in different wavelength ranges, with an SO_2 threshold of of 7×10^{17} molecules cm^{-2} as the criterion for valid data points.

below the volcanic plume. For an instrument with an elevation angle of 45°, the true CDs are underestimated by up to 80 %. However, even in these conditions the deviation of the BrO/SO_2 ratio is below 3.5 %. This shows that the BrO/SO_2 ratio is much more robust with regards to radiative transfer

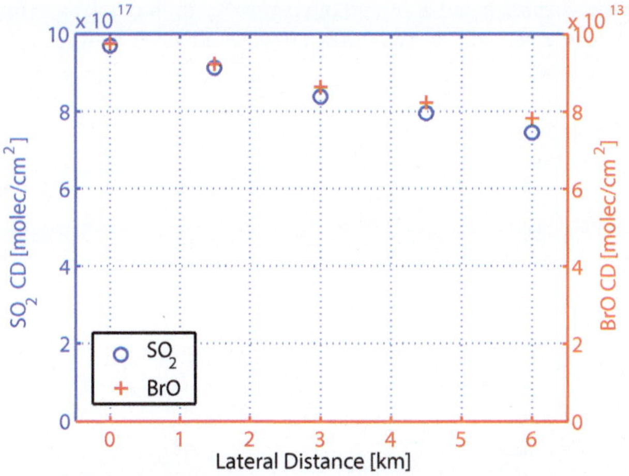

Figure 11. The influence of an increasing lateral distance between instrument and plume on the SO_2 and BrO CDs. For all calculations the volcanic plume was simulated 1.5 km above the instrument. The SO_2 column densities were simulated at 320 nm, the BrO column densities at 340 nm.

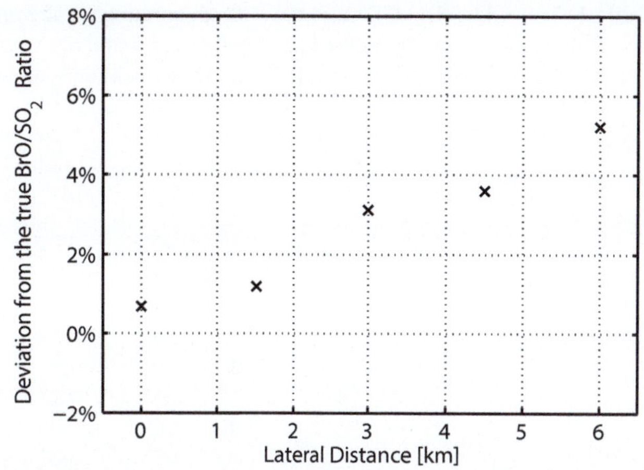

Figure 12. McArtim results for the influence of an increasing lateral distance between instrument and volcanic plume on the SO_2/BrO ratio. The ratio of the two trace gases is only slightly influenced by an increasing distance.

issues than individual SO_2 or BrO CDs, as one would expect intuitively.

As an additional verification step for the retrieved BrO/SO_2 ratios, both trace gases were evaluated in different wavelength ranges. As the alternative SO_2 evaluation range (326.8–353.5 nm) does not contain the strong SO_2 absorption bands, the DOAS fit error was found to be roughly 10 times higher than in our standard range. We found good correlation between the SO_2 CDs retrieved in the two evaluation ranges. The SO_2 CDs in the alternative wavelength range are $\approx 30\,\%$ higher than in the standard wavelength range. However, the alternative evaluation range should be treated with care because of the greater instability of the fitting algorithm due to the lower absorption of SO_2.

The BrO CDs retrieved in the two evaluation ranges are generally the same. The BrO fit error in the alternative wavelength range (327–347 nm) was found to be $\approx 30\,\%$ higher than in the standard range.

Time series of the BrO/SO_2 ratio were created for all combinations of the different wavelength ranges:

- *Set 1* SO_2: 314.8–326.8 nm, BrO: 330.6–352.75 nm; this are the standard evaluation ranges used in the rest of this work
- *Set 2* SO_2: 314.8–326.8 nm, BrO: 327–347 nm
- *Set 3* SO_2: 326.8–353.5 nm, BrO: 330.6–352.75 nm
- *Set 4* SO_2: 326.8–353.5 nm, BrO: 327–347 nm

For a clearer presentation only the running mean values for the four sets are shown at the bottom of Fig. 10. The same general trend is observed for all evaluation ranges, with a tendency of higher BrO/SO_2 ratios for set 1 and 2. These observations follow the previous findings on the effect of different evaluation ranges for SO_2. Evaluating SO_2 at larger wavelengths leads to higher CDs for two reasons: the influence of the light-dilution effect is lower at larger wavelengths, and the DOAS retrieval is less influenced by non-linearities that can occur for strong absorbers. The larger SO_2 CDs lead to lower BrO/SO_2 ratios when evaluating SO_2 in the alternative wavelength range.

3.3 Comparison with geophysical data and observations

We compared our obtained time series of BrO/SO_2 ratios with seismic activities as reported in Herrick (2012) and in the technical reports of the SGC – Servicio Geológico Colombiano[1] (formerly INGEOMINAS). Based on the SO_2 emission rate and number of volcano-tectonic (VT) and long-period (LP) seismic events, we divided the observation period in five different intervals.

In the first interval (A) between November 2009 (when the instruments where first installed) until mid-September 2010, we observe low average daily SO_2 emission rates below $3\,kg\,s^{-1}$. Gas emissions are too low to successfully retrieve BrO/SO_2 ratios. At the same time seismic activity (see Figs. 13 and 14) is very low, and the national park was open for tourists visiting Nevado del Ruiz.

Interval B, between mid-September 2010 and end of March 2011, is characterized by higher SO_2 emission rates up to $10\,kg\,s^{-1}$. Gas emissions are high enough to retrieve BrO/SO_2 ratios with an average value of $\approx 5\times10^{-5}$. The number of VT and LP events per month increases by an order of magnitude. At 6:30 a.m. on 5 November 2010, a first high volcanic column with a height of 900 m was reported by neighbours located more than 50 km from Nevado del Ruiz. On 11 December 2010 a first explosion was registered in the

[1] http://www.sgc.gov.co/lastaccessed:28.02.2014.

Table 1. SO_2 CDs, BrO CDs and the BrO/SO_2 ratios for the different cloud scenarios and a cloud optical density of 4 at 340 nm. (0) is a cloud-free atmosphere, (I) a cloud layer between 5.5 and 6.5 km, (II) a cloud layer between 5 and 5.5 km, (III) a cloud layer between 6.5 and 7 km, and (IV) an enhanced aerosol abundance coinciding with the volcanic plume.

		SO_2 [molec/cm^2] $\times 10^{18}$	BrO $\times 10^{14}$	BrO/SO_2 ratio $\times 10^{-4}$	Deviation [%]
0	90°	0.97	0.98	**1.006**	0.6
	45°	0.92	0.93	**1.014**	1.4
I	90°	1.09	1.09	**1.000**	0.0
	45°	1.04	1.05	**1.013**	1.3
II	90°	0.95	0.97	**1.028**	2.8
	45°	0.28	0.27	**0.976**	−2.5
III	90°	0.96	0.96	**1.000**	0.0
	45°	1.01	1.02	**1.008**	0.8
IV	90°	1.08	1.08	**0.999**	−0.1
	45°	1.07	1.09	**1.013**	1.3

Table 2. SO_2 CDs, BrO CDs and the BrO/SO_2 ratios for the different cloud scenarios and a cloud optical density of 12 at 340 nm. (0) is a cloud free atmosphere, (I) a cloud layer between 5.5 and 6.5 km, (II) a cloud layer between 5 and 5.5 km, (III) a cloud layer between 6.5 and 7 km, and (IV) a cloud layer coinciding with the volcanic plume.

		SO2 [molec/cm^2] $\times 10^{18}$	BrO $\times 10^{14}$	BrO/SO2 ratio $\times 10^{-4}$	Deviation [%]
0	90°	0.97	0.98	**1.006**	0.6
	45°	0.92	0.93	**1.014**	1.4
I	90°	1.37	1.36	**0.997**	−0.3
	45°	1.17	1.20	**1.023**	2.3
II	90°	0.81	0.83	**1.034**	3.4
	45°	0.22	0.22	**0.996**	−0.4
III	90°	0.99	0.99	**0.999**	−0.1
	45°	1.03	1.04	**1.006**	0.6
IV	90°	1.26	1.26	**0.999**	−0.1
	45°	1.20	1.22	**1.024**	2.4

centre of the active Arenas crater of Nevado del Ruiz at 12:51 local time.

Interval C, between April 2011 and mid-January 2012, shows SO_2 emission rates and BrO/SO_2 ratios similar to interval B. The number of LP events stays similar as well, but the number of monthly VT events drops to approximately half of the number in interval B. On 21 August 2011 another relatively high volcanic column with a height of 900 m was observed on the top of Nevado del Ruiz volcano. This period was accompanied by small explosions and some glacier avalanches around the main active Arenas crater.

Interval D starts at the beginning of February 2012 and ends with the explosion of Nevado del Ruiz on 30 June 2012. The SO_2 emission rate is generally largely increased with maximum values of the daily average emission rate up to 88 $kg\,s^{-1}$. In this interval the BrO/SO_2 ratio drops to lower values $\approx 1-2\times 10^{-5}$. At this time the number of VT events reaches similar values to interval B and the number of monthly LP events increases by an order of magnitude. Ash emissions were observed at the end of February (Garzón et al., 2013). The time between March and May 2012 was described as dominated by magma intrusion processes by Garzón et al. (2013).

Interval E starts after the explosion on 30 June and lasts till the end of our observations. The SO_2 emission rates slightly decrease (maximum daily averages of 45 $kg\,s^{-1}$), and the BrO/SO_2 ratio recovers to values of $\approx 5\times 10^{-5}$. However, another drop of the BrO/SO_2 ratio can be observed from mid-April 2013 until the end of our observations (June 2013). The number of LP events is lower again, similar to periods B and C. The number of VT events is characterized by high variations. The highest number of VT events coincides with another decrease of the BrO/SO_2 ratio. During this period ash emissions can be observed occasionally.

We can summarize that there are changes of the BrO/SO_2 ratio that coincide with variations of seismic signals. At the

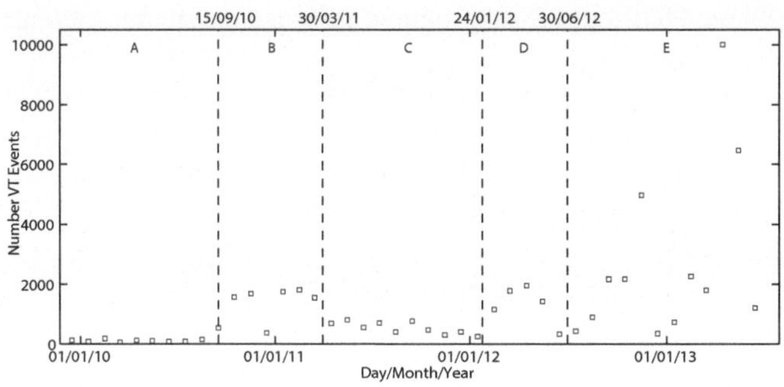

Figure 13. Number of volcano tectonic events per month. Data are taken from the SGC reports from http://www.sgc.gov.co/.

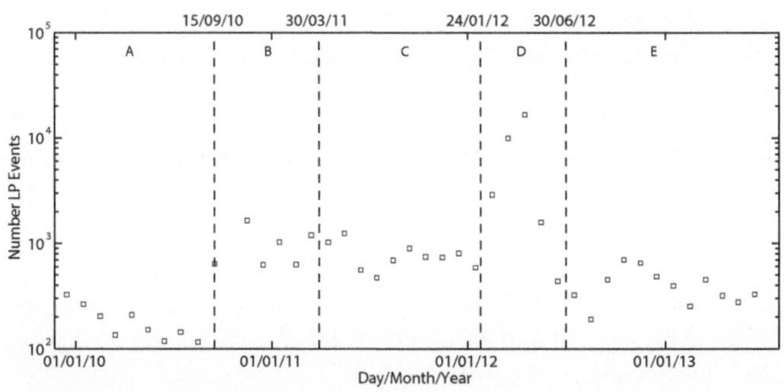

Figure 14. Number of long period seismic events per month. Data are taken from the SGC reports from http://www.sgc.gov.co/.

same time as the first drop of the BrO/SO_2 ratio (Interval D), the SGC reports indicate a large increase in the number of LP events. During the second drop of the BrO/SO_2 ratio, the number of VT events increases.

A robust interpretation of the observed BrO/SO_2 ratios is difficult with our current knowledge. Only few measurements of solubility of bromine have been carried out (e.g. Villemant and Boudon, 1999, Bureau et al., 2000, and 2010, Balcone-Boissard et al., 2010). No information can be found whether bromine exsolves from the magma earlier than sulfur. We therefore briefly discuss both possibilities. Assuming that bromine exsolves later than sulfur, we could identify interval B and C as a system in some kind of intermediate equilibrium with a BrO/SO_2 ratio of $\approx 5 \times 10^{-5}$. In case of a pressure decrease, we then expect first a decrease in the BrO/SO_2 ratio caused by the exsolution of sulfur. If the pressure decreases further at a certain point, bromine becomes exsolved as well and the BrO/SO_2 increases again, which can be observed in the data at the transition between intervals D and E.

However, the observed pattern can also be interpreted reasonably assuming that bromine exsolves from the magma before sulfur as suggested in Bobrowski and Giuffrida (2012). In this case we would expect an increase in bromine emissions (and therefore of the BrO/SO_2 ratio) when the pressure starts to decrease. This cannot be excluded nor confirmed from our data set. The BrO/SO_2 ratio could have increased in interval A, where general gas emissions were too low to exceed the detection limit. Small increases of the BrO/SO_2 ratio can be observed at the beginning of interval B and in the middle of interval C. We can also not exclude (with our time resolution) that an increase of the BrO/SO_2 ratio was too fast to be detected. The seismic signals already indicate changes of the volcanic system in interval B and C. As the pressure decreases further (Interval D), a relatively larger fraction of sulfur exsolves from the magma. As magma generally contains more sulfur than bromine, this leads to a decrease of the BrO/SO_2 ratio.

Another factor affecting the relative speciation of BrO/SO_2 in the plume is the effect of magma ascent rate on the efficiency of segregation of volatile species. Balcone-Boissard et al. (2010) found that for felsic magmas Cl, Br and I are extracted efficiently at high pressure during slow magma ascent, but remained in the melt during explosive degassing. Although the magmatic composition of Nevado del Ruiz is not felsic, a similar dynamical effect could be present in this volcano. The drop in the BrO/SO_2 ratio preceding the explosive phase of the eruption may thus indicate acceleration of the magma from the reservoir, affecting in different ways the exsolution of the two species.

4 Discussion of results and outlook

We have for the first time demonstrated that BrO column densities can be automatically evaluated from spectra routinely recorded by the NOVAC type 1 instruments of the NOVAC network. This proof of concept shows the possibility of achieving a 5-year or even longer time series of BrO/SO_2 ratios at many (if not all) volcanoes observed by the NOVAC network. The instruments with which the data were analysed in this study have exactly the same design as any other NOVAC type 1 instrument. In fact 2 years of data (January 2011–December 2012) at Galeras, Colombia, were also evaluated. While in general showing too low SO_2 levels for a successful BrO detection, the BrO DOAS retrieval error was comparable to the error from the instruments at Nevado del Ruiz. Since Nevado del Ruiz was chosen for this study due to the interesting activity and the good continuous data set, not for superior instrument performance, we have no indications that the other instruments from NOVAC should perform differently. Furthermore, there are other networks (i.e. the Flame network at Stromboli, Italy; Burton et al., 2009) that use comparable spectrometers, which could give further insight into volcanic BrO/SO_2 ratios. Evaluating BrO and SO_2 in different wavelength ranges and radiative transfer calculations showed that the BrO/SO_2 ratio is more robust against influences of radiative transfer than the trace-gas column densities.

The large NOVAC data set already gives the possibility to determine other influences than changes of the volcanic activity on the BrO/SO_2 molar ratio. BrO is not primarily emitted by volcanoes but forms from HBr when volcanic gases are released into the atmosphere (e.g. Bobrowski et al., 2003; Oppenheimer et al., 2006, von Glasow, 2010). While this potentially makes it more difficult to use the BrO/SO_2 ratio as an indicator of volcanic activity, Bobrowski and Giuffrida (2012) have found that the BrO/SO_2 molar ratio increases only in the first minutes after release into the atmosphere. The authors found that, at a distance of approximately 5 km after release into the atmosphere, the BrO/SO_2 molar ratio is constant (of course depending on the wind speed). The vast amount of available data in the NOVAC database allows the further investigation of influences like the residence time of the volcanic gas in the ambient atmosphere by estimating the time of transport from the volcanic vent to the point of measurement, or meteorological influences on the BrO/SO_2 molar ratio, as well as its dependence on height and latitude.

The BrO evaluation scheme presented in this manuscript was implemented for automatic operation to re-process the archived data. In the future, it is planned to be implemented into the routine evaluation for the NOVAC observatories. Besides the automated evaluation there are chances to further improve the quality of the BrO/SO_2 ratios.

One possible pitfall in the DOAS evaluation of volcanic gas measurements is the requirement of a reference spectrum free of volcanic gas. For instance, this problem is discussed in Salerno et al. (2009), where the authors state that wide volcanic gas plumes, covering the complete range of viewing directions, are frequently observed. This problem is probably most frequently observed during periods of low wind speeds, when meaningful gas emission rate measurements are difficult anyways. Therefore, it is not the main concern for SO_2 emission rate measurements. However, using gas-free reference spectra can improve the BrO retrieval. An approach similar to that described by Salerno et al. (2009) for the evaluation of SO_2 without a reference spectrum could be implemented. The background spectrum for the evaluation is modelled on the basis of an extraterrestrial solar reference spectrum (Chance and Kurucz, 2010). Unfortunately, while working well for volcanic SO_2, the approach of using modelled spectra does not yield sufficient accuracy for the BrO evaluation at this time.

However, this method could be used to identify "plume-free" spectra, by their SO_2 signatures, which subsequently can be used as reference spectra to evaluate the data for both BrO and SO_2. If all spectra from a particular scan of the sky contain significant SO_2 column densities, one could, for example, use a reference spectrum from a different scan or day, with similar atmospheric conditions and a similar instrument temperature.

Another possible approach to further improve the accuracy of our algorithm is taking temperature effects of the instruments, which were discussed in the Methods section, into account. This could further help to improve the quality of data evaluation.

The results presented in this work are – although there is still ample potential for improvements – encouraging and show that data recorded by NOVAC can be used to routinely perform measurements of the BrO/SO_2 molar ratio as an additional tracer of volcanic activity. The sample data set presented here already demonstrates that the BrO/SO_2 showed significant changes several weeks prior to major eruptive events. The variations of the BrO/SO_2 ratio occur at a similar time as increased seismic activity. If BrO/SO_2 ratios could be considered as an additional indicator of changes in the volcanic activity, even earlier warnings and discussion about potential risk could be initiated. However, more data are necessary to confirm this behaviour – but they might be soon available considering the large database of NOVAC.

Acknowledgements. The authors would like to thank Mike Burton for editing and two anonymous reviewers for their comments on the manuscript, which greatly helped to improve it. We would like to thank the European Commission Framework 6 Research Program for funding of the NOVAC project. We kindly acknowledge the NOVAC partners from the Colombian Geological Survey (formerly INGEOMINAS), especially the FISQUIM Research Group and the technical staff at the Manizales Volcanological Observatory, which permanently works for the maintenance of the energy supply for the ScanDOAS systems located around the main crater of Nevado del Ruiz volcano. The authors would like to thank Tim Deutschmann for providing the radiative transfer model McArtim. N. Bobrowski thanks for financial support from the DFG project DFG BO 3611/1-1. The authors thank the DFG project DFG PL 193/14-1 for financial support.

Edited by: M. Burton

Appendix A:

We simulated synthetic spectra as outlined in Vogel (2011). Measurement and Fraunhofer reference spectra (FRS) were both simulated. The BrO content of the measurement spectrum was 1.5×10^{14} molecules cm^{-2}, and the SO_2 content was 1×10^{18} molecules cm^{-2}. The synthetic spectra are based on the Chance and Kurucz (2011) solar atlas. The Beer–Lambert law was applied on the wavelength grid of the solar atlas using stratospheric absorbers and in the case of the measurement spectrum additionally BrO and SO_2. Then the high-resolution spectra were convolved to a lower resolution with a Gaussian profile with varying FWHM. The FWHM was varied between 0.525 and 0.775 nm in steps of 0.025 nm. For each FWHM a set of measurement and FRS spectrum (convolved with the same Gaussian profile) were evaluated using the DOAS retrieval as outlined in the Methods section. The absorption cross sections were convolved with a Gaussian profile with a FWHM of 0.65 nm. This approach mimics the situation in NOVAC. While a constant FWHM is used for the convolution of the absorption cross sections, the instrument line shape of both spectra varies with temperature. Figure 6 shows the deviation of the BrO/SO_2 ratio from the BrO/SO_2 ratio retrieved with both spectra using the same instrument line shape. The results show that the error in the BrO/SO_2 ratio is below 15 % for variations of the instrument line shape of up to 0.125 nm.

References

Arellano, S. R.: Studies of Volcanic Plumes with Spectroscopic Remote Sensing Techniques, Licentiate thesis, Chalmers University of Technology, Göteborg, Sweden, 2013.

Balcone–Boissard, H., Villemant, B. and Boudon, G.: Behavior of halogens during the degassing of felsic magmas, Geochem. Geophy. Geosy., 11, Q09005, doi:10.1029/2010GC003028, 2010.

Bobrowski, N. and Giuffrida, G.: Bromine monoxide/sulphur dioxide ratios in relation to volcanological observations at Mt. Etna 2006–2009, Solid Earth, 3, 433–445, 2012.

Bobrowski, N. and Platt, U.: SO_2/BrO ratios studied in five volcanic plumes, J. Volcanol. Geoth. Res., 166, 147–160, 2007.

Bobrowski, N., Hönninger G., Galle, B., and Platt, U.: Detection of bromine monoxide in a volcanic plume, Nature, 423, 273–276, 2003.

Boichu, M., Oppenheimer, C., Roberts, T. J., Tsanev, V., and Kyle, P. R.: On bromine, nitrogen oxides and ozone depletion in the tropospheric plume of Erebus volcano (Antarctica), Atmos. Environ., 45, 3856–3866, 2011.

Bureau, H., Keppler, H. and Métrich, N.: Volcanic degassing of bromine and iodine: experimental fluid/melt partitioning data and applications to stratospheric chemistry, EARTH PLANET SC LETT, 183, 51–60, 2000.

Bureau, H., Foy, E., Raepsaet, C., Somogyi, A., Munsch, P., Simon, G., and Kubsky, S.: Bromine cycle in subduction zones through in situ Br monitoring in diamond anvil cells, GEOCHIM COSMOCHIM AC, 74, 3839–3850, 2010.

Burrows, J. P., Richter, A., Dehn, A., Deters, B., Himmelmann, S., Voigt, S., and Orphal, J.: Atmospheric remote-sensing reference data from GOME-2. Temperature-dependent absorption cross sections of O_3 in the 231–794 nm range. J. Quant. Spectrosc. Ra., 61, 509–517, 1999.

Burton, M. R., Caltabiano, T., Murè, F., Salerno, G., and Randazzo, D.: SO_2 flux from Stromboli during the 2007 eruption: results from the FLAME network and traverse measurements, J. Volcanol. Geoth. Res., 182, 214–220, 2009.

Carroll, M. R. and Holloway, J. R.: Volatiles in Magmas, Mineralogical Society of America, Washington DC, 517 pp., 1994.

Chance, K. and Kurucz, R. L.: An improved high-resolution solar reference spectrum for earth's atmosphere measurements in the ultraviolet, visible, and near infrared, J. Quant. Spectrosc. Ra., 111, 1289–1295, 2010.

Deutschmann, T., Beirle, S., Fried, U., Grzegorski, M., Kern, C., Kritten, L., Platt, U., Prados-Román, C., Puckite, J., Wagner, T., Werner, B., and Pfeilsticker, K.: The Monte Carlo atmospheric radiative transfer model McArtim: Introduction and validation of Jacobians and 3D features, Journal of Quantitative Spectroscopy and Radiative Transfer, 112, 1119–1137, 2011.

Fleischmann, O. C., Hartmann, M., Burrows, J. P., and Orphal, J.: New ultraviolet absorption cross-sections of BrO at atmospheric temperatures measured by time-windowing Fourier transform spectroscopy, J. Photoch. Photobio. A, 168, 117–132, 2004.

Galle, B., Oppenheimer, C., Geyer, A., McGonigle, A. J. S., Edmonds, M., and Horrocks, L.: A miniaturised ultraviolet spectrometer for remote sensing of SO_2 fluxes: a new tool for volcano surveillance, J. Volcanol. Geoth. Res., 119, 241–254, 2003.

Galle, B., Johannson, M., Rivera, C., Zhang, Y., Kihlmann, M., Kern, C., Lehmann, T., Platt, U., Arellano, S., and Hidalgo, S.: Network for Observation of Volcanic and Atmospheric Change (NOVAC) – a global network for volcanic gas monitoring: Network layout and instrument description, B. Volcanol., 115, D05304, doi:10.1029/2009JD011823, 2010.

Garzon, G., Silva, B., Narvaez, A., Chacon, Z., and Galle, B.: Assessment of SO_2 emissions from three Colombian active volcanoes (2007–2012), GEOCHANGE: Problems of Global Changes of the Geological Environment, vol. 2, edited by: Khalilov, E., Science Without Borders, London, 6–14, 2013.

Grainger, J. F. and Ring, J.: Anomalous Fraunhofer line profiles, Nature, 193, 762, doi:10.1038/193762a0, 1962.

Hermans, C., Vandaele, A. C., Fally, S., Carleer, M., Colin, R., Coquart, B., Jenouvrier, A., Merienne, M.-F.: Absorption cross-section of the collision-induced bands of oxygen from the UV to the NIR, in: Weakly Interacting Molecular Pairs: Unconventional Absorbers of Radiation in the Atmosphere, edited by: Camy-Peyret, C. and Vigasin, A., Springer, Netherlands, 193–202, 2003.

Herrick, J., Bulletin of the Global volcanism Network (August 2012), vol. 37, 8. last accessed on 27 Nov 2013 from http://www.volcano.si.edu/reports_bgvn.cfm, 2012.

Hörmann, C., Sihler, H., Bobrowski, N., Beirle, S., Penning de Vries, M., Platt, U., and Wagner, T.: Systematic investigation of bromine monoxide in volcanic plumes from space by using the GOME-2 instrument, Atmos. Chem. Phys., 13, 4749–4781, doi:10.5194/acp-13-4749-2013, 2013.

Kelly, P. J., Kern, C., Roberts, T. J., Lopez, T., Werner, C., and Aiuppa, A.: Rapid chemical evolution of tropospheric volcanic emissions from Redoubt Volcano, Alaska, based on observations of ozone and halogen-containing gases, J. Volcanol. Geoth. Res., 259, 317–333, 2012.

Kern, C.: Spectroscopic measurements of volcanic gas emissions in the ultra-violet wavelength region, Ph.D. Thesis, Institute of Enviromental Physics, University of Heidelberg, 2009.

Kraus, S.: DOASIS, A Framework Design for DOAS, Ph.D. Thesis, University of Mannheim, 2006.

Meller, R. and Moortgat, G. K.: Temperature dependence of the absorption cross sections of formaldehyde between 223 and 323 K in the wavelength range 225–375 nm, J. Geophys. Res-Atmos., 105, 7089–7101, 2000.

Menyailov, I. A.: Prediction of eruptions using changes in composition of volcanic gases, B. Volcanol., 39, 112–125, 1975.

Noguchi, K. and Kamiya, H.: Prediction of volcanic eruption by measuring the chemical composition and amounts of gases, B. Volcanol., 26, 367–378, doi:10.1007/BF02597298, 1963.

Oppenheimer, C., Tsanev, V. I., Braban, C. F., Cox, R. A., Adams, J. W., Aiuppa, A. Bobrowski, N., Delmelle, P., Barclay, J., and McGonigle, A. J. S.: BrO formation in volcanic plumes, Geochim. Cosmochim. Ac., 70, 2935–2941, 2006.

Pennisi, M. and Le Cloarec, M.-F.: Variations of Cl, F, and S in Mount Etna's plume, Italy, between 1992 and 1995, J. Geophys. Res-Sol Ea., 103, 5061–5066, 1998,.

Pinardi, G., Roozendael, M. V., and Fayt, C.: The influence of spectrometer temperature variability on the data retrieval of SO2., In NOVAC second annual activity report, NOVAC consortium, 44–48, 2007.

Platt, U. and Stutz, J.: Differential Optical Absorption Spectroscopy – Principles and Applications, Physics of Earth and Space Environments, 1st Edn., Springer, Berlin, Heidelberg, New York, 597 pp., 2008.

Salerno, G. G., Burton, M. R., Oppenheimer, C., Caltabiano, T., Tsanev, V. I., and Bruno, N.: Novel retrieval of volcanic SO_2 abundance from ultraviolet spectra, J. Volcanol. Geoth. Res., 181, 141–153, 2009.

Vandaele, A. C., Hermans, C., Simon, P. C., Carleer, M., Colin, R., Fally, S., Merienne, M., Jenouvrier, A., and Coquart, B.:, Measurements of the NO_2 absorption cross-section from 42 000 cm^{-1} to 10 000 cm^{-1}(238–1000 nm) at 220 K and 294 K, J. Quant. Spectrosc. Ra., 59, 171–184, 1998.

Vandaele, A. C., Hermans, C., and Fally, S.: Fourier transform measurements of SO_2 absorption cross sections: I I.: Temperature dependence in the 29 000–44 000 cm^{-1} (227–345 nm) region, J. Quant. Spectrosc. Ra., 110, 2115–2126, 2009.

Villemant, B. and Boudon, G.: H_2O and halogen (F, Cl, Br) behaviour during shallow magma degassing processes, EARTH PLANET SC LETT, 168, 271–286, 1999.

Vogel, L.: Volcanic plumes: Evaluation of spectroscopic measurements, early detection and bromine chemistry, Ph.D. thesis, Ruperto Carola University of Heidelberg, Germany, available at: http://www.ub.uni-heidelberg.de/archiv/13219, 2011.

Vogel, L., Sihler, H., Lampel, J., Wagner, T., and Platt, U.: Retrieval interval mapping: a tool to visualize the impact of the spectral retrieval range on differential optical absorption spectroscopy evaluations, Atmos. Meas. Tech., 6, 275–299, 2013, http://www.atmos-meas-tech.net/6/275/2013/.

von Glasow, R.: Atmospheric chemistry in volcanic plumes, P. Natl. Acad. Sci. USA, 107, 6594–6599, 2010.

Wagner, T., Chance, K., Frieß, U., Gil, M., Goutail, F., Hönninger, G., Johnston, P., Karlsen-Tørnkvist, K., Kostadinov, I., Leser, H., Petritoli, A., Richter, A., Van Roozendael, M., and Platt, U.: Correction of the ring effect and I0 effect for DOAS observations of scattered sunlight, ESA Technical Report, 2002.

Wagner, T., Beirle, S., and Deutschmann, T.: Three-dimensional simulation of the Ring effect in observations of scattered sun light using Monte Carlo radiative transfer models, Atmos. Meas. Tech., 2, 113–124, doi:10.5194/amt-2-113-2009, 2009.

Wagner, T., Beirle, S., Dörner, S., Friess, U., Remmers, J., and Shaiganfar, R.: Cloud detection and classification based on MAX-DOAS observations, Atmos. Meas. Tech. Discuss., 6, 10297–10360, doi:10.5194/amtd-6-10297-2013, 2013.

AxiSEM: broadband 3-D seismic wavefields in axisymmetric media

T. Nissen-Meyer[1], M. van Driel[2], S. C. Stähler[3], K. Hosseini[3], S. Hempel[4], L. Auer[2], A. Colombi[5], and A. Fournier[6]

[1]Dept. of Earth Sciences, University of Oxford, South Parks Road, Oxford OX1 3AN, United Kingdom
[2]Institute of Geophysics, ETH Zurich, Sonneggstrasse 5, 8092 Zurich, Switzerland
[3]Dept. of Earth Sciences, Ludwig-Maximilians-Universität München, Theresienstrasse 41, 80333 Munich, Germany
[4]Institute of Geophysics, Universität Münster, Correnzstrasse 24, 48149 Münster, Germany
[5]ISTerre, Université Joseph Fourier, Grenoble, BP 53 38041 Grenoble CEDEX 9, France
[6]Institut de Physique du Globe, Sorbonne Paris Cité, Université Paris Diderot, UMR 7154 CNRS, 75005 Paris, France

Correspondence to: T. Nissen-Meyer (tarjen@earth.ox.ac.uk)

Abstract. We present a methodology to compute 3-D global seismic wavefields for realistic earthquake sources in visco-elastic anisotropic media, covering applications across the observable seismic frequency band with moderate computational resources. This is accommodated by mandating axisymmetric background models that allow for a multipole expansion such that only a 2-D computational domain is needed, whereas the azimuthal third dimension is computed analytically on the fly. This dimensional collapse opens doors for storing space–time wavefields on disk that can be used to compute Fréchet sensitivity kernels for waveform tomography. We use the corresponding publicly available AxiSEM (www.axisem.info) open-source spectral-element code, demonstrate its excellent scalability on supercomputers, a diverse range of applications ranging from normal modes to small-scale lowermost mantle structures, tomographic models, and comparison with observed data, and discuss further avenues to pursue with this methodology.

1 Introduction

Seismology currently enjoys transformative progress upon a simultaneous surge in instrumentation, software, and hardware. The dawn of high-performance computing and sophisticated numerical techniques to address seismic wave propagation in a physically robust and realistic manner has enabled seismologists to capture relevant physics of wave propagation in the seismic far field, and resolve structures for which direct comparisons with waveform data are feasible. This in turn opens doors to incorporating full waveforms into inversion algorithms, using, for instance, adjoint methods in conjunction with 3-D wave propagation (e.g., Tromp et al., 2005). Traditionally, global seismic tomography (consult Rawlinson et al., 2010, for a comprehensive summary) has been based upon ray theory utilizing traveltimes rather than full waveforms. Indeed, phase delays relate to wavespeed variations in a more robust manner than amplitude information. So why would numerical methods, within the realm of tomography at least, strive to capture the entire waveform? Most modern measurements of "traveltimes", such as cross-correlation (Nolet, 2008), time–frequency phase delays (Fichtner et al., 2008), or instantaneous phase (Bozdag et al., 2011) are based on waveforms, and therefore necessitate full wavefield modeling. Moreover, high-frequency waveform modeling (e.g., Thorne et al., 2007) is often employed to fit smaller-scale heterogeneities. Such studies are subject to significant tradeoffs, especially if secondary measurements such as traveltimes were used. Thirdly, accurate computation of the gradient of measurements with respect to model variations, often termed the Fréchet derivative, requires the convolution of a forward-propagating wavefield with a backward or adjoint wavefield, both of which need to be sampled in 3-D space and time (Nissen-Meyer et al., 2007a). For the purpose of this paper, let us postulate a desire for a method to deliver

1. *3-D wavefields* for realistic sources and structures,
2. across the *observable frequency band*,
3. at a *reasonable computational cost* for tomography.

We will now delve into some of these issues in more detail, and present our compromise for a solution that covers a significant, realistic and relevant fraction of these aspirations.

1.1 Effective Earth models and data

Spherically symmetric models are widely established as a common basis for Earth properties not only in seismology, but also as a bridge to mineral physics and geodynamics. This popularity stems not only from the relative simplicity in modeling 1-D structures, but largely from the fact that such laterally averaged models represent and fit a large majority of seismic (traveltime) data, as has been established since the traveltime tables by Jeffreys and Bullen (1940) and subsequent models such as PREM (Dziewonski and Anderson, 1981), IASP91 (Kennett and Engdahl, 1991), and ak135 (Kennett et al., 1995). Our understanding and interpretation of the Earth's interior has come a long way from the detection of its radial structure, and has been significantly fueled by means of seismic tomography (Rawlinson et al., 2010). On the global scale, 3-D tomographic models usually amount to a few percent of wavespeed perturbation from spherically symmetric models (Becker and Boschi, 2002), and behave close to linear in seismic traveltimes (Mercerat and Nolet, 2013).

Global tomographic models exhibit considerable agreement up to spherical harmonic degree 5 (Becker and Boschi, 2002; Auer et al., 2014), that is, for very large-scale structures, but often diverge at smaller length scales due to shortcomings such as insufficient data coverage and modeling. In this multi-scale context, it is important to remember that any discrete Earth model used in a numerical method is inherently upscaled (either by its own nature, or by porting to the discrete mesh), and at best a blurred rendition of reality. The challenge lies in tying the background model to both the desired frequency range and type of measurement extracted from the wavefield in a feasible, realistic manner.

Clearly, utilizing a maximal amount of broadband data is as desirable as capturing complexities in structure and wave physics. Even in times of a surge in data acquisition, source–receiver geometries are still largely controlled by continents, tectonic boundaries and the Northern Hemisphere. It thus seems desirable to seek a compromise for modeling between broad frequency ranges, realistic effective Earth models, while exploiting a maximal amount of usable data.

1.2 Numerical wave propagation

Unlike disciplines subject to more complex, non-linear physical systems such as fluid dynamics, the availability of mature, comprehensive seismic wave-propagation codes such as SPECFEM (e.g., Komatitsch and Tromp, 2002b) seems to have resolved most challenges in capturing the underlying physics of wave propagation. Instead, it appears to shift attention for achieving realistic wave propagation simulations towards:

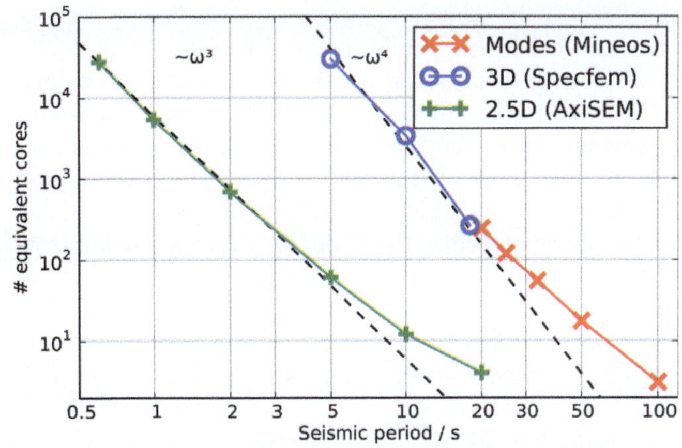

Figure 1. The cost of global wavefield simulations in "real time" (i.e., seismogram length equals wallclock simulation time) for different methods. Each data point is based on actual simulation times, and gives as a result the number of processors needed to achieve this, assuming perfect scalability. The cost of normal-mode solutions (Mineos, available from geodynamics.org) and wave propagation in 3-D domains (SPECFEM, geodynamics.org) scales with the seismic period to the fourth power, whereas the axisymmetric method (AxiSEM) scales to the third power. We calculate the cost estimation upon saving 10^6 spatial points, a moderate task to compute wavefields. This is especially noteworthy for the mode solution, whose cost scales directly with the number of saved spatial points.

- feasible choices for source and structure,
- usability of the method,
- availability of computational resources.

While all of these issues are common to any numerical method and not easily resolvable, the latter point is especially stringent. Worse still, for global seismology the prohibitive cost will remain a dominant limitation on the maximally resolved frequencies and a exponentially increasing number of usable data with millions of recorded waveforms (IRIS annual report 2011, www.iris.edu) for years to come. Full 3-D models in spherical geometry can be incorporated by spherical finite differences (Igel et al., 2002), or spectral-element methods (Komatitsch and Tromp, 2002a; Chaljub et al., 2007). As the computational cost scales with the frequency to the fourth power (three space, one time dimension), such comprehensive methods are still extraordinarily expensive for global-scale wave propagation at high resolution, and certainly more so if large numbers of simulations are needed as in most cases of geophysical interest. Alternative attempts such as high-order expansions of the Born series (Takeuchi et al., 2000) are similarly prohibitive for complex media and high frequencies. This is unfortunate, as especially the domain of seismic periods below 10 s (see Fig. 1) offers both a wealth of seismic data and a resolution harboring many open geophysical questions. The plot in Fig. 1 is an approximate, order-of-magnitude estimation of computational cost, and minor algorithmic optimization does not affect this overarching trend drastically. The slightly increased cost for AxiSEM above 10 s represents the fact that the thin crustal layers at these long periods start to dominate the smallest element size and thus increase the relative cost due to this geometric constraint on the global time step. This is not seen in SPECFEM, since intra-crustal layers are not explicitly meshed. Further commentary on Fig. 1 is given in Sect. 2.7.

Considering desirable features for the inversion such as comprehensive model sampling, uncertainty analysis, or probabilistic approaches, this represents not only a formidable challenge, but is essentially not computable even with most optimistic estimates of the evolution of computation on a decadal timescale, especially in 3-D.

Several strategies for speeding up numerical methods exist, focused on either the physical system or the implementation. Code optimization may exploit dedicated hardware infrastructures such as GPUs (Rietmann et al., 2012), or algorithmic tasks such as tensor–vector products (Nissen-Meyer et al., 2007b), irregular meshing (Zhu et al., 2009) or local time stepping. These approaches usually lead to a performance speedup of about 2–3 in total CPU time. Physics-based approximations often limit the frequency range either on the high end (as implicitly done due to the prohibitive cost in 3-D methods) or lower end (ray theory). Additionally, we commonly find cost reductions related to reduced dimensionality (e.g., 2-D, Zhu et al., 2009), rheology (e.g., acoustic wave propagation), or structural complexity by means of homogenization (Capdeville et al., 2013). Such approximations can lead to orders of magnitude faster codes, but need to be chosen carefully, depending on each application.

1.3 3-D waves in axisymmetric media

Several methods have been developed to accommodate various levels of complexity in background structures effectively. For spherically symmetric Earth models, normal-mode summation (Dahlen and Tromp, 1998) elegantly tackles the grave end of the spectrum including such effects as gravity and rotation (Dahlen, 1968). For higher frequencies, the direct-solution method (Kawai et al., 2006), GEMINI (Friederich and Dalkolmo, 1995), or Yspec (Al-Attar and Woodhouse, 2008) have proven efficient in delivering accurate seismograms. While in principle doable, all of these methods become computationally expensive if an entire wavefield is needed as for sensitivity kernels (Nissen-Meyer et al., 2007a), and do not allow for lateral heterogeneities. Axisymmetric finite difference methods (Toyokuni and Takenaka, 2006; Jahnke et al., 2008) may accommodate this effectively, but suffer various shortcomings such as approximate sources, lack of fluid domains and anisotropy (Jahnke et al., 2008), and high dispersion errors for large propagation distances of interface-sensitive phases such as surface or diffracted waves. However, recent advances include a full moment tensor, attenuation, and the Earth's center (Toyokuni and Takenaka, 2012).

The purpose of this paper is to introduce the axisymmetric spectral-element method for global wave propagation and the corresponding publicly available, production-ready code AxiSEM, which taps into parameter regimes that have been previously unavailable at similar computational cost. We motivate the relevance of these parameter regimes by various examples and present ideas for further extensions and applications. Exploitation of moment-tensor source and single-force radiation patterns allow the computational domain to be collapsed to a 2-D semi-disk, and the azimuthal third dimension is computed analytically. Radiation pattern symmetries require all sources to be located along the axis, and lateral heterogeneities are translated into a 2.5-D torus-like structure. Due to the dimensional reduction, global wave propagation at typical seismic periods can be tackled serially at workstations. Novel features in this manuscript with respect to the methodology already described in Nissen-Meyer et al. (2007b, 2008) include 2-D parallelization, scalability to > 8000 cores, benchmarks at 1 Hz and for normal modes, extensions to visco-elastic anisotropic media, fluid spheres, finite sources, axisymmetric structures, tomographic models, comparison with data, generic post-processing for arbitrary source–receiver settings, sensitivity kernels, and availability as an open-source code.

This paper is organized as follows. A methodological chapter briefly summarizes the mathematical background of our approach, delegating more details to previous publications, and focusing instead on practical matters such as scalability, runtime requirements, I / O, and code availability. Chapter 3 describes those source types that may be simulated with AxiSEM, ranging from moment tensors, single forces, to finite faults and stochastic sources. Chapter 4 similarly describes a range of background models to be discretized in AxiSEM, including anisotropy, intrinsic attenuation, classical 1-D Earth models, solar models, small-scale 2.5-D heterogeneities and tomographic cross sections, random media and explicit mesh representations of the crust and oceans (oceans are not part of the current code as of April 2014). Chapter 5 shows simulation results for a selection of the previously mentioned ranges of applicability, covering the entire seismic frequency spectrum, 3-D wavefield visualization, lowermost mantle structures, simulations for 2.5-D slices through a tomography model, comparison with observed data from lowermost mantle and core phases, and sensitivity kernels. A concluding chapter discusses the general applicability, limitations and an outlook for future developments.

AxiSEM is a mature methodology and code, able to address a number of intriguing scientific questions. As should be commonly known from any software implementation, the level of automatism for the applications listed here is diverse, and readers should refer to the manual of the most recent release version for up-to-date features of the code.

2 Methodology

The mathematical foundation and validation of spherically symmetric, solid–fluid lower-frequency settings is detailed in previous publications (Nissen-Meyer et al., 2007b, 2008). In this section, we only sketch key methodological concepts, while focusing on new additions and practical matters related to usability, functionality and applicability. Our approach accurately simulates 3-D wavefields in axisymmetric Earth models, and distinguishes itself by

1. decreasing the computational costs by orders of magnitude compared to the 3-D method by running in 2-D,
2. making no limiting assumptions about wave-propagation physics (except for very long-period effects such as rotation; see Sect. 4.7) or kinematic earthquake radiation.

It therefore falls in between traditional end members that are typically optimized for either end of the frequency spectrum (e.g., ray theory, normal-mode summation) and 3-D modeling, by not compromising on essential wave-propagation physics or the coverage of the entire recorded frequency band between 0.001 and 1 Hz. The efficiency gain is grounded upon assuming axisymmetric background models, which reduces the numerical cost to a 2-D domain, whereas the third dimension is tackled analytically. We shall forego detailed treatment of classical spectral-element methods to highlight the peculiarities associated with this axisymmetric setting.

2.1 Equations of motion

The 3-D integral (weak-form) elastodynamic equations of motion in the solid Earth $\oplus$ read

$$\overbrace{\int_{\oplus} \rho \boldsymbol{w} \cdot \partial_t^2 \boldsymbol{u}\, d^3\boldsymbol{x}}^{\text{mass term: } \mathbf{M}(\boldsymbol{u})} + \overbrace{\int_{\oplus} \nabla \boldsymbol{w} : \mathbf{C} : \nabla \boldsymbol{u}\, d^3\boldsymbol{x}}^{\text{stiffness term: } \mathbf{K}(\boldsymbol{u})} = \overbrace{\int_{\oplus} \boldsymbol{w} \cdot \mathbf{f}\, d^3\boldsymbol{x}}^{\text{source term: } \mathbf{F}(\boldsymbol{u})} \quad (1)$$

where $\boldsymbol{u}$ is the sought displacement vector, $\boldsymbol{w}$ a suitably chosen test vector, $\boldsymbol{f}$ the source term, ρ the mass density, and $\mathbf{C}$ the anisotropic fourth-order elasticity tensor with 21 independent parameters (consult Nissen-Meyer et al. (2007b) for details). It may be time dependent for intrinsic attenuation, in which case the double contraction : implies a convolution.

2.2 Axisymmetric dimensional collapse

As shown in Nissen-Meyer et al. (2007a), one may analytically separate radiation patterns into individual responses to each moment-tensor element M_{ij} factorized in azimuthal functions:

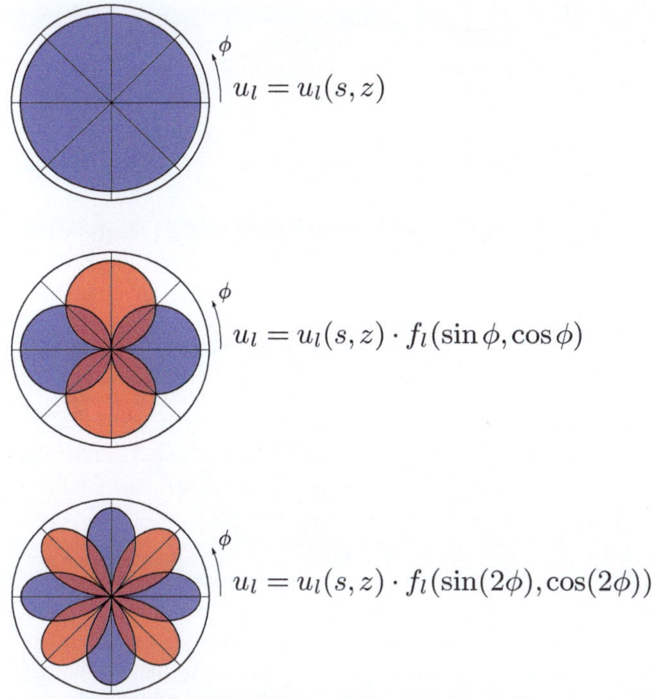

Figure 2. Radiation patterns for monopole (top), dipole (middle), and quadrupole angular orders of the respective moment tensor elements. The azimuthal radiation patterns encapsulated by f_l depend on multipole order m as well as component l, that is, no summation is implied by the above products.

$$\boldsymbol{u}_m(\boldsymbol{x}) = \begin{pmatrix} u_s(\tilde{\boldsymbol{x}}) \cos m\phi \\ u_\phi(\tilde{\boldsymbol{x}}) \sin m\phi \\ u_z(\tilde{\boldsymbol{x}}) \cos m\phi \end{pmatrix}, \quad (2)$$

where $m = 0, 1, 2$ are monopole, dipole, and quadrupole radiation types, respectively (Fig. 2), and $\tilde{\boldsymbol{x}} = (s, z) = (r, \theta)$ spans a two-dimensional domain (Fig. 3) by cylindrical (s, ϕ, z) or spherical (r, θ, ϕ) coordinates, respectively. This relation is accurate for axisymmetry in source $\mathbf{f} = \mathbf{f}(\tilde{\boldsymbol{x}})$ and structure $\rho = \rho(\tilde{\boldsymbol{x}}), \mathbf{C} = \mathbf{C}(\tilde{\boldsymbol{x}})$. After solving the set of 2-D problems, seismograms and wavefields at any location (s, ϕ, z) are obtained by multiplication with these azimuthal radiation factors in Eq. (2) during the post-processing stage (Sect. 2.5). Conceptually, 3-D integrals in $\oplus$ over any integrand ψ that contains azimuthal dependencies such as in Eq. (2) are then collapsed to 2-D integrals in D as

$$\int_{\oplus} \psi(\boldsymbol{x})\, d^3\boldsymbol{x} \Rightarrow \int_{D} \psi(\tilde{\boldsymbol{x}})\, d^2\tilde{\boldsymbol{x}}, \quad (3)$$

by evaluating the integration over ϕ analytically. This delivers solutions for the 3-D displacement vector $\boldsymbol{u}$ within a 2-D computational domain (Fig. 3). Symmetry about the axis (blue in Fig. 3) mandates all structural heterogeneities away

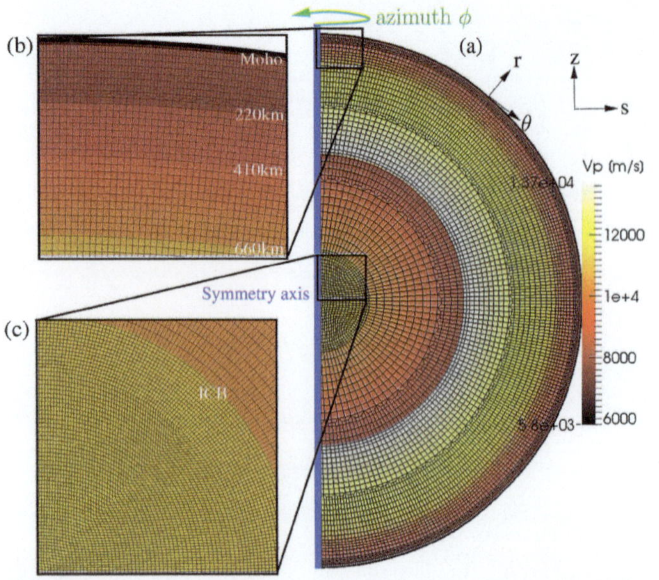

Figure 3. The 2-D computational domain D upon which the collapsed numerical system operates with a symmetry axis (blue). The method solves the three-dimensional equations of motion, but allows for an analytical representation within the azimuthal dimension (green). Sources and structure therefore obey axisymmetry with respect to the axis. Colors denote compressional velocities of the PREM model (Dziewonski and Anderson, 1981), and black lines an elemental mesh for a seismic period of 20 s. Zoom panels show a higher-resolution version (5 s) with upper-mantle discontinuities honored by the mesh **(b)**, as well as the rotated-coordinate meshing below the inner-core boundary (ICB) (Nissen-Meyer et al., 2008).

from it to adopt a torus-shaped, azimuthally invariant elongation, whereas the point source remains along the axis. Such lateral in-plane heterogeneities may prove useful for parameter studies at sufficiently high frequencies (see Sect. 4.4).

2.3 Spatial discretization

Finite element-based methods compute derivatives and integration upon reference coordinates. This entails a mapping $\tilde{x} = \tilde{x}(\boldsymbol{\xi})$ from a generic reference coordinate frame $\boldsymbol{\xi}$ to the physical domain $\tilde{x}$, represented by the Jacobian

$$\mathcal{J}(\boldsymbol{\xi}) = \left| \frac{\partial \tilde{x}}{\partial \boldsymbol{\xi}} \right|, \tag{4}$$

where $|.|$ is the determinant. This mapping is purely analytical for all element types of AxiSEM's automated mesh generator, with the exception of the cube at the center of the sphere (Nissen-Meyer et al., 2008). We then expand the wavefield within each elemental integral upon a basis of order N (typically 4, 5, 6) as

$$\boldsymbol{u}(\boldsymbol{\xi}) \approx \sum_{i,j=0}^{N} \boldsymbol{u}_{ij} l_{ij}(\boldsymbol{\xi}), \tag{5}$$

with two-dimensional Lagrange polynomials l_{ij} (Nissen-Meyer et al., 2007b). Partial derivatives $\partial_{\boldsymbol{\xi}} \boldsymbol{u}(\boldsymbol{\xi})$ are given by analytically differentiating l_{ij} along $\boldsymbol{\xi}$. These derivative operations are responsible for the bulk of the computational cost in typical spectral-element methods. Having performed these algebraic operations at the level of elements, these elemental contributions are gathered to define the discrete global stiffness $\mathbf{K}\boldsymbol{u}$ and mass terms $\mathbf{M}\boldsymbol{u}$. Our formulation with cylindrical coordinates leads to singularities of the type s^{-1} (Fig. 3) in the gradients at the symmetry axis. This is accommodated by a different basis compared to the interior domain, l'Hospital's rule (Fournier et al., 2005), and asymptotic expressions to accommodate boundary conditions (Nissen-Meyer et al., 2007a). By the choice of either kind of basis function, the mass matrix is exactly diagonal.

2.4 Temporal discretization

Such a discrete system leads to a set of ordinary differential equations in time, which may be rearranged as

$$\ddot{\boldsymbol{u}}(t) = \mathbf{M}^{-1} \left(\mathbf{f}(t) - \mathbf{K}\boldsymbol{u}(t) \right). \tag{6}$$

This system is conveniently solved by various explicit time-evolution schemes such as second-order Newmark, or higher-order symplectic schemes (Nissen-Meyer et al., 2008) up to eighth-order accuracy. Note that for the case of solid–fluid domains, the time stepping becomes a combined system of these two domains, which need to be linked appropriately to transmit waves across the solid–fluid interface (Chaljub and Valette, 2004).

2.5 Post-processing: summation, rotation, filtering

Unlike most seismic wave-propagation codes, AxiSEM requires a crucial sequence of post-processing steps to retrieve the full solution; see Eq. (2). While this may seem to be an undesirable additional burden, it represents a high level of flexibility, leaving a maximum amount of parameter choices to this post-processing step instead of having them fixed for the bulk simulations. For instance, one does not need to decide on the source mechanism, source–time function, filtering, instrument response, and receiver components at the time of the actual simulation, but can defer this to post-processing. The only necessary geophysical choices at the time of the simulation are source depth, receiver distances, maximal frequency, and background model. Figure 4 depicts an example of an automated output from post-processing to be read by Google Earth, containing source (red dot) and receiver (yellow pins) locations. Each receiver pin is linked to an image of the corresponding post-processed seismograms, and source information is provided as text (see Fig. 4). Section 5.3 sketches a generalization of this post-processing that fully exploits its flexibility in the context of solving the forward problem once and for all, deferring the choice of the source–receiver geometry to post-processing as well.

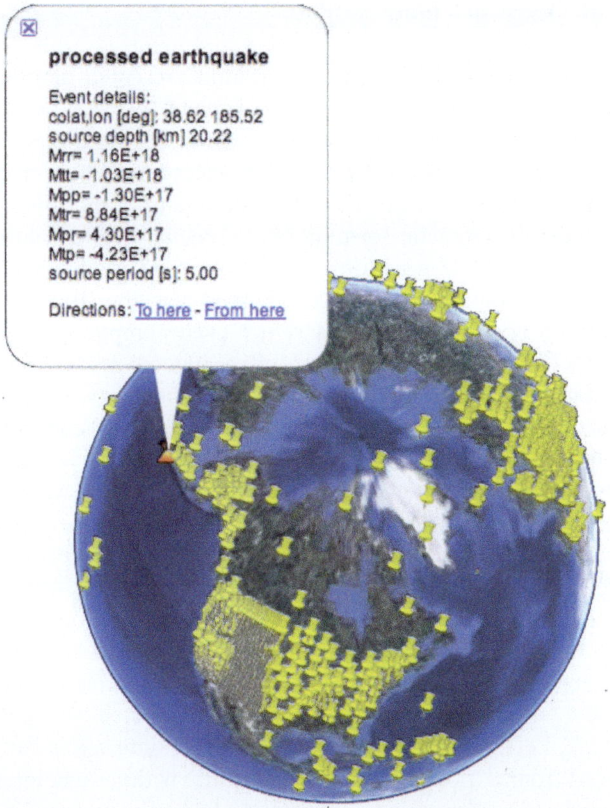

Figure 4. A Google Earth rendition of the source–receiver geometry used in an AxiSEM simulation. A generic output of the post-processing embedded within AxiSEM, this *kml* file contains earthquake parameters (red dot) and actual seismogram images as links at the station locations (yellow pins).

Recasting the 3-D equations of motion into a suite of 2-D problems yields a system of four independent wave equations to represent all six independent elements of the moment tensor ($M_{rr}, M_{\phi\phi}, M_{\theta\theta}, M_{r\theta}, M_{r\phi}, M_{\theta\phi}$) separately (see Fig. 7). This collapse from six to four independent systems honors azimuthal redundancy between the dipoles $M_{r\theta} \sim M_{r\phi}$ as well as quadrupoles $(M_{\theta\theta} - M_{\phi\phi}) \sim M_{\theta\phi}$ (Nissen-Meyer et al., 2007a). Consequently, AxiSEM simulations are by construction always given for each individual element of the moment tensor. The task of summing to a full moment tensor is described in Sect. 3.1. Additional features of post-processing are rotation from a pole-centric to an actual source–receiver geometry, bandpass filtering, convolution with a source–time function, rotation to arbitrary seismogram component systems, choice between displacement and velocity seismogram. A similar set of operations applies to 3-D wavefield visualizations. Users may for this visualization case also specify rendering perspectives, wavefield components, 3-D cuts, and hypersurface extractions within post-processing. In effect, this allows for in situ visualization and merges seismic trace analysis with visualization on the fly.

Figure 5. A typical mesh decomposition for the PREM model running at a dominant period of 9 s on 96 cores. Load balancing is exact, and the arbitrary permissible multiplication between the number of horizontal and radial slices guarantees flexibility for adapting numerical settings to existent hardware infrastructures.

2.6 Parallelization

At frequencies around 1 Hz, the required run-time memory (roughly 20GB) for 1 Million elements exceeds the typical memory of contemporary cluster cores. Also, as can be deduced from Fig. 1, the CPU time-to-solution becomes prohibitively lengthy if the system is simulated on a single core (although possible). We thus incorporated a generic, automated 2-D domain decomposition into $2N_\theta N_r$ domains, where $N_\theta N_r$ represent positive integers for the number of latitudinal and radial slices, respectively (see Fig. 5 for an example with 96 cores). This guarantees the simultaneous realization of the three crucial factors for scalability: (1) a minimal amount of neighboring domains (maximally eight), (2) minimal interfaces size (i.e., length of messages), and (3) exact load balancing. The non-blocking, asynchronous message-passing implementation is entirely hidden behind the computation of the stiffness term, which will be seen in the excellent scaling in the next section.

2.7 Performance & scaling

The reduction of 3-D wave propagation to a 2-D computational domain is reflected by the method's performance compared to 3-D methods (Fig. 1). This equally holds true against methods that are extremely efficient and fast in delivering singular seismograms such as normal-mode summation or DSM, but whose computational cost depends on the amount of desired output locations. To compute sensitivity kernels for the inverse problem, one needs to save the entire

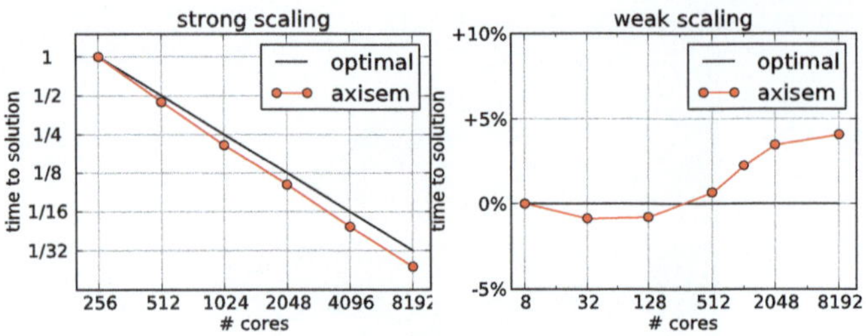

Figure 6. Scalability of AxiSEM on a Cray XE6 at CSCS (Switzerland). Left: Strong scaling, i.e., fixed global problem size (8 Million 2-D elements) as a function of the number of used cores communicating via the message-passing interface, for 12 000 time steps. AxiSEM scales super-optimally, which is mostly due to the more efficient usage of run-time memory if less memory is used per core. Right: Weak scaling, i.e., fixed problem size per core (1000 elements) for 1000 time steps. The desired constant time-to-solution is exceeded by 4 % for > 8000 cores, in which case communication is not entirely hidden behind the computation of the stiffness terms.

space–time wavefield everywhere, and hence such dependencies become inefficient especially when moving to higher resolutions. Figure 1 gives a flavor of the computational task along a typical range of global-scale seismic periods, quantified in terms of the required amount of CPU cores to achieve real-time simulations (seismogram time equals CPU wall-clock time), assuming perfect scalability. Figure 6 shows the new implementation of 2-D parallelization and strong as well as weak scaling results on a Cray XE6 supercomputer installed at CSCS, Switzerland. In both cases, the performance is excellent: strong scaling (fixed global degrees of freedom) is even super-optimal due to efficient memory usage. Weak scaling shows a slightly sub-optimal behavior at 96 % for > 8000 cores, indicating that message passing and parallelization are essentially hidden within the code. It is noteworthy to recognize that AxiSEM has little run-time memory, and applications at the high-frequency end benefit from vast multi-core systems mainly to reduce wall-clock time, unlike 3-D seismic methods, which are often memory bound.

As in any (2-D) time-domain discrete method, it is important to recognize that half the dominant period takes about 8 times longer if the seismogram length is fixed: the mesh is about 4 times larger, and the time step about twice as small. Note that monopole source types run faster than dipoles and quadrupoles due to their sparser stiffness terms (Nissen-Meyer et al., 2007b).

As a rough estimate, each 2-D element occupies 1.5 wavelengths and about 2.5 kB, and for seismic periods of 5 s approximately 400 000 elements are needed. The code requires about eight microseconds simulation time per time step and element.

2.8 Excessive input/output

Spectral-element methods of the kind presented here have excellent scalability properties in general (Fig. 6). The bottleneck, especially when moving to higher resolution and larger parallelization, lies in disk access, which is necessary for saving wavefields and seismograms at run time. For storage of synthetic seismic data, especially for a database of precomputed waveforms (see Sect. 5.3), platform independence of the data is needed. However, the storage format of Fortran binary files is not even compiler independent. To ensure true platform independence, AxiSEM fully supports the widely accepted NetCDF4 (http://www.unidata.ucar.edu/netcdf/) format to store seismograms and wavefields, but users may also revert to Fortran binary if desired. A NetCDF4 file is a container in which very large variables (e.g., wavefields) as well as single scalar values (e.g., general simulation information) can be stored. The format allows transparent compression of the variables using the SZIP algorithm (Shu Yeh et al., 2002), which saves around 50 % of hard drive space for a typical seismic wavefield with respect to generic binary format. The container character of a NetCDF file means that direct access to selected data is possible, i.e., small amounts of data such as time series can be read from a (potentially very large) file, without loading the whole file into memory. The code allows the storage of all simulation output in one self-contained NetCDF4 file, which facilitates the handling of a large number of simulation results, for example in parameter studies.

NetCDF4, which is based on the HDF5 format, allows for parallel writing into one file. Since this makes use of parallel file systems very efficiently, it might provide significant performance gains in the next generation of supercomputers. In the current generation however, the installation of parallel NetCDF4 is not generally reliable yet. Therefore, and to increase compatibility with older machines, AxiSEM uses a serial round Robin scheme for writing data to disk. All processors buffer their respective wavefield output locally. After a set number of time steps, one instance spawns a new thread and transfers its wavefield buffer to it. This new thread then opens the output file and compresses and writes the buffer to disk, while the original processor continues to simulate the wavefield. This non-blocking IO scheme has been tested to work well up to 224 parallel instances, in that wavefield storage marginally affects CPU time and performance.

2.9 Implementation and availability

The AxiSEM code is written in Fortran2003 combined with MPI message passing, requiring corresponding compilers. Optional additional packages are NetCDF4 (Rew and Davis, 1990) for improved I / O, fftw3 (Frigo and Johnson, 2005) for post-processing in the frequency domain, TauP (Crotwell et al., 1999) for traveltime picks, paraview for visualization of *vtk*- and *xdmf*-based wavefields, Matlab for visualization of record sections, gnuplot for creating seismogram image files, Google Earth for visualization of source–receiver geometries and seismograms on the sphere, and python wrappers for streamlined input/output and linkage to ObsPy (Beyreuther et al., 2010). The Fortran2003 code is divided into a Mesher utilizing OpenMP, a solver utilizing the message-passing interface (MPI) for communication between separate domains, and extensive post-processing for ease of visualization, filtering, source–time functions, various receiver component systems, and moment-tensor solutions. The number and meaning of input parameters for AxiSEM are kept at a generic, streamlined minimum, providing a robust basis in an effort to reduce failure. This is amended by a comprehensive number of sanity checks prior to the time loop, including critical tests upon mesh configurations, source, model, receiver setting parallelization, discrete volume and mass of the Earth, accuracy of internal surfaces, numerical quadrature, mass matrix and boundary terms.

AxiSEM is available through a release version with GPL license from www.axisem.info, and comes with no guarantee of functionality or support, but each version contains a detailed manual, examples, nightly builts and a tractable subversion control system as well as an existent user base.

3 Seismic sources

In global seismology, it is customary to rely on the point-source approximation and corresponding moment tensors $\mathbf{M}^p$ (not to be confused with the mass matrix $\mathbf{M}$ in the last section). The implementation of indigenous earthquake sources or single forces located at $\boldsymbol{x}_p$ is detailed in Nissen-Meyer et al. (2007b). We use temporal Dirac delta functions acting at time t_p in the simulations, such that a displacement time series is obtained by convolving a source–time function $S^p(t)$ (incorporated into a time-dependent moment-tensor term as $\mathbf{M}^p(t) = \mathbf{M}^p S^p(t)$), with the Green tensor solution $\mathbf{G}^p$:

$$\boldsymbol{u}^p(\boldsymbol{x},\omega) = \mathbf{M}^p(\omega) : \nabla_p \mathbf{G}^p(\boldsymbol{x},\omega), \tag{7}$$

where we reverted to frequency domain ω for concise notation, and ∇_p denotes spatial differentiation with respect to the source coordinate (no summation implied). Here, we focus on basic necessary post-processing operations to obtain the response to a full moment tensor, the extension to finite kinematic faults, stochastic sources (as for example in noise seismology or helioseismology), and the problem of handling Dirac delta functions $\delta(\boldsymbol{x})$ in a discrete world.

3.1 Moment-tensor and single forces

To obtain the response to a full moment tensor (e.g., CMT catalog, www.globalcmt.org), one applies posterior summation honoring the respective radiation patterns along the azimuth along with the convolution:

$$\boldsymbol{u}^p(\boldsymbol{x},\omega) = \sum_{m=1}^{4} \tilde{M}_m^p(\phi,\omega)\tilde{\mathbf{G}}_m^p(\tilde{\boldsymbol{x}},\omega), \tag{8}$$

where $\tilde{\mathbf{G}}_m^p$ represent 2-D vectorial Green's responses to each simulation m for point source p, and $\tilde{M}_m$ read

$$\tilde{M}_1(\phi) = M_{rr}, \tag{9}$$

$$\tilde{M}_2(\phi) = \left(M_{\theta\theta} + M_{\phi\phi}\right)/2, \tag{10}$$

$$\tilde{M}_3(\phi) = M_{r\theta}\cos\phi + M_{r\phi}\sin\phi, \tag{11}$$

$$\tilde{M}_4(\phi) = \left(M_{\theta\theta} - M_{\phi\phi}\right)\cos 2\phi + M_{\phi\theta}\sin 2\phi. \tag{12}$$

Only these four independent types of radiation patterns exist (monopoles $\tilde{M}_1$ and $\tilde{M}_2$, dipole $\tilde{M}_3$, and quadrupole $\tilde{M}_4$) in this axisymmetric framework. For a full earthquake moment tensor, four independent simulations are thus undertaken to account for the six moment-tensor elements M_{ij}, whereas single forces (as needed for Lamb's problem, ambient noise, impacts, or adjoint wavefields) require two simulations (monopole vertical force, dipole horizontal force) to account for the three components (Nissen-Meyer et al., 2007b).

Figure 7 depicts an example of the individual displacement solutions for each $\tilde{M}_m$ (quadrant on the left), and the final sum for a response to the full 2011 M9 Tohoku earthquake, recorded at station BILL (Eastern Siberia) at 33° distance. Note that the summed trace bears little resemblance to any of the generic solutions to radiation patterns (Eqs. 9–12).

3.2 Finite faults

Kinematic rupture over a fault plane can be modeled as a discrete sequence of point sources distributed across the fault plane, each of which may have individual moment tensors, magnitudes and source–time functions to mimic a time-dependent slip. AxiSEM is well positioned for an efficient incorporation of such finite faults: due to the rotational symmetries outlined above, the number of simulations for an arbitrary fault is simply given by its number of discrete depth points. The solution for finite-fault displacements may be written in terms of the solution to individual point-source solutions $\boldsymbol{u}^p$:

$$\boldsymbol{u}(\boldsymbol{x},t) = \sum_p \boldsymbol{u}^p(\boldsymbol{x},t). \tag{13}$$

Note that the dependence on the point-source locations $\boldsymbol{x}^p$ exists for the moment tensor $\mathbf{M}^p$ (by means of radiation pattern and source–time function) and the Green tensor $\mathbf{G}^p$ in Eq. (7), requiring separate solutions to the wave equation for each location $\boldsymbol{x}^p$. A significant shortcut can be made in the case of spherically symmetric media by saving seismograms at "all" distances and applying rotational properties to the Green tensor. As such, all laterally distributed points $\boldsymbol{x}^p$ are accommodated within one simulation, and only the

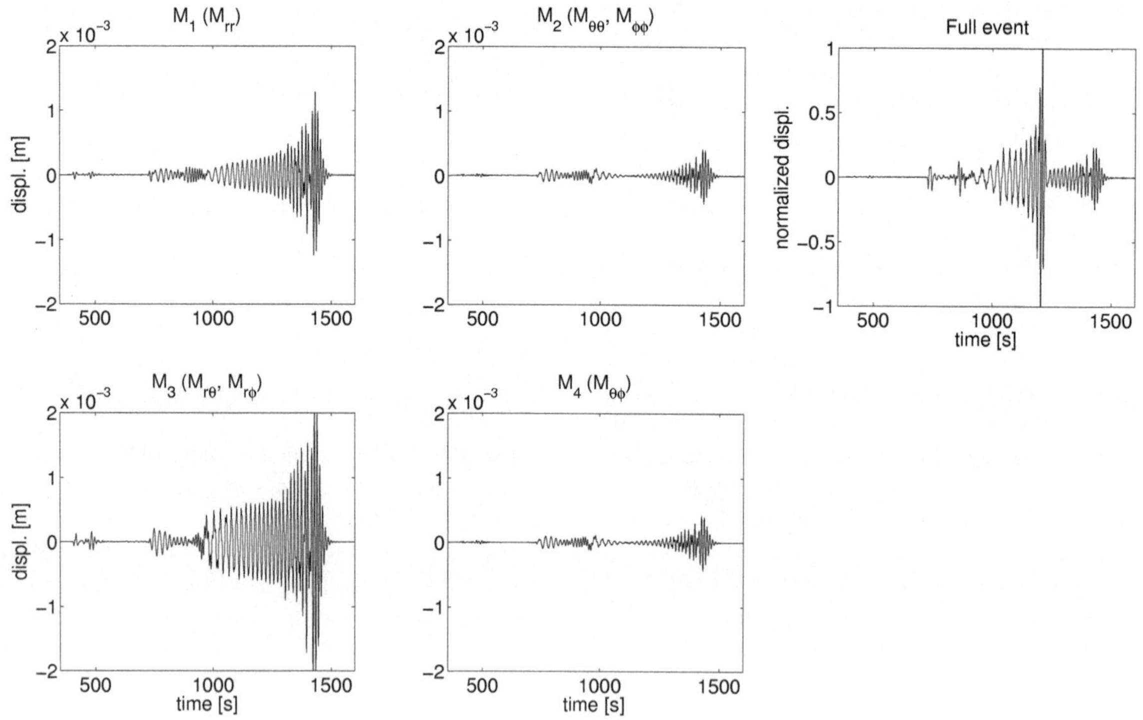

Figure 7. Left quadrant: The four time series upon the generic moment-tensor types (see Eqs. 9–12). Right: Summation to the full seismogram for the 2011 M9 Tohoku (point-source) event. Plotted is the displacement in the s direction (i.e., perpendicular to the symmetry axis, see Fig. 3), with a dominant period of 10 s recorded at station BILL (East Siberia) at 33° distance.

discrete depths need to be honored by separate simulations. This is advantageous, as most finite-fault models are mainly distributed laterally, and only require a few depth samplings. This allows for considerable flexibility should one wish to change certain properties of the fault model without conducting new simulations. In light of the common problem of local minima in (non-linear) source inversions, this offers an efficient engine for performing comprehensive studies on the behavior of different fault models and methodologies (Page et al., 2011). The modeling of finite sources is thereby largely delegated to post-processing (see Sect. 5.3), such that existent AxiSEM databases can simply be applied to finite sources as well, and finite faults can be naturally embedded within any application of AxiSEM with little additional computational effort.

3.3 Stochastic sources

The rotational invariance also facilitates applications of spatially distributed stochastic sources such as ambient noise generated by ocean–continent interactions or crustal scattering (e.g., Boué et al., 2013), or random pressure fluctuations in the Sun's interior (Gizon and Birch, 2002). Similar to finite faults, one simulates point sources at the relevant number of depths (for ocean–continent ambient noise, this is one depth) and relegates the spatial distribution and stochastic time–frequency behavior to post-processing for rotations and filtering, respectively.

This helps not only for generating a diffuse wavefield for structural imaging, but also for inverting for ambient-noise source locations.

3.4 Discrete Dirac delta distribution

It is desirable to simulate Green's functions, as they offer flexibility with respect to filtering and source–time functions after the simulation. The "source–time function" for the simulation is then a Dirac delta "function", which, from a rigorous perspective, is meaningless in any discretized system. To retain its attractive properties as the "source–time function" to generate Green's functions, one instead utilizes a triangle function that obeys integral properties of the Dirac distribution. Should one wish to extract a downsampled time series from a simulation of this kind, then the "width" of the delta function must be adjusted to guarantee the tradeoff between (1) its delta-function characteristics and (2) sufficient sampling below Nyquist frequency. This is automatically computed in the code, depending on the sampling and period ranges.

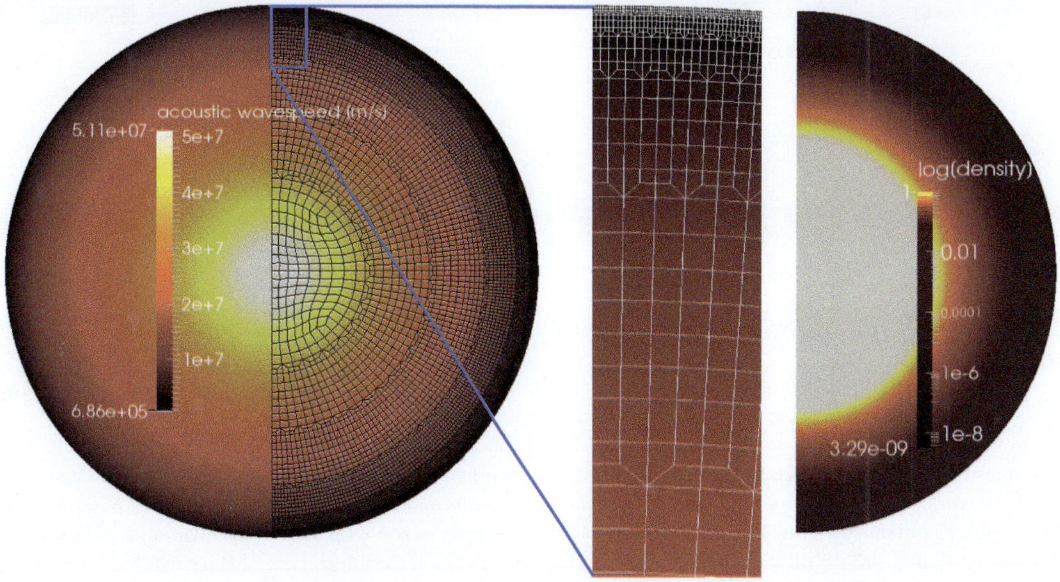

Figure 8. A mesh for the Sun's interior that accommodates the radial structure of the Sun for frequencies up to 5 mHz. It honors acoustic wavespeed variations of the Sun across two orders of magnitude (left), leading to seven coarsening (doubling) layers. Density (right) varies by eleven orders of magnitude, but does not affect the meshing process so long as these variations are smooth at the scale of elements. Such a mesh represents the basis for wave propagation and imaging the solar interior utilizing stochastic noise excitation within the framework of time–distance helioseismology.

4 Structural properties

The definition, discretization and implementation of background models are some of the most critical aspects for accurate wave propagation. Amongst the decision factors are

- the scale lengths of structural variations, and their feasible upscaling from a potentially smaller-scale model,
- merging diverse models of source and structure,
- sharp versus smooth variations,
- local reliability and resolution, for instance in global tomographic models.

Uncertain choices amongst these points may lead to an entirely wrong model and consequently useless wave-propagation results. Discretization and meshing in finite element-based methods usually strive to replicate all sharp boundaries of the model. Apart from algorithmic limitations in meshing arbitrary hexahedral elements, any failure to mesh a desired interface leads to false solutions.

AxiSEM offers an internal meshing algorithm that optimally honors any discontinuities in spherically symmetric Earth models, as well as arbitrary discrete spheres of any radial distribution of solid and fluid domains. Axisymmetric structures that are invariant in the azimuthal direction may then be superimposed onto the background mesh on-the-fly in the solver. These structures can be of pre-defined shape types, or arbitrarily superimposed by interpolation of discrete external grids (see Sect. 4.4).

4.1 Spherically symmetric models

Spherically symmetric models such as PREM, IASP91, and ak135 are automatically incorporated into AxiSEM. The code also allows for flexible inclusion of arbitrary 1-D structures in the meshing process, such that other Earth models such as those based on mineral physics can be easily accommodated. One may also apply the methodology to other planetary bodies (e.g., Moon, Mars, Europa; not shown here) including purely fluid media to facilitate acoustic wave propagation (a computational shortcut still popular in the exploration industry), which drastically reduces computational cost.

As a curious case of extreme medium variations readily discretized by our methodology, we reach out for our central star: the Sun is a giant fluid sphere subject to turbulent redistribution of masses, magnetic field variations, and acoustic body waves (Gizon and Birch, 2002). The background structure of the Sun covers many orders of magnitude in density due to its huge size and gravitational force, and about two orders of magnitude in acoustic wavespeeds. Only the radial wavespeed gradient matters for meshing, thus it is easily possible to adapt the AxiSEM mesh to the Sun, as seen in Fig. 8.

Surface boundary conditions for acoustic waves (i.e., vanishing pressure) pose no technical difficulty, but are not yet included in the first code release and shall be added in a timely future version.

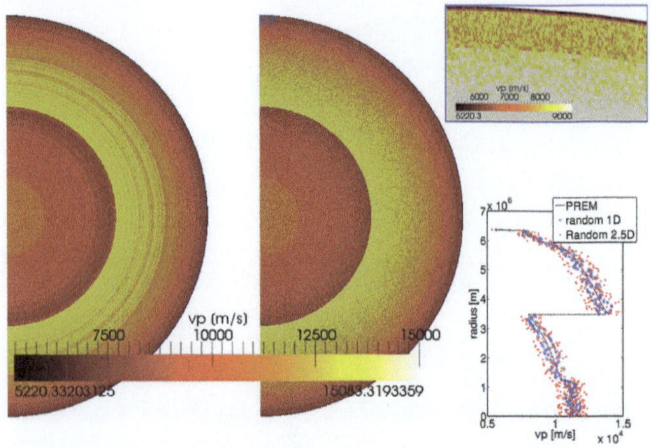

Figure 9. Two examples of random wavespeed variations superimposed onto PREM. Left: Variations with depth only. Middle: Variations in 2.5-D. Right (top): Zoom into the crust for the 2.5-D random medium. Right (bottom): Radial profile of these two realizations with respect to PREM.

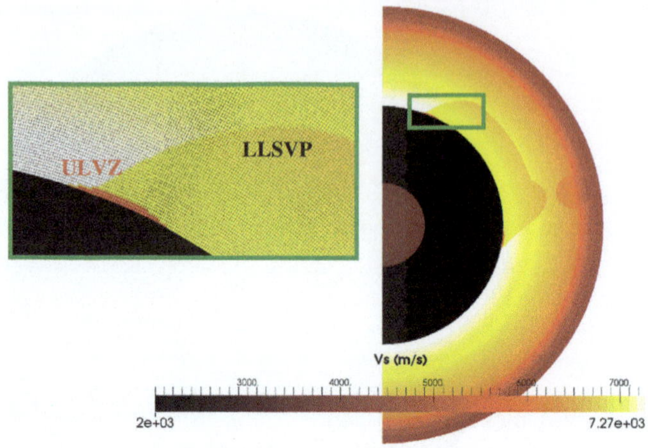

Figure 10. An example of various lateral heterogeneities, representing realistic deep-mantle structures projected onto the source–receiver plane with azimuthal invariance. The large volume in yellow denotes a Large Low-shear wave Velocity Province (LLSVP), flanked by two exaggerated ultra-low velocity zones (50 km height, 10 % P velocity decrease, 20 % S velocity decrease, 10 % density increase) underlying a detached uprising in the mid-mantle. The implementation is done by assigning laterally heterogeneous properties to the coefficients of the basis functions, as commonly done in high-order spectral-element methods (Peter et al., 2011) so long as elements are sufficiently small to capture variations in a smooth manner.

4.2 Crustal variations

Crustal thickness variations from 6–8 km (oceans) to 60–80 km (continental shields) owing to lithospheric composition have a significant imprint on those seismic phases that are sensitive to shallow structure, such as surface waves that traverse the crust to large distances. Additionally, covering 70 % of Earth's surface, the oceans are also a contributor to wavefield modifications, even though most seismometers are installed on land.

The computational efficiency of a 2-D numerical method allows for sufficiently small elements to mesh the crust explicitly, which is also necessary for wavelengths in the range of crustal thicknesses. This is also true for the actual oceans, which may be discretized by actual fluid-domain elements, instead of resorting to a loading equivalent (Komatitsch and Tromp, 2002b) or a homogenized crust (Fichtner and Igel, 2008). Similar to axisymmetric structures, this divides the sphere into oceanic and continental pole-centric rings. Discretized oceans are not available in the first code release, but will be added in the near future.

4.3 Random media

Spherically symmetric or axisymmetric variations in properties can be as general as desired in the method, including random media variations, so long as they are sufficiently smooth and mildly deviatoric. In Fig. 9, we show two types of such random variations, perturbing either radial structure (left) or 2.5-D structure by maximally 10 % velocity variations. Wave propagation through such complex structures can deliver useful insight into wave effects as a function of spatial scale dependence, scattering and homogenization properties, or the relation between structural heterogeneities and seismic measurements (Baig and Dahlen, 2003).

4.4 Localized heterogeneities

The next level of complexity in structural properties is represented by axisymmetric media, which may have arbitrary variations in the source–receiver plane, but are invariant in the perpendicular azimuthal direction. Especially in high-frequency regimes, 3-D edge effects from large-scale structures may become less dominant given the decreased ratio between seismic and structural wavelengths, such that Fresnel zones reside within the azimuthal extent of an anomaly. Axisymmetry can then represent a tangible basis for waveform modeling of unknown arrivals, precursors, undetected arrival delays, and oblique reflectors. The only neglected part of the wavefield compared to 3-D background models are 3-D wave effects from off-plane structures such as 3-D elastic lense focusing and off-plane back-scattering. All other scenarios in which structures vary preferentially in the source–receiver direction are respected, as for instance wave propagation through certain configurations for subduction zones, or forward scattering of small-scale lowermost mantle structures. Fig. 10 shows one example of lateral heterogeneities implemented in AxiSEM, including a Large Low-Shear Wave Velocity Province (LLSVP), e.g. (Romanowicz and Gung, 2002), an Ultra-Low Velocity Zone (ULVZ), e.g. (Rost, 2013), and a disconnected uprising (e.g., Zhao, 2001).

These variations are sufficiently smooth to be picked up by the elemental basis functions within the spectral-element mesh (see left panel). The inclusion of such lateral heterogeneities can be done by functional parameterization as shown here, but also by discrete external models. Such abritrary models are incorporated via a KD-tree (Kennel, 2004) search for nearest neighbors and interpolation, and therefore allows for any shape and complexity. The accuracy of wave propagation through such models is governed by the scale of heterogeneity versus finite-element size, in that strong variations on short spatial scales tend to be smoothed in the discrete model.

4.5 Anisotropy

In a spherically symmetric scenario, the most complex anisotropy is transverse isotropy, with five independent parameters. In axisymmetry, we may incorporate the full elasticity tensor with triclinic symmetry and 21 independent parameters, so long as the anisotropy does not vary within the azimuthal direction of the Fresnel zone. While logically mandated, this theoretical fact is in itself intriguing: An individual source–receiver wave senses only anisotropic variations within a suffiently narrow azimuthal range of sensitivity. If, however, a higher complexity of anisotropy is present but varies at a scale larger than the sensitivity of the traveling wave, then this level of large-scale complexity is not extractable from singular seismograms but instead represents an effective image of the actual structure. van Driel and Nissen-Meyer (2014a) provide a detailed analysis and implementation strategy for anisotropic wave propagation in axisymmetry, including proofs related to the multipole expansion for the presence of general anisotropy in the axisymmetric environment.

4.6 Attenuation

Intrinsic attenuation or visco-elastic damping is a natural property of the bulk real-Earth structure at relevant frequency ranges of seismic wave propagation. Although models for the quality factor Q (inverse damping) of the mantle show little agreement and the origin of damping may be inseparably coupled with elastic small-scale scattering, seismic attenuation on waveforms is a well-detected and significant phenomenon. We implemented an improved methodology based on coarse-grain memory variables with negligible additional computational cost compared with purely elastic wave propagation. This further includes attentuative bulk and shear deformation, five relaxation mechanisms, a combined linear/non-linear approach to identify optimal sets of parameters for the range of realistic Q, and analytical time stepping. Details on this new implementation, which is applicable to any higher-order finite-element method, are described in a separate paper (van Driel and Nissen-Meyer, 2014b).

4.7 Lack of ellipticity and rotation

The Earth's radius differs, depending on latitude, by up to 40 km between poles and Equator. For reasons of axial symmetry, AxiSEM does not allow waves to propagate from a non-polar point source through a pole-centric ellipsoid. To account for ellipticity a posteriori, three options are suggested: (1) phase correction, (2) epicentral distance correction, (3) Born perturbation theory. Phase-specific ellipticity corrections may be applied by shifting waveforms according to predicted traveltime shifts. This is useful only if individual phases are assessed, such as in most cases of tomography, and phase-specific waveform modeling. Epicentral distance correction may be conducted by recalculating receiver coordinates to account for the difference between purely spherical and ellipsoidal geometries, similar to the standard method in traditional tomography (Kennett and Gudmundsson, 1996). Finally, Born theory may be applied by assuming ellipticity (including internal interfaces) to act as a boundary perturbations to the spherical model domain. This way, entire seismograms are accounted for jointly. This approach has not been implemented or tested at this point. Rotation of the polar axis can in principle be incorporated into AxiSEM, but only for a polar source, which clearly is a rather unique case of rotation. On the scale of interest where rotation comes into play (above periods of 100 s), one could devise a torus-shaped, off-axis source in case its azimuthal radiation is of lesser significance – as may be the case for free oscillations. This would be a field of further study and implementation. In summary, such effects grow into a visible and recordable first-order issue for rather specific cases of seismic data analysis concerned with pathological body-wave paths and very long-period seismology.

5 Wave-propagation applications

Our methodology and the actual AxiSEM code are production-ready and may be used to tackle a diverse range of applications. Here, we sketch some of these, ranging from basic validation against reference solutions across the frequency spectrum and indefinite solutions to wave propagation, 3-D wavefield visualization, lowermost mantle heterogeneities, tomographic models, comparison with recorded data, and sensitivity kernels. All examples are deliberately disconnected as a showcase for the diverse range of applications.

5.1 High-frequency body waves

Previous publications on an early version of this method and implementation showcased the accuracy by comparison against normal-mode summation (Nissen-Meyer et al.,

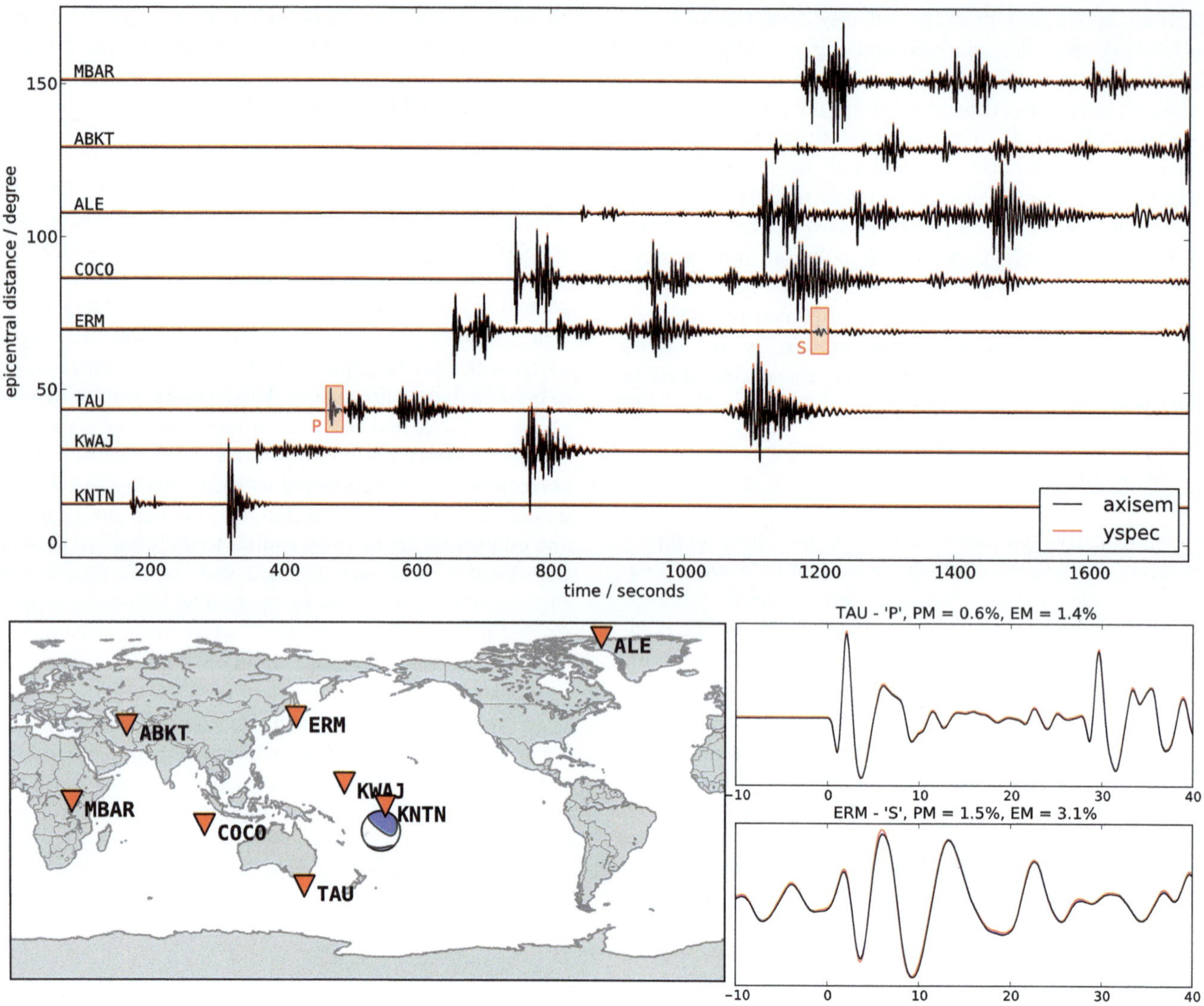

Figure 11. High-frequency validation (1 Hz dominant frequency) between AxiSEM and YSpec. Top: Record section for vertical displacements of an M4.1 event in Tonga (depth: 126 km), recorded at the stations shown on the map (bottom left) as red triangles. The background model is PREM, including anisotropy and attenuation, and the traces are filtered between seismic frequencies of 0.1–1 Hz, i.e., at the limit of recordable signals in global seismology. The traces from AxiSEM and Yspec are virtually indistinguishable. The zoom sections for individual seismograms (bottom right) in *P* and *S* waves (red boxes) represent phases that traveled 500 and 1200 wavelengths, respectively. Time in these panels is normalized to the ray-theoretical phase arrivals (Crotwell et al., 1999), and includes phase (PM) and envelope misfits (EM) measured following (Kristeková et al., 2009).

2008). Normal-mode summation is difficult to achieve at high frequencies due to the computational cost in generating mode catalogs, as well as numerical issues related to determining the eigenfrequencies. Instead, we now use an alternative frequency-domain reference solution capable of covering the entire relevant frequency band from 0.001 to 1 Hz ("YSpec" Al-Attar and Woodhouse, 2008). Figure (11) shows a record section and some details for both AxiSEM and YSpec modeling results in an anisotropic, visco-elastic PREM model for a Tonga event at 126 km depth, simulated at a dominant frequency of 1 Hz, i.e., at the limit of teleseismic detection of body waves. The fit is excellent for all phases and distances; the two solutions are indistinguishable almost everywhere out to 1600 propagated wavelengths. Minor differences are amplitude differences, and most probably due to a cut-off in the summation done in YSpec.

To our knowledge, this is the first accurate validation of two completely independent methods for anisotropic, viscoelastic media at such a high frequencies.

5.2 Free oscillations

At the grave end of the spectrum, free oscillations dominate and have revealed a great deal about Earth's internal

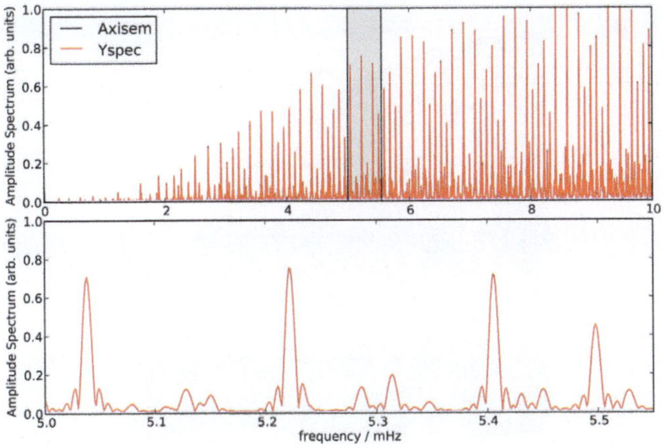

Figure 12. Amplitude spectra simulated by AxiSEM and YSpec for the PREM model (Dziewonski and Anderson, 1981) for frequencies below 0.02 mHz. The time-domain solution provided by AxiSEM extended over 48 h time series using 1.7 Million time steps. While amplitude spectra do not exhibit issues related to numerical dispersion, the fit between these two different methods is remarkable.

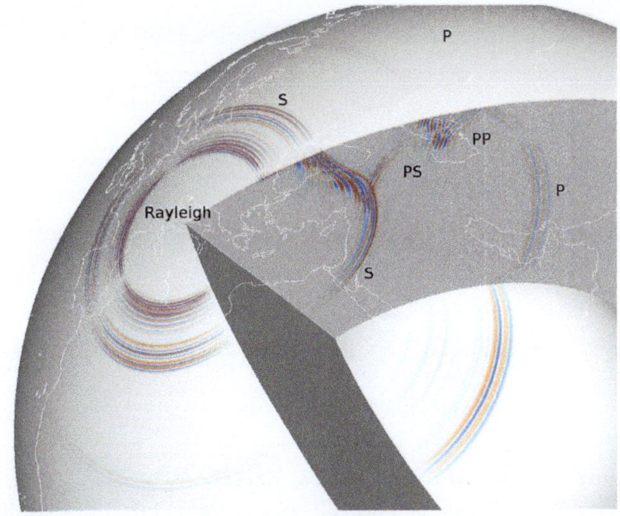

Figure 13. Snapshot of a 3-D wavefield emanating from a strike-slip event in Italy after 400 s. The background model is the isotropic, anelastic PREM, and the simulation is done at a dominant period of 10 s. Note the effect of the radiation pattern on the wavefield in 3-D. Similar snapshots are automatically generated in the post-processing of AxiSEM. A movie is available as supplementary material.

structure, in particular Earth's density structure (Dahlen and Tromp, 1998). We strive to provide a numerical method applicable to wave propagation across the observable frequency band. We thus compare amplitude spectra stemming from AxiSEM in a simulation over 48 h, 1.7 Million time steps, against YSpec in Fig. (12). The fit is again excellent, which is not trivial considering that time-domain numerical methods are exposed to steadily growing numerical dispersion errors with increasing numbers of propagated cycles. To the best of our knowledge, this is the first direct benchmark between time- and frequency-domain methods for free oscillations of the Earth, even if for a spherically symmetric, non-rotating, non-gravitating Earth model. Phase spectra, which may be more informative for actual studies with normal modes, are shown in (van Driel and Nissen-Meyer, 2014a).

5.3 Instantaneous forward solutions

The reduced dimensionality of AxiSEM opens doors to simulating the entire response due to a given background model once and for all, for all possible source–receiver choices. This seemingly daunting task is rendered tractable by the rotational properties of the displacement vector Eq. (2), such that seismograms only need to be stored along the one-dimensional distance range 0–180° for sources at a range of depths. This is computable. The remaining problem lies in deciding on a discrete sampling for source depth and receiver spacing to mimic continuous coverage. In the case of depths, this may be defined upon depth uncertainties in different earthquake catalogs, and in the case of receivers by choosing the closest one or interpolating upon epicentral distance uncertainty levels.

The computation of such a once-and-for-all solution can be conducted by taking into account the reciprocity of Green's function, resulting in only two simulations: one with a vertical and one with a horizontal single force, upon which the strain tensor needs to be stored for all realistic earthquake depths 0–660 km at all distances. This reciprocity shortcut is fueled by the fact that AxiSEM carries the full 3-D wavefield automatically, as opposed to reflectivity, DSM, Yspec or normal-mode solutions for which the number of saved seismogram locations factors into the computational cost. The problem is thereby shifted from CPU time to hard-drive storage. The permanent storage for the entire parameter space spanning all source–earthquake configurations and several Earth models is feasible (tens of terabytes). Queries to such databases (such as a record section of arbitrary source–receiver geometries, filters, source-time functions, and a range of spherically symmetric Earth models), can be completed within minutes by means of the same kind of post-processing as done upon the AxiSEM code. This can be tremendously beneficial in studies that need to sample a large range of parameters such as source inversion problems, especially in a probabilistic framework (Stähler and Sigloch, 2013).

5.4 Wavefield visualization

Most of the applications in this section, as well as in the literature, rely upon seismogram analysis. However, one of the major benefits of this method is the availability of the full global 3-D space–time wavefield, for both research and

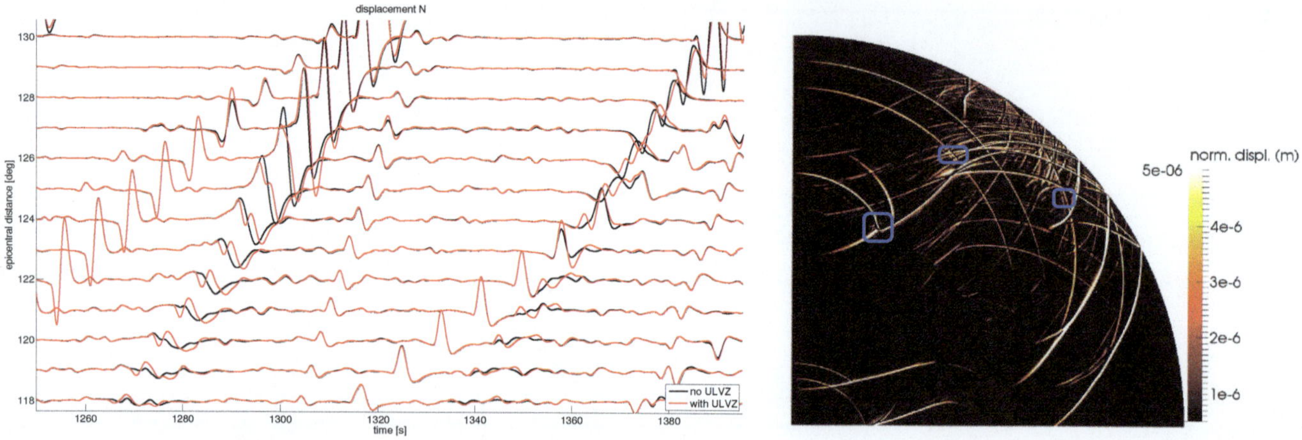

Figure 14. A forensic application of AxiSEM to the in-plane detectability of an azimuthally invariant representation of two adjacent structures: a "ULVZ" near an "LLSVP" (see model in Fig. 10). Left: Seismograms for a model with a ULVZ as in Fig. 10 (red traces), and one exactly the same but without the ULVZ (black traces). The underlying Earth model is the isotropic PREM (Dziewonski and Anderson, 1981), dominant period 2 s. The N-displacement record sections are at considerably large epicentral distance ranges. Right: Wavefield snapshot of the same simulation with ULVZ, at time 604 s. Blue quadrants denote those parts of the wavefield that are most affected by the presence of the ULVZ (in comparison to a similar plot for the simulation without ULVZ).

teaching purposes (Thorne et al., 2013). This is possible only due to the collapse to 2-D at run time and the convenience of on-the-fly extraction of the 3-D radiation upon azimuthal factors. In practice, this means that one may save the entire 2-D wavefield in space and time, and then subsequently decide on any moment tensor, summation, source–receiver geometry, and rendering choices. Figure 13 depicts a snapshot of a typical simulation of a strike-slip event in Italy with a dominant period of 10 s and isotropic, anelastic PREM background model. Note the characteristic dispersion in the surface wave train, the large amplitudes in the PP phase, and the 3-D radiation pattern. A movie of this setting is available in the supplementary material. Such visualization may offer complementary insight into complex propagation patterns beyond singular trace analysis, in particular as they can be devised as differential wavefields for diagnostic purposes in tracing the influence of changes in model parameters.

5.5 Wave propagation through in-plane heterogeneities

As mentioned in Sect. 4.4, we may readily insert lateral heterogeneities and compute 3-D wavefields upon those torus-like structures. To neglect the imprint of the torus-shaped azimuthal invariance, seismic frequencies should be chosen such that they represent local wavelengths that are smaller than the expected azimuthal extent of a 3-D heterogeneity to be modeled. In the regime of 1 Hz, this is warranted for a number of examples, and various lowermost mantle structures have been studied and constrained by waveform modeling for decades (e.g., Igel and Weber, 1996; Cottaar and Romanowicz, 2013). Constraining the geometry and structural composition of these features is crucial for understanding the thermo-chemical dynamical regime of the Earth's deep interior. In many previous applications, such structures have been modeled as azimuthally invariant, with source–receiver plane heterogeneity using approximations at frequencies below 0.25 Hz to study core-mantle boundary scattering (Thomas et al., 2000), D″-layer (Thorne et al., 2007), and LLSVP structure (Sun et al., 2007). Fig. 14 displays seismograms and wavefield snapshots for the model in Fig. 10. The record sections highlight phases and distances at which the existence of a ULVZ may be tested, possibly with array methods and stacking. The wavefield snapshots represent a complementary diagnostic for differential studies, from which the most significant imprints can be traced back to the surface, affecting phases such as PcP, ScS, and $SPKS$.

Modeling of such lateral heterogeneities taps into a regime of wave propagation that offers a grasp of wave effects at a resolution and computational cost that is difficult to achieve with alternative methods. Users should however, as always, be cautioned to recognize the structural assumptions imposed on such in-plane features: these may either approximate actual 3-D structures well (if the above-mentioned scale separation is warranted), or act as an upper bound of waveform effects (by means of azimuthal overestimation), but conversely they may also neglect elastic focusing and thus underestimate 3-D effects.

5.6 Tomographic models

Global models derived from tomography can also be approximated by an in-plane rendition with AxiSEM, in that they usually deviate only mildly and smoothly from spherically symmetric Earth models. Just as in the previous example, wavefield effects captured by this methodology are those that obey forward scattering, whereas true 3-D-medium

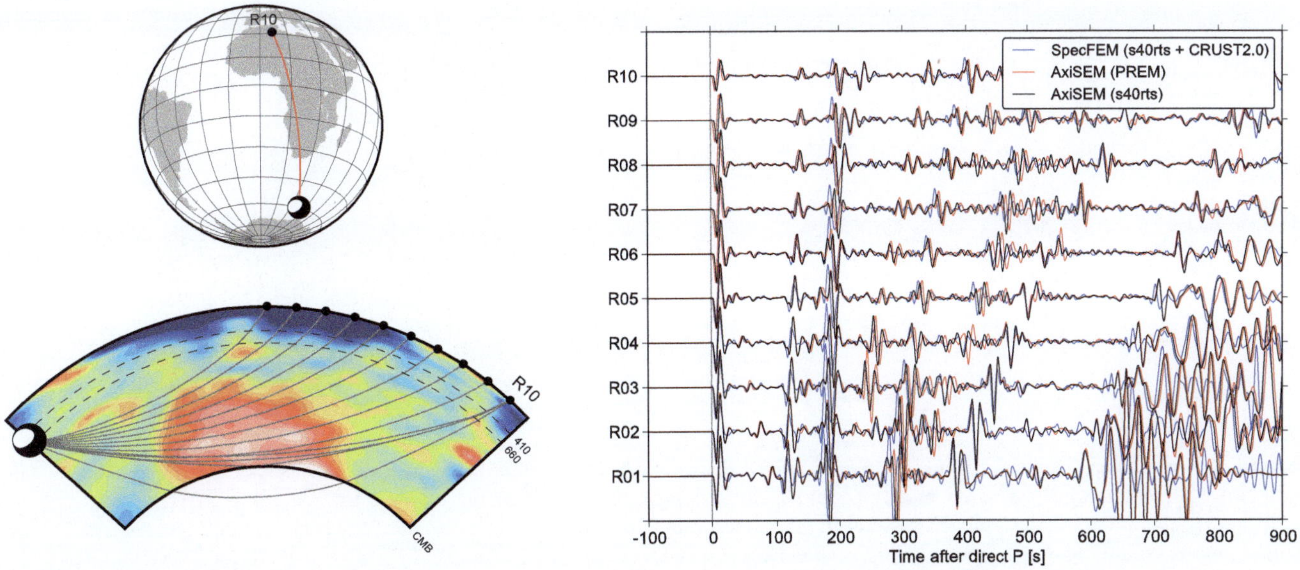

Figure 15. Modeling 3-D wave propagation through a 2.5-D version of tomographic model S40RTS (Ritsema et al., 2011) (bottom left) for an event near Antarctica (left top). Right: Seismograms filtered at 10 s from the receivers denoted in the cross section (bottom left) for the 1-D model, the 2.5-D tomographic model, and SPECFEM synthetics through S40RTS and CRUST2.0 (Bassin et al., 2000), aligned with the *P*-wave arrival time.

effects such as off-plane scattering are neglected. Of course, this azimuthal invariance does not represent our nature's dimensionality, but mimics a substantial sub-set of those data that are actually used for waveform modeling or tomography rather well and at a cost many orders of magnitude below that of simulating a 3-D domain. This can be seen in Fig. 15, a comparison between synthetic modeling through a PREM, an in-plane collapse of tomographic model S40RTS (Ritsema et al., 2011), and SPECFEM synthetics for S40RTS and CRUST2.0 (Bassin et al., 2000). As seen by the waveforms in Fig. 15, many phases are largely insensitive to the added complexity in these models, partly due to the smooth nature of tomographic models (as mandated by their inversion technique). Direct body waves and other early arrivals barely notice the different models, whereas later arrivals and surface waves exhibit considerable differences, most of which can be attributed to the crustal layer. The overall imprint of crustal variations overrides that of the tomographic model. The neglected effects such as 3-D back-scattering may indeed not contribute all that significantly to resultant seismograms, but this is subject to further parameter-space studies. In general, this provides an efficient new approach should one wish to validate different tomographic models within a synthetic exercise, or modify local properties for a given source–receiver geometry. The actual incorporation of tomographic models is trivial in AxiSEM for any model that is given by discrete cartesian, spherical grids, or spherical harmonics.

5.7 Relating to data

The ultimate raison d'être of any seismic modeling is its capability to relate to actual observed data, at least in some useful fraction of the generally impenetrable overall parameter space. Here, we showcase a comparison of waveforms at considerably high resolution (5 s) with observed data (Scheingraber et al., 2013). This resolution is at the cutting edge of supercomputing with 3-D methods (see Fig. 1), at a frequency range applicable to tomography, and also interesting for waveform modeling of relatively small-scale features in the lower mantle. The map (top right) shows the event and station locations (red triangles for PKiKP, blue for Pdiff). Filtering has been applied at 5–15 s (top) and 15–45 s (bottom). In the latter case, we included SPECFEM synthetics for the S362ANI tomographic model (Kustowski et al., 2008) and CRUST2.0 (Bassin et al., 2000), which are accurate to about 17 s (Tromp et al., 2010). AxiSEM synthetics are based on an inverted moment tensor and depth, whereas SPECFEM synthetics are taken from the IRIS database, i.e., calculated for GCMT. All synthetics have been convolved with an inverted source–time function. The phases have been aligned by frequency-dependent cross-correlation, forming the basis for tomographic inversions. The waveform differences between all three traces fall within a feasible range of conducting waveform tomography.

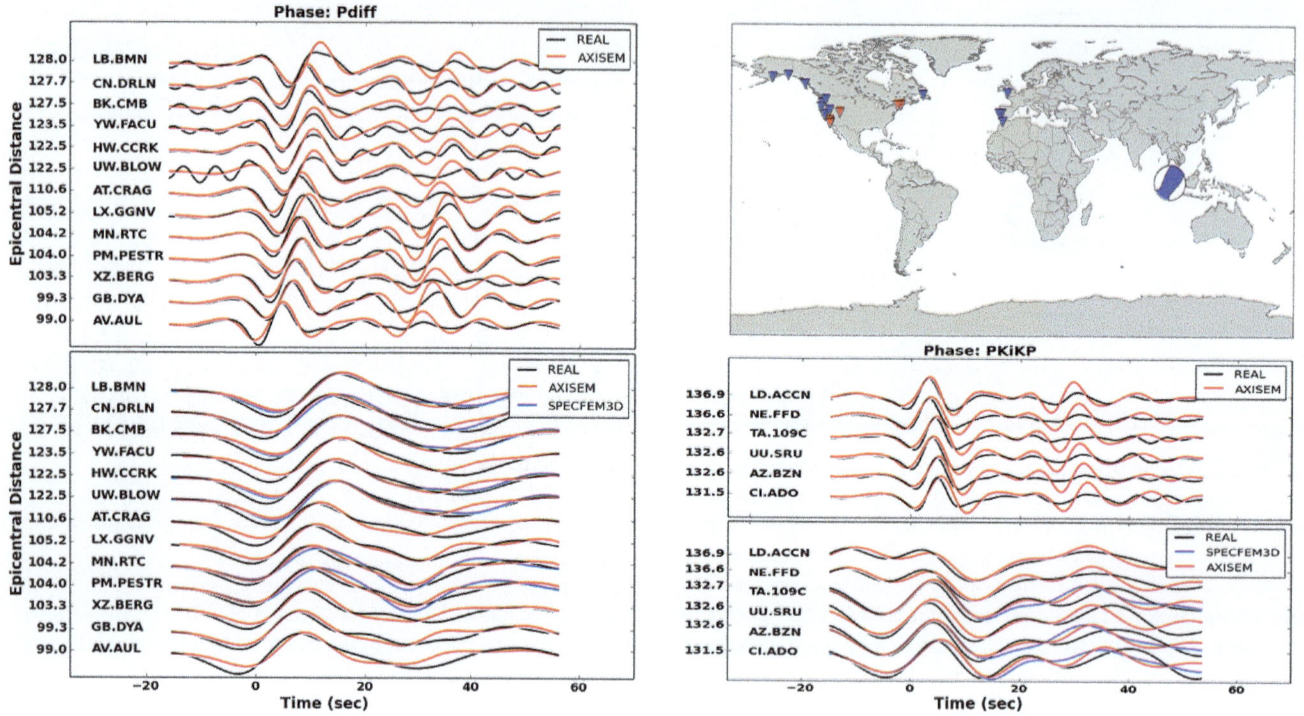

Figure 16. A comparison of AxiSEM synthetics with recorded data and SPECFEM synthetics (Tromp et al., 2010) for an Mw 7.5 2009 event in southern Sumatra at 82 km depth. Top right: Event-station distribution, where red triangles are for core-phase PKiKP, blue for the Pdiff CMB-diffracted phase. Left: Pdiff synthetics and observed data filtered at 5–15 s (top) and 15–45 s (bottom). In the latter case, SPECFEM synthetics are included, which are accurate down to 17 s. Bottom right: The same for PKiKIP. AxiSEM synthetics are simulated through a viscoelastic, anisotropic PREM model, SPECFEM synthetics through the S362ANI tomographic model (Kustowski et al., 2008), and both sets are shifted by cross-correlation traveltimes to align with the respective phases (left: Pdiff, right: PKiKP). Traveltime shifts are about 2–6 s (see main text).

The timeshifts based on the 13 Pdiff and 6 PKiKP measurements are

filter [s]	method	Δt(Pdiff) [s]	Δt(PKiKP) [s]
5–15	AxiSEM	2.5 ($\sigma^2 = 0.99$)	2.3 ($\sigma^2 = 0.5$)
15–45	AxiSEM	4.0 ($\sigma^2 = 0.89$)	3.4 ($\sigma^2 = 0.3$)
15–45	SPECFEM	3.7 ($\sigma^2 = 0.6$)	6.0 ($\sigma^2 = 0.37$)

Such comparisons include (as per usual) inevitable differences in processing such as event origin time and location, and source time function. However, it is noteworthy to recognize the waveform similarity confirming that wave propagation in spherically symmetric Earth models provides an excellent basis for broadband waveform tomography, in particular in the regime of periods below 10 s.

5.8 Wavefield sensitivity kernels

As a final example, we present the essential and possibly most intriguing application to time- and frequency-dependent sensitivity kernels (Nissen-Meyer et al., 2007a). The ability to store entire space–time wavefields from AxiSEM makes it principally trivial to compute time-dependent sensitivity kernels as a comprehensive basis for mapping seismograms to the Earth's structure, and thus the model-to-data operator for the tomographic inverse problem (Fuji et al., 2012; Colombi et al., 2014). As such, it logically extends existent ray-based or finite-frequency tomographies (which were based on approximate physics) by incorporating complete seismograms, arbitrary time and frequency windows as well as arbitrary wave effects such as triplicated phases from the mantle transition zone (Stähler et al., 2012), core-mantle diffraction (Colombi et al., 2012), or caustics. Figure (17) shows a sensitivity kernel for cross-correlation traveltimes with respect to compressional wavespeeds. This was computed with wavefields from AxiSEM, by time-integrating the velocity waveform of dominant seismic period 10 s within a 20 s time window around the P wave arrival time at 90° epicentral distance. The kernel exhibits considerable heterogeneity (partly due to saturated colorscales to highlight its complexity), notably missing the "donut hole" that is present for pure P wave kernels (Hung et al., 2000). This stems from the fact that our wavefield-based approach honors the large time window, which obscurs the purity of phase-based approaches, but correctly represents the measurement corresponding to the time window.

Figure 17. A sensitivity kernel computed with wavefields from AxiSEM, time-integrated (time window: 20 s) over the arrival of the direct *P* wave at 90° distance for a dominant period of 10 s. This denotes the "region of influence" in which this particular time window in the seismogram may "see" the compressional structure that deviates from the background model. Such kernels are not only the basis for waveform tomography, but may also aid in identifying obscure arrivals in the seismogram.

Time-dependent sensitivity (i.e., without integrating over the time window) is useful not only as a basis for tomography, but also in a forensic sense for detecting faint signals of sensitivity due to a given region or structure. Note that the (separate) calculation of sensitivity kernels is not part of the AxiSEM release, but will be added in the future as a separate package.

6 Conclusions

This paper presents a mature method and implementation for global seismic wave propagation across the seismic frequency spectrum. It describes crucial extensions with respect to the initial papers (Nissen-Meyer et al., 2007b, 2008) such as the inclusion of anisotropy, attenuation, lateral heterogeneities, finite sources, the basis for sensitivity kernels, and innovative visualization. The method is, to our knowledge, the first time-domain local numerical method successfully benchmarked against independent solutions across the entire frequency band recorded in global seismology, and exhibits excellent scaling in large multi-core systems. The code offers a diverse range of realistic applications in forward and inverse modeling and showcases promising comparisons with recorded data. The moderate computational cost allows for reaching any desirable frequency with moderate resources and storage of full space–time wavefields for sensitivity kernels.

The presented methodology is most accurate, efficient and useful in parameter regimes that are quite complementary to well-established, mature methods such as normal-mode summation (low-frequency seismology), 3-D numerical methods (with local basis functions) such as SPECFEM (3-D Earth models at intermediate frequencies), and asymptotic ray theory (high-frequency regime with potentially complex wavespeeds). As with any method, the realm of validity for AxiSEM is limited, approximative and blurred, and any application must be undertaken with caution despite the excellent and robust validation shown here and with the actual code available from http://www.axisem.info. This parameter space promises a diverse range of applications that were previously inconvenient, inexact, or unattainable due to limited computational resources or methodologies. Specifically, the main factor attributed to its efficiency in a 2-D computational domain is the availability of space–time wavefields for axisymmetric, viscoelastic anisotropic media and realistic earthquake sources.

We have touched upon a few key applications, far and away from explaining or validating each one of them. Rather, the purpose of this paper is to present a new open-source methodology and scan its usability specifically in those directions that we deem most benefitting from this modeling tool. Details on specific applications and implementation are found in other publications to be submitted, and the state of reliable features in the code should always be consulted in its concurrent manual.

All limitations of the methodology are by construction related to the existence of the symmetry axis, which mainly translates into neglecting true 3-D media (effects such as 3-D off-plane scattering and focusing) and realistic Earth rotation. All other limitations (lack of ocean layer, gravity and topography) mentioned here or in the code reflect the current stage of the algorithm, but pose no fundamental restriction.

6.1 Future additions

Current and future extensions of the presented methodology include low-frequency effects like gravitation (Chaljub and Valette, 2004), internal and external topography, and a local-scale version of the method. Sensitivity kernels upon AxiSEM also deliver the basis for scattering solutions to wave propagation, which may then allow for considering mild effects of 3-D (Born) scattering, which can be applied to both 3-D volumetric and boundary topography. AxiSEM-generated wavefields may also be injected into a small 3-D box of local 3-D heterogeneities in a hybrid sense (Tong et al., 2014). This will allow for the consideration of teleseismic wavefields to travel locally through 3-D heterogeneities (Masson et al., 2013), for instance beneath a dense seismic array above a tectonically active region such as USArray in the western USA or the Pyrenees (Monteiller et al., 2012). It may furthermore be useful to attempt a cost–accuracy benefit analysis across various wave-propagation codes that cover a sensible overlapping parameter space. This is a non-trivial task, as efficiency depends highly on the actual problem at hand (frequency range, distance range, number of sources, number of receivers, multi-scale models, source complexity, solid–fluid domains, etc.).

Acknowledgements. Review comments by Guust Nolet and Nobuaki Fuji significantly improved the manuscript. Financial support has been provided by the EU Framework program 7 (QUEST network), and the "Petaquake" HP2C project funded by the Swiss National Fund. Simulations were done at ETH's computing facilities and CSCS in Switzerland, as well as HECToR (UK), whose support is appreciated. The contribution of AF is IPGP contribution 3514. SCS was funded by the Munich Centre for Advanced Computing (MAC), SH under grant DFG TH1530/3-1. We are indebted to Jean-Paul Ampuero, Lapo Boschi, Emmanuel Chaljub, Tony Dahlen, Andreas Fichtner, Domenico Giardini, Heiner Igel, Dimitri Komatitsch, Guust Nolet, Karin Sigloch, Frederik Simons, Christine Thomas and Jeroen Tromp for constructive ideas, suggestions, and support that greatly improved AxiSEM over the past years. The online resource (including source tarballs, manual, tutorial, contact details etc.) for AxiSEM is: http://www.axisem.info. Development of the code takes place on GitHub, the latest version can be downloaded under Linux or MacOS with the terminal command: git clone https://github.com/geodynamics/axisem.git.

Edited by: G. Saccorotti

References

Al-Attar, D. and Woodhouse, J. H.: Calculation of seismic displacement fields in self-gravitating earth models–applications of minors vectors and symplectic structure, Geophys. J. Int., 175, 1176–1208, 2008.

Auer, L., Boschi, L., Becker, T., Nissen-Meyer, T., and Giardini, D.: Savani: A variable resolution whole-mantle model of anisotropic shear velocity variations based on multiple data sets, J. Geophys. Res., 119, 3006–3034, 2014.

Baig, A. M. and Dahlen, F. A.: Traveltimes of waves in three-dimensional random media, Geophys. J. Int., 153, 467–482, 2003.

Bassin, C., Laske, G., and Masters, G.: The Current Limits of Resolution for Surface Wave Tomography in North America, Eos, 81, 897 pp., 2000.

Becker, T. W. and Boschi, L.: A comparison of tomographic and geodynamic mantle models, Geochem. Geophys. Geosys., 3, 168, doi:10.1029/2001GC000168, 2002.

Beyreuther, M., Barsch, R., Krischer, L., Megies, T., Behr, Y., and Wassermann, J.: ObsPy: A Python Toolbox for Seismology, SRL, 81, 530–533, 2010.

Boué, P., Poli, P., Campillo, M., Pedersen, H., Briand, X., and Roux, P.: Teleseismic correlations of ambient seismic noise for deep global imaging of the Earth, Geophys. J. Int., 194, 844–848, 2013.

Bozdag, E., Trampert, J., and Tromp, J.: Misfit functions for full waveform inversion based on instantaneous phase and envelope measurements, Geophys. J. Int., 185, 845–870, 2011.

Capdeville, Y., Stutzmann, E., Wang, N., and Montagner, J.-P.: Residual homogenization for seismic forward and inverse problems in layered media, Geophys. J. Int., 194, 470–487, 2013.

Chaljub, E. and Valette, B.: Spectral element modeling of three dimensional wave propagation in a self-gravitating Earth with an arbitrarily stratified outer core, Geophys. J. Int., 158, 131–141, 2004.

Chaljub, E., Komatitsch, D., Vilotte, J.-P., Capdeville, Y., Valette, B., and Festa, G.: Spectral-element analysis in seismology, Adv. Geophys., 48, 365–419, 2007.

Colombi, A., Nissen-Meyer, T., Boschi, L., and Giardini, D.: Seismic waveform sensitivity to global boundary topography, Geophys. J. Int., 191, 832–848, 2012.

Colombi, A., Nissen-Meyer, T., Boschi, L., and Giardini, D.: Seismic waveform inversion for core-mantle boundary topography, Geophys. J. Int., doi:10.1093/gji/ggu112, 2014.

Cottaar, S. and Romanowicz, B.: Observations of changing anisotropy across the southern margin of the African LLSVP, Geophys. J. Int., 195, 1184–1195, 2013.

Crotwell, H. P., Owens, T. J., and Ritsema, J.: The TauP Toolkit: Flexible seismic travel-time and ray-path utilities, Seismol. Res. Lett., 70, 154–160, 1999.

Dahlen, F. A.: The Normal Modes of a Rotating, Elliptical Earth, Geophys. J. Roy. Astr. Soc., 16, 329–367, 1968.

Dahlen, F. A. and Tromp, J.: Theoretical Global Seismology, Princeton University Press, Princeton, 1998.

Dziewonski, A. and Anderson, D.: Preliminary Reference Earth Model, Phys. Earth Planet. Inter., 25, 297–356, 1981.

Fichtner, A. and Igel, H.: Efficient numerical surface wave propagation through the optimization of discrete crustal models – a technique based on non-linear dispersion curve matching (DCM), Geophys. J. Int., 173, 519–533, 2008.

Fichtner, A., Kennett, B. L. N., Igel, H., and Bunge, H.-P.: Theoretical background for continental and global scale full-waveform inversion in the time-frequency domain, Geophys. J. Int., 175, 665–685, 2008.

Fournier, A., Bunge, H.-P., Hollerbach, R., and Vilotte, J.-P.: A Fourier-spectral element algorithm for thermal convection in rotating axisymmetric containers, J. Comp. Phys., 204, 462–489, 2005.

Friederich, W. and Dalkolmo, J.: Complete synthetic seismograms for a spherically symmetric Earth by a numerical computation of the Green's function in the frequency domain, Geophys. J. Int., 122, 537–550, 1995.

Frigo, M. and Johnson, S. G.: The design and implementation of FFTW3, Proc. IEEE, 93, 216–231, 2005.

Fuji, N., Chevrot, S., Zhao, L., Geller, R. J., and Kawai, K.: Finite-frequency structural sensitivities of short-period compressional body waves, Geophys. J. Int., 190, 522–540, 2012.

Gizon, L. and Birch, A. C.: Time-Distance Helioseismology: The Forward Problem for Random Distributed Sources, Astroph. J., 571, 966–986, 2002.

Hung, S.-H., Dahlen, F. A., and Nolet, G.: Fréchet kernels for finite-frequency traveltimes—II. Examples, Geophys. J. Int., 141, 175–203, 2000.

Igel, H. and Weber, M.: P-SV wave propagation in the Earth's mantle using finite differences: Application to heterogeneous lowermost mantle structure, Geophys. Res. Lett., 23, 415–418, 1996.

Igel, H., Nissen-Meyer, T., and Jahnke, G.: Wave propagation in 3D spherical sections: Effects of subduction zones, Phys. Earth Planet. Inter., 132, 219–234, 2002.

Jahnke, G., Thorne, M. S., Cochard, A., and Igel, H.: Global SH-wave propagation using a parallel axisymmetric spherical finite-difference scheme: application to whole mantle scattering, Geophys. J. Int., 173, 815–826, 2008.

Jeffreys, H. and Bullen, K. E.: Seismological Tables, British Association for the Advancement of Science, London, 1940.

Kawai, K., Takeuchi, N., and Geller, R. J.: Complete synthetic seismograms up to 2 Hz for transversely isotropic spherically symmetric media, Geophys. J. Int., 164, 411–424, 2006.

Kennel, M. B.: KDTREE 2: Fortran 95 and C++ software to efficiently search for near neighbors in a multi-dimensional Euclidean space, Arxiv Physics Archive 0408067, http://arxiv.org/abs/physics/0408067, 2004.

Kennett, B. and Engdahl, E.: Traveltimes for global earthquake location and phase identification, Geophys. J. Int., 105, 429–465, 1991.

Kennett, B., Engdahl, E., and Buland, R.: Constraints on seismic velocities in the Earth from traveltimes, Geophys. J. Int., 122, 108–124, 1995.

Kennett, B. L. N. and Gudmundsson, O.: Ellipticity corrections for seismic phases, Geophys. J. Int., 127, 40–48, doi:10.1111/j.1365-246X.1996.tb01533.x, 1996.

Komatitsch, D. and Tromp, J.: Spectral-element simulations of global seismic wave propagation—I. Validation, Geophys. J. Int., 149, 390–412, 2002a.

Komatitsch, D. and Tromp, J.: Spectral-Element Simulations of Global Seismic Wave Propagation—II. 3-D Models, Oceans, Rotation, and Self-Gravitation, Geophys. J. Int., 150, 303–318, 2002b.

Kristeková, M., Kristek, J., and Moczo, P.: Time-frequency misfit and goodness-of-fit criteria for quantitative comparison of time signals, Geophys. J. Internat., 178, 813–825, 2009.

Kustowski, B., Ekström, G., and Dziewonski, A. M.: Anisotropic shear-wave velocity structure of the Earth's mantle: A global model, J. Geophys. Res., 113, 2156–2202, 2008.

Masson, Y., Cupillard, P., Capdeville, Y., and Romanowicz, B.: On the numerical implementation of time-reversal mirrors for tomographic imaging, Geophys. J. Int., 196, 1580–1599, doi:10.1093/gji/ggt459, 2013.

Mercerat, E. and Nolet, G.: On the linearity of cross-correlation delay times in finite-frequency tomography, Geophys. J. Int., 192, 681–687, 2013.

Monteiller, V., Chevrot, S., Komatitsch, D., and Fuji, N.: A hybrid method to compute short-period synthetic seismograms of teleseismic body waves in a 3-D regional model, Geophys. J. Int., 192, 230–247, doi:10.1093/gji/ggs006, 2012.

Nissen-Meyer, T., Dahlen, F. A., and Fournier, A.: Spherical-earth Fréchet sensitivity kernels, Geophys. J. Int., 168, 1051–1066, 2007a.

Nissen-Meyer, T., Fournier, A., and Dahlen, F. A.: A 2-D spectral-element method for computing spherical-earth seismograms–I. Moment-tensor source, Geophys. J. Int., 168, 1067–1093, 2007b.

Nissen-Meyer, T., Fournier, A., and Dahlen, F. A.: A 2-D spectral-element method for computing spherical-earth seismograms–II. Waves in solid-fluid media, Geophys. J. Int., 174, 873–888, 2008.

Nolet, G.: A Breviary of Seismic Tomography: Imaging the Interior of the Earth and the Sun, Cambridge University Press, Cambridge, 2008.

Page, M., Mai, P. M., and Schorlemmer, D.: Testing Earthquake Source Inversion Methodologies, Eos, 92, 75 pp., 2011.

Peter, D., Komatitsch, D., Luo, Y., Martin, R., Le Goff, N., Casarotti, E., Le Loher, P., Magnoni, F., Liu, Q., Blitz, C., Nissen-Meyer, T., P., B., and Tromp, J.: Forward and adjoint simulations of seismic wave propagation on fully unstructured hexahedral meshes, Geophys. J. Int., 186, 721–739, 2011.

Rawlinson, N., Pozgaya, S., and Fishwick, S.: Seismic tomography: A window into deep Earth, Phys. Earth Planet. Inter., 178, 101–135, 2010.

Rew, R. and Davis, G.: NetCDF: an interface for scientific data access, Computer Graphics and Applications, IEEE, 10, 76–82, 1990.

Rietmann, M., Messmer, P., Nissen-Meyer, T., Peter, D., Basini, P., Komatitsch, D., Schenk, O., Tromp, J., Boschi, L., and Giardini, D.: Forward and Adjoint Simulations of Seismic Wave Propagation on Emerging Large-scale GPU Architectures, in: Proceedings of the International Conference on High Performance Computing, Networking, Storage and Analysis, SC'12, IEEE Computer Society Press, Los Alamitos, CA, USA, http://dl.acm.org/citation.cfm?id=2388996.2389048, 1–11, 2012.

Ritsema, J., Deuss, A., van Heijst, H. J., and Woodhouse, J. H.: S40RTS: a degree-40 shear-velocity model for the mantle from new Rayleigh wave dispersion, teleseismic traveltime and normal-mode splitting function measurements, Geophys. J. Int., 184, 1223–1236, 2011.

Romanowicz, B. and Gung, Y.: Superplumes from the core-mantle boundary to the lithosphere: Implications for heat flux, Science, 296, 513–516, 2002.

Rost, S.: Deep Earth: Core-mantle boundary landscapes, Nature Geosci., 6, 89–90, 2013.

Scheingraber, C., Hosseini, K., Barsch, R., and Sigloch, K.: ObsPyLoad: A Tool for Fully Automated Retrieval of Seismological Waveform Data, Seismol. Res. Lett., 84, 525–531, 2013.

Shu Yeh, P., Xia-serafino, W., Miles, L., Kobler, B., and Menasce, D.: Implementation of CCSDS Lossless Data Compression in HDF, in: Earth Science Technology Conference, 2002.

Stähler, S. C. and Sigloch, K.: Fully probabilistic seismic source inversion – Part 1: Efficient parametrisation, Solid Earth Discuss., 5, 1125–1162, doi:10.5194/sed-5-1125-2013, 2013.

Stähler, S. C., Sigloch, K., and Nissen-Meyer, T.: Triplicated P-wave measurements for waveform tomography of the mantle transition zone, Solid Earth, 3, 339–354, doi:10.5194/se-3-339-2012, 2012.

Sun, D., Tan, E., Helmberger, D., and Gurnis, M.: Seismological support for the metastable superplume model, sharp features, and phase changes within the lower mantle, Proc. Natl. Ac. Sci., 104, 9151–9155, 2007.

Takeuchi, N., Geller, R., and Cummins, P.: Complete synthetic seismograms for 3-D heterogeneous Earth models computed using modifed DSM operators and their applicability to inversion for Earth structure, Phys. Earth Planet. Inter., 119, 25–36, 2000.

Thomas, C., Igel, H., Weber, M., and Scherbaum, F.: Acoustic simulation of P-wave propagation in a heterogeneous spherical Earth: Numerical method and application to precursor energy to PKPdf, Geophys. J. Int., 141, 307–320, 2000.

Thorne, M., Lay, T., Garnero, E., G., J., and Igel, H.: Seismic imaging of the laterally varying D″-region beneath the Cocos Plate, Geophys. J. Int., 170, 635–648, 2007.

Thorne, M., Crotwell, H., and Jahnke, G.: An educational resource for visualizing the global seismic wavefield, Seismol. Res. Lett., 84, 711–717, 2013.

Tong, P., Chen, C.-w., Komatitsch, D., Basini, P., and Liu, Q.: High-resolution seismic array imaging based on a SEM-FK hybrid method, Geophys. J. Int., 197, 369–395, 2014.

Toyokuni, G. and Takenaka, H.: FDM computation of seismic wavefield for an axisymmetric earth with a moment tensor point source, Earth Planets Space, 58, e29–e32, 2006.

Toyokuni, G. and Takenaka, H.: Accurate and efficient modeling of global seismic wave propagation for an attenuative Earth model including the center, Phys. Earth Planet. Inter., 200-201, 45–55, 2012.

Tromp, J., Tape, C., and Liu, Q.: Seismic tomography, adjoint methods, time reversal and banana-doughnut kernels, Geophys. J. Int., 160, 195–216, 2005.

Tromp, J., Komatitsch, D., Hjörleifsdóttir, V., Liu, Q., Zhu, H., Peter, D., Bozdag, E., McRitchie, D., Friberg, P., Trabant, C., and Hutko, A.: Near real-time simulations of global CMT earthquakes, Geophys. J. Int., 183, 381–389, 2010.

van Driel, M. and Nissen-Meyer, T.: Seismic wave propagation in fully anisotropic axisymmetric media, Geophys. J. Int., submitted, 2014a.

van Driel, M. and Nissen-Meyer, T.: Optimized visco-elastic wave propagation for weakly dissipative media, Geophys. J. Int., submitted, 2014b.

Zhao, D.: Seismic structure and origin of hotspots and mantle plumes, Earth Planet. Sci. Lett., 192, 251–265, 2001.

Zhu, H., Luo, Y., Nissen-Meyer, T., Morency, C., and Tromp, J.: Elastic Imaging and Time-Lapse Migration Based Upon Adjoint Methods, Geophysics, 74, WCA167–WCA177, 2009.

Permissions

All chapters in this book were first published in Solid Earth, by Copernicus Publications; hereby published with permission under the Creative Commons Attribution License or equivalent. Every chapter published in this book has been scrutinized by our experts. Their significance has been extensively debated. The topics covered herein carry significant findings which will fuel the growth of the discipline. They may even be implemented as practical applications or may be referred to as a beginning point for another development.

The contributors of this book come from diverse backgrounds, making this book a truly international effort. This book will bring forth new frontiers with its revolutionizing research information and detailed analysis of the nascent developments around the world.

We would like to thank all the contributing authors for lending their expertise to make the book truly unique. They have played a crucial role in the development of this book. Without their invaluable contributions this book wouldn't have been possible. They have made vital efforts to compile up to date information on the varied aspects of this subject to make this book a valuable addition to the collection of many professionals and students.

This book was conceptualized with the vision of imparting up-to-date information and advanced data in this field. To ensure the same, a matchless editorial board was set up. Every individual on the board went through rigorous rounds of assessment to prove their worth. After which they invested a large part of their time researching and compiling the most relevant data for our readers.

The editorial board has been involved in producing this book since its inception. They have spent rigorous hours researching and exploring the diverse topics which have resulted in the successful publishing of this book. They have passed on their knowledge of decades through this book. To expedite this challenging task, the publisher supported the team at every step. A small team of assistant editors was also appointed to further simplify the editing procedure and attain best results for the readers.

Apart from the editorial board, the designing team has also invested a significant amount of their time in understanding the subject and creating the most relevant covers. They scrutinized every image to scout for the most suitable representation of the subject and create an appropriate cover for the book.

The publishing team has been an ardent support to the editorial, designing and production team. Their endless efforts to recruit the best for this project, has resulted in the accomplishment of this book. They are a veteran in the field of academics and their pool of knowledge is as vast as their experience in printing. Their expertise and guidance has proved useful at every step. Their uncompromising quality standards have made this book an exceptional effort. Their encouragement from time to time has been an inspiration for everyone.

The publisher and the editorial board hope that this book will prove to be a valuable piece of knowledge for researchers, students, practitioners and scholars across the globe.

List of Contributors

V. K. Karastathis
National Observatory of Athens, Institute of Geodynamics, Lofos Nymfon, P.O. Box 20048, 11810 Athens, Greece

E. Mouzakiotis
National Observatory of Athens, Institute of Geodynamics, Lofos Nymfon, P.O. Box 20048, 11810 Athens, Greece

A. Ganas
National Observatory of Athens, Institute of Geodynamics, Lofos Nymfon, P.O. Box 20048, 11810 Athens, Greece

G. A. Papadopoulos
National Observatory of Athens, Institute of Geodynamics, Lofos Nymfon, P.O. Box 20048, 11810 Athens, Greece

T. Burschil
Leibniz Institute for Applied Geophysics, Stilleweg 2, 30655 Hannover, Germany

T. Beilecke
Leibniz Institute for Applied Geophysics, Stilleweg 2, 30655 Hannover, Germany

C. M. Krawczyk
Leibniz Institute for Applied Geophysics, Stilleweg 2, 30655 Hannover, Germany

F. Sodoudi
Helmholtz Centre Potsdam, GFZ German Research Centre for Geosciences, Telegrafenberg, 14473 Potsdam, Germany

A. Brüstle
Faculty of Geosciences, Ruhr-Universität Bochum, Universitätsstr. 150, 44801 Bochum, Germany

T. Meier
Institute of Geosciences, Christian-Albrechts Universität zu Kiel, Otto-Hahn-Platz 1, 24118 Kiel, Germany

R. Kind
Helmholtz Centre Potsdam, GFZ German Research Centre for Geosciences, Telegrafenberg, 14473 Potsdam, Germany
Department of Earth Sciences, Freie Universität Berlin, Malteserstr. 74–100, 12249 Berlin, Germany

W. Friederich
Faculty of Geosciences, Ruhr-Universität Bochum, Universitätsstr. 150, 44801 Bochum, Germany

EGELADOS working group
Ruhr-University Bochum, Institute of Geology, Mineralogy and Geophysics, Bochum, Germany

I.Koulakov
Trofimuk Institute of Petroleum Geology and Geophysics SB RAS, Prospekt Koptyuga, 3, 630090, Novosibirsk, Russia
Novosibirsk State University, Novosibirsk, Russia, Pirogova 2, 630090, Novosibirsk, Russia

G. Maksotova
Trofimuk Institute of Petroleum Geology and Geophysics SB RAS, Prospekt Koptyuga, 3, 630090, Novosibirsk, Russia
Novosibirsk State University, Novosibirsk, Russia, Pirogova 2, 630090, Novosibirsk, Russia

S. Mukhopadhyay
Department of Earth Sciences, IIT Roorkee, Roorkee, India

J. Raoof
Department of Earth Sciences, IIT Roorkee, Roorkee, India

J. R. Kayal
School of Oceanographic Studies, Jadavpur University, Kolkata 700032, India

A.Jakovlev
Trofimuk Institute of Petroleum Geology and Geophysics SB RAS, Prospekt Koptyuga, 3, 630090, Novosibirsk, Russia
Novosibirsk State University, Novosibirsk, Russia, Pirogova 2, 630090, Novosibirsk, Russia

A.Vasilevsky
Trofimuk Institute of Petroleum Geology and Geophysics SB RAS, Prospekt Koptyuga, 3, 630090, Novosibirsk, Russia
Novosibirsk State University, Novosibirsk, Russia, Pirogova 2, 630090, Novosibirsk, Russia

F. Braz Machado
Departamento de Ciências Exatas e da Terra, Universidade Federal de São Paulo, Diadema, SP09920-540, Brazil

E. Reis Viana Rocha-Júnior
Departamento de Física da Terra e do Meio Ambiente, Instituto de Física, Universidade Federal da Bahia, Salvador, BA40170-115, Brazil

L. Soares Marques
Instituto de Astronomia, Geofísica e Ciências Atmosféricas, Universidade de São Paulo, Sao Paulo, SP05508090, Brazil

A. J. Ranalli Nardy
Departamento de Petrologia e Metalogenia, Instituto de Geociências e Ciências Exatas, Universidade Estadual Paulista, Rio Claro, SP13506-900, Brazil

M. Lupi
ETH Zurich, Department of Earth Sciences, Zurich, Switzerland

S. A. Miller
Steinmann Institue, Geodynamics/Geophysics, University of Bonn, Bonn, Germany

V. Baptiste
Géosciences Montpellier, Université Montpellier 2 & CNRS, CC 60, Place E. Bataillon, 34095 Montpellier cedex 5, France

A. Tommasi
Géosciences Montpellier, Université Montpellier 2 & CNRS, CC 60, Place E. Bataillon, 34095 Montpellier cedex 5, France

S. Gregersen
Geological Survey of Denmark and Greenland, Ostervoldgade 10, 1350 Copenhagen K, Denmark

P. H. Voss
Geological Survey of Denmark and Greenland, Ostervoldgade 10, 1350 Copenhagen K, Denmark

A. C. F. Luchetti
Instituto de Geociências e Ciências Exatas, Universidade Estadual Paulista, Rio Claro, SP, 13506-900, Brazil

A. J. R. Nardy
Instituto de Geociências e Ciências Exatas, Universidade Estadual Paulista, Rio Claro, SP, 13506-900, Brazil

F. B. Machado
Departamento de Ciências Exatas e da Terra, Universidade Federal de São Paulo, São Paulo, SP, 09920-540, Brazil

J. E. O. Madeira
Departamento de Geologia da Faculdade de Ciências (GeoFCUL), Instituto Dom Luiz, Universidade de Lisboa, Lisboa, Portugal

J. M. Arnosio
Instituto Geonorte, Facultad de Ciencias Naturales, Universidad Nacional de Salta, Salta, Argentina

W. Friederich
Ruhr-University Bochum, Institute of Geology, Mineralogy and Geophysics, Bochum, Germany

L. Lambrecht
Ruhr-University Bochum, Institute of Geology, Mineralogy and Geophysics, Bochum, Germany

B. Stöckhert
Ruhr-University Bochum, Institute of Geology, Mineralogy and Geophysics, Bochum, Germany

S. Wassmann
Ruhr-University Bochum, Institute of Geology, Mineralogy and Geophysics, Bochum, Germany

C. Moos
Ruhr-University Bochum, Institute for Computational Engineering, Bochum, Germany

J. E. Kendrick
School of Earth, Ocean and Ecological Sciences, University of Liverpool, Liverpool, L69 3GP, UK

Y. Lavallée
School of Earth, Ocean and Ecological Sciences, University of Liverpool, Liverpool, L69 3GP, UK

K.-U. Hess
Department of Earth and Environmental Sciences, Ludwig-Maximilians-Universität, Theresienstr. 41, 80333 Munich, Germany

S. De Angelis
School of Earth, Ocean and Ecological Sciences, University of Liverpool, Liverpool, L69 3GP, UK

A. Ferk
Department of Earth and Environmental Sciences, Ludwig-Maximilians-Universität, Theresienstr. 41, 80333 Munich, Germany
Central Institute for Meteorology and Geodynamics, 1190 Vienna, Austria

H. E. Gaunt
Rock & Ice Physics Laboratory, Department of Earth Sciences, University College London, Gower Street, London, UK

P. G. Meredith
Rock & Ice Physics Laboratory, Department of Earth Sciences, University College London, Gower Street, London, UK

D. B. Dingwell
Department of Earth and Environmental Sciences, Ludwig-Maximilians-Universität, Theresienstr. 41, 80333 Munich, Germany

R. Leonhardt
Central Institute for Meteorology and Geodynamics, 1190 Vienna, Austria

A. Brüstle
Ruhr-University Bochum, Institute of Geology, Mineralogy and Geophysics, Bochum, Germany

L. Küperkoch
BESTEC GmbH, Landau, Germany

T. Meier
Christian-Albrechts-University Kiel, Institute of Geosciences, Kiel, Germany

S. Lamara
Ruhr-University Bochum, Institute of Geology, Mineralogy and Geophysics, Bochum, Germany

C. Haldar
National Geophysical Research Institute (CSIR), Uppal Road, Hyderabad-500007, India

P. Kumar
National Geophysical Research Institute (CSIR), Uppal Road, Hyderabad-500007, India

M. Ravi Kumar
National Geophysical Research Institute (CSIR), Uppal Road, Hyderabad-500007, India

L. Bonini
Dipartimento di Scienze della Terra e dell'Ambiente, Università di Pavia, Pavia, Italy

D. Di Bucci
Dipartimento della Protezione Civile, Rome, Italy

G. Toscani
Dipartimento di Scienze della Terra e dell'Ambiente, Università di Pavia, Pavia, Italy

S. Seno
Dipartimento di Scienze della Terra e dell'Ambiente, Università di Pavia, Pavia, Italy

G. Valensise
Istituto Nazionale di Geofisica e Vulcanologia, Rome, Italy

P. Lübcke
Institute of Environmental Physics, University of Heidelberg, Heidelberg, Germany, Max Planck Institute for Chemistry, Mainz, Germany

N. Bobrowski
Institute of Environmental Physics, University of Heidelberg, Heidelberg, Germany

S. Arellano
Department of Earth and Space Sciences, Chalmers University of Technology, Gothenburg, Sweden

B. Galle
Department of Earth and Space Sciences, Chalmers University of Technology, Gothenburg, Sweden

G. Garzón
FISQUIM Research Group, Laboratory Division, Colombian Geological Survey, Cali, Colombia

L. Vogel
Institute of Environmental Physics, University of Heidelberg, Heidelberg, Germany

U. Platt
Institute of Environmental Physics, University of Heidelberg, Heidelberg, Germany

T. Nissen-Meyer
Dept. of Earth Sciences, University of Oxford, South Parks Road, Oxford OX1 3AN, United Kingdom

M. van Driel
Institute of Geophysics, ETH Zurich, Sonneggstrasse 5, 8092 Zurich, Switzerland

S. C. Stähler
Dept. of Earth Sciences, Ludwig-Maximilians-Universität München, Theresienstrasse 41, 80333 Munich, Germany

K. Hosseini
Dept. of Earth Sciences, Ludwig-Maximilians-Universität München, Theresienstrasse 41, 80333 Munich, Germany

S. Hempel
Institute of Geophysics, Universität Münster, Correnzstrasse 24, 48149 Münster, Germany

L. Auer
Institute of Geophysics, ETH Zurich, Sonneggstrasse 5, 8092 Zurich, Switzerland

A. Colombi
ISTerre, Université Joseph Fourier, Grenoble, BP 53 38041 Grenoble CEDEX 9, France

A Fournier
Institute de Physique du Globe, Sorbonne Paris Cité, Université Paris Diderot, UMR 7154 CNRS, 75005 Paris, France